Wissenschaftsethik und Technikfolgenbeurteilung
Band 21

Schriftenreihe der Europäischen Akademie zur Erforschung
von Folgen wissenschaftlich-technischer Entwicklungen
Bad Neuenahr-Ahrweiler GmbH
herausgegeben vo

T0216722

Springer

Berlin
Heidelberg
New York
Hong Kong
London
Milan
Paris
Tokyo

D. Solter · D. Beyleveld · M. B. Friele · J. Hołówka
H. Lilie · R. Lovell-Badge · C. Mandla · U. Martin
R. Pardo Avellaneda

Embryo Research in Pluralistic Europe

With 17 Figures and 26 Tables

 Springer

Editor of the series

Professor Dr. Carl Friedrich Gethmann
Europäische Akademie GmbH
Wilhelmstraße 56, 53474 Bad Neuenahr-Ahrweiler, Germany

For the authors

Professor Dr. Davor Solter
Institut für Immunbiologie
Max-Planck-Institut
Postfach 1169, 79011 Freiburg, Germany

Editing

Friederike Wütscher
Europäische Akademie GmbH
Wilhelmstraße 56, 53474 Bad Neuenahr-Ahrweiler, Germany

ISBN   978-3-642-05798-4

Bibliographic information published by Die Deutsche Bibliothek
Die Deutsche Bibliohek lists this publication in the Deutsche Nationalbibliografie; detailed bibliographic data is available in the Internet at <http://dnb.ddb.de>.

Springer-Verlag is a part of Springer Science+Business Media
springeronline.com

© Springer-Verlag Berlin Heidelberg 2010
Printed in Germany

Coverdesign: deblik, Berlin

Printed on acid-free paper   62/3020hu – 5 4 3 2 1 0 –

Europäische Akademie

zur Erforschung von Folgen wissenschaftlich-technischer Entwicklungen
Bad Neuenahr-Ahrweiler GmbH

The Europäische Akademie

The *Europäische Akademie zur Erforschung von Folgen wissenschaftlich-technischer Entwicklungen GmbH* is concerned with the scientific study of consequences of scientific and technological advance for the individual and social life and for the natural environment. The Europäische Akademie intends to contribute to a rational way of society of dealing with the consequences of scientific and technological developments. This aim is mainly realised in the development of recommendations for options to act, from the point of view of long-term societal acceptance. The work of the Europäische Akademie mostly takes place in temporary interdisciplinary project groups, whose members are recognised scientists from European universities. Overarching issues, e.g. from the fields of Technology Assessment or Ethic of Science, are dealt with by the staff of the Europäische Akademie.

The Series

The series "Wissenschaftsethik und Technikfolgenbeurteilung" (Ethics of Science and Technology Assessment) serves to publish the results of the work of the Europäische Akademie. It is published by the Academy's director. Besides the final results of the project groups the series includes volumes on general questions of ethics of science and technology assessment at well as other monographic studies.

Acknowledgement

The project Embryo Research was supported by the Federal Ministry of Education and Research (Bundesministerium für Bildung und Forschung).

Foreword

The Europäische Akademie Bad Neuenahr-Ahrweiler GmbH is concerned with the study of the consequences of scientific and technological advance for individual and social life and therefore, not least with the study of consequences of recent developments in life-sciences and medical disciplines. The Europäische Akademie intends to contribute to finding a rational way for society to deal with the consequences of scientific development. This aim is mainly realised by developing recommendations for options to act focussing on long-term social acceptance. The work of the Europäische Akademie mostly takes place in interdisciplinary project groups, whose members are recognised scientists from European universities and other independent research institutes.

In light of the recent development on molecular biology and reproductive medicine and the fact that the research as well as the clinical conduct is not restricted to national borders but is increasingly taking place on an international or at least European level, the Europäische Akademie set up a project on "Embryo Research in Pluralistic Europe" in January 2001. Experts from biology, jurisprudence, sociology and philosophy from different European countries were brought together to discuss current scientific development in reproductive and regenerative medicine, current public attitudes towards embryo research and related topics in the nine largest European countries, the rationality and implications of the moral arguments used in the debate and how to regulate new bio-medical techniques, considering the differences between already existing national legislation and public attitudes in the context of the project of harmonisation of the European Union.

I hope that this memorandum will attract international attention and thus will be able to contribute to moral understanding and legal harmonisation in the European integration process.

The members of the project-group have taken on the demands of an interdisciplinary project for more than two years. I would like to thank the members of the project-group and their chairman, Professor Davor Solter, for this commitment. Special thanks also go to Minou Bernadette Friele, M.A., who co-coordinated the project on behalf of the Europäische Akademie.

Bad Neuenahr-Ahrweiler, August 2003   Carl Friedrich Gethmann

Preface

Based on the results of a workshop on "Embryo Experimentation in Europe", in January 2001 the Europäische Akademie set up a project group to consider several aspects of research on human embryos as it is at present and as it will be performed in Europe. Numerous conferences and projects have been dealing with the scientific aspects of human embryo experimentation. However, we felt that the international dimensions of this increasingly borderless scientific endeavour had not been properly considered. Our goal was to place these complex scientific issues within the framework of equally complex currently existing legal solutions often created as a delayed response to fast-changing scientific reality. The legal response of each nation-state to scientific technology dealing with such an emotionally charged issue as reproduction and embryo experimentation will obviously be shaped by its past legal history and convention, by the general attitude of its citizens and by prevailing moral ethics. In a complex multinational unity in the making as Europe is today, the solution – legal and otherwise – of such a difficult issue is further complicated by often conflicting national laws and attitudes. While dealing with these issues, we tried to keep in mind the special case of Europe and to always pose the question whether European harmonisation in this scientifically, legally and philosophically very novel and unpredictable field is possible and desirable or if it is better left in the sphere of each individual member state.

The work of a small project group like ours faces many challenges and we often felt the need for further expert advice and help. We were very fortunate in being able to consult numerous colleagues who provided the insight and knowledge we were sorely lacking and who pointed out important issues we would most certainly have missed. We profited immensely from the advice of the following colleagues and would thus like to thank them for their co-operation, their helpful contributions and critique: Peter Braude, Mirentxu Corcoy, Detlev Ganten, Ferdinand Hucho, Barbara Knowles, Hans Lehrach, Jens Reich, Bettina Schöne-Seifert and Michael Selgelid. We would particularly like to thank David Fraser for providing important information on different national laws and regulations concerning embryo research and for drafting large parts of the legal appendix and Felix Thiele for considerably contributing to our discussions. For acting as reviewers of our book, we would like to thank: Dieter Birnbacher, Bernhard Böhm, Anne McLaren, Sheila McLean, Julian Savulescu and Baroness Mary Warnock. We would also like to thank Friederike Wütscher for the editorial work in preparing the text for print.

Freiburg, August 2003   Davor Solter

Table of Contents

Introduction

Recent advances in techniques and understanding in the fields of genetics, embryology and reproductive biology have opened up exciting and potentially valuable ways to treat a variety of medical and social problems. These range from new options in infertility treatment and preimplantation genetic diagnosis to embryonic stem cell-based therapies for debilitating diseases. However, these new approaches involve the manipulation of human gametes, embryos and embryonic cells, and could also permit more contentious uses, such as reproductive cloning and genetic manipulation. They have therefore stimulated a heated political debate as to whether research on human embryos that leads to their destruction should be permitted and, if the answer is yes, to what extent and for which aims such experiments and actions are morally permissible and socially desirable. Is it justifiable to destroy a potential human being in order to develop stem cell lines for therapeutic purposes? Is it acceptable to pre-select one's offspring by preimplantation diagnosis (PGD)? And if yes, should PGD only be allowed in order to eliminate traits that are clearly linked to severe genetically inherited diseases and disabilities or also in order to select for other, medically irrelevant, genetic traits? And should we seek to allow therapeutic cloning, even though the development of this technique might also allow human reproductive cloning?

In view of the fact that research in these areas is often carried out in multinational projects and scientists as well as patients are increasingly ready to bypass regulative, social and moral boundaries by going abroad, these questions obviously cannot be dealt with on a mere national level. Under the given circumstances policy makers will have to bear in mind debates that have taken place in other countries and take into account the co-existence of different legislative solutions. Due to their shared political features and the European unification process, this particularly applies to European debates. In view of the numerous interdependencies of European states and the project of European unification one might even have to consider Europe-wide harmonisation of embryo research related laws. Would harmonisation in these areas not be desirable in order to ensure that the presumable or actual interests of embryos, patients and scientists are equally respected throughout all European countries? The answer to this question is just as difficult to provide as the answers to the questions mentioned above.

Current policies, as well as public opinion on the legitimacy of research on human embryos, are clearly linked to the differing social, legal, and moral traditions of the nations in which they are rooted. In fact, this cultural pluralism is an important trait of the European Union. It is, as a European Commission staff working paper states, "mirroring the richness of its tradition and adding a need for mutual respect and tolerance." Further it is stated, that "(r)espect for different philosophical, moral or legal approaches and for diverse cultures is implicit in the ethical dimension of building a democratic European society." (2.3. Pluralism and European ethics; Commission Staff Working Paper – Report on Human Embryonic Stem Cell Research, Brussels, 3.4. 2003, SEC (2003) 441).

In other words, political decisions on embryo research are not made in a vacuum. Current regulations, their constitutional framework and their underlying principles need to be taken into consideration when discussing legislation for as yet unregulated or only insufficiently regulated areas. Otherwise new laws would be not only likely to cause inconsistencies with respect to previously established national or European regulations but – and in democratic societies, this is even more important –would be likely to lack wide acceptance and support of the general public. Thus sufficient knowledge about and respect for widespread public opinions and beliefs is essential if adequate and effective political decisions are to be made, It is important to note, however, that the mere existence of public opinions and moral stances does not allow us to deduce their legitimacy – particularly not if they are to be enforced by criminal law. History has far too often shown that political majorities were wrong about 'absolute demands' of morality. Often these demands were based on various illegitimate claims and/or on false beliefs. If political decisions are to be acceptable in the long term, it is thus crucial to ensure that they are also based on the best scientific information available and that the normative implications of this information are considered as thoroughly as possible. In spite of differences which may provide good reasons for regulating morally contentious issues like embryo research on a national instead of a supranational or international level, this demand for rationality can be assumed to be acceptable to all democratic European societies.

Obviously the authors of this study are not in a position to provide the final answers concerning the moral status of human embryos or on how they should be handled. Neither can we prescribe any particular national or European policy. However, with this study we hope to contribute to a rationalisation of the political and public debate by providing relevant scientific, legal and social information essential for an informed decision. Furthermore we have examined – we hope impartially – various moral and ethical stances in response to embryo experimentation and have tried to identify criteria which could help in reaching a rational political solution of these possibly not solvable moral problems.

Summary

It may be difficult for many people to accept experiments on human embryos for various ethical, moral and religious reasons. On the other hand it is not likely to be acceptable to others, or to many societies at large, if whole areas of research are rejected before a proper understanding of the issues at stake and without fully realising the implications that political and legal decisions in this field may have for all the different parties concerned. The pros and cons must first be carefully balanced. We therefore begin this book with a chapter that explores embryo research from a scientific perspective. This is not purely scientific and it is important to stress that each topic covered in this book relates to and has been informed by all the others. We then take a look at the current state of regenerative medicine and stem cell research, as these are hotly debated areas that impinge on embryo research in several ways. The next chapter is concerned with the laws governing embryo research where we survey and comment on the various approaches adopted by different countries, especially within Europe, and ask if it is possible to move towards some common regulatory mechanism. Chapter 4 covers public attitudes towards embryo research and includes results from a new survey of opinions across Europe specifically conducted for this project. The final chapter is a philosophical treatment of the arguments for and against embryo research where we propose a rational basis for decision making in this field. We then close the book with a list of recommendations, which can be considered as our most pertinent conclusions.

Scientific and Clinical Aspects of Research on Human Embryos

Two broad justifications can be put forward for experimentation using human embryos. One deals with projects aimed at addressing basic biological questions concerning human development. Provided that any suitable alternative experimental approaches exist, one would certainly avoid using human embryos per se. Such alternatives are the embryos of other animals or cultured cells. However, we have to bear in mind that some aspects, and possibly quite important ones, are unique for the development of each mammalian species. This brings us to the second justification, which applies to clinical practice. If the information to be obtained in an experiment is of crucial medical significance, be it reproductive medicine, teratology, fetal surgery or the clinical application of cloning procedures, it will eventually have to be confirmed by experiments using human embryos or cell lines derived from them, such as embryonic stem cells. Many of these experiments would not result in any evident benefit to the embryos under study; indeed these will often be destroyed. Some experiments might even necessitate the creation of embryos and their subsequent destruction. On the other hand, the results of this work may well lead to benefits for other embryos and the live-born offspring that would develop from them, as well as to their parents and other patients undergoing treatment for infertility or disease.

Much current research on human embryos responds to the wish to have choice in reproductive outcomes, in particular to overcome infertility or genetic defects in the embryo, which result in non-viable or severely genetically injured offspring. These wishes are generally seen to be benign, though the techniques developed for such purposes might also be used in more contentious areas such as sex-selection and genetic manipulation that cannot be justified by health considerations. But, as is asked in Chapter 1, "Should we throw the baby out with the bath water?"

The use of IVF (In Vitro Fertilisation) and ICSI (Intra Cytoplasmic Sperm Injection) is widely accepted in most societies today. As early human embryos are frequently of poor quality, with a reduced chance of developing to term, these techniques generally rely on superovulation to ensure sufficient numbers of good quality embryos for implantation. However, very few embryos can be transferred into the uterus. So an increasing number of spare embryos (or embryos in a pronuclear stage) are discarded or stored frozen in case they are required for future attempts at pregnancy. Since unlimited storage of supernumerary embryos is usually not practical for economic reasons, and/or because they are no longer required by the couple, embryos that were not implanted either have to be destroyed without any further use or they can be used for research, which also results in their destruction. If IVF is permitted, and this is the situation throughout all the countries investigated in this study, it has to be recognised that it inevitably involves destruction of embryos and, critically, that it required, and still requires, research on embryos to perfect the technology. The latter is an imperative unless one is happy to accept inefficiency and waste. A considerable fraction of actual and prospective research on human embryos is exclusively designed to develop methods to improve, facilitate or make reproduction possible. Provided that IVF is acceptable, some of the arguments presented against research and associated procedures, including preimplantation genetic diagnosis (PGD) or the derivation of Embryonic Stem (ES) cells, need to be evaluated in view of these facts.

With regard to PGD, it must be borne in mind that this technique is not primarily a means to overcome problems of infertility (by finding embryos with the best chance to implant successfully in the womb) but to produce offspring without a certain genetic make-up, one that would lead to a genetic disease. Its use therefore promotes considerable debate about where to draw the line. How serious does a genetic disease need to be in order to legitimise a decision not to implant an embryo? What if the aim is to choose an embryo of a particular genotype, in order that the newborn child can be used as a donor to save the life of another already suffering from a genetic disease?

Even more controversial are purposes such as sex-selection, positive selection of desired traits, or even genetic enhancement. It is obvious that moral questions arising from these new technical possibilities cannot be answered by merely pointing to biological facts. However, biological knowledge can help us to avoid becoming submerged in merely theoretical debates, which frequently raise unrealistic hopes and fears. For example, not only do we know almost nothing about the genes that control the characteristics many parents would like to see in their children (intelligence, beauty, artistic skills, etc), but these characteristics are presumably regulated by a large number of different genes, and it would be impossible to select an embryo that carries all of the favourable alleles. Furthermore, every individual's genome comprises a random mixture of 'positive' and 'negative' alleles for different genes and selecting for a slight increase in one desired characteristic may easily result in loss of others. Genetic determinism is very difficult or impossible to apply to complex traits, and certainly not in a way that would enable us to predict the development of a child created using the techniques of IVF and PGD.

Even if we disregard selection as a way to obtain embryos with a particular genetic make-up, should we be concerned about the possibility of altering genetic make-up during development? This could be done in theory by the introduction of transgenes, containing one or a few genes, or artificial chromosomes that might contain many. If these led to heritable changes they would fall under the term germline gene therapy: are there circumstances when this should be permitted? We should remember that we generally strive to give our children advantages in life.

And what about nuclear transfer (cloning)? The debate on human reproductive cloning mainly concerns the question of whether it is categorically wrong to clone a human being for reproductive purposes. If one regards clones as "postponed" twins, and twins are usually not considered to live worse lives than other individuals despite the fact that they share their genetic make-up with somebody else, this categorical denial of another method of human procreation seems astonishing. Indeed, arguments based on religion or gut feelings are sometimes strong enough to abandon a practice that might be acceptable in theory. This is true even if the techniques should not be attempted for purely practical reasons. Reproductive cloning is a good example of this because it would involve considerable risks for the mother and the resulting offspring, while at the same time there seem to be no genuinely pressing needs for the procedure, certainly none that would make the risks and the costs involved worthwhile.

Embryo research has always been relevant for regenerative medicine, but recent developments have made it even more so. Present therapies using artificial tissue and organ substitutes have many drawbacks. Allogeneic transplantation is not feasible for all organs, infections and malignancies occur from the use of immunosuppressive agents and there is a permanent shortage of organs, so that a significant number of patients die while waiting for a suitable graft. Thus the need for new treatment approaches can hardly be overestimated. It is the expectation of many researchers that in the non too distant future the use of stem cells in combination with tissue engineering technologies will enable the development of bioartificial tissue and organs and thereby overcome some of the current problems. Stem cell based therapies also offer potential cures for many more types of disease or trauma than can be treated by any conventional graft. Moreover, knowledge gained from studying stem cells, both outside and inside the body, can help us to understand both normal biology and mechanisms leading to disease. Similarly, knowledge gained from studying embryonic development can inform us how to best use stem cells in regenerative medicine.

On the other hand, the derivation of human Embryonic Stem (ES) cells involves the destruction of potential human beings and their instrumentalisation, i.e. they are produced to be destroyed for the benefit of another human being. Thus research using ES cells is considered by some to be controversial and there are those who argue that it should be prohibited – in particular since many see an adequate alternative in the use of adult stem cells, cord blood or other cell sources.

It has been argued that adult stem cells possess a remarkable plasticity which can transform adult neural stem cells into several blood cell types and vice versa. If the patients' own cells can be used to replace lost or damaged tissues, should not all research using ES cells, and in particular therapeutic cloning, be declared illegal? In order to put the reader in a position to compare and evaluate the various approaches, we present in this book not only current and anticipated studies based on ES cells but also discuss cell replacement therapies based on alternative cell sources. The latter, which are already in use in some instances, include various somatic differentiated cells, adult stem and progenitor cells, cord blood stem cells and fetal cells.

At present, each approach has advantages as well as disadvantages. For example, while allowing autologous grafts is less contentious, there are serious scientific doubts about the pluripotentiality of adult stem cells. For most desired cell types it is not known if these stem cells can fulfil the necessary requirements as regards proliferation and differentiation capacity. Further major obstacles for the clinical use of adult stem cells are the unsolved questions of stem cell identification, isolation, expansion, differentiation and functional integration.

In contrast to adult cells, ES cells seem to have the capacity to propagate indefinitely in culture, and to give rise to any desired cell type. On the other hand, this ability to differentiate is hard to direct and ways of sorting out the required cell type for a particular therapy need to be developed. Moreover, ES cells, which can persist even in well differentiated cultures, could develop into teratocarcinomas in vivo. Cells would have to be selected with absolute certainty before injections are administered to humans. Also, most, if not all, of the existing human ES cell lines are unsuitable for clinical applications since they were cultured on feeder layers of mouse fibroblasts. This imposes a risk of contamination of the ES cells by feeder-derived endogenous viruses. Some countries have laws stipulating that it is only possible to use ES cell lines established before a particular date. At present these laws might rationally merely serve as a moratorium. They would have to be altered if progress, i.e. the invention of novel clinically applicable therapies, were not to be thwarted.

In addition, the use of established ES cells for therapeutic purposes would still require immunosuppression. "Therapeutic cloning" (otherwise known as cell nuclear replacement or CNR) would overcome this latter problem as a nucleus from one of the patient's own somatic cells is transferred into an enucleated egg, which is then allowed to develop to a blastocyst from which ES cells could be isolated, manipulated in vitro, induced to differentiate and finally the appropriate cell types returned to the patient. However exciting this approach may seem, it imposes additional problems, some of which might not be easy to overcome. Certainly, the time required for ES cell cultivation would be too long for injuries and diseases requiring immediate treatment.

Human ES cells in general, and those made via "therapeutic cloning" in particular, may find more immediate uses in studying genetic defects or for screening potential pharmaceuticals, etc. In the meantime, a great deal of research will have to be done to achieve the goal of producing cell-based therapies that are safe and reliable enough to enter the clinical trial phase. This research can neither be done using one particular type of stem cell nor can it be done only on those cell lines that have been established so far. The most pragmatic approach will be to pursue research down as many avenues as possible, including both adult and embryonic stem cells.

Biological problems might eventually be surmountable. The problematic moral core issue that a potential human being has to be destroyed to gain ES cells will remain and further ethical questions, e.g. that therapeutic cloning could encourage reproductive cloning, need to be considered in legal, sociological and ethical terms.

The Regulation of Embryo Research

We can distinguish three general regulative approaches regarding research on human embryos or research and conduct closely related to such research: (1) a more liberal approach such as in the UK, one of the few countries in the EU that has legislation which explicitly permits embryo research; (2) a restrictive approach like that in Germany, which does not allow any research not directly benefiting the embryo itself; and (3) a compromise of these two positions like in Spain, where research is permitted under specifically defined circumstances. In order to gain some understanding of the factors that would affect endeavours towards a common European legislation, Chapter 3 examines in some depth the history and details of regulation of embryo research and related fields, such as in vitro fertilisation, preimplantation genetic diagnosis and abortion, in the three countries mentioned above. This is complemented by an overview and discussion of the reasons behind the various regulations in European countries and in those non-European countries which are particular important for the European debate. In order to elucidate possible regulative approaches, a detailed overview on regulations of embryo research and related issues is presented in an appendix at the end of the book. Some non-European countries (such as Australia, Canada and the USA) are organised by a federal structure or have been posed with the issue of harmonisation themselves, so we also consider whether they might serve as a model for a European solution.

Regulation of Embryo Research in the United Kingdom, Spain and Germany

Even before considering the possibility of research, it is apparent that legislation already differs significantly between countries in defining the human embryo. Definitions range from 'a living group of cells resulting from fertilisation not implanted in a women's body' to a distinction between 'pre-embryos' as fertilised eggs until 14 days or implantation and the 'embryo' as existing from 14 days to two and a half months. And some laws even avoid definitions altogether.

Obviously definitions can influence the regulative scope of the law to a considerable extent, by, for example, affecting the permitted time limits on use of embryos for research or by determining what falls within the scope of the law. For instance, human embryos created by cell nuclear replacement (CNR) are not created by 'fertilisation', and hence might be thought not to fall within the definition of embryo as it is given in the act of the British Human Fertilisation and Embryology Authority (HFEA). However, UK courts have now held that CNR embryos are within the scope of the Act, on the grounds that a purposive interpretation of the Act requires the Act to cover cells capable of developing into born human beings regardless of how these are created.

Other issues regarding the enactment of regulations might be more challenging. This applies in particular if they result from certain dynamics of power in political decision making processes. For example, what happens if none of the three broad views on the moral status of the embryo, which happen to be pro-life, proportionality or libertarian, get sufficient majority support, but a majority only can be formed by an alliance between the proportionality camp and either of the other two? Regulatory scrutiny is the most probable outcome of such situations since the liberal camp is likely to be able to compromise in the direction of regulatory scrutiny, while the pro-life camp can not be expected to compromise its ideals, especially if its stance is rooted in religious belief. Alternatively, proponents of what we call a regulatory position may argue that it is better to permit embryo research in at least some situations. They might do so on the ground that a position of regulatory oversight involves active recognition of the inherent value of having decisions made in a transparent and accountable manner. Indeed the regulatory scrutiny position in the UK is partly to be seen as result of different alliances and of the pragmatic style that characterises the conduct of British public life including its common law legal system.

The situation is different in other countries, in particular in Germany. The primary set of regulations governing the treatment of human embryos, the *Embryonenschutzgesetz* (Embryo Protection Law – ESchG), has been developed according to the regulative requirements of the federal system of Germany, and thereby in accordance to the requirements given in its constitution. Given the rather recent scientific developments in stem cell research, the ESchG was supplemented in mid 2002 by a Stem Cell Act (StZG). Such an additional act became necessary because the German ESchG did not allow for embryo research that would lead to the destruction of embryos when it was passed in 1990. When considering how regulations can have an international context, the Stem Cell Act is of particular interest as it prohibits the production of human embryonic stem cell lines in Germany or by German scientists working abroad. Nevertheless it allows such cell lines to be imported as long as they have been produced abroad under certain conditions and prior to a specified date. The problems occurring from regulation of this sort in terms of biological needs and safety have already been mentioned. Further problems result in terms of consistency of the law. Germany has one of the most prohibitive legislations on embryo research, giving the highest priority to the protection of the embryo. Is it not inconsistent to take advantage of research conducted by others under less restrictive regulations? Additional consistency problems occur from the fact that embryos are better protected in vitro than in vivo. While the German law on abortions (which is to be found in the Criminal Code § 218a) left terminations of pregnancies unpunished on the basis of so-called "maternal indication", the EschG does not permit the screening of an embryo before implantation (i.e. PGD) even with regard to the very same defects. The practice of medical tourism, widely known in other areas such as abortion, is now also being used for PGD. A considerable number of couples who are facing a high probability that their child will be affected by a genetic disease, and who can afford the costs, request the service where it is available, e.g. in Belgium.

The German situation can be seen to result partly from its constitutional background. Such backgrounds must be kept in mind when discussing possible harmonisation of embryo research on a European level, especially when regulation in other countries, for example, the UK, is based and organised on an entirely different structure.

The Spanish legislation on embryo research and related issues is also based on constitutional law, but the relevant act, the Assisted Reproduction Techniques Act (Law 35/1988 sobre Técnicas de Reproducción Asistida) is less prohibiting than the German laws. It permits certain research activities on human (pre-)embryos as long as these are not conducted beyond the fourteenth day of development or implantation has taken place. The postimplantation embryo again is protected by a special law, regulating the donation and use of human embryos, fetuses and cells, tissues or organs obtained from them. In both cases, whether the research is proposed to be using pre-embryos or postimplantation embryos, written informed consent of the donors has to be obtained and the research has to be carried out in Health Centres and by legalised and authorised teams. In these, as well as in some other aspects, the Spanish law is much closer to the British regulation than to that in Germany. In contrast to the British law the Spanish law, again, also distinguishes between viable and non-viable embryos insofar as in vitro embryos shall only be authorised for diagnostic, therapeutic or preventive purposes, i.e. for research benefiting the embryo itself. Other kinds of research are restricted to embryos that are considered to be non-viable, such as aborted embryos. This law also applies on PGD related issues. It permits advice being given against the transfer of a (pre-)embryo if a hereditary disease which cannot be treated has been detected, since the legal protection guaranteed by the constitution is less for the pre-embryo than it is for the embryo in utero.

Is European Harmonisation of Law Possible?

In each of these three countries, scientific advisory committees are still dealing with questions of embryo research that might need political, i.e. legislative solutions. Do the rather few common features found amongst all the regulations adopted by different European countries in the field of embryo research give a sufficient basis to consider harmonisation? Or do we have to admit that the moral, social, and economic factors that currently exist, render non-harmonisation the optimal solution with respect to the interests of both those countries that currently prohibit research on embryos and those that permit it? In countries that prohibit destructive embryo research, moral or religious arguments, to the effect that human life is sacred from the moment of conception, appear to be the main reasons for prohibition. An additional factor, especially in the German case, arises from misgivings or regrets over historical events. On the other hand, those who do not share these values and/or do not share the misgivings, argue for a permissive stance. There are many arguments. For instance, they may point out the loss of scientific prestige that could follow prohibition. They are accompanied by those who argue that it is hypocritical to use the products of embryo research if one is not prepared to carry the research out oneself. There is also the perception that the moral objections seem to be not so overriding for prohibitive countries if they are still, in general, prepared to associate with permissive countries.

In permissive countries, the embryo has only a proportional status which renders it possible to weigh its worth against the possible benefits for others that might result from its use in research. Consequently these countries are only likely to consider prohibiting embryo research under fairly significant economic or political pressures. Accordingly, non-harmonisation seems to have an advantage in that cultural sovereignty will allow high moral principles to be maintained in prohibitive countries while economic and prestige advantages can be gained by permissive countries.

Despite these arguments it might well turn out that some form of positive action is required if non-harmonisation is to be maintained. For example, it might require some mechanism to reduce the competitive disadvantage that the prohibition countries would be faced with. Because the constitution of the European Union makes it difficult for EC legislation to levy taxes on Member States, and funding strategies that might be adopted will not prevent researchers obtaining funds elsewhere, we conclude that harmonisation in the direction of permission is the likely direction in which the EU will move. In this context it is important to note also that the impact of Newly Associated States (NAS) is not likely to facilitate the situation, as they usually lack the economic power to influence European legislation. Rather they might be susceptible to adopting a permissive stance in order to attract research investment. This may happen even if they lack the necessary culture of public debate on the issues and that the debates that have taken place on subjects such as abortion have evinced a generally strong 'pro-life' attitude that would be contradicted by a permissive stance in embryo research related legislation.

Attitudes and Knowledge on Biotechnology, Cloning, and Stem Cells

The complexity of the issues posed by current scientific research has led to diverse opinions, images, and beliefs, which often produce contradictions and confrontations both at the individual and social level. These pose difficulties for any attempt at a common European policy.

There is public resistance to some scientific advances. While many blame this on scientific illiteracy or a growth of neo-luddite attitudes, the reality is more complex. On the basis of a systematic study of European attitudes towards science, the scientific community, embryo research and regulation, and value judgements on 'the natural', life, and risk perception, it is apparent that resistance to certain areas of research is qualified and we are able to understand the attitudinal structures underlying these feelings. Furthermore, this data allows us to form a view of the aspirations held by Europeans with respect to science.

While the scientific community enjoys high levels of legitimacy, and public confidence and expectations about future scientific developments remain strong, the public demonstrates a growing ability to differentiate amongst areas of research, applications, and treatments such as basic/pure research, applied research, and clinical practice. Thus, general optimism does not seem to determine attitudes towards specific applications, which, as the data shows, rely to a much greater extent on images, fears, stereotypes and beliefs than on knowledge and information.

Cloning, irrespective of whether it is conducted for therapeutic or reproductive purposes, most clearly triggers these "images of fear", and "the gene" stands as a cultural symbol of everything that is 'natural' and vulnerable to manipulation for illicit or 'unnatural' ends. Such fears, along with the belief that by engaging in human therapeutic cloning we are going down "the slippery slope", which will inevitably lead to human reproductive cloning, have placed cloning in the centre of the public debate regarding the life sciences.

Attitudes towards embryo research, on the other hand, are most clearly conditioned by moral-religious reservations. These have led to an enormously wide variety of views on the moral status of the embryo both across Europe and within each country.

Nevertheless, moral judgements are not abstract matters governed by uniform principles. Rather, they arise in relation to specific scenarios, such as the choice between the use of spare embryos and the creation of embryos for research. Survey data show that Europeans clearly favour the former, although important 'minorities' in many countries oppose both options or have not yet formed an opinion on this issue. These 'minorities' are particularly important in Catholic countries, with the exception of Spain, while other countries share more flexible and permissive views. PGD techniques are regarded in a similar way. The prevention of hereditary diseases tends to deflate possible moral reservations, while the idea of selection of characteristics for aesthetic or non-health related reasons, triggers a strong rejection on moral grounds.

At the national level, to the extent that the regulation of embryo experimentation might be a function of each society's perception of the status of the embryo, one would expect a tendency toward restrictive rules in Austria, Germany, Italy and Poland. Conversely, a more permissive regulatory stance could be expected, in descending order, in Denmark, the UK, and the Netherlands. Spain and France represent the middle ground, with more uncertain positions that are susceptible to being swayed one way or another depending on other variables and factors (these mainly being the "voice" of various organised stakeholders, from religious groups and institutions to political parties, patient associations and scientific and medical institutions).

At the individual level, three basic positions with regard to embryo research can be distinguished: (a) positions favouring prohibition of any embryo research, (b) positions favouring that research as well as clinical practice remain unrestricted, and (c) positions whose proponents neither accept a full prohibitive nor a full permissive stance. Even from the public's point of view, a complete legal rejection or approval of embryo research would be difficult to accept before a serious attempt is made at balancing the potential benefits and risks.

The dominant aspiration is a common regulatory framework, despite the significant differences in perceptions, opinions, and value judgements. In the medium term, the public's perceptions play a significant role in the success or failure of public policies and regulations. Thus a consensus needs to arise from a rational and inclusive public debate, which, despite media attention, has not yet developed.

To date, the public debate on these issues seems to have been dominated by cloning, from which the majority of Europeans do not perceive any clear benefits that outweigh the risks and challenges to the genetic integrity of both humans and animals that it poses. If a constructive consensus in relation to embryo research is to be reached, the debate must change to provide accurate and complete information that enables both individuals and society to make their own critical assessment of the benefits and risks.

However, in so far as public attitudes serve to guide legislation on Biotechnology, a common European regulation permitting the use of embryos left over from artificial reproduction treatments seems possible. Any serious attempt to regulate embryo experimentation at the European level has to take into account the markedly pluralistic character of Europe today, both inside each society and among societies: not only are there different religious beliefs, with different degrees of sensitivity to issues regarding the moral condition of the embryo, but also large minorities who don't seem to have any particular religious affiliation and who have fewer reservations about embryo experimentation for biomedical purposes.

Philosophical Analysis of Argumentative Strategies

Though public opinion obviously needs to be respected when political decisions are taken, it cannot be regarded as a sufficient reason for adopting a certain policy. The reasoning behind it must be considered and evaluated. Thus, a philosophical analysis of the current political stances and the arguments used in the debate is necessary.

In academic, as well as in public debate, we can generally distinguish three fundamentally different positions: (a) positions favouring prohibition of any embryo research, (b) positions favouring unrestricted research as well as clinical practice, and (c) positions that demand regulatory oversight of such matters. In essence, the prohibitive point of view insists that embryos are, or have the potential to become, human beings. Therefore they deserve full life protecting rights from the moment of conception. Most prominently this view is defended by the Catholic Church. From this initial premise it follows that any kind of embryo research that does not benefit the embryo itself is unacceptable. This applies to both the extraction of ES cells and PGD, as the latter can be expected to be followed by an abortion if the embryo shows genetic defects. This position consistently does not distinguish between the production of ES cells by using embryos that have been discarded in the course of IVF treatments and embryos created for research purposes. Rather it is stated that each human embryo should be allowed to develop into a child. This position, and its metaphysical as well as scientific assumptions, is strongly objected to by those who favour legally unrestricted research options and those who favour permission of embryo research under certain, usually legally defined, conditions. Proponents of completely unrestricted embryo research may adopt a position according to which no institution should be allowed to impose its moral opinions on other people – particularly if these moral opinions are based on traditional values that might not be shared by everyone. They hold that whatever is technically feasible will be tried eventually. However, this stance is not likely to be acceptable. Though an embryo is not a person, it nevertheless should not be treated as mere property but demands special consideration, not because of metaphysical assumptions resulting in moral values that one does not have to share, but at least because of the legitimate interests that other people may have, in particular the resulting children. Interest conflicts between parents or researchers and the offspring resulting from medical interventions can easily occur, for example, if a child would have a reasonable suspicion that it was produced for instrumental reasons only, not for its own sake but merely to save another child's life by being an appropriate tissue donor. Similar considerations arise regarding the use of ES cells, in particular regarding possible harm to sentient persons. Neither unrestricted freedom of choice nor full prohibition is likely to be an acceptable stance for research that interferes with human reproduction or uses embryonic cells for the benefit of others. Rather it seems to be necessary to balance the pros and cons of the various settings of research. A position that tries to avoid the shortcomings of each of the other positions might be called a regulatory stance. In order to avoid unnecessary and/or unjustifiable harm such a position might require, for example, that the use of embryos for medical and scientific purposes is regulated and controlled by law in accordance with the aforementioned harm principle and in accordance with legitimate interests of parties concerned. In accordance with such a regulatory stance, procedures need to be developed that allow 'hard cases' to be dealt with, such as cases in which gamete donors disagree about the further use of their embryos. Here it needs to be taken into account that parents have a special relationship to their offspring and therefore deserve an extended right to a say in such matters. Also, if an embryo is going to be implanted and thus is planned to become a person with certain capacities (sometimes referred to as capacities of a rational agent in contrast to those of a mere member of the species Homo sapiens), its presumed future interests should be respected even if it is yet to show the relevant features necessary to have interests.

In the public debate as presented in the mass media as well as in parliamentary debates, some arguments and argumentative strategies are particularly common. These rely on taboos, make use of redefinitions, emotions or rhetorical language, refer to the embryo's potentiality, dignity or presumed rights, use slippery slope arguments and arguments referring to requirements of caution or justice. These arguments and argumentative strategies need to be evaluated if the debates are not to be unnecessarily hindered by false assumptions or conclusions.

Fears associated with taboos or cultural abhorrence are frequently not products of reason and often deflect from the considerations that are necessary for a rational decision. The same applies to attitudes associated with what is often called 'naturalism'. It is often stated that abortion, IVF, stem cell therapy and embryo selection by prenatal or preimplantation screening are unacceptable practices because they are not 'natural'. Such attitudes, which sometimes are aligned with concepts of human dignity and/or theological world views that demand that men shall not intervene in divine providence, are again not sufficiently rational. Human interventions into natural processes have frequently been advantageous. Since this is not necessarily the case, evaluation of new developments is still required. So-called 'slippery slope' arguments suggest that such interventions should not be made all. According to these arguments, however, it must be assumed, that though a practice A is acceptable, once it is allowed, it will inevitably be followed by an unacceptable practice B. But such an assumption is seldom based on evidence. Of course, there are many troubling cases in which it may be difficult to decide whether an act is acceptable. However, laws are regularly passed to distinguish between (morally) acceptable and criminal behaviour and to prevent the latter.

Obviously accepting A does not always necessitate the acceptance of B.

Procedural Solutions and the European Convention on Human Rights

The last part of our study aims at giving some tentative advice to handle the problem, that the current debate is seemingly deadlocked. This situation is mainly caused by different views about the inherent moral status of the human embryo. Quite often different views derive from different moral theories. It does not seem to be possible to find a consensus about which theory is the 'right' one. Thus, if moral disputes about embryo research are to be 'managed' in a fully rational manner, this management must at some point be procedural. The essence of a procedural solution to a moral dispute is that it requires persons to accept decisions as binding even if they disagree with them morally. This might be possible if the persons involved have moral reasons to accept even those decisions they disagree with morally, which will happen if not reaching a decision is regarded as threatening more important values than the values at stake in the primary dispute. Successful procedural solutions at this level, thus, presuppose a set of core values between the involved parties that is shared between their competing moral theories. The extent to which such a consensus-set exists, limits the efficacy of procedural solutions. It is thus necessary to determine whether any core values exist that rational persons who assent to that core will not be willing to sacrifice because they disagree over matters that they must, if they are sincere and rational, admit that others who are sincere and rational can disagree with them about. Indeed such a consensus currently seems to exist in the Human Rights Conventions, especially the European Convention on Human Rights (ECHR), to which all EU Member States are party. While there is nothing explicit in the ECHR about embryos, it contains some indications that the moral status of embryos has to be viewed differently from that of a person (defined as a rational agent). This is indicated by the fact that many of the rights of the ECHR, which are civil and political rights, cannot be exercised by embryos. Even rights that could in principle be held to apply to the embryo (such as the Article 2 right to life) have not been extended by the European Court of Human Rights to the unborn. However, this does not mean that duties to the unborn cannot be inferred from the Convention rights. It is at least arguable that they can be derived using precautionary and proportionality reasoning.

But, in the final analysis, and regardless of whether issues of embryo research or similar biopolitical questions are to be solved on a European or national level, we ought to address our institutional structures and ask whether or not they are designed to maximise accountability and transparency, reviewability, and reason giving. At present we, above all, need to establish adequate safeguards against conflicts of interests and outright corruption and ensure that decisions on these matters are conducted in good faith and competently.

Zusammenfassung

Die Forschung an menschlichen Embryonen stößt vielfach auf grundsätzliche Ablehnung – ethische bzw. moralische sowie religiös motivierte Gründe werden gegen eine solche Forschung vorgebracht. Die Ablehnung erweist sich aber häufig als unvereinbar mit den Ansprüchen anderer und oft auch dem gesamtgesellschaftlichen Interesse, problematische Forschungsbereiche erst dann aufzugeben, wenn ein angemessenes Verständnis der in Frage stehenden Sachverhalte und eine realistische Abschätzung der Folgen erreicht ist, die etwaige politische und rechtliche Entscheidungen für alle betroffenen Parteien hätten. Erst nachdem die Sachzusammenhänge geklärt und alle wesentlichen Aspekte sorgfältig abgewogen sind, sollte über die Akzeptabilität oder Nicht-Akzeptabilität solcher Forschung entschieden werden.

Die Darstellung der relevanten biologischen und klinischen Zusammenhänge bildet daher einen ersten Schwerpunkt der vorliegenden Studie (Kapitel 1). Dabei wird nicht zuletzt auch der Stand der Forschung in der regenerativen Medizin und der Stammzellforschung in den Blick genommen (Kapitel 2). – Wissenschaftliche Fakten erlauben jedoch noch keine Rückschlüsse auf die Akzeptabilität politischer Entscheidungen. Einen weiteren Schwerpunkt bildet daher die Analyse der unterschiedlichen rechtlichen Regulierungen der Embryonenforschung in Europa, verbunden mit der Frage, ob eine Harmonisierung der Gesetzgebung auf europäischer Ebene möglich ist (Kapitel 3). Eine Antwort auf diese Frage wird nicht nur davon abhängen, welche Spielräume die bereits bestehenden Regulierungen für ein solches Harmonisierungsbemühen lassen, sondern auch davon, ob sich in der öffentlichen Meinung der jeweiligen Länder hinreichende Gemeinsamkeiten finden lassen. Im Rahmen der Studie wurde daher eine umfassende europaweite Umfrage zu Haltungen und zum Wissen in Fragen neuer medizinisch-technischer Entwicklungen wie der Embryonenforschung durchgeführt. Die Ergebnisse werden hier vorgestellt und diskutiert (Kapitel 4). Der letzte Teil will für die anstehenden politischen und rechtlichen Entscheidungen eine rationale Grundlage anbieten. Hierfür werden insbesondere auch die unterschiedlichen politischen Ansätze und die in der Debatte verwendeten Argumente und argumentativen Strategien einer kritischen philosophischen Prüfung unterzogen (Kapitel 5). Die Studie mündet in eine Liste von Handlungsempfehlungen, die die zentralen normativen Konklusionen noch einmal übersichtlich zusammenführt (Kapitel 6).

Wissenschaftliche und klinische Aspekte der Forschung an menschlichen Embryonen

Zurzeit findet wissenschaftliche Forschung an Embryonen v.a. in zwei Bereichen statt: einerseits in der reproduktionsmedizinischen Forschung, die auf die Überwindung ungewollter Kinderlosigkeit und die Vermeidung genetisch bedingter Schädigungen zielt, andererseits in der regenerativen Medizin, die die Entwicklung neuer Gewebe- oder Organersatztherapien vorantreiben will. Die dem forscherischen Handeln dabei primär zu Grunde liegenden Zwecksetzungen – ungewollt kinderlosen Paaren zu helfen und Krankheit, Leid und Tod zu vermeiden – sind in aller Regel moralisch unbeanstandet. Moralisch problematisch sind vielmehr die für die Erreichung der Zwecke eingesetzten *Mittel*, wenn für die Forschung die Zerstörung von Embryonen in Kauf genommen wird, sowie die Möglichkeit, die Forschungs*ergebnisse* für andere als die genannten primären Zwecke zu verwenden. Solche weiterreichenden Zwecke werden etwa in der pränatalen Geschlechtswahl gesehen oder in der genetischen Manipulation von Embryonen ohne klare medizinische Indikationen. Daher ist es für eine angemessene ethische Beurteilung der Embryonenforschung unabdingbar, sie nicht lediglich isoliert, sondern im Lichte bereits etablierter und weithin akzeptierter Verfahren zu betrachten, die zwar auf die Erreichung derselben primären Zwecke gerichtet sind, aber hierfür andere Mittel einsetzen und nicht die Gefahr bergen, dass etwaige Forschungsergebnisse für die genannten nicht akzeptierten weiterreichenden Zwecke einsetzbar wären.

Mit Bezug auf die *reproduktionsmedizinische Forschung* wird so vor allem die In-Vitro-Fertilisation (IVF) als Alternative angeführt. Sie ist inzwischen in den meisten Gesellschaften als Mittel zur Überwindung ungewollter Kinderlosigkeit weitgehend akzeptiert und gesetzlich reguliert. Diese Tatsache ist für die weitere Diskussion insbesondere der Frage nach dem einzusetzenden Mittel insofern interessant, als zum einen die Entwicklung der IVF sehr wohl ebenfalls Embryonenforschung erfordert hat (und im Zuge der Optimierung dieser Technik noch weiter erfordert), und insofern zum anderen die Anwendung dieser Technik regelmäßig mit der Zerstörung von Embryonen einhergeht: Nicht alle der im Rahmen einer IVF-Behandlung erzeugten Embryonen können zum Zwecke einer Schwangerschaft übertragen werden, und es ist in der Regel nicht möglich, die überzähligen Embryonen zeitlich unbefristet aufzubewahren – überzählige, d.h. nicht transplantierbare Embryonen müssen letztlich zerstört bzw. aufgegeben werden. Angesichts dieser Tatsache scheint die Frage berechtigt, ob nicht deren vorherige Verwendung im Rahmen der biomedizinischen Forschung gestattet werden sollte.

Aber auch bzgl. der Gefahr eines über die primäre Zwecksetzung hinausgehenden Einsatzes der Forschungsergebnisse ist die IVF keineswegs harmlos: Ein im Zusammenhang mit der IVF gegenwärtig breit diskutiertes Problem liegt in den inzwischen bereitstehenden Möglichkeiten der Präimplationsdiagnostik (PGD). Diese zielt gerade nicht auf die bloße Überwindung ungewollter Kinderlosigkeit, sondern auf die Wahl und Steuerung der genetischen Ausstattung eines Embryos, indem etwa Embryonen, bei denen eine genetisch bedingte Erkrankung diagnostiziert werden konnte, nicht zur Reife gebracht werden. Hier aber sind die Grenzen des rechtfertigbaren und des nicht-rechtfertigbaren Einsatzes der Technik keineswegs von vornherein klar: Sollen nicht alle, aber wenigstens einige der diagnostizierbaren genetischen Eigenschaften eine Entscheidung gegen das Transferieren des betroffenen Embryos rechtfertigen, dann ist die Aufgabe einer klaren Abgrenzung gestellt. Dabei steht – entgegen dem durch die öffentlichen Debatten erzeugten Eindruck – gar nicht einmal die Befürchtung im Vordergrund, durch die Selektion auch nicht krankheits-relevanter genetischer Prädispositionen könnten in Zukunft „Designerbabys" geschaffen werden. Solche Bedenken erweisen sich aus wissenschaftlicher Sicht als weitgehend unbegründet, insofern die Möglichkeit einer gezielten genetischen Beförderung von Eigenschaften wie Intelligenz, Schönheit oder künstlerischer Begabung als sehr begrenzt gelten muss. Charakteristika wie diese werden durch eine so große Zahl verschiedener Faktoren gesteuert, dass die Möglichkeit der gezielten Auswahl eines Embryos, dessen Genom solche favorisierten Eigenschaften aufweist, praktisch nicht gegeben ist. Abgrenzungsprobleme erzeugen vielmehr andere Anwendungen der PID, wie etwa die Möglichkeit, Embryonen mit dem Ziel zu selektieren, einen geeigneten Zell- oder Organspender für ein bereits vorhandenes Kind zu erzeugen. Einsatzmöglichkeiten wie diese unterstehen keineswegs der pri-mären Zwecksetzung der Reproduktionsmedizin, ungewollte Kinderlosigkeit zu überwinden, haben aber bereits Eingang in die medizinische Praxis gefunden.

Die genetische Ausstattung eines Embryos lässt sich heute auch, zumindest theo-retisch, mittels Keimzelltherapie oder Zellkern-Transfer (Klonen) manipulieren. In der hierüber geführten Debatte wird vor allem die Frage diskutiert, ob nicht das reproduktive Klonen von Menschen aus kategorischen Gründen zu verbieten sei, insofern die Erzeugung eines menschlichen Wesens, für das von vornherein fest-steht, dass es genetisch identisch sein wird mit einem anderen, eine Verletzung des Prinzips der Menschenwürde darstelle. Dass eine so grundlegend andere Form der Fortpflanzung kategorische Ablehnung gerade aus diesem Grund erfährt, muss jedoch erstaunen: Unter dem Aspekt ihrer genetischen Identität lassen sich Klone biologisch wie zeitlich versetzte Zwillinge betrachten. Selbst dann aber, wenn man einen starken genetischen Determinismus unterstellte, wäre nicht anzunehmen, dass Zwillinge nur deshalb, weil sie genetisch mit ihrem Zwilling identisch sind, prinzi-piell ein schlechteres Leben führen als ihre Mitmenschen. Tragfähige Gründe für die Ablehnung des reproduktiven Klonens sind daher nicht so sehr mit Blick auf die Eigenschaften des im Sinne des Verfahrens perfekt gelungenen Resultats – des Klons – zu suchen als vielmehr in den Risiken des Verfahrens selbst: Erfahrungen im Bereich des reproduktiven Klonens bei Tieren weisen auf beträchtliche Risiken für den Klon selbst und dessen biologische Mutter hin, die als hinreichend angesehen werden dürfen, das reproduktive Klonen beim Menschen – zumindest beim gegen-wärtigen Entwicklungsstand der Technik – aus praktischen Gründen abzulehnen. Da es ferner bis auf Weiteres keineswegs klar ist, auf welchen Bedarf überhaupt mit dem Mittel des reproduktiven Klonens reagiert werden soll, sind die hohen Risiken und Kosten, die mit der weiteren Entwicklung dieser Technik verbunden sind, nicht zu rechtfertigen.

In der *regenerativen Medizin* erhoffen sich viele Forscher, dass sich in nicht all zu ferner Zukunft unter Verwendung embryonaler Stammzellen in Kombination mit dem sogenannten Tissue-Engineering, künstliche Gewebe und Organe entwickeln lassen, mit denen die gegenwärtigen Probleme der Transplantationsmedizin wie etwa der permanente Mangel an geeigneten Transplantaten oder die mit der Ver-wendung von Immunsuppressiva zusammenhängenden Infektionsrisiken überwunden werden können. Darüber hinaus erwartet man sich neue Erkenntnisse über die Entstehungsbedingungen bisher unbehandelbarer Erkrankungen und damit entscheidende Fortschritte in der Therapeutik. Auch für die regenerative Medizin gilt, dass die forscherisch verfolgten Zwecke in der Regel unstreitig sind, die dafür einzusetzenden Mittel jedoch kontrovers diskutiert werden, insofern sie die Erzeugung menschlicher embryonaler Stammzellen und die Zerstörung potentiellen menschlichen Lebens oder dessen Instrumentalisierung einschließen. Viele plädieren daher für ein vollständiges Verbot jedweder Embryonenforschung, die auf die Gewinnung menschlicher Stammzellen zielt.

Dabei wird oft darauf verwiesen, dass mit der Gewinnung adulter Stammzellen oder der Verwendung von Nabelschnurblut u.ä. eine moralisch weniger bedenkliche Alternative zur Verwendung embryonaler Stammzellen zur Verfügung stehe. Die Plastizität adulter Stammzellen, also deren Fähigkeit, sich in verschiedene Zelltypen umzuformen, biete dabei die Möglichkeit, patienteneigenes Zellmaterial als Ausgangsmaterial von Ersatztherapien zu verwenden und so die Forschung auf der Basis embryonaler Stammzellen oder das sogenannte therapeutische Klonen überflüssig zu machen. Tatsächlich weist jedoch die Verwendung adulter Stammzellen ebenso wie die Verwendung embryonaler Stammzellen Probleme auf, die eine klare und moralisch eindeutige Entscheidung für eine von beiden zum gegenwärtigen Zeitpunkt verbieten.

So bestehen etwa begründete Zweifel über eine hinreichende Pluripotentialität adulter Stammzellen. Zudem ist für viele der avisierten Zelltypen bisher nicht bekannt, ob adulte Stammzellen, Zellen aus Nabelschnurblut o.ä. über die Proliferationsfähigkeiten verfügen, die für deren erfolgreichen therapeutischen Einsatz notwendig wären. Nicht zuletzt muss die klinische Anwendung angesichts bislang ungeklärter Probleme der Stammzellidentifikation, der Isolation, Expansion, der Differenzierung und der funktionalen Integration der Zellen als fraglich gelten.

Demgegenüber scheinen embryonale Stammzellen über ein unbegrenztes Proliferationspotential in vitro zu verfügen – sie lassen sich offenbar in jeden gewünschten Zelltyp differenzieren und dabei leichter von anderen Zelltypen unterscheiden, so dass sie für patientenspezifische Therapien aufbereitet werden können. Andererseits ist aber die Differenzierung von embryonalen Stammzellen schwer zu steuern. Es erweist sich daher als notwendig, allererst sichere Methoden zu entwickeln, mit denen sich die für die Therapie geeigneten Zellen auszusortieren lassen. Darüber hinaus muss als besonders problematisch gelten, dass embryonale Zellen nach ihrer Injektion zur Bildung von Teratokarzinomen neigen – auch hier setzt der klinische Einsatz eine Beherrschung dieses Problems voraus. Wegen dieser Risiken sind ferner – insgesamt oder zumindest in großen Teilen – die gegenwärtig verfügbaren Zelllinien nicht für den klinischen Einsatz geeignet: Da sie in der Regel auf Mäusefibroblasten kultiviert worden sind, kann eine Kontamination mit exogenen Viren nicht ausgeschlossen werden. Die Etablierung neuer Zelllinien unter Verwendung neutraler Medien wäre daher unumgänglich. Die Verwendung vorhandener embryonaler Stammzellen wäre auch insoweit nicht unbedenklich, als hierfür ein weiterer Einsatz von Immunsuppressiva erforderlich wäre.

Zumindest theoretisch böte das therapeutische Klonen eine Möglichkeit, dieses Problem zu umgehen, insofern dabei der benötigte Zellkern einer patienteneigenen Körperzelle entnommen würde. In eine entkernte Eizelle transferiert ließe sich – wenn möglich wiederum unter Verwendung einer patienteneigenen Fibroblaste als Kulturmedium – eine Blastozyste entwickeln, aus der embryonale Stammzellen isoliert werden könnten. Diese wiederum müssten dann in vitro so manipuliert werden, dass sie sich in hinreichender Zahl differenzieren, um dabei auch die zur Therapie erforderlichen Zellen hervorzubringen, die schließlich in den Körper des Patienten zurück transferiert werden sollen. So vielversprechend dieser Ansatz auch erscheint – seine Realisierung sieht sich einer Vielzahl technischer Probleme gegenüber, von denen einige als nahezu unlösbar eingestuft werden. Vor allem erfordert die Durchführung der beschriebenen Prozedur einen beträchtlichen Zeitrahmen, der in vielen klinisch relevanten Fällen schlicht nicht zur Verfügung steht.

Vorläufig werden daher embryonale Stammzellen, insbesondere auch die auf dem Wege des therapeutischen Klonens gewonnenen, eher bei der Untersuchung genetischer Defekte oder beim Screening potentieller Pharmazeutika eine Rolle spielen. Sollen sie in Zukunft auch die Grundlage sicherer und verlässlicher therapeutischer Verfahren bilden, dann setzt dies einen großen forscherischen Aufwand voraus. Eine solche Forschung wird aber keinesfalls aussichtsreich betrieben werden können, solange sie auf die Verwendung nur eines bestimmten Zelltyps oder bereits vorhandener Zelllinien beschränkt wird. Unter forschungspragmatischen Gesichtspunkten scheint es daher geboten, verschiedene Ansätze nebeneinander zu verfolgen und dabei sowohl von adulten als auch von embryonalen Stammzellen auszugehen.

Wie immer es aber um die Lösbarkeit der beschriebenen biologisch-medizinischen Probleme steht – der moralisch problematische Kern der Debatte, dass für die Gewinnung embryonaler Stammzellen potentielles menschliches Leben zerstört werden muss, bliebe davon unberührt. Die angemessene Behandlung dieser und anderer ethischer Probleme wie z.B. die Möglichkeit, dass eine Freigabe des therapeutischen Klonens dem reproduktiven Klonen Vorschub leisten könnte, erfordern auch die Einbeziehung rechtswissenschaftlicher, soziologischer und philosophischer Aspekte.

Rechtliche Regulierungen der Embryonenforschung

Grundsätzlich lassen sich drei regulative Ansätze zum Umgang mit der Forschung an menschlichen Embryonen unterscheiden: (1) eher liberale Ansätze, wie bspw. in Großbritannien, eines der wenigen Länder in der EU, die Embryonenforschung ausdrücklich erlauben; (2) restriktive Ansätze, wie bspw. in Deutschland, wo Forschung an Embryonen nur dann erlaubt ist, wenn sie dem Embryo selbst unmittelbar zu Gute kommt und schließlich (3) differenzierende Ansätze wie bspw. in Spanien, das Embryonenforschung unter ganz bestimmten Bedingungen erlaubt, unter anderen verbietet. Eine nähere Betrachtung der Realisierung dieser drei Ansätze in den genannten Ländern unter Einbeziehung der jeweiligen historischen Hintergründe verspricht einigen Aufschluss über die Aspekte, die für eine gemeinsame europäischen Lösung entscheidend sein werden. Dabei ist auch ein Seitenblick zu werfen auf verwandte Bereiche wie etwa die Gesetzgebung zur In-Vitro-Fertilisation, zur Präimplantationsdiagnostik und zum Schwangerschaftsabbruch.

Regulationen der Embryonenforschung in Großbritannien, Spanien und Deutschland

Bereits bei der Bestimmung des Objekts der zu regulierenden Forschung, bei der Definition dessen also, was in den Rechtstexten als Embryo bezeichnet werden soll, bestehen erhebliche Unterschiede zwischen den europäischen Staaten. Das Spektrum reicht von Bestimmungen wie „a living group of cells resulting from fertilisation not implanted in a woman's body" bis zur Unterscheidung zwischen einem „Prä-Embryo", worunter die befruchtete Eizelle bis zum 14. Tag nach der Befruchtung oder bis zur Implantation verstanden wird, und dem Embryo selbst als der weiteren Entwicklungsphase zwischen dem 15. Tag und 2 und 1/2 Monaten. Die Gesetzgebung einiger Länder vermeidet sogar insgesamt nähere Bestimmungen.

Selbstverständlich haben solche Begriffsbestimmungen erhebliche Auswirkungen auf den regulativen Rahmen, der durch die Gesetzgebung gesetzt werden soll, indem sie etwa bereits einen Zeitrahmen für erlaubte Forschung setzen oder festsetzen, welche Forschungsobjekte überhaupt durch das Gesetz geschützt werden und welche nicht. Lebende Zellgruppen etwa, die durch Kernaustausch erzeugt werden, entstehen nicht durch „fertilisation" und fallen so im strengen Sinne nicht unter die oben zitierte Bestimmung der British Human Fertilisation and Embryology Authority (HFEA). Gleichwohl haben englische Gerichte jüngst entschieden, dass auch so erzeugte Zellen unter die Bestimmung fallen, insofern die Definition bei sachangemessener Deutung sämtliche Zellen einschließe, die sich – unabhängig von ihrer Genese – zu menschlichen Individuen entwickeln können. – Eine solche Flexibilität macht deutlich, dass mit den Unterschieden in den begrifflichen Bestimmungen dem Bemühen um eine Harmonisierung der Rechtsprechung auf europäischer Ebene keine prinzipiellen Hindernisse entgegenstehen.

Entscheidendere Probleme sind im Zusammenhang mit den Haltungen und Überzeugungen zu sehen, die jeweils hinter den verschiedenen Regulierungsansätzen stehen und die im Rahmen der politischen Entscheidungsfindung ihre eigene Dynamik entwickeln können. So etwa korrespondieren den drei Regulierungsansätzen drei grundsätzlich verschiedene Auffassungen bzgl. des moralischen Status eines menschlichen Embryos: (1) die libertäre Auffassung, der zufolge Embryonen keinerlei speziellen Schutzes bedürfen und daher Regelungen überhaupt nicht notwendig sind; (2) die pro-life-Auffassung, der zufolge Embryonen einen prinzipiellen, keinerlei Abwägung zugänglichen Lebensschutz genießen sollen; (3) die Proportionalitätsauffassung, der zufolge sich das Maß des Lebensschutzes durch Abwägung mit anderen Rechtsgütern bestimmt. Erhält nun in der Beratung keine dieser drei grundsätzlich verschiedenen Überzeugungen eine politisch hinreichende Unterstützung durch eine Majorität, bildet sich vielmehr nur eine Allianz zwischen den Vertretern der Proportionalitätsauffassung und einem der beiden anderen Lager, so ist als Ergebnis eine regulative Politik zu erwarten, also eine Politik, in der anders als von Vertretern der libertären Auffassung gefordert, Regelungen irgendwelcher Art für die Embryonenforschung getroffen werden: Der Vertreter einer libertären Auffassung wird mutmaßlich zu Kompromissen bereit sein und gegen seine grundsätzliche Auffassung einer Regulierung zustimmen – für ihn ist es besser, wenigstens in einigen Fällen Embryonenforschung freigeben zu können als ein generelles Verbot zu riskieren. Zudem wird auch er den Gewinn an Transparenz und eindeutiger Zurechenbarkeit, der mit einer expliziten Regulierung einhergeht, begrüßen. Demgegenüber wird von Vertretern der pro-life-Auffassung eher nicht erwartet werden können, dass sie gegen ihre prinzipiellen Überzeugungen in Kompromisse einwilligen – dies um so weniger, je mehr diese Überzeugungen in religiösem Glauben verankert sind.

Tatsächlich kann die in Großbritannien etablierte Politik der geregelten Prüfung zum Teil als Ergebnis wechselnder Allianzen zwischen den Positionen gesehen werden, was nicht zuletzt auch durch den pragmatischen Stil im politischen und juristischen Handeln ermöglicht wurde.

Die Situation in anderen Staaten, insbesondere auch diejenige in Deutschland, weist hierzu deutliche Unterschiede auf. So stand z.B. die Formulierung des deutschen Embryonenschutzgesetzes (EschG), deutlich unter den Erfordernissen der föderalen Struktur Deutschlands und den Vorgaben des deutschen Grundgesetzes. Mit Blick auf neuere Entwicklungen in der Stammzell-Forschung wurde es 2002 um ein Stammzellgesetz (StZG) ergänzt. Ein solches ergänzendes Gesetz hatte sich als notwendig erwiesen, da das 1990 verabschiedete EschG grundsätzlich keinerlei Embryonenforschung zuließ, die mit der Zerstörung von Embryonen einhergeht. Es verbietet die Herstellung von Stammzellen in Deutschland sowie die Beteiligung deutscher Forscher an deren Herstellung im Ausland, gestattet aber den Import von Zelllinien, die unter bestimmten Bedingungen und vor einem bestimmten Zeitpunkt im Ausland entstanden sind. Regelungen wie diese sind, indem sie von vorneherein transnationale Aspekte beinhalten, von besonderem Interesse, wenn es um Fragen der internationalen Harmonisierung von Regulierungen der Embryonenforschung geht. Vor allem wäre dabei zu prüfen, ob Bestimmungen wie diese überhaupt konsistent erlassen werden können: Deutschland hat eine der prohibitivsten Rechtsprechungen im Bereich der Embryonenforschung überhaupt. Dabei steht klar der individuelle Schutz des Embryos im Vordergrund. Es ist aber mit dieser Zielsetzung nicht verträglich, dass man zugleich Möglichkeiten schaffen will, von der Forschung zu profitieren, die andere unter weniger restriktiven Regelungen betreiben. Ein weiteres Konsistenzproblem besteht darin, dass durch die genannte Gesetzgebung für den Embryo in vitro stärkere Schutzbestimmungen errichtet werden als für den Embryo in vivo. Während die deutsche Gesetzgebung § 218a StGB den Schwangerschaftsabbruch von Strafe freistellt, wenn dadurch nach ärztlicher Erkenntnis und unter Berücksichtigung der Lebensumstände Gefahren für die seelische oder körperliche Gesundheit der Schwangeren abgewendet werden, verbietet das EschG das Screening von Embryonen vor der Implantation auch dann, wenn eine Implantation mit Risiken dieser Art einherginge. Dies führt etwa dazu, dass viele Paare, die bei einer geplanten In-Vitro-Fertilisation solche Risiken fürchten, sich für die Durchführung einer Präimplantationsdiagnostik ins benachbarte Ausland begeben, wo andere gesetzliche Regelungen gelten.

Die deutsche Situation ist zum Teil durch die verfassungsrechtlichen Vorgaben zu erklären. Die durch solche konstitutionellen Vorgaben gesetzten Rahmenbedingungen sind stets zu berücksichtigen, wenn über eine europäische Harmonisierung nachgedacht wird. Allerdings darf die Verbindlichkeit und die Eindeutigkeit solcher Vorgaben auch nicht überschätzt werden. So haben sich etwa in Spanien eher pragmatische und weniger restriktive Lösungen gegen die kategorischen Alternativen durchgesetzt, obwohl auch hier eine verfassungsrechtliche Ordnung im Hinter-

grund steht. Das Gesetz 35/1988 „sobre Técnicas de Reproducción Asistida"
erlaubt bestimmte Forschungen an überzähligen menschlichen (Prä-)Embryonen,
solange diese noch nicht über den vierzehnten Tag ihrer Entwicklung hinaus oder
bereits implantiert sind. Bereits implantierte Embryonen werden durch ein weiteres
Gesetz geschützt, das den Umgang mit Embryonen und aus Embryonen gewonnenen Materialien regelt, darunter auch die Spende von Zellen, Geweben und Orga-
nen. Danach ist z.B. für die Forschung an (Prä-)Embryonen als auch an bereits
implantierten Embryonen stets die schriftlich niedergelegte informierte Einwilligung der Gametenspender einzuholen – und auch dann ist die Forschung nur in entsprechend lizensierten Gesundheitseinrichtungen durch entsprechend lizensierte
Teams zulässig.

In den genannten und weiteren anderen Momenten steht die spanische Regulierung der britischen sehr viel näher als der deutschen. Anders als die britische jedoch
unterscheidet die spanische Gesetzgebung zwischen lebensfähigen und nicht-
lebensfähigen Embryonen: Embryonen in vitro sollen generell nur für diagnosti-
sche, therapeutische oder präventive Zwecke zur Verfügung stehen, also nur für solche Forschung, die dem betroffenen Embryo selbst zu Gute kommt. Darüber
hinausgehende forscherische Aktivitäten sind auf solche Embryonen beschränkt,
die für nicht-lebensfähig gelten müssen, wie etwa abgetriebene Embryonen. Diese
Regelungen gelten etwa auch für Aktivitäten, die im Zusammenhang mit der Präimplantationsdiagnostik stehen. So ist es dem Arzt in Spanien erlaubt, von einem
Transfer des (Prä-)Embryos abzuraten wenn eine erbliche, nicht heilbare Krankheit
diagnostiziert wird, da der rechtliche Schutz des Embryo in vitro hinter dem des
Embryo in utero zurücksteht.

Ist eine europäische Harmonisierung möglich?

Bieten die wenigen Gemeinsamkeiten der unterschiedlichen nationalen Regulierungen eine hinreichende Basis, um ernsthaft über eine europäische Harmonisierung
im Bereich der Embryonenforschung nachzudenken? Oder ist angesichts der viel-
fältigen nationalen Differenzen in moralischen, sozialen und ökonomischen Fragen
eher eine Beibehaltung der nationalen Unterschiede sowohl im Interesse der restriktiv als auch der eher liberal verfahrenden Länder vorzuziehen?

In Staaten mit restriktiver Gesetzgebung geben v.a. moralische und religiöse
Argumente den Ausschlag dafür, dass das Leben vom Moment der Empfängnis an
als vollständig schützenswert erachtet wird. Weitere bestimmende Faktoren können,
wie bspw. im Falle Deutschlands, in besonderen historische Bedingungen liegen.
Demgegenüber plädieren diejenigen, die die moralischen oder religiösen Einstel-
lungen und etwaigen historisch bedingten Befürchtungen nicht teilen, für liberalere
Regulierungen, indem sie etwa den drohenden Verlust an wissenschaftlichem Prestige ins Feld führen oder auf das Missverhältnis verweisen, das zwischen der Nut-
zung von Resultaten aus der Embryonenforschung in anderen Ländern und dem
Mangel an Bereitschaft zu eigener Forschung besteht. Auch wird bezweifelt, dass
die moralischen Bedenken restriktiv verfahrender Staaten tatsächlich von so funda-
mentaler Bedeutung seien wie behauptet, da sie doch nichts an der Bereitschaft
ändern, sich mit Embryonenforschung freizügig gestattenden Staaten zu assoziie-
ren. Nach einer solchen liberaleren Auffassung besäße der Embryo lediglich einen

relativen Status, der sich aus der Abwägung mit dem potentiellen Nutzen seiner forscherischen Verwendung für andere ergibt. Konsequenterweise werden in Fragen
der Embryonenforschung liberal verfahrende Staaten ein Verbot nur unter entspre-
chendem ökonomischen oder politischen Druck erwägen.

In einer solchen Konstellation scheint der Verzicht auf eine Harmonisierung Vor-
teile für alle zu haben: Die kulturelle Souveränität der restriktiv verfahrenden Staaten würde bewahrt, so dass sie uneingeschränkt an ihren moralischen Prinzipien
festhalten könnten, während liberal verfahrende Staaten von den ökonomischen
Vorteilen, die die Forschung eröffnet, und vom Zugewinn an wissenschaftlichem
Prestige profitieren könnten.

Trotz solcher Argumente darf davon ausgegangen werden, dass zumindest einige
positive Regelungen auch dann gerechtfertigt sind, wenn grundsätzlich von einer
Harmonisierung abgesehen werden soll. So etwa mögen sich Regelungen als notwendig erweisen, die eventuelle wettbewerbliche Nachteile restriktiv verfahrender
Staaten auszugleichen im Stande sind. Insofern aber die Verfassung der Europäischen Union es kaum gestattet, Steuern von den Mitgliedstaaten zu erheben, und
mögliche Förderungsstrategien nur sehr eingeschränkt als Steuerungsinstrument in
Frage kommen da die Forscher ja durch nichts gehindert sind, auch anderweitig
nach Fördermöglichkeiten Ausschau zu halten, muss es als eher wahrscheinlich
gelten, dass sich die Gemeinschaft in Richtung auf eine Harmonisierung bewegt,
die dann freilich einen eher freizügigen Charakter haben wird. Dabei wird auch der
Einfluss der neuen assoziierten Staaten nicht unerheblich sein. Diese verfügen im
allgemeinen zwar nicht über die ökonomische Macht, den Gang der europäischen
Gesetzgebung wesentlich zu beeinflussen. Es ist jedoch zu erwarten, dass sich
gerade deshalb bei den Administrationen dieser Länder eine freizügigere Haltung
durchsetzt, um entsprechende Forschungsinvestitionen zu befördern. Diese Prog-
nose kann auch für diejenigen Staaten aufrechterhalten werden, in denen kein entsprechender kultureller Rückhalt zu erwarten ist, ja selbst dann, wenn sich ganz im
Gegenteil in ähnlich gelagerten Diskussionen, etwa in der Abtreibungsfrage,
gezeigt hat, dass in der Mehrheit eher eine ablehnende Haltung vorherrscht.

Haltungen und Wissen in Fragen der Biotechnologie, des Klonens und der Stammzellforschung

Die Komplexität der modernen wissenschaftlichen Forschung geht zunehmend mit
konträren, teilweise auch einander widersprechenden Haltungen, Vorstellungen und
Überzeugungen über die Forschung selbst und der mit ihr verbundenen Chancen
und Risiken einher. Die dadurch erzeugten Konflikte auf individueller ebenso wie
auf gesellschaftlicher Ebene stellen alle Versuche einer gemeinsamen europäischen
Politik vor schwierige Aufgaben. Insbesondere ist auch der offensichtliche Wider-
stand breiter Teile der Öffentlichkeit gegen den wissenschaftlichen Fortschritt ernst
zu nehmen. Dabei handelt es sich nämlich keineswegs lediglich um blinde Maschi-
nenstürmerei aus einem Mangel an wissenschaftlicher Informiertheit, wie gelegent-
lich behauptet wird – die Wirklichkeit ist ungleich komplexer. Die im Laufe des
Projekts erstellte systematische Erhebung der Haltungen europäischer Bürger
gegenüber der Wissenschaft, der Wissenschaftlergemeinschaft, der Embryonenfor-

schung und der bestehenden Regulierungsansätze, ihrer Werturteile über das „natürliche Leben" und ihrer Wahrnehmung von Risiken erlaubt deutlichere Schlussfolgerungen über die Gründe des verbreiteten Widerstands gegenüber der Embryonenforschung und der dahinter stehenden Strukturen.

Grundsätzlich genießen Wissenschaftler nach wie vor eine hohe Reputation, und das Vertrauen der Öffentlichkeit in zukünftige wissenschaftliche Entwicklungen ist ebenso ungebrochen hoch wie die Erwartungen. Zudem zeigt sich bei größeren Teilen der Öffentlichkeit eine zunehmende Fähigkeit, zwischen einzelnen Bereichen der Forschung und der Anwendung der Forschungsergebnisse zu unterscheiden sowie zwischen reiner Forschung, angewandter Forschung und klinischer Praxis. Es ist daher kein Widerspruch, wenn einer generell optimistischen Haltung skeptische Haltungen in Bezug auf spezifische Anwendungen gegenüberstehen, die, wie die erfassten Daten belegen, eher von vagen Vorstellungen, diffusen Ängsten, Stereotypen und vorgefassten Meinungen bestimmt sind als durch Wissen und Information.

Insbesondere die Haltung gegenüber dem Klonen, gleich ob es sich dabei um das therapeutische oder das reproduktive Klonen handelt, ist durch angstbesetzte Vorstellungen bestimmt. Dabei wird die Rede von „dem Gen" vielfach wie ein kulturelles Symbol für alles das verwendet, was „natürlich" ist und der Manipulation für illegitime und „unnatürliche" Zwecke schutzlos ausgeliefert. Solche Ängste, verbunden mit der Befürchtung, dass die Einführung des Klonens für therapeutische Zwecke zu einem Dammbruch führen und die Einführung des reproduktiven Klonens beim Menschen unausweichlich nach sich ziehen könnte, haben das Klonen in das Zentrum der gegenwärtigen öffentlichen Debatte gerückt. Andererseits sind – verglichen mit anderen Forschungsbereichen – die Haltungen gegenüber der Embryonenforschung am deutlichsten durch moralisch-religiöse Grundüberzeugungen geprägt. Damit einher geht eine enorme Diversität der Haltungen sowohl zwischen den europäischen Staaten als auch innerhalb der jeweiligen Gesellschaften. Besonders die Frage nach dem moralischen Status des menschlichen Embryos wird in sehr unterschiedlicher Weise beantwortet.

Die faktischen moralischen Urteile sind keine durch systematische Überlegungen oder gar aus einem einheitlichen Prinzip stringent hergeleite Erkenntnisse. Sie bilden sich vielmehr kontextabhängig und immer im Zusammenhang mit bestimmten Szenarien aus, etwa der Situation, in der die Entscheidung ansteht zwischen einer Forschung an überzähligen Embryonen und an solchen, die eigens zu Forschungszwecken hergestellt wurden. Der Datenlage nach bevorzugen Europäer klar die Forschung an überzähligen Embryonen, auch wenn sich wichtige „Minoritäten" deutlich gegen beide Optionen aussprechen oder noch keine Überzeugung in dieser Frage gebildet haben. Diese „Minoritäten" sind besonders in überwiegend katholischen Ländern (mit Ausnahme von Spanien) von Bedeutung, wohingegen in anderen Länder eher die differenzierenden und liberalen Haltungen zum Einsatz der Embryonenforschung vorherrschen.

Sofern man einen direkten Zusammenhang zwischen der gesellschaftlichen Auffassung über den moralischen Status des Embryos und der Weise der Regulierung der Embryonenforschung unterstellen darf, ist mit restriktiven Regulierungen in Österreich, Deutschland, Italien und Polen zu rechnen. Eher permissive Regulierungen darf man in Dänemark, Großbritannien und den Niederlanden erwarten. Spanien und Frankreich stehen in der Mitte, insoweit dort je nach den in Frage stehenden Aspekten eher restriktive oder eher permissive Regelungen zu erwarten sind. Aus Sicht der Allgemeinheit ist dabei aber in der Regel keinesfalls ein vollständiges Verbot der Embryonenforschung zu akzeptieren, bevor nicht ernsthafte Versuche unternommen worden sind, die Chancen und Risiken gegeneinander abzuwägen.

Trotz der signifikanten Unterschiede in der öffentlichen Meinung, in den Haltungen und Werturteilen ist die Befürwortung eines gemeinsamen europäischen Rahmens für die Embryonenforschung vorherrschend. Insofern mittelfristig der Erfolg oder Misserfolg einer politischen Entscheidung wesentlich auch von der öffentlichen Meinung abhängen wird, sollte daher ein zumindest weitgehender Konsens über die wesentlichen Fragen angestrebt werden. Dieser aber kann nur erreicht werden, indem eine rationale und vollständig öffentliche Debatte geführt wird. Eine solche hat bisher, trotz aller Aufmerksamkeit der Medien in Fragen der modernen biomedizinischen Forschung, noch nicht in hinreichendem Maße stattgefunden.

Die gegenwärtige Debatte ist stark von den Diskussionen über das Klonen dominiert, dessen Risiken nach Auffassung der überwiegenden Mehrheit der Europäer in keinem Verhältnis zum erwartbaren Nutzen stehen. Um einen konstruktiven Konsens in den zentralen Fragen der Embryonenforschung möglich zu machen ist es hier erforderlich, präzise und vollständige Informationen bereitzustellen, so dass Individuen und Gesellschaften zu einer eigenen kritischen Abschätzung und Bewertung der Chancen und Risiken gelangen können. Bereits jetzt scheint jedoch eine weitgehend permissive gesamteuropäische Regelung zur Forschung an überzähligen, bei künstlicher Befruchtung nicht implantierten Embryonen möglich. Stets jedoch müssen Ansätze zu einer gesamteuropäischen Regelung den ausgeprägten Pluralismus sowohl innerhalb der einzelnen Gesellschaften als auch der Gesellschaften als Ganzer berücksichtigen – nicht nur der Vielzahl der religiösen Überzeugungen mit ihrer je eigenen Aufmerksamkeit für die Belange des menschlichen Embryos, sondern auch den Überzeugungen derer, die keinem Glauben anhängen und jedenfalls keine kategorischen Vorbehalte gegen eine Embryonenforschung für biomedizinische Zwecke hegen, gilt es mit einer solchen Regelung gerecht zu werden.

Philosophische Analyse verbreiteter Argumente und argumentativer Strategien

Unerachtet der Tatsache, dass die öffentliche Meinung für politische Entscheidungen eine offensichtlich beachtliche Bedeutung hat, kann sie doch kein hinreichendes Kriterium für die Richtigkeit einer bestimmten Politik sein. Soll eine politische Entscheidung nicht nur eine (mehr oder minder allgemein) faktisch akzeptierte, sondern eine akzeptable Entscheidung sein, dann sind auch die hinter den Überzeugungen und Wertungen stehenden Gründe und Argumentationen in den Blick zu nehmen. So etwa wird von Befürwortern einer restriktiven Politik häufig betont, dass Embryonen ein besonderer moralischer Status zukomme insofern sie menschliche Wesen sind bzw. das Potential haben, sich zu Menschen im Vollsinne zu entwickeln. Aufgrund dieser Eigenschaft stehe ihnen vom Zeitpunkt der Kernverschmelzung an dasselbe uneingeschränkte Recht auf Leben und Unversehrtheit zu wie geborenen Menschen. Ein prominenter Vertreter dieser Position ist etwa die katholische Kirche. Aus dieser (pro-life-)Position ergibt sich unmittelbar, dass alle Forschung, die nicht dem direkten Nutzen des jeweiligen Embryos dient, gegen die elementarsten Rechte verstößt und insofern zu verwerfen ist. Zu verbieten wäre danach also etwa jedwede Gewinnung von Stammzellen oder die Präimplantations- und Pränataldiagnostik, insofern diese ja mit dem Ziel eingesetzt werden, bei entsprechender Indikation von der Implantation abzusehen bzw. einen Schwangerschaftsabbruch einzuleiten. Vertreter dieser Position unterscheiden konsistenterweise auch nicht zwischen aus Behandlungen mit IVF hervorgehenden überzähligen Embryonen und Embryonen, die eigens zu Forschungszwecken erzeugt werden, insofern jedem Embryo unabhängig von seiner Genese grundsätzlich das Recht zustehe, sich zum vollwertigen Menschen zu entwickeln.

Diese Position sieht sich nicht zuletzt auch wegen ihrer starken metaphysischen Voraussetzungen dem Widerspruch sowohl der Befürworter einer uneingeschränkten Forschungsfreiheit als auch der Befürworter einer beschränkten, unter bestimmten rechtlich festgesetzten Bedingungen aber feizugebenden Embryonenforschung ausgesetzt. Vertreter einer grundsätzlichen und umfassenden Liberalisierung der Embryonenforschung etwa entgegnen, dass es keiner Institution gestattet sein darf, die Berücksichtigung bestimmter moralischer Werthaltungen auch von denjenigen zu erzwingen, die diese Haltungen nicht teilen – dies gelte insbesondere dann, wenn die entsprechenden Haltungen allein auf nicht weiter begründeten traditionellen Wertvorstellungen beruhen, die nicht unbedingt von jedermann geteilt werden müssen.

Das Argument ist jedoch wenig überzeugend: Es enthält die verborgene Prämisse, dass alles, was nicht Person ist, eine beliebig für jedermann verfügbare Sache darstellt. Diese Alternative besteht aber so nicht: Es bedarf keineswegs des Rückgriffs auf metaphysische Annahmen, um die Beschränkung der Verfügbarkeit von etwas beliebigem, das nicht Person ist, zu rechtfertigen – in zahllosen alltäglichen Situationen akzeptieren wir vielmehr eine solche Beschränkung bereits dann, wenn andere legitime, auf die in Frage stehende Sache gerichtete Ansprüche und Interessen bekunden. Analog kann bereits ein Abrücken von einer grundsätzlich liberalen Regelung der Embryonenforschung geboten sein, wenn auf den Embryo andere als bloß forscherische Interessen gerichtet sind. Solche Interessen müssen dabei nicht einmal explizit geäußert werden, sondern dürfen anderen unter plausiblen Annahmen auch zugeschrieben werden – so etwa auch gerade den aus manipulierten Embryonen bei entsprechender Förderung sich entwickelnden Kindern. So sind Interessenskonflikte zwischen Forschern und Eltern auf der einen und den in Frage stehenden werdenden Kindern ohne Weiteres denkbar in Fällen, in denen etwa das Kind mit dem begründeten Verdacht aufwachsen müsste, nicht um seiner selbst willen, sondern aus rein instrumentellen Gründen ge- bzw. erzeugt worden zu sein, z.B. zu dem Zweck, als Gewebespender für ein Geschwister zu dienen.

Die argumentative Lage lässt weder eine völlig unreglementierte Forschung noch ein vollständiges Verbot jeglicher Embryonenforschung als akzeptable politische Lösung vermuten. Vielmehr wird eine Lösung anzustreben sein, die es erlaubt, die jeweiligen Vorzüge und Nachteile verschiedener Forschungs- und Therapieansätze abzuwägen und die Chancen einer solchen Forschung nach Möglichkeit zu nutzen, die Risiken aber zu vermeiden. Eine solche Position kann als regulatorische Position bezeichnet werden. Die Umsetzung eines solchen Ansatzes erfordert, dass sowohl die Forschung als auch die klinische Praxis durch eine Gesetzgebung geregelt und kontrolliert werden, die die Vermeidung bzw. Minderung von Schädigungen oder Risiken ermöglicht ohne dabei die legitimen Interessen der beteiligten Partein zu vernachlässigen. Ergänzend sind Prozeduren zu entwickeln, die in schwierigen Fällen eine für alle Seiten akzeptable Lösung herbeizuführen erlauben. Ein solcher schwieriger Fall wäre etwa gegeben, wenn die Gametenspender in Konflikt geraten über die weitere Entwicklung ihrer cryopräservierten Embryonen. Hier kann die besondere Bindung, die zwischen den Gametenspendern als genetischen Eltern und dem Gezeugten besteht, ein besonderes Mitspracherecht und eine stärkere Berücksichtigung ihrer Interessen gegenüber denen der Forscher und anderer beteiligter Parteien rechtfertigen. Auch die Interessen, die einem zu implantierenden Embryo mit Blick auf sein künftiges Leben zugeschrieben werden dürfen, verdienen besondere Berücksichtigung – auch wenn er zum gegenwärtigen Zeitpunkt noch nicht über diejenigen Eigenschaften oder Fähigkeiten verfügt, die wir für das Zuschreiben von Interessen üblicherweise voraussetzen.

Die öffentlichen Debatten, wie sie etwa in den Massenmedien oder in der parlamentarischen Beratung geführt werden, sind z. Zt. stark von Argumentationen und Argumentationsstrategien bestimmt, in denen die Erzeugung von Tabus, gezielte semantische Verschiebungen, eine suggestive Metaphorik und insgesamt die Verwendung emotionalisierender Sprache und rhetorischer Strategien vorherrschend sind. Von Embryonen wird als potentiellen Personen gesprochen, von der Würde des Embryos und den ihm zukommenden Rechten ist viel die Rede, Dammbrüche werden heraufbeschworen, Gerechtigkeit wird eingefordert und vor den dramatischen Folgen eines allzu leichtfertigen Umgang mit Risiken gewarnt. Soll in den zur Lösung der Probleme auch weiterhin anstehenden Debatten der Zugang zu rationalen und wohlinformierten Konsensen nicht durch unzutreffende Vorraussetzungen und durch inkorrekte Schlüsse verstellt werden, dann sind solche Argumentationen und Argumentationsstrategien einer kritischen Prüfung zu unterziehen.

So etwa besitzen Tabus und kulturell bedingte Voreingenommenheiten in der Regel keinen rationalen Kern oder haben ihre für frühere Gesellschaftsformen rekonstruierbare Bedeutung durch den kulturellen Wandel verloren, blockieren aber oft die notwendige vorurteilsfreie und von Emotionen weitestgehend freie Debatte. Auch die gerade in bioethischen Debatten vorherrschenden Natürlichkeitsargumente finden häufig eine eher durch kulturell bestimmte Intuitionen geleitete Zustimmung, halten aber einer rationalen Prüfung in der Regel nicht stand. So erfahren medizinische Techniken wie IVF, Stammzelltherapie oder die Selektion untherapierbar erbgeschädigter Embryonen durch Präimplantations- oder Pränataldiagnostik eine häufig strikte Ablehnung gerade unter Hinweis auf ihre „Unnatürlichkeit". Oft stehen dabei geistesgeschichtliche Traditionen im Hintergrund wie etwa das neuzeitliche Verständnis von der Unverletzlichkeit der – die Gleichheit aller und damit die Abwehrrechte des Bürgers gegen den Zugriff absoluter Herrscher begründenden – natürlichen Würde des Menschen oder die theologische Vorstellung vom Lauf der Natur als der göttlichen Vorsehung, in die einzugreifen dem Menschen nicht erlaubt ist. Man muss dabei nicht diese Traditionen in Frage stellen, um Natürlichkeitsargumente als letztlich nicht triftig zu erweisen: „Unnatürlich" ist jedes Resultat menschlichen Handelns, mit dem der homo faber sich gegen Bedrohungen durch „die Natur" zu schützen sucht – und „unnatürlich", aber in der

Regel wünschenswert, ist damit auch jeder Versuch, eine Krankheit zu heilen. Richtig ist aber auch, dass mit der immer weiter fortschreitenden Entwicklung von Wissenschaft und Technik die Folgen des naturverändernden Handelns immer weniger überschaubar werden und entsprechend einer möglichst frühzeitigen rationalen Abschätzung, einer genauen Beobachtung und kritischen Beurteilung bedürfen.

Die wiederum vor allem im Zusammenhang mit Fragen der Bioethik kursierenden Dammbruch-Argumente machen von der skeptischen Prämisse Gebrauch, dass die Möglichkeit, korrigierend in einmal eingeleitete Entwicklungen einzugreifen, in den streitigen Fällen gar nicht besteht: In vielen Fällen wird es zwar sicher so sein, dass die Einführung einer neuen Technik, die mit Blick auf ihre primären Folgen als unproblematisch oder sogar als hochgradig wünschenswert gelten muss, allmählich und über viele Zwischenschritte vermittelt ein bisher bestehendes und akzeptiertes Tabu aushöhlt und – nachdem der Damm einmal gebrochen ist – zur Einführung von Techniken oder zur Etablierung von Praktiken führt, die als inakzeptabel gelten müssen. Auch hier gilt: Bei jeder Einführung neuer Techniken, für die die Gefahr eines Dammbruchs besteht, besteht auch das Gebot der Technikfolgenforschung, damit möglichst vorausschauende – etwa auch strafrechtliche – Regelungen getroffen werden können, die die inakzeptablen Folgen zu verhindern im Stande sind, zugleich aber den Zugang zu den erwünschten Folgen nicht verstellen. Sollte sich bei einer solchen Prüfung erweisen, dass der Einführung einer prima facie harmlosen Technik mit nachgerade naturgesetzmäßiger Zwangsläufigkeit und unabhängig von jedweder normativen Rahmengebung notwendig auch die Etablierung einer inakzeptablen Praxis folgen müsste, dann wäre natürlich dem unter diesen Bedingungen triftigen Dammbruchargument zu folgen und die Einführung der Technik zu verhindern.

Prozedurale Lösungen und die europäische Menschenrechtskonvention

Der letzte Teil der Studie versucht, die zeitweilig wie festgefahren wirkenden Debatten konstruktiv wieder in Gang zu bringen und eine Strategie zu entwickeln, wie mit den scheinbar unüberwindlichen Gegensätzen in den Auffassungen über den moralischen Status des Embryos diskursiv umgegangen werden kann. Diese Gegensätze beruhen im Wesentlichen auf den Unterschieden in den kulturell geprägten Weltverständnissen und den darauf gebildeten unterschiedlichen ethischen Konzeptionen. Dabei erscheint es als aussichtslos, eine Übereinkunft darüber herstellen zu wollen, welche der unterschiedlichen ethischen Konzeptionen denn „die Richtige" genannt zu werden verdient.

Unter solchen Ausgangsbedingungen sind prozedurale Strategien zu entwickeln, die es trotz aller fundamentalen Überzeugungsgegensätze erlauben, wenigstens bis zu einem gewissen Grade Übereinkunft über das gemeinsame Vorgehen zu erzielen. Der Kerngedanke eines solchen prozeduralen Ansatzes besteht darin, dass auch denjenigen, die inhaltlich mit einer bestimmten Entscheidung nicht übereinstimmen, aber den Prozeduren, die zu dieser Entscheidung geführt haben, explizit zugestimmt haben, abverlangt werden kann, dass sie die Verbindlichkeit der Entscheidung akzeptieren. Dies ist immer dann möglich, wenn die Diskursteilnehmer auch einen Diskursausgang, der ihren moralischen Überzeugungen entgegensteht, zu akzeptieren bereit sind, um nicht mit anderen, grundlegenderen moralischen Überzeugungen in Konflikt zu geraten. Erfolgreiche prozedurale Lösungen setzen dementsprechend die Existenz gewisser grundlegender moralischer Überzeugungen voraus, die von den beteiligten Parteien bei aller Differenz in den inhaltlichen Fragestellungen und den dahinter stehenden ethischen Konzeptionen gleichermaßen geteilt werden. Prozedurale Strategien sind mithin gerade in dem Maße effizient, in dem eine solche Übereinstimmung auf der Ebene grundsätzlicher moralischer Überzeugungen besteht oder hergestellt werden kann. Es ist daher notwenig zu prüfen, inwieweit in Bezug auf die Embryonenforschung solche Voraussetzungen bestehen.

Ein Kernbereich an grundlegender moralischer Überzeugungen scheint tatsächlich in Form der bekannten Menschenrechtskonventionen und hier insbesondere in der Menschenrechtskonvention des Europarates, der alle Mitgliedstaaten beigetreten sind, gegeben zu sein. Obwohl dort keinerlei expliziten Bestimmungen über den Umgang mit menschlichen Embryonen formuliert sind, finden sich in der Konvention doch Hinweise darauf, dass der moralische Status von Embryonen nicht als identisch angesehen wird mit dem moralischen Status von Personen (im Sinne rationaler Agenten). Dies zeigt sich bereits darin, dass die meisten der in der Konvention festgeschriebenen Bürgerrechte und politischen Rechte gar nicht von Embryonen ausgeübt werden können. Auch solche Rechte, die im Prinzip auf Embryonen angewandt werden könnten, etwa das im Artikel 2 formulierte Recht auf Leben, sind durch den Europäischen Gerichtshof für Menschenrechte als dem über die Einhaltung der Menschenrechtskonvention wachenden Organ nicht auf das ungeborene Leben ausgedehnt worden. Dies bedeutet jedoch nicht, dass die Menschenrechtskonvention dem ungeborenen menschlichen Leben keinerlei Schutz gewähren würde. Gewisse Schutzrechte lassen sich vielmehr aus der auferlegten Pflicht zur Vorsicht und Verhältnismäßigkeit gewinnen.

Insgesamt gilt, dass unabhängig davon, ob die Regelung der Forschung an menschlichen Embryonen oder ähnliche biopolitische Aufgaben auf nationaler oder internationaler Ebene gelöst werden sollen, der Blick auf die institutionellen Strukturen gerichtet werden muss. Auf beiden Ebenen sind Strukturen zu schaffen, die ein Maximum an Zurechenbarkeit und Transparenz, an Überprüfbarkeit und rationaler Stringenz bieten. Vordringliches Ziel ist dabei die Schaffung adäquater prozeduraler Strategien zur Moderation von Interessenskonflikten und zur Verminderung des Risikos, dass Entscheidungen durch Korruption beeinflusst werden. Solche Strukturen sollen gewährleisten, dass die Entscheidungen in den anstehenden Fragen nach bestem Wissen und so rational wie möglich getroffen werden können.

Empfehlungen

Die nachfolgend zusammengestellten Empfehlungen zum forscherischen Umgang mit menschlichen Embryonen, zur Gestaltung gesetzlicher Forschungsregulierung und von Maßnahmen, die zur Vorbeugung inakzeptabler Entwicklungen geeignet sind, sind aus einem international und interdisziplinär geführten Diskurs hervorgegangen. Die Abschätzung der Chancen und Risiken der Embryonenforschung erfolgt auf der Basis des aktuellen Forschungsstandes. Die Abwägung und normative Beurteilung beruht auf einer Analyse der positiven Regulierungen in unterschiedlichen Staaten sowie der in der vorliegenden Studie angestellten soziologischen und ethischen Erwägungen. Sie bilden die rechtfertigende Grundlage dieser Empfehlungen.

(Terminologische Vorbemerkung: Im Folgenden bezieht sich der Ausdruck ‚Embryo', sofern nicht im Einzelfall Abweichungen explizit markiert sind, stets auf die Entwicklungsphase bis zum vierzehnten Tag nach der Fertilisation oder vor der Ausbildung des Primitivstreifens.)

> 1. Menschliche Embryonen sind von anderen menschlichen Biomaterialien zu unterscheiden: Sie verdienen erhöhten Respekt und sollten mit Vorsicht behandelt werden.

Diese Forderung basiert auf den folgenden Überlegungen:

1.1 Der besondere Status menschlicher *Embryonen* leitet sich aus ihrem Entwicklungspotential her. Dieses Potential besteht in ihrer prinzipiellen Fähigkeit, alle Zelltypen des Körpers auszubilden und/oder sich zu einem menschlichen Lebewesen zu entwickeln. Je höher das Potential einzuschätzen ist, desto größerer Wert sollte ihnen zugeschrieben werden. So ist etwa Embryonen, die das Ergebnis von Fertilisation sind, unter Umständen ein höherer Wert zuzuschreiben als solchen, die mittels parthenogenetischer Aktivierung erzeugt wurden. Ihr Wert ist auch deshalb als hoch anzusetzen, weil ihre Verfügbarkeit eines hohen technischen Einsatzes bedarf (N.B.: Embryonen anderer Spezies kommt in beiderlei Hinsicht ebenfalls ein Wert zu). Hätten Embryonen keinerlei Entwicklungspotential, wäre ihnen auch kein Wert zuzuschreiben, wären sie leicht verfügbar, wäre der ihnen zuzusprechende Wert geringer. De facto wird aus diesem Grund der schwer zu isolierenden weiblichen Eizelle ein höherer Wert zugesprochen als dem männlichen Spermium, obwohl keines von beiden ohne das andere die Fähigkeit besitzt, sich zu einem menschlichen Lebewesen zu entwickeln. Der Wert der Eizelle wird auch insofern höher angesetzt, als ihr Zytoplasma Faktoren enthält, die für die frühe Entwicklung des Embryos oder für die Reprogrammierung nach dem Zellkerntransfer erforderlich sind. Ferner rechtfertigt sich ihre Höherbewertung auch dadurch, dass sie, wenn sie aktiviert oder fertilisiert werden, über ein wahrnehmbares Potential verfügen.

1.2 Eingriffe in die Entwicklung menschlicher Embryonen können Schädigungen hervorrufen. Vor jedem Eingriff sind daher in einem gründlichen Risiko-Nutzen-Vergleich die Risiken gegen den zu erwartenden Chancen abzuwägen. Die angemessene Durchführung dieser Abwägungsprozedur ist zu kontrollieren und vor der Öffentlichkeit zu dokumentieren. Die Reaktionen der Öffentlichkeit sind zu beobachten.

1.2.1 Eingriffe in die Entwicklung von Embryonen können direkte Ursache sein für die Beeinträchtigung der Lebensqualität des sich aus dem Embryo entwickelnden Lebewesens. Die anzunehmenden Interessen des menschlichen Lebewesens, zu dem sich – zumindest seiner Potenz nach – der zu reproduktiven Zwecken erzeugte Embryo entwickeln kann, gilt es zu schützen. Auch wenn nicht von aktuellen Interessen des Embryos die Rede sein kann, so kann doch bei zu hohem Risiko für dessen zukünftige Interessen die Unterlassung des Eingriffs geboten sein. Mit Blick auf das Verhältnis von Chancen und Risiken scheinen viele Techniken, etwa die Gametenspende, IVF und die Cryokonservierung von Embryonen harmlos, während ICSI und PGD mit geringen Risiken einhergehen. Das reproduktive Klonen ist hingegen – jedenfalls beim gegenwärtigen Stand der Technik – mit sehr hohen Risiken behaftet und bietet wenig bis gar keinen Nutzen.

1.2.2 *Eizellspenderinnen* können unter den besonderen Umständen einer reproduktionsmedizinischen Behandlung oder auch auf Grund ihrer schwierigen finanziellen und sozialen Lage in eine kritische Situation geraten, die sie in besonderer Weise dem Risiko einer Ausbeutung ihrer reproduktiven Fähigkeiten aussetzt (N.B.: Dies gilt auch für den Verkauf anderer Körperteile. Dabei gilt die Eizellspende als stärker risikobehaftet als die Spende von Haaren, schwächer als die Spende von Nieren). Hier sind eine gründliche Aufklärung und die Vorschrift zur Einholung einer informierten Einwilligung von wesentlicher Bedeutung.

1.2.3 Entscheidungen über den Umgang mit menschlichen Embryonen können unerwünschte Nebeneffekte auf andere *Familienmitglieder* und/oder die *Gesellschaft* als Ganze haben. Betroffene Familien könnten beispielsweise Diskriminierungen und Anfeindungen ausgesetzt sein.

1.2.4 Entscheidungen in Fragen der rechtlichen Regulierung des Umgangs mit menschlichen Embryonen können *soziale* und *politische* Nebeneffekte haben, etwa die Erzeugung von Inkonsistenzen oder Spannungen, die das Gesamtsystem der Regulierungen zur biomedizinischen Forschung und Praxis gefährden.

---

2. Embryonenforschung ist moralisch zulässig und kann sogar moralisch geboten sein, wenn die legitimen Interessen von Personen, die unter bislang unbehandelbaren Erkrankungen oder Schädigungen (einschließlich der Unfruchtbarkeit) leiden, sowie die Interessen von Forschern, die Grundlagen- oder klinische Forschung betreiben mit dem Ziel, solchen Patienten zu helfen, das generelle Interesse der Gesellschaft am Schutz von Embryonen überwiegen.

2.1 Die Interessen der Parteien sind gemäß der Forderung eines vorsichtigen Umgangs mit Embryonen gegeneinander abzuwägen. Rationale Argumente sprechen gegen die Gleichsetzung des moralischen Status früher menschlicher Embryonen und erwachsener Personen. Entsprechend ist menschlichen Embryonen nicht der volle Schutzstatus zu gewähren, den erwachsene Personen für sich beanspruchen dürfen. Biologische Faktoren wie ihre Zugehörigkeit zur Spezies Homo Sapiens oder das Vorhandensein bestimmter genetischer Merkmale sind zwar eine notwendige, nicht aber bereits eine hinreichende Bedingung dafür, dass Embryonen dieselben Schutzrechte zukommen wie Personen, die über Empfindungsfähigkeit, die Fähigkeit, Interessen auszubilden oder die Fähigkeit, selbständig zu handeln verfügen.

2.2 Das Gesagte gilt jedoch allein für solche Embryonen, die noch keine der moralisch relevanten Fähigkeiten aufweisen und die nicht zu reproduktiven Zwecken implantiert werden sollen. Die mutmaßlichen (zukünftigen) Interessen der dann unter normalen Bedingungen sich aus den Embryonen entwickelnden Personen müssen durch alle Stadien ihrer Entwicklung hindurch geschützt werden. Experimente, die die normale Entwicklung des Embryos gefährden könnten und damit möglicherweise der Verwirklichung dieser zukünftigen Interessen entgegen stehen, sind nur aus solchen Gründen zulässig, die wiederum mutmaßlich in den zukünftigen Interessen des betreffenden Embryo selbst liegen.

2.3 Das Verständnis der Biologie des Menschen ist in der Regel eine notwendige Bedingung für die methodische Entwicklung erfolgreicher Strategien zur Behandlung von Krankheiten oder von Unfruchtbarkeit. Eine Rechtfertigung der biologischen Forschung, die ein solches Verständnis bereitstellen soll, ist aber nicht nur aus solchen instrumentellen Gründen möglich, sondern auch mit dem Hinweis darauf, dass Forschung generell zum menschlichen Selbst- und Weltverständnis beiträgt. Die Rechtfertigbarkeit steht dabei aber stets unter der einschränkenden Bedingung, dass das forscherische Handeln nicht gegen elementare moralische Forderungen verstößt, wie sie etwa oben für den Umgang mit dem Embryo formuliert wurden. Diese Einschränkung stellt aber die Rechtfertigbarkeit von Experimenten an Embryonen nach dem vierzehnten Entwicklungstag oder nach deren Implantation grundsätzlich in Frage.

2.4 Überzählige Embryonen, die im Verlauf einer IVF-Behandlung von den Gametenspendern aus freiem Willen zu Forschungszwecken gespendet werden und die anderenfalls vernichtet werden müssten, sollten der Forschung zur Verfügung stehen, solange diese den üblichen wissenschaftlichen Standards entspricht. Eine solche Regelung würde nicht nur den wissenschaftlichen Erfordernissen entgegenkommen, sie würde auch von der Mehrheit der Europäer mitgetragen werden.

2.5 Die Erzeugung von Embryonen eigens für Forschungszwecke wird nicht selten als moralisch fragwürdig erachtet. Auch stehen dem etliche praktische Probleme entgegen wie etwa die geringe Zahl der verfügbaren und für Forschungszwecke gespendeten Eizellen (auch wenn sich dies in Zukunft ändern könnte). Die Herstellung von Embryonen zu Forschungszwecken sollte aber dann und nur dann gestattet sein, wenn dies im Hinblick auf die erwartbaren wissenschaftlichen und medizinischen Resultate für ein spezielles Forschungsvorhaben gerechtfertigt erscheint und wenn nachgewiesen werden kann, dass die Verwendung überzähliger Embryonen mangels Zahl oder Eignung nicht möglich ist. Regelungen für solche Fälle sollten flexibel gestaltet werden, damit auf weitere Fortschritte in der biomedizinischen Forschung reagiert werden kann. Sollte sich etwa in Zukunft die Möglichkeit eröffnen, Eizellen aus embryonalen Stammzellen in vitro zu erzeugen, dann wäre ggf. die Verwendung von Embryonen, die mit Hilfe dieser Eizellen gezeugt wurden, der Verwendung überzähliger Embryonen vorzuziehen.

2.6 Der Einsatz des reproduktiven Klonens sollte – zumindest bis auf Weiteres – nicht gestattet werden. Die Gründe für dieses Verbot sind nicht so sehr kategorisch-moralischer als vielmehr pragmatischer Art: Die bisherige Erforschung des Klonens bei Tieren zeigt, dass neben übermäßigen Verlusten von Embryonen die Zahl der abnormal geborenen Klone unverantwortlich hoch ist.

> 3. Eine Beschränkung der Forschung auf die Verwendung adulter Stammzellen oder bereits gewonnener Stammzelllinien ist nicht empfehlenswert.

3.1 Die Behauptung, alle mit embryonalen Stammzellen angestrebten Forschungsziele und klinischen Anwendungen ließen sich auch unter Verwendung adulter Stammzellen oder bereits etablierter Stammzelllinien realisieren, ist nicht bewiesen. Entscheidungen darüber, Zellen welcher Art für ein bestimmtes Forschungsvorhaben verwendet werden sollen, sind im Hinblick auf dessen spezifischen Bedingungen zu treffen. Sind gleiche Ergebnisse zu erwarten, sollte der sozial weniger belastete Ansatz gewählt werden. Ist dies nicht der Fall, sollte diejenige Option gewählt werden, die die besseren Ergebnisse verspricht.

3.2 Die Gewinnung neuer menschlicher Stammzelllinien erweist sich aus einer Vielzahl von Gründen als erforderlich und sollte erlaubt sein. Beispielsweise ist es spätestens bei Eintritt der Forschung in die klinische Phase erforderlich, Voraussetzungen zu schaffen, die menschliche embryonale Stammzellen auf menschlichen Fibroblasten als Nährboden oder besser noch auf einem nährstofffreien Medium zu isolieren, zu klonen und/oder zu züchten erlauben, um möglichen Übertragungen spezies-spezifischer endogener Viren vorzubeugen. Um dies zu erreichen sind umfassende Studien an isolierten humanen embryonalen Stammzellen nötig, die nicht durchgeführt werden können, wenn die Forschung auf die Verwendung einiger willkürlich gewählter und bereits etablierter Stammzelllinien beschränkt wird.

3.3 Menschliche embryonale Stammzellen können sich als ein wertvolles Mittel erweisen, um die Erforschung genetisch bedingter Krankheiten in vitro voranzutreiben, um Screenings für die Suche nach Pharmazeutika effizient zu gestalten und Toxizitäts-Tests durchzuführen. Menschen reagieren – oft in Abhängigkeit von ihrer divergierenden genetischen Ausstattung – in unterschiedlichster Weise auf pharmakologische Agenzien. Embryonale Stammzellen könnten bei der Erforschung solcher Abhängigkeiten von Nutzen sein. Für beide Forschungsbereiche wäre die gezielte Erzeugung solcher embryonaler Stammzellen erforderlich, die spezifischen Individuen oder Genotypen korrespondieren.

3.4 Die unter 3.3 genannten Gründe könnten eine erste Rechtfertigung liefern für den Einsatz von Klonierungstechniken für die Gewinnung menschlicher embryonaler Stammzellen (das sogenannte therapeutische Klonen). Möglicherweise wäre dies der einzige Weg, eine embryonale Stammzelllinie (oder überhaupt irgendeine Zelllinie) mit einem bestimmten genetischen Defekt oder einer ganz bestimmten Genkombination herzustellen. Ist der therapeutische Nutzen embryonaler Stammzellen erwiesen, so könnte sich ein solches Verfahren auch als Strategie empfehlen, passende Gewebe bereitzustellen und Abstoßungsreaktion auf implantierte embryonale Stammzell-Derivate vorzubeugen.

3.5 Aus dem Dargelegten folgt, dass die Entwicklung von Klonierungstechniken zum Zweck der Gewinnung menschlicher embryonaler Stammzelllinen als akzeptables Forschungsziel anerkannt werden sollte. Regulierungsansätze zur Embryonenforschung sollten Vorkehrungen enthalten, dass zumindest zu einem späteren Zeitpunkt die Erforschung und der Einsatz solcher Techniken unter bestimmten Bedingungen freigegeben werden können.

> 4. Um zu gewährleisten, dass Forschung an menschlichen Embryonen nur in Übereinstimmung mit den zum Schutz der Interessen der beteiligten Parteien erlassenen Bestimmungen stattfindet und um zu verhindern, dass wissenschaftliche und ökonomische Ressourcen verschwendet werden, sind strenge Sicherheitsbestimmungen zu etablieren.

4.1 Um diese Forderung zu erfüllen, sollte Forschung an menschlichen Embryonen nur stattfinden dürfen, wenn

(i)   zuvor die informierte Einwilligung der Gametenspender eingeholt wurde;
(ii)   sie in lizensierten Institutionen und durch als kompetent ausgewiesene Wissenschaftler betrieben wird;
(iii)   sie zuvor durch ein wissenschaftlich-ethisches Beratungsgremium genehmigt wurde, das zu prüfen im Stande ist, ob die Forschung mit den Erfordernissen guter klinischer Praxis und den gesetzlichen Bestimmungen übereinstimmt.

4.2 Um zu gewährleisten, dass die reproduktive Freiheit der Gametenspender respektiert wird und um ihre mögliche Ausbeutung in IVF-Behandlungen und bei der Gewinnung von Eizellen zu Forschungszwecken zu verhindern, sollte die informierte Einwilligung der Spender mindestens bezüglich folgender Fragestellungen eingeholt werden:

(i)   Wie lange sollen die Embryonen aufbewahrt werden?
(ii)   Dürfen die Embryonen für Forschungszwecke verwendet werden?
(iii)   Für welche Forschungszwecke dürfen die Embryonen verwendet werden?

4.3 Nur lizensierten Institutionen und Kliniken soll es gestattet sein, Forschung an menschlichen Embryonen durchzuführen.

4.4 Menschliche embryonale Stammzellen sind keine Embryonen, sondern Zelllinien. Ihre Verwendung muss nicht in derselben Weise reguliert werden wie die Verwendung von Embryonen. Gleichwohl sollte aber ggf. auch bei der Arbeit mit Zelllinien die Zustimmung der lokal verantwortlichen Ethikkomitees eingeholt werden. Fiele die Rückübertragung embryonaler Stammzellen in den menschlichen Embryo unter die Bestimmungen zur Embryonenforschung, wäre eine Genehmigung dieses Verfahrens unwahrscheinlich. Die Verwendung embryonaler Stammzellderivate zu therapeutischen Zwecken hingegen fiele unter die Zuständigkeit anderer Regelungen, beispielsweise unter die entsprechenden Transplantationsgesetze.

4.5 Unabhängige wissenschaftlich-ethische Beratungsgremien sind einzurichten, die in ähnlicher Weise arbeiten wie die bereits breit etablierten Gremien im Bereich der Tierforschung.

4.6 Auf europäischer Ebene sind Organisationen zu schaffen, die in die allgemeinen Entscheidungsprozesse eingebunden sind – insbesondere wissenschaftlich-ethische Beratungsgremien, die für die Lizensierung von Institutionen und Forschungsvorhaben verantwortlich sind. Dies sollte möglichst frühzeitig geschehen, um zu gewährleisten, dass Entscheidungen in den moralisch sensiblen For-

schungsbereichen und insbesondere dort, wo wie in der Embryonenforschung
wertvolle Ressourcen in Frage stehen, mit der erforderlichen Kompetenz und
aus bestem Wissen getroffen werden. Die Gremien sollten so organisiert sein,
dass ein Höchstmaß an Zurechenbarkeit und Transparenz gewährleistet ist.
Geeignete Maßnahmen zur Sicherung der Qualität von Begutachtungsverfah-
ren, zur geregelten Moderation von Interessenkonflikten und zur Verhinderung
von Korruption sind vorzusehen. Ferner sollten die einzurichtenden Organisationen in der Lage sein, Informationen über die Embryonenforschung und ihre
Ergebnisse in breiter Form anzubieten.

**5. Auf politischer Ebene sind prozedurale Lösungen und die wissen-
schaftliche Bildung der Bevölkerung zu fördern**

5.1 Politische Entscheidungen in moralisch sensiblen Bereichen wie etwa der
Embryonenforschung müssen den „Common Sense" so weit wie möglich res-
pektieren – eine Politik, die die öffentliche Meinung und mehrheitlich vertre-
tene Haltungen nicht respektiert, würde die Unterstützung aufs Spiel setzen,
die sie benötigt um überhaupt wirksam zu sein.

5.2 Da bei einem großen Teil der europäischen Bevölkerung die Haltungen zur
Embryonenforschung nicht so sehr auf Wissen basieren als vielmehr auf angst-
besetzten Vorstellungen, Stereotypen und Glaubensüberzeugungen bzgl. des
moralischen Status und der Rechte des Embryo, sollte die Öffentlichkeit vor
allem über die biologische Natur des Embryo unterrichtet werden. Um den
moralischen Belangen gerecht zu werden sollte die öffentliche Debatte über
Embryonenforschung so rational wie möglich geführt werden.

5.3 Jedwede politische Entscheidung in pluralistischen Gesellschaften muss in
Übereinstimmung mit legitimierten Verfahren getroffen werden, so dass eine
konsistente Politik gewährleistet ist, die von allen Bürgern respektiert werden
kann.

5.4 Hierbei ist zu beachten, dass Respekt gegenüber einer prozeduralen Lösung nur
dann erreichbar ist, wenn die beteiligten Personen Gründe haben, die politische
Entscheidung auch dann zu akzeptieren, wenn sie mit ihren jeweiligen moralischen Überzeugungen nicht übereinstimmt. Diese Bedingung ist am ehesten
erfüllt, wenn Entscheidungen auf den besten verfügbaren wissenschaftlichen
Erkenntnissen basieren und wenn in den entsprechenden Verhandlungen die
Interessen aller beteiligten Parteien berücksichtigt wurden. Darüber hinaus
müssen für den Erfolg prozeduraler Lösungen die Konsequenzen für den Fall,
dass gar keine Entscheidung getroffen wird, von den Beteiligten als Infragestel-
lung solcher moralischer Überzeugungen wahrgenommen werden, die grundlegender sind als diejenigen, die sie bei einer Zustimmung zu einer anderen als
der von ihnen gewünschten Entscheidung hintanstellen müssten. Erfolgreiche
prozedurale Lösungen setzen einen Bestand von Grundüberzeugungen voraus,
der zwischen den beteiligten Parteien trotz ansonsten konträrer ethischer Konzeptionen geteilt wird.

6. Zukünftige Gesetzgebung im Bereich der Embryonenforschung sollte sowohl wohlinformiert als auch flexibel sein. Darüber hinaus gilt, dass auch wenn die Regulation der Embryonenforschung zum gegenwärtigen Zeitpunkt allein auf nationaler Ebene möglich ist, der Gesetzgeber dennoch die Möglichkeit einer internationalen Harmonisierung offenhalten oder sich auf die Folgen einer abweichenden Haltung einrichten sollte.

6.1 Nach den Ergebnissen der für die vorliegende Studie erfolgten Meinungsumfrage ist eine Entwicklung hin zu einer gemeinsamen europäischen Politik ein klar formuliertes Anliegen der überwiegenden Mehrheit der europäischen Bürger. Die EU sollte diese Präferenz beachten, wenn über die Politik und Gesetzgebung im Bereich der Embryonenforschung beraten wird.

6.2 Zum gegenwärtigen Zeitpunkt verfügt die EU über keine legislativen Kompetenzen, Festsetzungen allein auf ethischer Grundlage zu treffen. Regulierungen im Bereich der Embryonenforschung fallen nur dann in den Kompetenzbereich der EU, wenn der gemeinsame Markt oder Fragen der Gesundheit betroffen sind (letzteres ist nicht auf Emryonenforschung in vitro anwendbar). Gleichwohl kann die EU Schritte in Richtung einer künftigen Harmonisierung der Gesetzgebung im Bereich der Embryonenforschung einleiten, indem sie etwa auf der Grundlage einer Analyse der Streitfragen grundsätzliche Leitlinien für Ethikkomitees oder für wissenschaftliche Vereinigungen auf europäischer Ebene erlässt.

6.3 Es besteht ökonomischer und sozialer Druck in Richtung auf eine permissive Harmonisierung. Demzufolge muss handeln, wer eine Nicht-Harmonisierung wünscht. Die Entscheidung gegen eine Harmonisierung sollte aber womöglich mit der Schaffung von Regelungen einhergehen, die es erlauben, Wettbewerbsnachteile derjenigen Staaten, die eine prohibitive Haltung gegenüber der Embryonenforschung einnehmen, zu reduzieren.

6.4 Gesetzgebung im Bereich der Embryonenforschung sollte im allgemeinen einen eher regulativen als einen prohibitiven Charakter haben. Solche Gesetzgebung ist etwa mit der Herangehensweise Großbritanniens und derjenigen Staaten verwirklicht, die Embryonenforschung zwar generell verbieten, dennoch aber mit Blick auf die speziellen Erfordernisse entsprechende Genehmigungen erteilen.

6.5 Die Regulierungen sollten hinreichend flexibel sein, um auf neue Entwicklungen bzgl. der technischen Möglichkeiten und der wissenschaftlichen Erkenntnis, aber auch bzgl. der Haltung der Öffentlichkeit angemessen reagieren zu können. Die Entwicklungen im Zusammenhang mit der Biomedizin werden in naher Zukunft mutmaßlich zu schnell voranschreiten als dass die Entwicklung des Rechts auf den üblichen Bahnen der Gesetzgebungsverfahren erfolgen können wird.

1 Theoretical and Practical Possibilities in Human Embryo Experimentation

Introduction

In this chapter we consider scientific and clinical aspects of research on human embryos. We review the current state of the art, highlighting some instances where research in the past was particularly important, but we also give speculations as to what might be feasible in the not too distant future. It is not possible to cover all topics in depth; this would be several books in length. Instead we mostly give a broad overview raising issues that we feel are relevant to the theme of the book. However, we have gone deeper into some specific areas, such as preimplantation genetic diagnosis and cloning, partly because they illustrate general concepts, but also because these topics are usually considered to be the most contentious.

Much current research on human embryos is driven not by a desire to understand basic mechanisms, which is better approached in model systems, but in response to our wish to have choice in reproductive outcomes. Our wishes are mostly benign. In general, we would like to be able to have children, to choose when to have them and how many, and of course we would prefer them to be normal and healthy. It so happens that human reproduction is notoriously bad in comparison with that of most animals. Many people are infertile, those that are fertile have little natural control over this, and the frequency of embryo loss and abnormal outcomes is extremely high. All these areas therefore provide the focus for most current research. There are a number of spin-offs however. We do get to understand how early human embryos work in comparison with those of other animals. This helps with our basic understanding of mechanisms and is enlightening with respect to the human condition. Most would consider this knowledge beneficial. On the other hand, many of the methods that are developed could be applied in areas that are more contentious. These range from sex-selection to genetic manipulation and cloning. But should we throw the baby out with the bath water?

Before any rational decisions can be reached, it is necessary to appreciate the reasons why research is carried out, taking into account some of the biological problems that have to be overcome together with the techniques currently available and their limitations. We will not address any of the psychosocial aspects that drive individuals to make often very difficult choices, suffice to say that the desire to have children is sometimes so overwhelming that there is little room for rationality. Before delving into the research, we begin with an outline of the major problems facing human reproduction.

1.1
The Human Problem

1.1.1
Infertility

Up to 70 million couples worldwide have turned to the use of assisted reproduction technologies (ART) to overcome their problems with infertility. This corresponds to more than 2% of the world population, but this greatly underestimates the total number of individuals with fertility problems, as these technologies are not readily available to all. It is likely that the true incidence approaches 10%. Of course, many women who have problems of producing eggs can be treated simply by hormone stimulation, which, as a procedure on its own, is quite successful. It also has few complications, except for the critical need to monitor for, and avoid, pregnancies with multiple embryos. Since 1978, more than one million children worldwide have been conceived using ART, mostly involving in vitro fertilization (IVF), either mixing sperm and eggs or with the use of intracytoplasmic sperm injection (ICSI) (Palermo et al. 1992; Steptoe and Edwards 1978). In some countries the technologies have become very popular, for example, ART accounted for approximately 1 out of 150 children born in the USA in 1999 (Schultz and Williams 2002). These numbers are impressive, clearly indicating the problem to be faced and giving an impression that the technologies are successful. The latter is a false impression, however, as it does not take into account the number of IVF cycles attempted, problems of multiple births and other complications, and there are still many couples that fail to have successful pregnancies.

In about 40% of infertile couples the infertility is due at least in part to the male, most often to a low sperm count, abnormal motility or sperm morphology. There are many causes ranging from known, specific single gene defects to purely environmental factors (see Table 1 for some examples). There has been much concern about the latter, in particular with a class of pollutants termed endocrine disruptors. These mimic the steroid hormones involved in controlling reproductive function and have been proposed as an explanation for an apparent decline in sperm counts. But a recent WHO report on endocrine disruptors was rather inconclusive: its strongest recommendation being that more research is needed (www.who.int/pcs/emerg_site/edc/global_edc_TOC.htm).

ICSI is commonly used to overcome male infertility. This involves the direct injection of a mature sperm, or in some cases an immature stage, for example, a round or elongating spermatid, directly into the oocyte cytoplasm. It is a sophisticated technique, involving micromanipulation, but with human oocytes it is a relatively easy and robust method. (Paradoxically, it is much more difficult in mice where the oocyte membrane is more fragile.) Because of this, it is now often used in IVF, even when the infertility is not due to the male. However, despite its apparent success, ICSI is still thought by some to be a contentious technique. Some of the problems are discussed below.

Female infertility also has many causes (see Table 2 for some examples). These again range from single gene defects to environmental factors, with damage caused to the reproductive tract by infections being quite common. An important contribu-

Table 1.1 Causes of male infertility

---

1.   Defects of the hypothalamus-pituitary-gonadal axis resulting in low gonadotrophins (FSH, LH)

   (i)   *Specific gene defects:*
          Kallmann's Syndrome (Anosmin 1)
          Adrenal Hypoplasia Congenita (DAX1)
          Cushing's syndrome (Adrenal Hypoplasia)
          Congenital adrenal hyperplasia (CAH)

   (ii)   *Acquired:*
          Pituitary tumours
          Meningitis
          Trauma

2.   Testicular failure

   (i)   *Chromosomal abnormalities:*
          Klinefelter's Syndrome (XXY)
          XO males (XYO)
          XX males (sporadic)
          Autosomal translocations, deletions or duplications

   (ii)   *Specific gene defects:*
          Cryptorchidism (e.g. INSL-3)
          Persistent Müllerian duct syndrome (AMH or AMHRII)
          Myotonic muscular dystrophy
          Y-linked spermatogenesis genes (e.g.DAZ)

   (iii) *Germ cell development and migration defects, e.g. Steel, W, BMP4*

   (iv)   *Acquired:*
          Mumps
          Radiotherapy
          Testicular tumours

3.   Reproductive tract development or obstruction (sperm are usually present)

   (i)   *Congenital:*
          Cystic Fibrosis
          Absence of tract (e.g. CAVD, Zinner's syndrome)

   (ii)   *Acquired:*
          Tuberculosis
          Gonococcal or chlamydial infections
          Parasite infections (e.g. Filariasis)
          Surgical trauma (e.g. from prostate removal)
          Vasectomy

Notes: Some examples of specific genes that are affected are given in brackets. There may be sufficient sperm or immature spermatid stages present to allow ICSI in some cases. In others, such as XX males, spermatogenesis is blocked at early stages and no germ cells survive. Data obtained from several sources, notably Harper et al. (Harper et al. 2001)

Table 1.2 Causes of Female Infertility

1.	**Genetic disorders affecting oocyte number or quality**	
	(i)	*Germ cell development and migration defects* Specific gene defects, e.g. in : Steel, W, BMP4
	(ii)	*Gonadal agenesis or dysgenesis and sex determination/differentiation disorders* Chromosome defects, such as Turner's Syndrome, DSS Specific gene defects, e.g. in LHX9, WT1, SF1, SRY, AR
	(iii)	*Premature ovarian failure (reduction in oocyte number)* Chromosome abnormalities including translocations and deletions. Specific gene defects, e.g. FOXL2
	(iv)	*Ovulatory disorders* Polycystic ovaries Pituitary defects, e.g. those that affect GNRH release Adrenal defects, e.g. Congenital adrenal hyperplasia (CAH)
2.	**Genetic or non-genetic disorders preventing ovary function, conception or pregnancy**	
	(i)	Congenital absence of the oviducts or uterus
	(ii)	Tubal blockage
	(iii)	Endometriosis
	(iv)	Ovarian cancer
	(v)	Chemotherapy and other medical interventions.
	(vi)	Hysterectomy
	(vii)	Immunological problems, e.g. antibodies against sperm
	(viii)	Medical conditions in which pregnancy could be life threatening

Table 2, notes: It is clear that in some cases oocytes can be recovered, used for IVF treatment and returned to the women's own uterus, in other cases a surrogate mother is required, while in others it is necessary to have oocyte donation. Data obtained from several sources, notably Harper et al. (Harper et al. 2001)

tory factor to the statistics is the increasing number of women who delay attempts to become pregnant until relatively late. IVF is an obvious solution to many of the problems.

1.1.2
Embryo Loss

It is difficult to obtain accurate figures for normal rates of embryo loss in humans, but all indications suggest it is very high. From studies on embryos produced by IVF, about 70% are abnormal. Of course this figure may not represent the natural situation. In part this is because hormone treatment will have been used to induce superovulation. This is likely to increase the number of abnormal oocytes, perhaps by rescuing some follicles which would have otherwise been lost through a natural process termed atresia (Davis and Rosenwaks 2001; Ertzeid and Storeng 2001; Van Blerkom et al. 1997; and see below). Women receiving IVF treatment tend on average to be older than those able to conceive naturally, which will also increase the chances of chromosome abnormalities. Moreover, many of the couples seeking IVF have fertility problems and there will be a sampling bias in that the more normal looking embryos are selected for transfer and therefore not analyzed. Nevertheless, many of the transferred embryos will fail to implant and will be lost without being noticed. Of those embryos that survive to implant in the uterus, there is good evidence to suggest that 31% miscarry shortly afterwards. Subsequent losses are not so dramatic, but they do occur throughout gestation, and probably account for another 5–10% of conceptuses (Hassold and Hunt 2001). (These miscarriages can also compromise the mother especially when they occur close to term.) It is generally accepted that only about one in five fertilized eggs will give a viable human baby, whereas in many mammals, such as mice, the rate approaches 100%.

Certain types of genetic abnormality are a frequent cause of embryo loss. Humans should have 46 chromosomes comprising 22 pairs of autosomes and a pair of sex chromosomes, the latter being XX in females and XY in males. However, it is estimated that 10–30% of fertilized human eggs have the wrong number (Hassold and Hunt 2001; Hunt and Hassold 2002). Furthermore, chromosomal abnormalities account for at least 50% of all miscarriages and 5–10% of stillbirths. Of these, 60% are aneuploid, with either an extra chromosome (trisomy) compared to the normal complement, or one less (monosomy). The remainder include polyploids, where there is one or more extra sets of chromosomes (as in triploids and tetraploids), translocations, where two chromosomes are joined to one another, inversions and deletions. It follows from the rates of fetal loss that aneuploidy is the most common genetic problem in humans. The only trisomies that are not lethal are those involving chromosomes 13, 18 and 21 (the latter responsible for Down's syndrome) and the sex chromosomes (X and Y). Together these make up the single largest cause of mental retardation and developmental disabilities. All monosomies are lethal except for XO, which leads to Turner syndrome (although many of these also die in utero). As a comparison to highlight the extraordinarily high frequency of chromosomal abnormalities in human embryos, the rate of aneuploidy in early mouse embryos is just 1–2% (depending on strain) (Hassold and Hunt 2001; Hunt and Hassold 2002; Yuan et al. 2002).

Deletions can be large, and therefore similar in effect to monosomies where many genes are present in only one copy, or they can affect only a single gene. In each case a resulting abnormal phenotype can be due to haploinsufficiency, where it is necessary to have two active copies of a gene, or to the aberrant juxtaposition of sequences to give either abnormal gene expression or an abnormal gene product. Similarly, while translocations and inversions may not involve the loss of any DNA, genes at the breakpoints may be disrupted to give either loss or gain of function. For certain genes that show "parental imprinting" and are normally expressed only from either the maternally or the paternally inherited copy, any mutation affecting the active copy can be detrimental. These genes are scattered throughout the genome and may be responsible for some of the abnormal phenotypes associated with aneuploidy (Cox et al. 2002; Reik et al. 2001; Solter 1998b).

Aneuploidy arises most often from defects in meiosis, the special cell divisions that occur during the formation of sperm and eggs (or oocytes). The first meiotic division (MI) is preceded by DNA synthesis to replicate the chromosomes, homologues then pair and exchange genetic material through crossing over (recombination), before separating to two daughter cells, each of which then divide again without new DNA synthesis (MII) to produce four cells each with a haploid set of chromosomes (23 in human). Spermatogenesis begins at puberty and occurs throughout adult life due to the presence of a stem cell population residing in the testes (the spermatogonia). These stem cells give rise to committed precursor cells that divide several times by mitosis before eventually entering meiosis. Each of the four daughter cells, the products of meiosis, gives rise to a round spermatid. These then differentiate into mature sperm in a complex process termed spermiogenesis. However, female meiosis is different in several respects. First, the germ cells all enter meiosis in the fetal ovary, so the total number of potential oocytes is already fixed before birth. They then arrest in MI just after recombination, but before separation of the chromosomes (disjunction). They are then frozen in this state until they either die (atresia) or are chosen to grow, develop further (maturation) and are ovulated. It is only at this point that they complete the first meiotic division. This is asymmetric, with most cytoplasm and one (diploid) set of chromosomes giving rise to the oocyte, while the other set of chromosomes ends up within a much smaller cell, known as the first polar body. The latter plays no role in subsequent development. The second meiotic division does not occur until after the oocyte is fertilized by a sperm (or artificially activated). Again, this is an asymmetric division, with one (this time haploid) set of chromosomes remaining in the fertilized egg, the other within the second polar body (which again plays no role in subsequent development).

It has been shown that 90% of all chromosomal abnormalities have a maternal origin and most of these can be traced back to an MI error in which the chromosomes failed to separate correctly (non-disjunction) (Hassold and Hunt 2001; Hunt and Hassold 2002). Moreover, the frequency of aneuploidy increases dramatically with maternal age. This is perhaps not surprising given that the oocytes can be arrested in MI for up to 50 years. It seems that the machinery for correctly separating the chromosomes, the spindle, slowly degrades, increasing the likelihood of non-disjunction events. However, this is a hotly debated topic and one in which there is a considerable research effort. While some of this research can be done on gametes prior to fertilization, to study MII and subsequent events requires artificial activation of the egg or fertilization. Moreover, while much work is done in animals, particularly mice, because these have such a low rate of aneuploidy in comparison to humans, these are poor models. Furthermore, it is possible that human-specific differences in the spindle are the real cause of the high rate of aneuploidy and the effect of maternal age (Hewitson et al. 2002). This requires experiments on human material.

Another proposed cause of aneuploidy is defective mitochondria within the oocyte (Palermo et al. 2002; Yin et al. 1998). These provide the "energy" for cellular processes including that of meiosis. The mitochondria might also become compromised with age. Transfer of some cytoplasm from an egg of a younger woman into that of an older woman has been used in attempts to cure this problem (Barritt et al. 2001a; Cohen et al. 1998). Several babies have been born in the USA after this procedure, but because it involves transferring mitochondrial DNA, and will therefore result in embryos with three genetic parents, it has been considered a form of gene therapy and at least temporarily banned. It is also an area where much more research is needed, both to understand the nature of defects in mitochondria and how they arise, whether they are really involved in generating aneuploidy, as well as the consequences of heteroplasmy (more than one type of mitochondria within a cell).

Other genetic reasons for embryo loss will include single gene defects and more complex genetic effects where particular combinations of mutant alleles will compromise viability of the embryo. There is no a priori reason to suppose these will occur more frequently in humans than in other animals. However, because of our altruistic society and health care, the 'mutation load' carried by humans is likely to be much larger than in wild populations of animals, where the chances of reaching breeding age and transmitting mutant alleles will be lower.

There is much experimental work done on the role of specific genes in embryonic development. Indeed, "functional genomics", the main theme of research in the "post-genome" era, is largely a continuation of what developmental geneticists have been doing for almost a century. Much of this is carried out using animal models, notably mice, chick, frogs and fish, as well as invertebrates, such as the fruitfly *Drosophila* and the roundworm *Caenorhabditis elegans*. There is a wide range of techniques available to scientists using these species, many of which could not be applied to humans or it would be unethical to do so, since it would involve considerable risks for patients (see 5.2.9 on caution). However, many important developmental genes have been found directly through the study of human mutations. It is clear that we share many such genes with other animals and they function in similar ways. In general developmental processes seem remarkably well conserved. However, there must be differences (otherwise we would look like mice or fruitflies). Many of these are likely to be due to more or less subtle changes in the way the genes are regulated, allowing novel combinations of gene products, altered timing, etc. It is therefore important to be able to compare and contrast gene expression between humans and other animals and between normal and mutant embryos. A certain amount of research involving human embryos, in particular fetal material obtained from abortions, is carried out with this in mind. Human fetal tissue banks and atlases of gene expression data have been established to help with this effort (Bullen et al. 1998; Fougerousse et al. 2000). The further relevance of genetic defects to embryo research will be discussed in more detail under preimplantation genetic diagnosis.

While most early embryonic failure is likely to be due to defects in the embryo itself, a smaller proportion of cases will involve deficiencies in the maternal environment or a mismatch between the two. Appropriate hormonal priming is essential to prepare the uterus for implantation and there are many reasons why this may go awry. This could include pollutants or pharmaceutical agents in addition to nutritional status, etc. We know very little about non-genetic reasons for embryo loss. But, in short, there are many reasons why the human embryo might fail and these are all subjects of research.

If we are to be rational, experimentation on human embryos should perhaps be considered in the light of this dramatic natural loss of early human embryos. Most go unnoticed and we do not grieve for them. Moreover, while all research will necessarily involve destruction of embryos, this is also true of IVF itself. IVF is generally accepted in most societies, but if it is necessary to use superovulation to ensure sufficient good quality embryos for implantation, as happens today, then it is likely that there will be spare embryos. These are stored frozen for future attempts at pregnancy, used for research, or they have to be destroyed. In some IVF clinics, or for some patients, economics dictates that embryo storage is too expensive, so this is not always an option. In these cases the embryos are usually destroyed, especially if the use of spare embryos for research is not possible. This can be because parental consent is not granted, because the clinic is not associated with a research group or because the research team is not able to handle the embryos at the time they become available. It follows that the ability to store frozen embryos as efficiently as possible is important both for reproductive purposes and for research. Cryopreservation therefore becomes a topic of research in its own right. In several countries, for instance in the United Kingdom and in Spain (see Chapter 3, Case Study 1 and 2) there is usually a time limit on how long embryos can be stored frozen. These could be a valuable source of embryos for research, but issues of consent often make this impossible. These comprise the largest class of embryos that are destroyed.

Frozen embryos sometimes seem to be given a different status to those that are freshly made by IVF. For example, the Australian Government reached a decision on 5 December 2002 to allow the use of human embryos to derive embryonic stem cells (ES cells), but only from frozen embryos. Moreover, these had to be frozen prior to 5 April 2002, the date of a previous government agreement on embryo research. There are about 70,000 such embryos, but this does not take into account their quality, the remaining length of time that they can be stored or issues of consent, so the number actually available for making cell lines will probably be far fewer. German regulations concerning the importation of embryonic stem cell lines are similar, as will be discussed below (see Chapter 3, Case Study 3).

In the UK it is also possible, with appropriate licensed approval from the HFEA, to create embryos specifically for research (see: Chapter 3). This option has not been used very often, although some argue that it may be necessary to do so for embryonic stem cell derivation, and of course it would be an essential part of therapeutic cloning, although the latter does not involve fertilization. In many countries, notably the USA, where there is little or no regulation of IVF practice or embryo research, it is difficult to obtain figures for the numbers of embryos made by IVF or details of their fate. For the UK, the HFEA keeps accurate records. These show that for the nine years between 1991 and 2000, a total of 925,747 embryos were created by IVF, of which only 118 were created specifically for research. More than 50,000 babies were born (after transfer of 2–3 embryos). A total of 53,497 embryos were used for research, while 294,584 were destroyed (many after 5 years storage) and the remainder was still frozen (see: http://www.hfea.gov.uk/. For further regulations see: Appendices).

1.2
Experiments Designed Exclusively to Develop Methods to Improve, Facilitate or Make Reproduction Possible

1.2.1
In Vitro Fertilisation: All Possible Clinical Indications

(i)   To develop better methods of hormonal stimulation
(ii)   To develop better methods of egg recovery
(iii) To develop better methods of *in vitro* fertilization
(iv) To develop better methods of *in vitro* culture
(v)   To develop better methods of introducing the embryo into the uterus of the biological or foster mother
(vi) To develop better methods of gamete and embryo storage

All these method developments are going on, but often in a cursory fashion. Although much work is published, and some is of high quality, many results are little more than anecdotal. The prevailing attitude amongst some IVF clinics seems to be that the methods as they exist at present are probably good enough: the success rate is clearly sufficient to encourage couples to attempt IVF and to make sufficient profits for the clinics[1]. However, this is clearly not the case, especially as clinics in academic settings are actively researching ways to improve each of the areas mentioned above. These experiments are obviously hampered by problems such as the number of available embryos. A similar experiment in mice, designed to test different culture media for example, would typically use hundreds of embryos to obtain statistically valid comparisons, a number that would be very hard if not impossible to obtain on one day for a study on human embryos. Another problem is that of genetic diversity. Again with mice, the use of inbred strains can greatly reduce variability of outcome. This in turn allows statistically meaningful results to be generated with fewer embryos than would otherwise have to be used with an outbred population, as with humans. Finally, in some countries where there is a requirement to preserve the viability of the embryos and to give them a chance to develop further after transfer into the mother's uterus, research of this type is very difficult, if not impossible.

The ethical and philosophical concerns are difficult to estimate, but they are not great. The principle of IVF and other methods of assisted reproduction seem to be broadly accepted and the techniques are now carried out in most countries. There is some concern, however, when untested protocols are introduced without prior experiment. In most branches of clinical practice it is usual to carry out experiments first, to define parameters, improve conditions and test safety. Only then are the

---

[1]   In a survey of 48 countries Collins (2002) reported that there were on average 289 IVF/ICSI cycles per million people per year, ranging from 2 in Kazachstan to 1657 in Israel. There was a strong correlation with the quality of health services. In 2002, it is estimated that the average cost per IVF/ICSI cycle in the USA will be US$ 9,547 or US$ 58,394 per live birth, whereas in 25 other countries these figures would be US$ 3,518 or US$ 22,048 respectively. This can be related to national costs of $1.00 per capita, but for individual couples it translates to 10% of annual household expenditure in European countries and to 25% in Canada and the USA.

methods adopted into the clinic. Sometimes in IVF it seems that the clinical outcome is an integral part of the experiment. For example, intracytoplasmic sperm injection (ICSI) was introduced into IVF practice with almost no prior knowledge of its likely success or safety. It is fortunate that it seems on the whole to be both reliable and safe (Bonduelle et al. 2002; Van Steirteghem et al. 2002).

Many aspects of research on human embryos are therefore concerned directly with improving the efficiency of all the different procedures that are required to obtain a successful pregnancy using ART. It is clearly important to the patient to reduce the number of attempts; each cycle of hormone treatment and surgery is physically and mentally traumatic, as well as expensive (see section 1.3 for discussion of some details of IVF procedures). Also, while currently the simplest way to improve the chances of success is to implant more than one embryo, this often leads to multiple gestation pregnancies. Compared to singletons, these pregnancies result in a significantly increased incidence of pregnancy complications, low birth weight and learning disabilities (Hack and Fanaroff 1999). They can paradoxically also lead to higher economic costs, as managing risky pregnancies and intensive care for premature babies are both expensive. Clearly the best solution would be to implant only a single embryo, but this requires not only the most favourable conditions possible, but also an ability to identify the most suitable embryos for implantation. At present, this can only be done retrospectively, or with some help from PGD. Indeed aneuploidy screening is frequently used solely for this purpose (see section 1.3 and Table 3a and Wilton 2002). However, because of the problems of mosaicism, even the latter is far from a guarantee (see 1.1.2 and 1.3). Irrespective of genotype the criteria for what constitutes either a good or a poor quality embryo are still relatively imprecise. Any approach to address this requires research on spare embryos, where development to blastocyst stages is essentially the only relevant assay (but see below in section 1.2.2). At some point, however, a correlation has to be made as to how well the embryos will develop after implantation. This will always involve some risk, but it is perhaps remarkable how few problems have occurred. This is testament to a lot of careful research that has taken place in the past, plus the courage of researchers, clinicians and especially their patients.

Specific areas of research that are of particular scientific interest and/or raise ethical issues include the following:

1.2.2
Embryo Culture

It is clear that the whole field of IVF depends on research that was done to define appropriate culture conditions. Since the original success of Steptoe and Edwards (Steptoe and Edwards 1978) there have been substantial improvements in our understanding of the requirements of human preimplantation embryos and a corresponding improvement in methods allowing efficient *in vitro* fertilisation and culture to blastocyst stages. The latter is still a problem with usually no more than about 50% developing as blastocysts and even these rates are not achieved in many centres. The relatively poor development *in vitro* of human embryos compared to that of other mammals may reflect the high frequency of aneuploidy, etc. as discussed above (1.1.2). Newly developed sequential media allow many more embryos to reach blastocyst stages and to generally develop more consistently (Gardner and Schoolcraft 1998; Van Winkle 2001), however, there has been as yet no convincing improvement in rates of pregnancy following transfer (Braude 2001). Again, this probably reflects the chromosomal abnormalities. On the other hand, it is possible that we are still missing something essential from the medium. Rates of development and of cell proliferation are never quite that found *in vivo*. There are changes in glycogen metabolism and often an increased metabolic rate. These could all have consequences for development of human embryos and/or for the efficiency of techniques of ART, PGD, etc.

While further research is still needed, there is some risk that changes even in something as apparently simple as the culture medium can have unexpected consequences for the newborn child or resulting adult. For example, "large-calf syndrome" is a culture-induced effect leading to abnormal growth of cattle embryos (and similar problems are sometime seen in sheep). It seems to be due to abnormal programming of genes within the preimplantation embryo that can have long lasting effects on the activity of those genes in subsequent development (Khosla et al. 2001; Leese et al. 1998; McEvoy et al. 2000; Sinclair et al. 2000; Young et al. 2001). Dramatic effects on growth have not been seen with other animals, such as mice, however, more or less subtle changes in gene expression have been noted (Reik et al. 2001). The latter includes imprinted genes, developmentally important genes and even liver enzymes. Similar problems are just beginning to be noticed and described in humans (Couzin 2002). A recent report suggested an increased incidence of retinoblastoma in children born after *in-vitro* fertilization in the Netherlands (Moll et al. 2003). At present the mechanism or cause of such an increase is unknown and it is even possible that it represents a statistical artefact (BenEzra 2003). However, it is something to be aware of, especially as there could be effects on gene expression that do not have such obvious effects on phenotype, but which could still be relevant to normal health.

Given all the issues outlined above, it is likely that further improvements in rates of pregnancy from cultured embryos will depend on ways to select those that are normal. For example, it may be possible to assay the culture medium to assess the metabolic state of the embryo (Houghton et al. 2002; Leese 2002), which ought to provide a measure of its viability. As mentioned above, aneuploidy screening is already being used to look for chromosome defects, but eventually it may also be possible to use a biopsy to determine gene expression profiles.

It is thought that failure of the embryo to implant may contribute significantly to the overall failure rate of IVF (and natural pregnancy). Some of this is due to problems with maternal hormone status, which can sometimes be corrected by careful monitoring and hormone replacement. In many other cases, it will be due to a failure of the interactions between the embryo and the uterus and subsequent cellular and developmental processes that allow implantation. Most research on these topics is being done with animal models (Paria et al. 2002), but the placenta appears to be a rapidly evolving structure, and the human placenta is very different from that of many of the animals used in research (such as mice).

With respect to other types of method development, the results of research at both ends of pregnancy may reduce the time that an embryo needs to be implanted in the uterus. For example, there are attempts to use cultured endometrial cells (from the lining of the oviducts and uterus) as a substrate for improving preimplantation culture and for defining better conditions for implantation (Lee et al. 2001; Xu et al. 2001b). These could in theory also be used to investigate early postimplantation development *ex-utero*. This is unlikely to allow normal development to proceed very far as this requires a correct three-dimensional structure and an appropriate supply of oxygen, nutrients, hormones and other factors, which necessitates an appropriate vascular system and the efficient exchange mechanism of the placenta. However, there is a reasonable assumption that the endometrial cells may provide critical factors that are missing for embryo development and/or implantation and one goal will be to isolate such factors.

Similarly, there is research to improve methods of keeping premature babies alive. These include liquid ventilation methods, where oxygen-rich fluids are pumped into the lungs, and attempts to make a so-called artificial placenta, essentially a system to allow oxygen exchange. While giving better survival of premature babies, it is thought they could not allow survival of foetuses of less than 20 weeks (Knight 2002).

It seems extremely unlikely that it will be possible to do without the uterus and placenta altogether. It is relatively simple to obtain reasonable development of postimplantation mouse embryos for about two to three days in culture for any period up until the time when the placenta becomes critical (at about 12 days). This is significant for a rapidly developing species such as the mouse, where gestation is only about 20 days. If legal, it might be possible to obtain survival of human early postimplantation stage embryos for similar periods of time, but their slower rate of development and large size would almost certainly compromise normal development very quickly. It would also be almost impossible to reimplant such embryos into a uterus (Beddington 1985).

1.2.3
Intracytoplasmic Sperm Injection (ICSI)

ICSI is designed to solve problems of infertility caused by low sperm number or defective or immobile sperm, however it is frequently used for other reasons, notably for PGD (see section 1.3). In the case of ICSI, a single sperm from males with too few (oligospermy), or sperm that are otherwise incapable of fertilization, or even immature stages such as a single round or elongating spermatid, is injected directly into the cytoplasm of the oocyte thus providing the possibility for paternal genetic continuity (Kahraman et al. 1998; Ogura and Yanagimachi 1995; Van Steirteghem et al. 1993). The use of spermatid stages in humans (ROSI or ESCI) is quite inefficient, however, prompting some doubts about its safety as well as its usefulness as a method (Aslam et al. 1998; Vanderzwalmen et al. 1998). Nevertheless it is clear that spermatozoa do not need to be morphologically normal or even alive in the conventional sense to participate in embryo development; as long as they have intact genomes they are able to produce normal offspring. Freeze-dried sperm have even been shown to work in the mouse (Kusakabe et al. 2001). When ICSI is used to treat infertility, the underlying defect is frequently unknown; moreover very little or no experimental work preceded the introduction of the technique into clinical practice. Indeed it is quite surprising that ICSI works so well as the whole sperm is injected straight into the egg cytoplasm. This avoids all the usual checks and balances required for fertilization and introduces sperm components, such as the plasma membrane and acrosomal contents, which would normally remain outside. For further discussion around this topic see Yanagimachi (Yanagimachi 2001).

It is likely that ICSI and related methods will perpetuate the genetic defects causing the problem in the first place. This is already apparent for some cases of male infertility due to abnormal genes on the Y chromosome, which have been inherited by male offspring from ICSI (Komori et al. 2002; Mulhall et al. 1997; Oates et al. 2002). Some concern has also been expressed with respect to recent data on early mouse development, where a correlation has been noted between sperm entry point and axis development of the resulting embryo. (The axes refer to the orientation of the embryo and the breaking of symmetry in what is originally a sphere.) The formation of the "fertilization cone" at the sperm entry point is due to the reorganization of cytoskeletal elements. This does not occur after microinjection. However, it appears that mammalian development is highly regulative, and there is little evidence that these structures, or indeed any early indicators of axis development, are important for subsequent embryogenesis (Davies and Gardner 2002; Piotrowska and Zernicka-Goetz 2001; Piotrowska and Zernicka-Goetz 2002).

It is now becoming apparent that assisted reproduction, be it merely *in-vitro* fertilization or ICSI, poses an increased risk for major birth defects and/or chromosomal abnormalities (Hansen et al. 2002) and also for the occurrence of syndromes associated with imprinting failures (DeBaun et al. 2003; Ørstavik et al. 2003). It is unlikely that these increased risks will reduce the demand for procedures requiring embryo manipulation *in vitro* (IVF, ICSI, ooplasm injection, preimplantation genetic diagnosis). However, the increased awareness of possible dangers should spur further research aimed at understanding the underlying mechanisms and at determining which measures can be taken to avoid or minimize the risks involved.

1.2.4
Injection of Ooplasm

It has been suspected that some problems of infertility are caused by defective oocytes, especially when IVF or ICSI fails and there is no evidence of sperm abnormality. There have been attempts to overcome this by rescuing the cytoplasm of the "bad" oocyte by injection of cytoplasm from presumably "good" oocytes obtained from anonymous donors (Barritt et al. 2001a; Cohen et al. 1998). However, there is no evidence that cytoplasm is at fault and indeed no evidence that the method works. It has been considered as genetic modification of embryos as it involves introduction of donor mitochondrial DNA and is not approved as a technique in the UK and has been stopped in the USA.

Some experimental data suggest that introduction of cytoplasm from other oocytes may in fact cause problems with nucleo-cytoplasmic interactions taking place at the beginning of development. Other problems may result from mitochondrial dysfunction or heteroplasmy or from a disrupted spindle, which would affect cell division. Significant amounts of work in experimental animals should be done before these methods can be evaluated and accepted – if indeed it can be demonstrated that there is a problem in the first place.

1.2.5
Embryo and Gamete Freezing

The ability to store frozen embryos and gametes has been critical to many aspects of IVF research and clinical practice. A human pregnancy after cryopreservation of an 8-cell embryo was first reported in 1983 (Trounson and Mohr 1983) and successful births followed not long after (Cohen et al. 1985; Downing et al. 1985) from frozen cleavage or blastocyst stage embryos. These cases had followed several years of research to establish appropriate conditions for freezing human embryos, which in turn had followed pioneering research in animals, e.g. mice (Whittingham 1974; Whittingham et al. 1972). Since the late 1980's it has become almost routine to freeze spare human embryos. The ability to store embryos means that several attempts at re-implantation can occur without the need for repeated cycles of superovulation, while the hormone treatments required for the former will not be disturbed by those required for the latter. The storage of embryos superfluous to requirements for reproduction also provides a ready source of embryos for research, assuming that parental consent is given.

Many other techniques discussed below, such as autologous ovary transplants, also critically depend on reliable methods of freezing tissue. Freezing of embryos, gametes and gonadal tissue are therefore all topics of continuing research (Bredkjaer and Grudzinskas 2001; Porcu et al. 2000; Wininger and Kort 2002).

While solving many practical problems, having embryos in storage does itself create a number of ethical dilemmas. These are mostly obvious, but some are quite contentious. For example, how long should they be stored, who pays for this, what should happened to them if they are unwanted or if the parents cannot be contacted? Should consent be granted at the time they are frozen or does it need to wait until the research project is known in detail? We do not discuss these here, but some issues of this sort are considered in Chapter 5.

1.2.6
Other Techniques

(i) Spermatogonial transfer. This has been shown to work in rodents, where the germ line stem cells are transplanted from one testis to another (usually depleted of endogenous spermatogonia) and shown to be functional (Brinster 2002; Ogawa et al. 2000). The donor cells can be stored frozen and still function and it will even work to a limited extent across species barriers. This method could be used to restore fertility to an infertile male, but with the loss of genetic continuity (less so if a brother was used as the donor). However, the method could also be used for autologous grafts, for example to restore fertility in patients having to undergo radiotherapy. A related technique, recently shown to work in mice, is to graft pieces of testis from a donor mouse to subcutaneous sites on the backs of immunocompromised host mice (Honaramooz et al. 2002). Spermatogenesis could occur normally in these grafts, giving rise to sperm able to fertilise eggs *in vitro*. The testis grafts could also come from other species, such as pigs, although spermatogenesis was less normal in these cases. Such techniques could in theory be useful in cases where the infertility is due to low levels of gonadotrophins, although a suitable host has to be found.

(ii) Ovary transplants. These are used frequently in rodents as a means of maintaining a mutant strain where the females are unable to reproduce (due to abnormalities of the reproductive tract or behaviour or because of early lethality). Restoration of fertility with cryopreserved ovarian grafts has also been successful in various animals. The method has been attempted in several women who have had ovarian material removed and stored frozen prior to chemo- or radiotherapy. As an autologous graft it would pose few ethical problems. In theory (barring immune problems) it would also be possible to graft from one woman to another (as is commonly done in animals). After maturation in the host, the donor's oocytes could then be used for IVF and returned to the donor, or the host could act as a surrogate mother. If the ovary has been removed prior to cancer therapy, there is a risk of transmitting malignant cells, with either an autologous or heterologous graft. Nevertheless, ovarian transplantation could prove to be clinically useful for women at risk for premature ovarian failure, especially if this is due to defects of gonadotrophins. There is also research on using xenotransplantation to mature primordial follicles, but the efficacy and especially the safety of this method needs further investigation. (See Hovatta and Kim et al. (Hovatta 2000; Kim et al. 2001) for further discussion on these topics.)

(iii) In vitro maturation of germ cells. It has been possible to mature oocytes in mice, and in some farm animal species, beginning with primary follicles (a very immature stage). It could be used to obtain large numbers of oocytes at a specified time. These could be used subsequently for IVF to provide embryos for research or for "therapeutic cloning", etc. The donors could be women at any age and include those undergoing hysterectomy, but in theory it may also be possible to use ovaries from aborted fetuses. These methods could greatly increase the numbers of oocytes and embryos available for research or for assisted reproduction (egg donation). Methods need to be improved and validated for safety (e.g. karyotype integrity) in animal models. Some work is being done on storage and maturation of primary follicles from humans, but this is still very experimental.

There is also a considerable effort to obtain spermatogenesis *in vitro*, beginning with spermatogonia. Except for some reports (in the mouse) that involve transformed cells, it has been difficult to obtain either proliferation or differentiation of the stem cells. More needs to be understood about the relationships between the germ cells and the surrounding supporting cells in the testis.

Given that mouse ES cells are able to give rise to sperm and oocytes in chimaeras, it may eventually be possible to derive mature gametes directly from ES cells maintained entirely *in vitro*. This would offer prospects of infertility treatments for men and women via therapeutic cloning. It would also allow a ready supply of oocytes for the latter technique and generally, after IVF, of embryos for research. How would such a readily available source of embryos, with no living parents, affect the value we place on preimplantation human embryos?

(iv) Sperm sorting. The difference in size between the X and Y chromosomes confers a subtle difference in the density, and in the uptake of fluorescent dyes that intercalate with DNA or chromatin, between X and Y bearing sperm. These differences are sufficient to allow live sperm to be separated using a fluorescence activated cell sorter (FACS) and the two fractions used in artificial insemination, IVF

or ICSI to increase the likelihood of conceiving either an XX female or XY male offspring. The techniques are being applied with some considerable success with farm animal species, notably cattle. They are now also being used to sort human sperm. The Genetics and IVF Institute in Virginia, USA, has recently reported that they have helped to produce over 300 babies using sperm sorting techniques. They claim a success rate in obtaining the desired sex of up to 91%, although this rate is for females: for males it is more like 65%. In the majority of cases, this is being used simply to balance the sex within a family. Sex selection for purely social reasons has been considered unethical in the UK, although the HFEA has little jurisdiction over the procedure as it does not need to involve IVF. This attitude may have come from concern that sex selection would be used to choose only sons, however, so far in the USA there appears to be no bias. Consequently the HFEA are now re-surveying public attitudes. If the procedures were more reliable, sperm sorting could also be used for medical reasons, notably as an alternative to PGD for some X-linked diseases. In theory, it might also be possible to discriminate against some types of chromosomal abnormality, e.g. large duplications or deletions, but the methods would have to be improved significantly to allow this. Although techniques to quantitate fluorescent dyes have become very accurate, they have to be used with care as the UV light used to visualise the dyes can cause damage to DNA. Even if the DNA does not absorb the wavelengths that are used, the radiation can cause secondary damage. This could lead to an increased mutation load and to visible mutations in the offspring or their descendents. It is not clear how well this has been controlled for in the procedures currently being used with human sperm.

1.3
Prenatal and Preimplantation Genetic Diagnosis

1.3.1
Major Goals

(i)   Selection and elimination of embryos with a genetic defect, including specific gene mutations and chromosomal abnormalities
(ii) Selection and elimination of embryos with a non-genetic defect
(iii) Selection of embryos on the basis of sex
(iv) Selection of embryos on the basis of desirable characteristics (single gene, multiple genes)
(v)   Selecting an embryo for tissue type (HLA, etc.), such that the resulting child or cord blood, could be used to provide grafts to treat a sibling with, for example, leukemia. This is related to 1 and 4 above, but is not for the benefit of the child.

All these uses would obviously involve the destruction of embryos that do not meet the set criteria. If destruction of embryos is not an option, the use of these methods is questionable. Indeed, they are illegal in some countries (see Chapter 3 and Appendices). However, this is true even in countries where IVF is permitted. This is inconsistent as IVF also involves destruction of embryos (see section 1.1). Selection of embryos for the desired sex is relatively easy, and it is used when a

Table 1.3a Most common reasons for requesting PGD

Genetic risk and previous termination of pregnancy	21%
Genetic risk and objection to termination of pregnancy	36%
Genetic risk and fertility problems	26%
Genetic risk and sterilization	1%
Aneuploidy screening	14%
Other or unknown	12%

Some patients are in two categories.
Adapted from ESHRE Preimplantation Genetic Diagnosis Consortium (2002)

Table 1.3b PGD referrals according to indication

Specific locus abnormalities (N = 884)	
X-linked	18.8%
Autosomal recessive	18.6%
Autosomal dominant	16.3%
Mitochondrial	0.4%
Two indications	0.6%
Y-chromosome deletion	0.1%
Unknown	1.9%
Structural chromosomal aberrations (N = 647)	
Reciprocal translocation	16.2%
Robertsonian translocation	3.9%
Inversion	0.8%
Deletion	0.4%
Aneuploidy risk	16.0%
47, XXY, 47, XYY	1.6%
Sex chromosomal mosaicism	1.0%
Male meiotic abnormalities	0.9%
Other	0.3%
Unknown	0.5%
Social sexing (N = 30)	1.9%

Total No. of cycles = 1561. Cumulative data obtained from 1999 to 2001 from approximately 25 centres. 2001 was the first year to include social sexing. Adapted from ESHRE Preimplantation Genetic Diagnosis Consortium (2002)

genetic defect is sex-linked (i.e. located on an X or Y chromosome), but it tends to be frowned upon if it is to be used simply to obtain a child of a particular sex. There are, however, indications that this attitude may be changing. This was discussed above under sperm sorting, which although less reliable, is a much less invasive and simpler method. Nevertheless, PGD is being used for this purpose in some countries (see Table 3). This is despite strong objections from some practitioners of PGD (Ray et al. 2002).

Selection and elimination of embryos with a genetic defect is possible today for many diseases, especially if details of the mutation are known. Systematic screening for numerous mutations may be feasible in the future, although it would be difficult to select embryos for more than 2 or 3 traits at one time. The same is true for positive selection for a desired genetic characteristic, which is also possible now, but the range of options is likely to increase in the future with improvements in understanding of the genetics underlying the development of specific characteristics. The latter could include anything from height to skin colour or disease resistance. The idea of positive selection does seem to raise significant opposition and is considered morally questionable. Some put forward the counterargument that this would be no different from parents choosing a good school or other way of giving their children an advantage. Selection for or against may be especially contentious in some countries, notably Germany.

1.3.2
Prenatal Diagnosis

Prenatal Diagnosis (PND) involves biopsy of extraembryonic cell types (so as to minimize the risk of damage to the embryo), notably from amniotic fluid (amniocentesis) or placenta (chorionic villus sampling). On rare occasions, sampling of foetal blood or even of foetal tissues is performed, but the only valid reason for doing so is when a genetic test is not available for the disease. PND is combined with imaging of the foetus via ultrasound scans or MRI, which, depending on the circumstances can reveal specific phenotypes to help confirm the genetic abnormality. There is a slight risk of miscarriage, but the main disadvantage is that pregnancy is well on the way before the diagnosis and when the decision has to be made whether or not to terminate the pregnancy. Often there is a strong desire to have a child, so this is a wanted pregnancy, which can make the decision to terminate very difficult. However, as will be discussed in Chapter 3, Case Study 3 (Regulation in Germany) the use of PND is a controversial issue as the criteria whether to terminate pregnancy can be seen from completely different individual standpoints and, furthermore, abortion – though it remains unpunished – is regarded in Germany as a breach of law.

1.3.3
Preimplantation Genetic Diagnosis

Preimplantation Genetic Diagnosis (PGD) is similar in that it involves biopsy, but in this case it involves removal of a single cell, or sometimes two cells, from a cleavage stage embryo or several cells from a blastocyst (Hardy et al. 1990; Monk 1988; Van de Velde et al. 2000; Veiga et al. 1997). Because little time is available before a blastocyst has to be implanted in the uterus, blastocyst biopsy is rare. It is also possible to take the first and/or second polar bodies as the biopsy, however this only gives indirect information as to the genotype of the oocyte and none for the sperm, and it is only rarely used (Rechitsky et al. 1999; Verlinsky et al. 1996 and see

ESHRE Preimplantation Genetic Diagnosis Consortium (2002)). The diagnostic test is then made on the biopsy and only embryos believed to be unaffected by the genetic disorder are considered suitable for implanting into the uterus. To be useful PGD requires a sufficient number of preimplantation embryos to give a reasonable chance of finding at least one that is unaffected and of good quality. It therefore relies on ART techniques of superovulation and IVF (often by ICSI) as well as good embryo culture conditions. Nevertheless, it should be stressed that PGD is used to solve genetic problems and not problems of fertility per se. However, one should

Table 1.4a Specific locus diseases for which PGD has been requested

X-linked	*Autosomal dominant*
– Fragile X syndrome* (75)	– Myotonic dystrophy* (88)
– Duchenne and Becker's muscular dystrophy (69)	– Huntington's disease* (73)
	– Charcot-Marie-Tooth disease (20)
– Haemophilia (26)	– BRCA1, 2 (inherited cancer predisposition)
– Agammaglobulineamia	
– Alport syndrome	– Central core-disease
– Anderson-Fabry disease	– Crouzon syndrome
– Ataxia	– FAP-Gardner
– Autism	– Marfan's syndrome
– Barth syndrome	– Neurofibromatosis
– Coffin-Lowry syndrome	– Osteogenesis imperfecta I
– FG syndrome	– Osteogenesis imperfecta IV
– Goltz syndrome	– Stickler syndrome
– Granulomatous disease	– Tuberous sclerosis
– Hunter syndrome	
– Hydrocephalus	
– Hypohidrotic ectodermal dysplasia	*Autosomal recessive*
– Incontinentia pigmenti	– Cystic Fibrosis (109)
– Kennedy disease	– β-thalassaemia (53)
– Lesch-Nyhan syndrome	– Spinal muscular atrophy (50)
– Lowe syndrome	– Sickle cell anaemia
– Mental retardation (several genes)	– Tay-Sachs disease
– Oro-facial-digital syndrome type 1	– CDG1C
– Pelizaeus-Merzbacher disease	– Epidermolysis bullosa
– Proliferative disease	– Gaucher's disease
– Retinitis pigmentosa	– Hyperinsulinaemic hypoglycaemia
– Retinoschisis	– PHH1
– Vitamin D resistant rickets	– RhD blood typing
– Wiskott-Aldrich syndrome	– Acyl CoA dehydgrogenase deficiency
	– Familial adenomatous polyposis coli
	– Congenital adrenal hyperplasia

Table 1.4b Chromosomal abnormalities for which PGD has been requested

Female chromosome abnormalities	No of cycles	
45XX, t(13;14)	16	
45XX, t(14;21)	6	
46XX, t(1;10)(q44;q11.2)	2	
46XX, t(1;13)	2	
46XX, t(1;19)	2	
46XX, t(10;11)(p11.2;q23.3)	3	
46XX, t(11;17)	2	
46XX, t(11;22)(q23.3;q11.2)	2	
46XX, t(12;13)(q21.3;q32)	2	
46XX, t(13;22)	2	
46XX, t(14;18)(q11.2;q21.2)	3	
46XX, t(2;7)(q37.3;q34)	2	
46XX, t(5;11)(q34;p25)	2	
46XX, t(6;7)(p25;q11.2)	2	
46XX, t(9;13)(q12;p13)	2	
46XX, t(9;16)(q34.3;p12)	3	
plus 22 other translocations	1	each
Patient mosaic (XX<->XO)	1	
Patient 47 XXX	1	
Pseudocentric chromosome trisomy	8	
Male chromosome abnormalities	No of cycles	
45XY, t(13;14)	14	
45XY, t(14;15)(q10;q10)	3	
45XY, t(14;21)	7	
46XY, t(1;11)(p36.3;q13)	2	
46XY, t(1;17)(p34;q25)	2	
46XY, t(11;12)	11	
46XY, t(11;22)	3	
46XY, t(3;7)	2	
46XY, t(4;11)(q35.1;q15)	2	
46XY, t(6;11)(q23;p13)	2	
46XY, t(7;10)(q15.3;q23)	3	
46XY, t(8;12)(q24.1;q22)	3	
46XY, t(8;14)(q21.3;q31), inv(9)	3	
46XY, t(9;17)(q21;p11.2)	2	
46XY, t(Y;18)(q12;p11.2)	2	
plus 24 other translocations	1	each
DiGeorge syndrome	1	
Klinefelter (XXY)	10	
45, XY, psu dic (15;22)(p12;p12)	2	

Notes: X-linked diseases do not necessarily require a specific diagnostic test, whereas these are required for autosomal diseases and chromosomal disorders. In Table 4a, the numbers in brackets correspond to the 1999–2001 cumulative total of referrals for the most common diseases and relate to the data presented in Table 3a. *Triplet repeat disorders, which can be technically difficult to assay. The data in Table 4b refer to those collected in 2001 only. The most common chromosomes to be affected by translocations are: 14, 13, 11, 1 and 21. Data obtained from several sources, notably (ESHRE Preimplantation Genetic Diagnosis Consortium 2002; Geraedts et al. 1999; Geraedts et al. 2000; Geraedts et al. 2001; Harper et al. 2001).

take into account the recent reports of problems associated with ART (see 1.2.2 and 1.2.3) when deciding whether PGD is indicated and necessary.

PGD also requires sensitive and robust techniques of genotyping, especially as this is being done on only one or at best a few cells. This usually involves polymerase chain reaction (PCR) techniques to amplify the very small amounts of DNA, followed by molecular techniques to detect specific (known) mutations in the affected gene. Alternatively, fluorescence in situ hybridisation (FISH) is used to directly visualise specific genes, segments of chromosomes or entire chromosomes. This is most often done to look for X and Y chromosomes, in case of sex-linked gene disorders, or to test for aneuploidy or other chromosome abnormalities. With these sensitive techniques care must be taken to avoid contamination, which is one reason why ICSI is commonly used as it eliminates the risk of contaminating supernumerary sperm or sperm DNA.

It follows from the above that PGD usually requires prior knowledge of the genetic defect. Perhaps one or both of the parents suffer or have a high risk of developing the disorder, or they have close relatives who suffered, or the parents have already had one or more affected children or several miscarriages (see Tables 3 and 4). However, it is increasingly being used also for aneuploidy screening. This may include older women who have a greater chance of having children with abnormal chromosome number, e.g. trisomy 21 (Down's Syndrome). This is sometimes referred to as "aneuploidy risk" and it reflects the likelihood within the population as a whole of having an affected child, rather than any previous indication that the woman herself is at risk (Wilton 2002). It can also be used simply to improve the success of IVF, i.e. to help eliminate embryos that are unlikely to implant or develop. The demand for this is increasing. In some respects it is beneficial, as it may both increase the chance of success and reduce the need to transfer several embryos with the associated risk of pregnancies with multiple foetuses. However, it is a significant use of resources.

To work efficiently, PGD therefore depends on expertise and practical skills within several different disciplines, notably genetics, molecular biology and embryology, as well as on clinicians and genetic counsellors. The whole procedure must be well coordinated and it is clearly beneficial if all the steps occur in one centre. Once superovulation is initiated in the woman, clinicians and embryologists must be on hand to harvest oocytes, perform IVF or ICSI (the latter involving micromanipulation), culture the embryos, choose those suitable for testing at, for example, the 8-cell stage, and remove one blastomere from each (again using micromanipulation). The embryos then have to be cultured separately and allowed to continue development (although freezing is theoretically possible) while the biopsy is analysed by appropriate molecular genetic or cytological techniques to determine the genotype. Embryos thought to be free of risk are then selected and the remainder are either discarded or processed further to verify that they were indeed abnormal. There is little time between biopsy and the need to transfer the embryos into the patient's uterus. If pregnancy is established there is usually careful monitoring and often CVS or amniocentesis is performed to confirm the genetic status of the foetus.

1.3.4
Procedures Involved in IVF and PGD

For IVF the female's own reproductive hormone cycle is downregulated, follicle stimulating hormone (FSH) is then given to stimulate oocyte growth. This needs to be fairly aggressive for PGD to maximise the number of embryos. The patient is monitored with ultrasound and blood tests to measure estrogen levels. At the appropriate time, human chorionic gonadotrophin (HCG) is injected to trigger oocyte maturation and about 36 hours later the oocytes are collected by ultrasound-guided aspiration. After IVF or ICSI, the embryos are allowed to develop in culture, reaching 2–4 cells on day two and about 8 cells on day three. They will form blastocysts about two days later. Culture to these stages is often relatively inefficient, with usually only about 50% reaching blastocyst stages, and this is only achieved with experience and optimum culture conditions (see 1.2.2).

In each cycle of superovulation and IVF the number of embryos obtained is maximally about 16 and usually less. A number of these will develop abnormally and be discarded. With sufficient skill, the biopsy procedure itself should not physically damage the embryo, and success rates of 97% are cited. It is either done at a time when all blastomeres are thought to be equivalent, usually at the 8-cell stage, and where the loss of one will not compromise development, or at the blastocyst stage where a few mural trophectoderm cells are taken that would have made only a negligible contribution to the placenta. Nevertheless, a significant amount of research was done to get to this point, initially on mouse embryos and subsequently on human embryos, and research still continues on potentially better and alternative methods of biopsy. While biopsy is mostly successful, it is quite usual to have some ambiguous results with the genotyping. This is often because of the high frequency of aneuploidy and mosaicism that occurs naturally in human preimplantation embryos. It is usually considered too risky to implant these embryos, although some will have sufficient normal cells to allow development. From the remaining embryos (some of which may be of poor quality and unlikely to give pregnancies), it is necessary to choose those of appropriate genotype for transfer to the patient. The expected proportions will vary according to the nature of the genetic disorder, for example the aim may be to discard embryos that are homozygous for a recessive defect (25%), or embryos heterozygous for a dominant mutation or abnormal chromosome (50%) or all embryos that carry the mutation, whether heterozygous or homozygous (75%). It follows that the number of embryos remaining that are acceptable and likely to give pregnancies is often extremely low.

A single blastomere that has been removed for genotyping from an 8-cell embryo is highly unlikely to be able to undergo normal embryogenesis by itself. Although it could in theory give rise to any cell type in the body, it is too small on its own to allow a blastocyst or subsequent embryo to develop. This has not been tested in humans, but it is a clear result from experiments on animals, notably mice and sheep, with some data from rhesus monkeys (Chan et al. 2000). It is theoretically possible, however, that single blastomeres could be used to make an ES cell line. The embryos that are discarded as either being at risk or because of genotyping failure can also be used to derive ES cells. While some of these embryos may carry genetic defects, this is not necessarily relevant to the use of ES cells derived from them for research, indeed such ES cells may provide valid *in vitro* models of the disease. It also does not preclude their use clinically if the genetic defect is irrelevant to the disease being treated or if the genetic defect can be corrected (see Chapter 2).

After initial tests of viability (Hardy et al. 1990), PGD was first applied to allow sex selection following PCR diagnosis for Lesch-Nyhan, an X-linked disease (Handyside et al. 1990). FISH was introduced a little later (Griffin et al. 1994; Griffin et al. 1991) and polar body biopsy in 1996 (Verlinsky et al. 1996). By 1998, there were more than 40 centres in 17 countries performing PGD and it is now used for a wide range of single gene defects and chromosomal abnormalities (Table 4). It is difficult to gather statistics on its use and efficacy as each centre performing PGD can handle only a relatively small number of cases. However, attempts to do so within Europe and elsewhere are being undertaken by a European consortium (ESHRE PGD Consortium Steering Committee 2002; Geraedts et al. 1999).

Following PGD, rates of pregnancy to term are about 20–25% per cycle, which is very similar to that for standard IVF (ESHRE Preimplantation Genetic Diagnosis Consortium 2002). It perhaps ought to be higher as the parents are fertile. Misdiagnosis can occur, but it is relatively infrequent and techniques are constantly improving (Lewis et al. 2001). (By 1999 there were only about 6 cases reported to ESHRE, involving two for cystic fibrosis and one each for sexing, myotonic dystrophy, β-thalassaemia and trisomy 21). Prenatal diagnosis is usually advised to confirm a normal foetus, but many patients apparently decline for fear of losing a healthy pregnancy.

1.3.5
How Do We Decide when PGD Should Be Used and what Are Its Limits?

The use of PGD promotes considerable debate about where to draw the line. For example, how serious does the genetic disease need to be? Should it be used for late onset diseases? Should PGD be used for positive selection of desired traits or only to eliminate bad ones? Should it be used for aneuploidy screening on embryos from women older than 35, when the most likely outcome is Down's Syndrome? Should it be used simply for choosing the sex of an embryo? In the UK decisions like this are referred to the HFEA, who require details of each condition for which PGD is requested and the procedure cannot be carried out without a license. The latter is only granted if the HFEA are convinced of the medical importance and of the ability of the practitioners to perform the appropriate tests, provide counselling for the parents, etc. (for further details on the HFEA see Chapter 3, Case Study 2 and Appendices). Throughout the rest of the world, however, where no equivalent to the HFEA exists, the decisions are usually taken without the support of a good regulatory system by the individual clinicians and their collaborators. It is worth considering several case studies that emphasise some of the issues. The first two of these are taken from Braude (2001).

(i) A couple both suffer from achondroplasia, a dominant genetic condition characterised by extreme short stature, but no other problems. The chance of them conceiving an affected child is 50%, with a 25% chance of an unaffected child and a 25% chance of a homozygous dominant, which is lethal *in utero*. They request PGD to avoid having the latter, but unusually, they also request to have only heterozygotes replaced rather than the unaffected embryos, as they wish to have a child with achondroplasia like them. They believe that achondroplasia is a positive advantage that has enhanced their lives, partly through having to cope with the disability itself and with the perceptions of other people towards them. They also feel that an unaffected child may feel excluded and stigmatised amongst their family and friends through "suffering normal stature". They are also worried about obstetric complications if the woman was to be pregnant with an unaffected embryo.

In terms of who should take the decision, patient autonomy needs to be respected, but under the terms of the HFEA Act, the welfare of the child must be taken into account before agreeing to provide treatment. If the couple were to conceive naturally, which cannot be prevented by law, although they have a 25% risk of having a dead foetus or baby, the odds of them having a live born child with achondroplasia are 2:1. Fulfilling their wishes by PGD at least allows them to avoid the lethal condition.

It seems extraordinary to want to replace only affected embryos. However, there are clearly other conditions, such as congenital deafness, where the individuals concerned do not feel that they are suffering from a disability, on the contrary they feel special. It may not be so surprising, therefore, if the parents are happy for their children to inherit the same condition. Would we feel so alarmed if the above case had been about deafness, which is not a visible condition? But on the other hand, is deafness a condition that would warrant PGD?

(ii) Huntington's disease (HD) is a dominantly inherited, late onset disorder characterised by progressive neuropsychiatric problems and cognitive deterioration. There is no cure. The mean age of onset is 40 years and symptoms progress slowly with death occurring on average 15 years later. A woman aged 34 realised that she had a 50% chance of developing HD when her father was diagnosed with the disease some 6 years earlier. However, after counselling she declined to undergo predictive testing. This is quite common as many feel they could not cope with the news if it was bad. She and her husband decided to have children and opted for PGD to select against embryos carrying the HD allele. However, the woman specified that she did not wish to know the result of the diagnosis under any circumstances (non-disclosure) as long as only unaffected embryos are replaced. Fifteen embryos are tested during the PGD cycle and are all found to be unaffected. This makes it highly unlikely that she has HD. Three embryos are replaced, but she does not become pregnant. The woman requests a further cycle of IVF and PGD, which the clinicians now know is both unnecessary and costly.

This case highlights some of the problems associated with using PGD for late onset diseases. Is it reasonable to perform PGD to exclude embryos for a disease which only strikes after 40 years or so of normal life? What about the interests of the child if the mother was a carrier? The added complication of having to comply with a request for non-disclosure makes the case even more difficult as the clinicians have to explain what they have done without inferring the diagnosis. What if all the embryos they obtained were affected? See Braude et al. (1998) for a fuller discussion on this case.

(iii) In a well publicised case in the UK, the HFEA turned down an application for PGD for a couple whose son, Charlie Whitaker, suffers from Diamond-Blackfan anaemia. This is a very rare sporadic genetic disease caused by a new mutation that occurred in Charlie. Neither of his parents are carriers of the disease. Charlie's only chance of survival is to have a bone marrow transplant from a compatible donor. The Whitakers had a daughter since Charlie, but she was not an appropriate tissue match. They decide that they would like another child, but wish to use PGD to select an embryo of the correct HLA type, which should occur with a probability of 25%. However, the HFEA refused treatment because the Whitakers are not at risk of having another child with the same condition as Charlie: the diagnosis would have merely been for tissue typing. This decision of the HFEA had come only a short time after a similar case, that of the Hashmi family. This couple had also had a child affected with a blood disorder that could be cured by a bone marrow transplant. However, in this case the parents were carriers of the genetic disease. Therefore the Hashmis wanted PGD to avoid the genetic condition in their next child as well as to ensure that it would be a compatible donor for their first. In this case the HFEA approved the use of PGD because it would benefit the new child. Needless to say, the decision of the HFEA in the Whitaker case was met with some dismay, as the distinction between the two seemed rather subtle to many. PGD and tissue typing are unlikely to compromise the welfare of either the Hashmis' next child, or that of the Whitakers, but denying it in the latter case will have a deleterious effect upon Charlie Whitaker. The Whitakers are apparently seeking to have a child by PGD in the USA.

But should PGD be used in this way, to generate a child to save the life of another? Previously parents, hoping to win a "genetic roulette", simply decided to have another child who could prove to be a genetically compatible donor. Most recently in England, the parents of a child with beta thalassemia major received permission to use IVF and PGD to determine and produce the most suitable offspring as a bone marrow donor (Gross 2002). Using PDG parents' chances are obviously somewhat better than trusting to the genetic luck of the draw, but depending on the number of HLA alleles among the parents and whether the HLA of the affected child was a result of crossing over, it may prove somewhat difficult to find a suitable embryo. There are many reasons why couples decide to have more children. If saving the life of a sibling was the only reason (and it was not in the above cases), is it a bad one? On the other hand, how would the parents or the new child be affected if the transplant failed? Beside this aspect, it might be important to ask, how important it would be to the decision if the parents needed to use IVF anyway. If it were possible to obtain the appropriate cells (haematopoietic stem cells) for therapy from ES cells, then lines could have been derived from each of the early embryos at blastocyst stages (and the genotyping performed retrospectively on these), making it unnecessary to implant them in the mother. Given the choice, would the parents have opted to do this rather than have another child? Therapeutic cloning may have been a more suitable option in these cases. All the unsolved problems with therapeutic cloning notwithstanding, it may have represented, ethically and scientifically, a superior solution (see also 1.3).

With respect to PGD, we should also consider the practical limits to its use. In theory it should be possible to test for specific alleles at several or even many gene loci within the same embryo. However, there are practical (and statistical) limits on the number of characters (traits) that could be selected. For example, let us consider desired traits for which it is necessary to be homozygous, but for each of these the parents are heterozygous. For each trait, therefore, there is a one in four chance of an embryo being homozygous. For two traits together, the probability is one in 16, for three it is one in 64. For ten traits the chances of finding an embryo homozygous for the desired allele at all loci would be less than one in a million. As the number of embryos obtained with each cycle of superovulation is usually no more than 16, it follows that the chances are already low of being able to find an embryo with even just two of the desired characters. If one of the parents is homozygous for the trait or if heterozygosity is sufficient, then this will improve the chances, but it would still be practically impossible to select for more than three or four traits at once. Suggestions that it may be possible to have children selected for many different traits are therefore no more than science fiction (see Fukuyama 2002; Stock 2002). This is especially so when talking about characteristics like intelligence, ability to play music, etc., which are almost certainly dependent on the interaction of many genes.

1.4
Experiments Designed to Develop or Improve Methods of Contraception

A better understanding of fertilisation and early development (pre- and peri-implantation) of human embryos can be used to develop improved methods for preventing or controlling conception, implantation and pregnancy (Adjaye et al. 1998; Griffin et al. 1991). This is an important, although rather paradoxical avenue of research on human embryos. The research clearly has no benefit for the embryos being studied, and could be argued as being deleterious to embryos in general. However, most people accept that there is a need for alternative reliable methods of contraception. Rather than preventing ovulation, spermatogenesis, or physical barriers to fertilization, the types of contraceptive methods relevant here will be those designed to prevent fertilization, cleavage stage development or implantation. They could target anything from cell surface molecules to DNA replication. The methods used will involve IVF and embryo culture and are otherwise obvious.

1.5
Research to Obtain Basic Understanding of Early Human Embryogenesis

Much of this is related to and feeds into research on fertility, contraception, genetics and chromosomal abnormalities and general methodologies for ART. Work of this type has provided data on gene activity, DNA replication, chromosome structure and behaviour, cell division, apoptosis, determinants of cell shape, metabolic activity and stress responses (Braude et al. 1988; Capmany et al. 1996; Jurisicova et al. 1996; Leese 2002; Leese et al. 1993; Munne et al. 2000; Munne et al. 1994; Tesarik et al. 1988; Van Blerkom and Davis 1995). It has also allowed some work on the establishment of cell lineages and characterisation of markers of early cell types.

While in general terms early human development is similar to that in other mammals, this work reveals that early human embryos have several properties that make them distinct from other mammalian embryos. For example, human embryos often develop poorly in culture. They frequently show fragmentation and multinucleation (Pickering et al. 1995). There is a high level of aneuploidy (discussed above) and of mosaicism, where some cells (blastomeres) in an embryo may have a normal chromosome complement, whereas others in the same embryo may be abnormal. This has major problems for PGD (see above). Some of these problems may be due to inadequate conditions for oocyte retrieval (e.g. superovulation regimes), fertilisation or embryo culture. As discussed above, it is quite possible that the culture media is suboptimal and this is an important reason why research on embryo metabolism, and on growth factor requirements and signalling pathways is pursued.

Generally research on normal human development is limited to the preimplantation period, but as mentioned above (section 1.2.2), new culture methods may allow later development to be followed. This has clear moral constraints.

1.6
Experiments Designed to Develop Methods for the Alteration of the Genetic Makeup of the Human Embryo and Resulting Adult

1.6.1
Correction of a Genetic Defect

Any attempt to correct a genetic defect in an embryo is predicated on the ability to unambiguously identify such an embryo. Following identification, one has to be able to insert into the genome a piece of DNA that carries the replacement for the faulty or missing information. Both of these prerequisites may be very difficult, if not impossible. If the introduced DNA is to become part of the entire embryo, it has to be injected into the sperm or egg or into one of the pronuclei, which presupposes that the defective gamete or embryo can be identified and this may be impossible without destroying the gamete or embryo. It may be possible to determine the genetic status of the oocytes by examining the first polar body and taking into account the possibility of cross-over. As one cannot always wait for the results before *in-vitro* fertilization (some eggs may be frozen), all eggs would have to be treated individually, injected with DNA and sorted after the result of the analysis is known. It may be necessary to check them again at the 8-cell or blastocyst stage to determine in which embryos the DNA integration took place. All these elaborate manipulations are likely to reduce the success rate of subsequent development, possibly to zero. Moreover, one has to realize that integration of injected DNA per se does not guarantee the correct expression and function of the introduced gene. Because of random integration, the additional possibility of an introduced mutation cannot be excluded. For all these reasons it is very unlikely that simple transgenesis as outlined above will ever be used to correct a genetic defect in the embryo. If this is the only option, one would have to elect the simple alternative used today, namely discarding all embryos identified as homozygous mutants. This solution obviously depends on the quality of the preimplantation genetic diagnosis (PGD) and also does not reduce the number of carriers unless all heterozygous embryos were discarded as well (which would reduce the number of available embryos even further).

Beside the simple transgenic approach, we can envision certain other options, at least theoretically. One is the use of artificial chromosomes, and some early results have been presented (Knight 2001). This method would eliminate the risk of introducing mutations by DNA integration and would, at least theoretically, increase the likelihood of the introduced genes being correctly expressed. The difficulties associated with the identification of mutant embryos would remain. In addition, the artificial chromosome would have to behave as a normal chromosome during mitosis. Our present knowledge suggests that there is a possibility that single artificial chromosomes may interfere with meiosis and render the carrier sterile. A substantial amount of additional work is necessary before we can even contemplate the use of artificial chromosomes. As they might represent a way for genetic improvements, it is likely that investigation in this field will continue.

Another possible way of correcting a genetic defect in the embryo would involve nuclear transfer, which will be discussed later. One has to realize that the transgenic approach or the use of artificial chromosomes may solve the problem in the embryo itself, but if the embryo and the resulting individual is heterozygous, the transgene or artificial chromosome would separate from the mutation in half of the germ cells, so the entire process would have to be repeated in the next generation. If genetic screening becomes much easier, heterozygous individuals (carriers) would have the option to avoid having children with another carrier or to at least screen and discard all mutant and heterozygous embryos, thus reducing the mutation load to spontaneous mutations only. The dark side (and there always is one) could be that identified carriers of a serious genetic mutation may find it difficult to find a mate at all. The balance between the right to privacy versus the demand for divulging crucial genetic information to the future mate will be difficult to determine. The ability to identify and possibly correct our genetic defects is (as any other technological advance) going to complicate and not simplify our lives, but this is always the case when blind luck is replaced by the possibility of a voluntary option.

So far we have discussed germline gene therapy, i.e. the attempt to correct a mutation in all cells of the future organism with the possible proviso that such a correction will be passed on to future generations. One can also envision a more limited approach, namely somatic gene therapy – the prenatal or postnatal introduction of a normal gene or cells carrying the normal gene into the fetus or newborn. In this approach only those cells affected by the mutation would be targeted and the correction would presumably not be transmitted to the progeny of the patient (Cavazzana-Calvo et al. 2000). Although not directly relevant to our discussion, such gene therapy is not without problems. At the present time retroviral vectors are usually used as the best method of stably introducing a functioning gene into a large number of cells (May et al. 2000). A recent report (Li et al. 2002) indicates that such an approach could lead to the development of leukemia, and it is obvious that substantial additional risk assessment studies in animal models will have to be carried out before massive use of gene therapy targeting long-living stem cells is initiated. The selection of alternative vectors and the use of a so-called suicide gene, which would enable the destruction of all cells carrying the inserted construct, may become a mandatory part of somatic gene therapy.

1.6.2
Genetic Improvement

Provided that some or all of the methods outlined above become easy and reliable, we can envision that they will be used not only for the correction of a genetic defect but also for improving the genetic makeup of an individual. Some authors believe that this is not only desirable but also inevitable (Stock 2002), while some argue that it will result in an irrevocable change in human nature and thus lead to the total loss of human identity and dignity (Fukuyama 2002). The main question, however, should be: is it possible? Characteristics regulated by a single gene (eye colour, ear shape, disease resistance) may be suitable targets, though one could not envision much need or demand for such genetic alterations beside the possible interest in securing specific disease resistance if the disease in question is life-threatening and there is no available therapy (resistance to HIV infection may be a good example). Beyond single gene improvement we can envision two ways of systematically "improving" humans, both currently in very early, mostly theoretical stages, and neither generally acceptable or even permissible. One, and much the simpler, has already been discussed, and this involves embryo selection (see 1.3). Again, provided that genetic screening becomes much easier, we can imagine that PGD could be extended to "normal" embryos and, instead of analysing one gene, we could analyse hundreds. If that were the case, one could theoretically pick which embryos to implant and which to discard. Isolated cases of such practice in its most primitive form (selecting embryos on the basis of sex) are already going on at the present time. It is rather doubtful that this approach will be fruitful. Genetic screening will certainly be possible but the follow-up is questionable. First of all, we know very little about the genes that control the characteristics most parents would like to see in their children (intelligence, beauty, artistic ability, physical prowess) and it is very likely that these are regulated by a large number of genes. Moreover, even if we know exactly the responsible genes and how they work, it would be very difficult to do something with this knowledge. Each embryo would present a random mixture of "positive" and "negative" genes, and selecting for a slight increase in one desired characteristic may easily result in a slight decrease in another. As discussed above (end of section 1.3), there would never be sufficient numbers of embryos from which to choose one with more than a very few desired traits. As long as embryos represent a random reshuffling of their parental genes, it is unlikely that the ability to pick and choose among them will result in a significant improvement.

The aforementioned artificial chromosome would represent a more deliberate vehicle for improvement. Once identified, genes for intelligence and beauty could all be packed into one artificial chromosome and injected into the zygote. How would they interact with each other and with endogenous genes is an open question. In addition, most of our complex characteristics are likely to be significantly affected by small and unpredictable environmental influences that occur during development (Finch and Kirkwood 2000). It is entirely unclear as to how soon we can expect any serious attempt along these lines and whether we should spend a significant amount of time discussing the ethical acceptability of genetically improved humans.

1.6.3
Cloning

Cloning by means of nuclear transfer can be viewed as another way of deliberately controlling the genotype of an individual, and only these aspects will be discussed here. The use of nuclear transfer to produce embryonic stem (ES) cells for therapeutic purposes (therapeutic cloning) will be discussed separately.

Before we examine the possible usage of cloning as it applies to humans, which will be discussed more thoroughly below in Chapter 2, we have to briefly analyse the current state of the techniques and results that have been obtained. Several mammalian species have been successfully cloned, and one can reasonably assume that all mammals can be cloned provided that species-specific reproductive problems are resolved (Solter 2000). Although possible, cloning by nuclear transfer continues to be extremely inefficient, and it is unclear to what extent biological and/or technical problems contribute to the low success rate (Perry and Wakayama 2002; Solter 2000; Wilmut 2002). It may be too early to try and draw definite conclusions in this rapidly developing field, but some tentative inferences can be made. It is possible that technical improvements addressing various aspects of the cloning procedure (egg collection and selection, removal of the genetic material of the egg, introduction of a foreign nucleus, egg activation and culture) may increase the overall success rate, i.e. currently 1% of manipulated eggs develop to adults (Chung et al. 2002; Heindryckx et al. 2001; Ono et al. 2001). However, biological problems encompassed by the term "incorrect reprogramming" are likely to be much more serious and more difficult to solve. When one considers what is involved in successful cloning, it is certainly amazing that it works at all, no matter how rarely. In normal development the egg cytoplasm must, following fertilization, reprogram the genetic material of both egg and sperm with the resulting activation of the embryonic genome, and this has to occur in a limited time span (Solter et al. 2002). The egg cytoplasm is obviously designed to accomplish this reprogramming, and the fact that both sperm and egg genomes are inactive at the time of fertilization may contribute to its success. Following the transfer of a somatic cell nucleus, the egg cytoplasm encounters a transcriptionally active genome it is not designed to reprogram. Whatever pattern of gene expression specific for a particular somatic cell is present, it needs to be shut down and a pattern specific for the zygotic genome must be established. It is no wonder that the majority of clones does not succeed and fails before implantation. Further failures in reprogramming result in fetal and perinatal loss (Renard et al. 1999; Solter 1999) and phenotype abnormalities are observed in the majority of adult clones. These include obesity (Tamashiro et al. 2002), pneumonia and liver failure resulting in premature death (Ogonuki et al. 2002) and arthritis recently observed in Dolly, the first clone from adult cells (Williams 2002). Though there have been some reports suggesting that adult clones are phenotypically normal (Lanza et al. 2001), the preponderance of abnormalities suggest that no clone is entirely normal (Wilmut 2002). Similarly, examination of gene expression in clones (embryos, placentas and newborn) suggests that numerous imprinted and non-imprinted genes are incorrectly expressed, and this incorrect expression is usually accompanied by an irregular methylation pattern (Bourc'his et al. 2001; Daniels et al. 2000; Fairburn et al. 2002; Humpherys et al. 2002; Humpherys et al. 2001; Inoue et al. 2002; Kang et al. 2001; Ohgane et al. 2001; Rideout et al. 2001; Young et al. 2001). In view of these uncertainties, it is absolutely clear that human reproductive cloning should not be attempted. In addition to a tremendous loss of embryos, it is very likely that the clones would be born abnormal. Experiments aimed at improving the technology of cloning may change this situation somewhat, but the problem of incorrect genomic reprogramming will remain, and it is not at all clear how it could be resolved.

We can be optimistic, however, and assume that one day the risks of reproductive cloning will approach the, now negligible, risks of *in vitro* fertilization. When such a time comes, what can be envisioned as the proper application of the nuclear transfer technique aimed at producing an adult? The most often cited need is to give a sterile person or a couple the chance of their own genetic continuation. The replacement of a tragically lost child is also usually mentioned. Although understandable, these desires do not seem pressing enough especially considering the risks involved. However, a few possible and useful applications of the technology do exist, though they may not be reproductive cloning in the strict sense. One would address mitochondrial mutations which are obviously transmitted only through the female germline. One possible solution to such a problem would be the injection of donor egg cytoplasm into the oocyte or zygote carrying the mutated mitochondrial DNA (see 1.2.4). Ooplasmic transplantation leads to mitochondrial heteroplasmy and has been used to correct compromised oocytes from patients with repeated implantation failure of unknown origin (Barritt et al. 2001b). As an aside, this procedure has also resulted in a few fetuses with chromosomal abnormalities (Knight 2001), another indication that procedures used in assisted human reproduction need to be thoroughly investigated and tested in animal models before being applied in clinical practice (Naviaux and Singh 2001). Obviously, in the case of mutated mitochondrial DNA in oocytes, the best solution would be a complete exchange of the cytoplasm (Hesterlee 2001). This can be best accomplished by transferring both pronuclei following *in vitro* fertilization into a recipient enucleated oocyte or zygote (McGrath and Solter 1983). Although involving nuclear transfer, the procedure is not cloning in the strict sense, since the genome of the zygote was transferred and only one individuum can be born following the transfer.

Another similar, though slightly more complicated, scenario involves the progeny of parents both homozygous for the same recessive mutation causing a disease. Admittedly rare, this situation will inevitably result in all children being affected by the same disease. One possible solution involves IVF and the development of embryos to blastocyst in culture followed by the isolation of ES cells (see later) from these blastocysts. Once the ES cells have been derived, one could use the technique of homologous recombination (well-established for mouse ES cells) to repair one or both mutated alleles. The transfer of nuclei from such corrected ES cells into enucleated oocytes and the return of the developing embryos to the mother womb would complete the procedure which, if successful, would result in the birth of the couple's own genetic but now normal child (Solter 1999). Obviously we cannot forget the aforementioned problem connected with nuclear transfer, its low success rate and the likelihood of phenotypic abnormalities, all of these reported following the transfer of nuclei from ES cells in mice (Humpherys et al. 2001; Rideout et al. 2001). However, such a procedure may be ethically and theologically palatable to all foes of reproductive cloning since no embryo was destroyed and no individual was cloned, just normal reproduction with a few detours along the way.

1.7
Research Involving the Use of Embryos to Derive ES Cells for Therapeutic Purposes

Following the above discussion, which mainly dealt with experimentation carried out directly on human embryos, let us now expand the discussion and consider stem cells. This topic is mostly covered in Chapter 2 on Adult and Embryonic Stem Cells: Clinical Perspectives. However, several of the issues that are considered in Chapter 2 need also to be examined from the perspective of the embryo. The connection is both practical and conceptual, as the majority of stem cells can be isolated from embryos and fetuses, and understanding the function and biology of stem cells will significantly contribute to our understanding of the biology of the embryo.

Thus research on early human embryos is necessary to:

(i) improve efficiency of ES cell derivation;
(ii) improve methods of derivation with respect to safety for clinical use;
(iii) research the parameters required in early embryo culture and manipulation that confer maximal developmental potential of the resulting ES cells and/or appropriate ways to direct differentiation in required directions or to select for specific cell types.

It is not clear that any of the currently available cell lines has been derived in the best way. Perhaps a new method will be found that allows them to be established more efficiently and to reliably give rise to cells appropriate for a given therapy. It is clear with mouse ES cell lines, that some of the first ones to be established are not as useful as many of the more recent ones, and this correlates with a better understanding of their properties and how to grow them. But this will be a difficult question to answer for human ES cells as we do not have the ultimate test of their potential – the ability to make chimaeras. With experience, perhaps the use of microarray technology or screening for the appropriate expression of a set of specific gene markers will at least allow for some measure of reproducibility.

There is nevertheless a strong requirement to develop human ES cell lines that will meet standards appropriate for their use in a clinical setting (good manufacturing practice (GMP). With this in mind there is already research ongoing to develop feeder-free conditions and serum free media for ES cell derivation and maintenance. However, IVF is currently not done under GMP conditions, which already poses several logistical problems. Once routine methods have been developed to establish ES cell lines that are appropriate for clinical use, then further research on embryos for this purpose becomes less important. However, there will almost certainly be a continuous need to generate new cell lines.

The ES cell lines can be used to generate cells for therapeutic purposes, to understand some aspects of human embryonic development and the role of specific genes (*in vitro*), and for testing of pharmaceuticals. For the latter, ES cell lines carrying mutations which result in particular genetic diseases may be useful (see 1.3).

Embryonic stem cells were isolated from mouse blastocysts in the early eighties (Evans and Kaufman 1981; Martin 1981) and their utilisation revolutionized experimental approaches to mammalian developmental biology and genetics (Lovell-Badge 2001). Human embryonic stem (ES) cells were isolated from human blastocysts (surplus embryos from IVF programs) using essentially the same approach (Thomson et al. 1998), and biologically very similar human embryonic germ (EG) cells were isolated from the genital ridges of 5–9 week old human fetuses (Shamblott et al. 1998). Both ES and EG cell lines have since then been rather extensively studied: human ES cell lines can be cloned[2] (Amit et al. 2000) and the cloned cells propagated, the phenotype of ES and EG cells in terms of several markers have been established (Shamblott et al. 1998; Thomson et al. 1998) and their capacity to differentiate *in vitro* and *in vivo* analysed (Amit et al. 2000; Reubinoff et al. 2000; Schuldiner et al. 2000; Shamblott et al. 1998; Thomson et al. 1998). Though very important in their own right as a model of early human development, ES cells attracted much wider attention because they were seen by many as *the* unique promise in human medicine. They were hailed as the source of all differentiated cells, tissues and organs to be used in replacement therapy, the Holy Grail in human medicine and the scientific breakthrough of the decade. Now, several years after their isolation and with considerably more data available, we can examine to which extent these promises were real and are likely to be accomplished. At the outset it must be emphasized that there are numerous unresolved technical and biological problems (although it is difficult to draw a line between them) to be solved before we can expect the routine use of ES and EG cells in human medicine (Donovan and Gearhart 2001). We can easily conceptualize the ideal conditions and criteria that have to be established for the efficient use of human ES and EG cells in practice. These criteria include the ability to indefinitely propagate the cells in culture, to induce their differentiation into any desired cell type, to sort out from the mixture of different cell types the one needed for the particular purpose and to develop the best possible way of applying these cells to a patient (Solter and Gearhart 1999). Let us briefly examine how we stand on these criteria. The isolation and propagation of ES and EG cells required the use of a feeder layer (irradiated mouse fibroblasts were the most commonly used) (Shamblott et al. 1998; Thomson et al. 1998). Feeder-free growth of human ES cells has recently been reported (Xu et al. 2001a), however these cells required medium conditioned with mouse embryo fibroblasts. In addition, these feeder-free ES cells have initially been established using mouse feeder layers; the establishment of human ES cells without the use of feeders has not been reported. If the differentiated derivatives of human ES cells are to be used in clinical practice, it would make sense for them never to have been in contact with cells of any other species in order to avoid the possibility of transferring species-specific endogenous viruses. It will thus be necessary to establish conditions in which human ES cells are isolated, cloned and propagated using human (preferably the patient's own) fibroblasts as a feeder layer, which has recently been accomplished (Richards et al. 2002), or, even better, using a completely defined

---

[2]   Cloning in this context means taking a single cell and establishing a cloned cell line from it by placing this single cell in a separate culture dish.

medium without feeders. To further this goal, extensive research on the isolation of human ES cells needs to be done, and this work cannot be done if research is limited to certain arbitrarily chosen and already established cell lines, as is sometimes stipulated, e.g. by the German regulations (see Chapter 3, Case Study 3).

Differentiation of human ES and EG cells *in vivo* and *in vitro* follows similar pathways as that of mouse ES cells. Investigators used various culture conditions and different cell type-specific markers in order to achieve and demonstrate differentiation into endothelial cells (Levenberg et al. 2002), neural precursors (Reubinoff et al. 2001; Zhang et al. 2001), haematopoietic colony-forming cells (Kaufman et al. 2001) and several other cell types (Shamblott et al. 2001). The process of differentiation is regarded as somewhat haphazard – multiple cell types appear and the desired cell type must be sorted out from the majority of the other, often undefined, cell types. We are still very far from the ideal situation, namely of precisely defining the conditions in which the vast majority of ES or EG cells reproducibly differentiate into a single desired cell type. It is encouraging that the interest in the differentiation of human ES cells provoked a renewed interest in studying the same question using mouse ES cells. Mouse ES cells are less difficult to handle and the extent of their differentiation *in vivo* can be easily assessed, which is not the case with human ES cells. It is to be hoped that by defining the precise conditions for the differentiation of mouse ES cells (Brüstle et al. 1999; Eto et al. 2002; Kim et al. 2002; Lumelsky et al. 2001; Perlingeiro et al. 2001; Rathjen and Rathjen 2001; Yamashita et al. 2000), we will be able to translate these results to the differentiation of human ES cells. One of the issues which has to be resolved before using derivatives of ES cells in therapy is their potential to develop into tumors. Obviously ES cells themselves will develop into teratocarcinomas if injected, so before injection they would have to be sorted out with absolute certainty (Solter and Gearhart 1999). The use of both endogenous and transgenic (Eiges et al. 2001) markers would be necessary to accomplish this goal as would the ability to follow the injected cells (Lewin et al. 2000). Nevertheless, it may be prudent and necessary to introduce a suicide gene into ES cells so that they or their derivatives could be induced to die if necessary.

So far we have been discussing ES cell-based therapies in conjunction with *established* ES and EG cell lines, and the majority of the current work is orientated in this direction. However, this is not the most desirable approach for several reasons. An absolutely ideal cell and tissue replacement therapy would involve the use of patients' own cells, thus eliminating the problem of immune rejection (Drukker et al. 2002) and the need for immunosuppression and also avoiding any danger of infection from the grafted tissue. This approach is theoretically possible and has been repeatedly elaborated as so-called "therapeutic cloning" (Smith 1998; Solter 1998a, 1999; Solter and Gearhart 1999). To put it in the most simple terms – therapeutic cloning involves the transfer of a nucleus from the patient's own somatic cells into an enucleated egg, the growth of the resulting nuclear transfer embryo to a blastocyst and the isolation of ES cells (using the patient's own fibroblasts as a feeder layer). These ES cells could then be manipulated *in vitro*, the desired DNA changes introduced if necessary by homologous recombination, the cells induced to differentiate and the appropriate cell types then returned to the patient. Experimental confirmation that this approach is possible has recently been repeatedly described using mouse models (Kawase et al. 2000; Munsie et al. 2000; Rideout et al. 2002; Wakayama et al. 2001). Although therapeutic cloning may sound ideal for the desired purpose, there are definitely several problems connected with this procedure, some of which may not be solvable. One of the problems, which will be difficult to resolve, is the issue of time. Obviously, to complete the entire therapeutic cloning procedure a substantial length of time is required even in an ideal situation, and some clinical situations might not allow for this. For example, one of the anticipated uses of ES cell-based therapy is myocardial infarction. It is very likely that one of the conditions for the success of cell replacement therapy would be the creation of a suitable "empty space", which the introduced cells could then easily occupy. This would determine the ideal timing for the introduction of the cells and, in the case of myocardial infarction, the first week when the ischaemic muscle is dying seems the best time. There is no way that autologous ES cells can be prepared in that space of time and by the time they are ready, scarification of the heart muscle may have made replacement with fresh tissue impossible. Similar logic may apply to cases of acute liver failure or traumatic or infectious spinal cord damage. Therapy with autologous ES cells would be, by necessity, reserved for chronic conditions; but there are certainly enough of those to justify developing this therapeutic modality. Alternatively, one could identify individuals "at risk" and prepare their ES cells ahead of time. It is certain that therapeutic cloning will be very labour intensive and expensive but, considering the expenses currently incurred to care for patients suffering from chronic diseases (juvenile diabetes, Parkinson's disease) for which ES cell therapy may be usable, the expenses and labour may be justified and even economically advantageous. After all, we envision that ES-based therapy will actually cure and not merely treat these diseases.

The other problem with therapeutic cloning as evidenced from the "proof in principle" work in mice (Kawase et al. 2000; Munsie et al. 2000; Rideout et al. 2002; Wakayama et al. 2001) is low efficiency. Depending on different reports, 0.2–3.5% of reconstituted oocytes resulted in successful ES cell line derivations. Considering that this work was done in mice where extensive experience and unlimited material exist, one would anticipate an even worse success rate when using human material. It is also clear from the reported results that the success rate of the isolation of ES cell lines from cloned blastocysts is substantially lower when compared with normal blastocysts, again emphasizing the problem with reprogramming. It is possible that the reported low efficiency is mostly due to inexperience or technical problems which could be solved in the future. A recent and very controversial report dealing with the very limited development of human nuclear transfer embryos (Cibelli et al. 2001) further emphasizes this problem, though unconfirmed rumours suggest that a group of Chinese scientists have succeeded in producing cloned human blastocysts, but not ES cell lines. If the success rate of therapeutic cloning is not substantially increased, the entire approach may be theoretically possible but unpractical, since it will be impossible to secure the necessary number of human oocytes (Rossant 2002). One possibility would be to intensify research into the use of fetal and adult ovaries (obtained post mortem or during operation) as a source from which usable oocytes can be derived by *in vitro* growth and maturation (Klinger and De Felici 2002). One may even imagine that we will be able to control the differentiation of ES cells in culture into human germ cells and full-grown oocytes. This has been accomplished using mouse ES cells (Hübner et al. 2003), thus the shortage of oocytes necessary for therapeutic cloning may no longer be a problem.

Considering all these problems, we also have to start simultaneously looking for alternatives to therapeutic cloning. After all, the sole purpose of therapeutic cloning is to turn the patients' own cells into ES cells, and there may be other ways to accomplish this beside reprogramming using human oocytes.

We can envisage several possible approaches to reprogram the adult nucleus (Solter 1999; Surani 2001), and these include the use of non-human oocytes, fusion with ES cells and exposure to defined controlling factors. Early attempts to produce human ES cells by transferring adult human nuclei into rabbit or cow eggs were reported anecdotally (Marshall 1998) and never published. A recent, again anecdotal, report of a Chinese group (Abbott and Cyranoski 2001) producing ES cell lines from blastocysts derived by transferring human somatic cell nuclei into rabbit eggs appears more solid, but we will have to await the publication of these results. Fusion of a somatic cell with an ES cell can result in reprogramming and somatic cell hybrids resembling ES cells (Tada et al. 2001; Terada et al. 2002; Ying et al. 2002) were observed. Such hybrids are tetraploid or nearly tetraploid possessing chromosome complements from both parents, and it is unlikely that they can be used for cell replacement therapy. For that purpose one would have to fuse somatic cells with enucleated ES cells (cytoplasts) and produce so-called cybrids, which would hopefully contain the reprogrammed genome of the somatic cell. The use of ES cell cytoplasts is technically rather difficult and no results along these lines have been presented.

The final and, in many ways most desirable, option would be to expose somatic cells to a set of factors (signalling molecules, DNA binding proteins, transcription factor, chromatin remodelling molecules) which would turn them into ES cells. Some very preliminary, though promising, results are just appearing and one can expect intensive and probably productive efforts in that direction (Håkelien et al. 2002; McGann et al. 2001; Western and Surani 2002).

Many of the problems outlined above may not necessarily apply with another major goal of "therapeutic cloning", namely the use of ES cells derived from an individual, not to directly cure that individual, but as a means to study genetic diseases or genetic variation and its interaction with environmental factors *in vitro*. This is discussed briefly in Chapter 2. This goal is often ignored in debates, but it is very important and it may well be the only currently practical reason for carrying out the procedure in the context of human embryo research.

So far we have discussed the biological problems of therapeutic cloning, though there are also additional legal, ethical and religious objections (as discussed in Chapter 3 and 5) which may lead to it being declared illegal. The main objection is that it involves the destruction of potential human beings and that these human beings are being instrumentalized, i.e. produced to be destroyed, for the benefit of another human being. One possible way around these objections would be to use parthenogenetic and androgenetic blastocysts to produce ES cells. Parthenogenetic embryos are produced by activation of the egg but without fertilization, while androgenetic embryos would be derived from fertilization of an enucleated oocyte. Both of these can develop to blastocysts and ES cells can be derived from such embryos (Kaufman et al. 1983; Mann et al. 1990). In addition, ES cells have been derived from parthenogenetically produced monkey eggs (Cibelli et al. 2002) and even human parthenogenetic cleavage-stage embryos but not blastocysts have been reported (Cibelli et al. 2001). Although it is not absolutely certain that ES cell lines derived from parthenogenetic and androgenetic blastocysts would behave as those derived from normal blastocysts, they may still prove useful as their derivation should not cause any moral misgivings. Namely parthenogenetic and androgenetic embryos cannot develop much further than implantation and certainly cannot result in newborns because of imprinting (Solter 2001). Briefly, imprinting is a process which differentially marks the genome of male and female mammalian germ cells resulting in their differentiated functioning and in the absolute necessity for both of them to be present to secure normal development (Reik and Walter 2001). The use of androgenetic and parthenogenetic ES cell lines may be, from a scientific point of view, a second less favourable alternative, but it will illustrate, not for the first time, that applied science, like politics, is an act of the possible.

1.7.1
Will Clinical Application of ES-cells Be Necessary in View of the Availability of Adult Stem Cells?

One could argue that the consideration of fetal and adult stem cells does not properly fit into a book dealing with embryo experimentation. However, as discussed above, some aspects of the therapeutic applications of human ES cells definitely require the use and destruction of early human embryos while fetal and adult stem cells have been repeatedly proposed as ethically and morally painless alternatives. It is thus necessary to briefly discuss their therapeutic potentials as seen today. A more extensive analysis of adult stem cells is further presented in Chapter 2.

The existence of adult stem cells was always obvious or we would not be able to explain the continuous renewal our tissues undergo during our lifetime. Some of the stem cells (bone marrow, keratinocytes) have been extensively used in clinical practice and the use of neural stem cells has also been intensified (Blackshaw and Cepko 2002; Blau et al. 2001; Gage 2000; Rossi and Cattaneo 2002; Watt and Hogan 2000; Weissman 2000; Wu et al. 2002). However, two relatively recent developments again attracted considerable attention to adult stem cells and their possible usage. One not so surprising finding was the fact that many, if not all, adult organs possess stem cells and that these cells can be isolated, studied and moreover used in cell and tissue therapy. Examples using both human and mouse models include: liver (Braun et al. 2000), pancreas (Ramiya et al. 2000), brain (Roy et al. 2000), retina (Tropepe et al. 2000), heart (Anversa and Nadal-Ginard 2002; Kao 2001) and also situations where one type of stem cells shows a multipotential capacity as in the case of oligodendrocytes (Kondo and Raff 2000) and bone marrow stromal cells (Sekiya et al. 2002). These latter examples lead us to another recent and more surprising finding, namely the capacity of somatic stem cells to show remarkable plasticity. Among the first of the reported transdifferentiation cases was the observation that adult neural stem cells can give rise to several blood cell types (Bjornson et al. 1999). This brain into blood sensation opened a flood of similar reports, so obviously reciprocal cases of blood into brain, i.e. the observation of cells with neural phenotypes following the engraftment of adult bone marrow, were soon reported (Brazelton et al. 2000; Mezey et al. 2000). Further examples included the differentiation of bone marrow stem cells into hepatocytes (Alison et al. 2000; Lagasse et al. 2000), muscle stem cells into haematopoietic stem cells (Jackson et al. 1999) and bone marrow stem cells into myocardium (Orlic et al. 2001), which in this particular case resulted in the regeneration of the infarcted cardiac muscle.

In addition to these one-to-one transdifferentiation examples, several researchers observed cells with the capacity to contribute to multiple lineages. Bone marrow stromal cells, bone marrow haematopoietic cells and adult neural stem cells displayed a remarkable capacity to differentiate into many cell types *in vitro* and to colonize numerous organs *in vivo* where colonizing host cells expressed appropriate tissue-specific markers (Clarke et al. 2000; Krause et al. 2001; Pittenger et al. 1999). The existence of cells with even more remarkable pluripotent capacities has recently been reported, namely that of mesenchymal stem cells (Jiang et al. 2002; Prockop 2002) and of epidermal stem cells (Liang and Bickenbach 2002), both of which, when injected into mouse blastocysts, contributed to many adult tissues in the resulting chimaeras.

The finding of adult stem cells with such remarkable properties served many opponents of ES cell research and therapeutic cloning as a clinching argument to request that all research using ES cells, and especially therapeutic cloning, be forbidden and declared illegal. After all, if each one of us possesses cells which can be used to replace any cell in our body that needs replacement, what is the justification for destroying potential humans as is necessary in therapeutic cloning? Well, as many things too good to be true usually turn out to be not so true, there is serious doubt about the pluripotentiality of adult stem cells. Turning brain into blood (Bjornson et al. 1999) was among the first of these reports, and it now seems that this happens either extremely rarely or never (D'Amour and Gage 2002; Morshead et al. 2002). Using a very sensitive assay, the researchers injected 128×10^6 neural stem cells into 128 host mice and did not find a single contribution to haematopoietic repopulation (Morshead et al. 2002). The authors who initially reported that stem cells from skeletal muscle have a significant capacity to produce haematopoietic stem cells (Jackson et al. 1999) have now realized that this observation was due to the large number of haematopoietic stem cells present in adult muscle and that no transdifferentiation took place (McKinney-Freeman et al. 2002). Reports of bone marrow stem cells repopulating the brain with cells of neural phenotype (Brazelton et al. 2000; Mezey et al. 2000) could not be confirmed (Castro et al. 2002a; Wagers et al. 2002a) and this resulted in at times acrimonious and still unresolved confrontation (Blau et al. 2002; Castro et al. 2002b; Wagers et al. 2002b). Moreover, two recent reports (Tada et al. 2001; Terada et al. 2002) demonstrated that mouse ES cells can spontaneously fuse with other cell types *in vitro* and impart their phenotype on the hybrids. It is entirely possible (though not likely) that all or nearly all cases of transdifferentiation could be explained by such rare fusion events. One has to realize that a fusion event will result in a cell which will show both the markers of the origin of the donor cell and the differentiation marker of the host cell, so one could easily be misled into believing that the donor cell transdifferentiated (DeWitt and Knight 2002; Lemischka 2002; Wurmser and Gage 2002).

Should we thus conclude that adult stem cells are useless and that only ES cells can be a real basis for cell and tissue therapy? Of course not. We should conclude, as it should have been obvious from the beginning, that working in the hot crucible of media attention and in competition for money, patients and prizes, the slow and careful goes out of the window and widely exaggerated claims replace reality. At this point in time – and this has been repeated time and time again – we do not know nearly enough to made an educated guess as to which of the approaches will result in therapies of the future. Choosing only some of all possible options by legislative fiat (USA), or voluntary moratorium (as some scientists in Germany propose), goes against everything that sound science should support.

1.8
Conclusions

Two broad justifications can be put forward for experimentation using human embryos. One envisages experiments aimed at addressing basic biological questions concerning human development. Provided that any adequate alternative experimental approaches exist, one would certainly avoid using human embryos per se. The alternatives are the embryos of other mammalian species or cultured cells. Much of what we now know about early mammalian development resulted from work using rodent (mostly mouse) embryos, and it is likely that many basic molecular mechanisms are shared by all mammalian (and possibly non-mammalian) embryos and can be profitably investigated without any use of human embryos. The same is true for the investigation of feto-maternal interaction for which larger farm animals (mostly sheep) have been predominantly used. However, one has to realise that some aspects (possibly quite important in a given situation) are unique for the development of every mammalian species and, if the information is of crucial medical significance, it will have to be confirmed using human embryos. This brings us to the second justification, which applies to clinical practice. Be it reproductive medicine, teratology, fetal surgery and so on, there is a limit to how much we can learn by using experimental models. Any procedure or drug which is to be widely used in human medicine and in any way associated with reproduction will have to be tested on human embryos or pregnant women. The use of Thalidomide and its consequences is the most tragic example of how inadequate test results on experimental animals can be. Many of these tests will necessitate the creation of embryos and their subsequent destruction and some of the tests and experiments will not result in any evident benefit to the embryos and fetuses used in the test, though they should anticipate benefits for others.

It may be difficult to accept these premises for various ethical, moral and religious reasons, but this type of balancing benefits is sometimes unavoidable. It is thus extremely important to precisely understand the goal of each experiment involving the use of human embryos, the need for the resulting information and the likelihood of a valuable outcome and also to be absolutely certain that no alternative approach is available or possible.

Advances in genetics and developmental biology achieved in the last ten or so years have opened up possibilities which were previously relegated to science fiction, and it is likely that even more undreamed of advances will be forthcoming. We are thus in a very fluid situation in which it may be impossible to set down firm rules to guide our conduct in the field of human embryo experimentation. We can only hope that, by approaching the issue without prejudice, bias and hidden agenda and by being guided by logic and compassion, we will be able to balance apparently irreconcilable interests and to somehow enhance the well-being of humanity.

2 Adult and Embryonic Stem Cells: Clinical Perspectives

2.1 Introduction

Besides reproductive purposes, which we considered in Chapter 1, the second major purpose of embryo research is to be seen in the development of new tissue and organ replacement therapies on the basis of material derived from embryonic cells or fetal tissues. Since the use of embryos and fetuses for other than reproductive purposes has often been discussed as being not necessary since sufficient alternatives were assumed to exist, we will not only review the current state of art and highlight the prospects for the near future of replacement therapies based on embryonic or fetal cells but discuss also alternative approaches, such as the use of adult stem cells.

During the last century, modern medical developments led to dramatic progress in the treatment of congenital or acquired diseases resulting in a significant increase in life expectancy. These outstanding developments include the introduction of antibiotics and vaccines to evade and treat infectious diseases, optimised surgical and intensive care procedures, and replacement of tissues and limbs. Even in the case of complete organ failure, allotransplantation can save lives and restore quality of life for many years.

However, for a wide range of vital disorders we have only very poor treatment or have no cure at all. So far, artificial tissue and organ substitutes suffer many drawbacks. Artificial prostheses rarely achieve sufficient quality or long-term replacement of natural parts of the body, while problems of scarring or immune rejection arise frequently due to incompatibility of materials. Although allogeneic transplantation is now a well-established clinical procedure, it is not feasible for all organs, e.g. the brain, because of ethical considerations or technical limitations. Infections and malignancies due to the use of immunosuppressive agents represent other important drawbacks of clinical organ transplantation. Moreover, in spite of significantly improved immunosuppressive protocols graft rejection is still common. Recent results in transplantation research raise hopes that through induction of peripheral or central tolerance pharmacological immunosuppression may become unnecessary, but this is still a long way in the future (Waldmann 1999).

Another major problem in allotransplantation is the increasing shortage of organs. Since fewer organs than needed are available, waiting lists continuously grow and an increasing number of patients die while waiting for a suitable graft (figure 1). With regard to organ shortage, xenotransplantation, the transplantation of cells, tissues and organs from one species to another one, has to be mentioned as a therapeutic concept. For clinical use, pig organs would be practical in terms of size and physiology and may offer an essentially unlimited availability of organs of

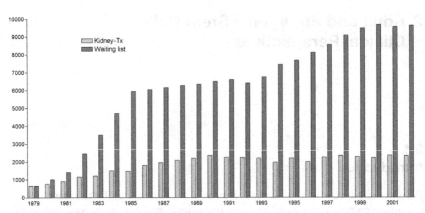

Figure 2.1 Number of kidney transplantations compared to the number of patients waiting for kidney transplantation in Germany 1979–2000

constant quality (Cascalho and Platt 2001; Logan 2000). However, it is currently unknown whether the vehement rejection of grafted pig organs and the tremendous physiological problems (e.g. interference with the coagulation system) can be managed (Platt 2000a; Platt 2000b) or whether the risk of infections with porcine pathogens is significant and if so, whether it can be avoided (Blusch et al. 2002).

In view of the problems discussed above, the employment of human stem cells offers totally new therapeutic perspectives for a variety of different diseases. For example, treatment of Morbus Parkinson or Alzheimer's disease might be possible by transplantation of neural stem cells or differentiated neuronal cells (Freed et al. 2001). The use of stem cells in combination with tissue engineering technologies may allow the development of bioartificial tissues and organs. When using autologous cells of the affected patient, which will not be immunologically rejected by the recipient, tissues and organs can potentially be produced in unlimited numbers.

During future decades, new technologies of cell transplantation and tissue engineering may help to overcome current therapeutic limits and may enable treatment of so far untreatable diseases. However, to finally reach these goals, many problems have to be resolved: At present it is unknown whether *in vitro* differentiation of embryonic or adult stem cells will result in cells functionally equivalent to the corresponding "normal" somatic cell type. Moreover, we still need to greatly improve our understanding of how organs normally develop in the embryo, how they acquire their structure and function and to develop better ways to achieve appropriate vascularisation, innervation if required and growth in three dimensions.

2.2
Regenerative Medicine: New Approaches for Tissue Repair

The term "Regenerative Medicine" has been applied recently as a way of highlighting several emerging therapeutic strategies that should allow regeneration or repair of lost tissues and organs. These include:

1. Application of growth factors or other factors regulating proliferation, migration, differentiation, and structural integration of cells and tissues
2. Gene therapy ⇒ cell stimulation to achieve *in vivo* expansion of somatic cells or the production of clinically relevant factors (e.g. insulin, dopamine)
3. Cell transplantation ⇒ regeneration of tissues and organs through application of cells expanded (and differentiated) *in vitro*
4. Tissue engineering for the construction *in vitro* of bioartificial structures

If it is assumed that the above methods will work, each of the first three approaches could result in regeneration of affected tissues. In contrast, the most sophisticated strategy, designated as "Tissue Engineering" may allow future replacement of lost or damaged tissues or organs.

2.2.1
Application of Factors Regulating Proliferation, Migration, Differentiation and Structural Integration

Cell proliferation ceases in many tissues and organs once they reach their adult size, however others hold an intrinsic regeneration capacity. This is due to the remaining proliferative potential of some differentiated somatic cell types (e.g. endothelial cells and fibroblasts), to the existence of organ specific stem cells (e.g. epidermal stem cells) or to the plasticity of non-tissue specific stem or progenitor cells (e.g. endothelial cell progenitor cells of the peripheral blood). Nevertheless, even in these tissues, regeneration is usually limited by the extent of injury, by the formation of scar tissues or by insufficient proliferative capacity of the necessary cell type. Moreover, the lack of appropriate chemotactic factors or guidance cues, as well as an inability to integrate into a structure, can compromise regeneration. Application of appropriate factors may improve the body's own regeneration capacity in the future. For example, bone marrow derived stem cells mobilised by granulocyte colony stimulating factor (GCSF) and/or stem cell factor (SCF) are proposed to allow improved regeneration after hepatic injury or revascularisation after myocardial infarction. Essential for such therapies is detailed knowledge of the cell types involved in regeneration, of factors resulting in mobilisation and proliferation of these cells, of chemotactic factors and of components necessary for structural integration.

2.2.2
Gene Therapy

Similar results to those obtained with purified factors, especially the induction or improvement of proliferation of cell types involved in tissue regeneration, can be achieved by gene therapy techniques. Candidate genes may be those encoding telomerase reverse transcriptase (Meyerson et al. 1997), cell cycle regulators (e.g. p21 or p27) (Brown et al. 1997; Karnezis et al. 2001), growth factors, growth factor receptors, or chemokines. Several promising results in animal models have been achieved, for example by transient immortalisation through telomerase expression in hepatocytes (Rudolph et al. 2000). Inhibition of gene expression e.g. of cell cycle inhibitors can be achieved by means of antisense strategies, ribozymes, siRNA or decoy oligonucleotides.

Besides defining the correct factors to be expressed or to be inactivated, the method of gene delivery is crucial. Oligonucleotides can be delivered efficiently by non-viral methods. Viral vectors that allow highly efficient gene transfer *in* vivo are currently the system of choice to overexpress transgenes. Viral gene transfer systems include vectors allowing transient expression (e.g. adenoviruses) or stable long-term expression (retroviruses, lentiviruses, adeno-associated viruses). Some vector systems including adenoviruses, lentiviruses or adeno-associated viruses even allow efficient transduction of resting cells. However, all viral vector systems suffer from certain limitations including insufficient safety features (they may promote adverse immune reactions (Marshall 1999) or may activate oncogenes (Hacein-Bey-Abina et al. 2003; Li et al. 2002)), limited insert size, or complex production. Therefore, future clinical gene therapy approaches may be based on non-viral systems copying certain essential features of viruses that are responsible for high infectivity and expression, but not those for pathogenicity.

2.2.3
Cell Transplantation

In contrast to the above strategies, cell transplantation therapies require *ex vivo* handling of the required cell type. The cell types to be transplanted have to be isolated and purified, for example from bone marrow, peripheral blood, or from skin biopsies. Subsequently, these cells can be expanded *in vitro*, genetically manipulated if necessary, or differentiated into the appropriate specific cell type. To develop cell transplantation strategies allowing clinical applications, certain prerequisites have to be established. First, there is the question of the proper cell type for each specific application. For some applications terminally differentiated cells may be useful, but most applications will probably require stem or progenitor cells. However, for most desired cell types, it is not known if adult stem cells or cord blood stem cells can match the requirements regarding proliferation and differentiation capacity. In theory, embryonic stem (ES) cells, which have relatively high rates of proliferation, can be grown perhaps indefinitely and are pluripotent, could be used to derive any cell type. But whether these embryonic stem cells will be required may well depend on the specific application.

Beyond this, appropriate methods for efficient cell delivery and robust engraftment will be required. Whereas efficient cell delivery might be achieved with relative ease for most applications, the question of functional engraftment and structural integration into pre-existing tissues will be critical. This is especially so for cell types participating in nerve conduction or mechanical processes. Thus far, the critical question of proper structural integration has not been addressed sufficiently well.

2.2.4
Tissue Engineering

The aim of tissue engineering is the *in vitro* generation of bioartificial tissues and in the future of whole organs, suitable to replace those that are damaged, functionally impaired, or missing. In most medical fields, tissue engineering approaches are being applied with the aim of developing biologically active prostheses. The stage of development extends from very basic research (e.g. kidney), via preclinical stages (e.g. heart valves) up to clinical application, (notably skin). As for cell transplantation, appropriate cell sources are a major prerequisite for tissue engineering. In principle, all demands for the proper cell types described above apply also to tissue engineering, including knowledge on differentiation potential of the cell types to be applied. In contrast to cell transplantation, however, the generation of functional tissue will not be done by the body itself, but has to be done *in vitro*. This may be relatively easy for simple tissue types such as connective tissue. For more complex structures such as renal tissue it will require extensive knowledge of how the tissues form during embryonic development. This will include knowledge of cell to cell and cell to matrix interactions. Furthermore, cell signalling during tissue formation and within existing tissues has to be understood in detail. All tissue structures include matrix substances as well as cellular components. The composition, structure and mechanical properties of the extracellular matrix of certain tissues can be quite special. Depending on the function of the tissue type referred to, the extracellular matrix has to be more or less stable or elastic. Matrix fibres can be oriented in one-, two- or three-dimensional patterns to fulfil their specific requirements. Therefore, novel three-dimensional culture conditions have to be developed, which consider the influence of mechanical stress, oxygen partial pressure and/or pulsatile liquid flow. Current matrices include collagen, fibrin, or other components of natural extracellular matrix. In addition, synthetic polymers such as PFTE or biologically degradable metals, growth factors and short synthetic peptides (e.g. RGD-peptides) are utilised. Some groups have attempted to use animal tissues to produce suitable matrices (Steinhoff et al. 2000). Different technologies have been developed to acellularize these tissues. The biochemical composition of the resulting biological matrix is very similar to the native structures and has the strong advantage that the matrix is available in the desired three-dimensional structure. After seeding with human cells, the animal matrix is replaced by human matrix, finally resulting in a completely humanised structure.

One of the major challenges may be the development of vascularised tissue, which can be surgically connected to existing blood vessels. The lack of technologies required to do this is currently the most serious obstacle in generating tissue structures of more extended dimensions. So far it has only been possible to supply tissue structures not exceeding a few millimetres with sufficient oxygen and nutrients. To overcome this hurdle, strategies employing angiogenic factors or the seeding of endothelial progenitors are currently under investigation. However, neovascularisation or neoangiogenesis is not sufficient to generate tissues that need to be surgically connected to the blood circulation. To achieve this goal it will be essential to development of matrices with appropriately preformed tubular networks that will become arteries and veins. The development of bioartificial tissues or even organs may be the most sophisticated task in regenerative medicine and will need close collaboration between molecular biologists, cell biologists, biotechnologists, clinicians, chemists, and physicists. While generation of relatively simple structures such as cartilage or connective tissue seems feasible, it is currently unknown whether more complex structures such as kidney tissues can be generated *in vitro*. Table 2.1 depicts an overview on the present state of the art of tissue engineering. For a comprehensive overview on the current state and perspective of tissue engineering, the articles of Vacanti (Stock and Vacanti 2001) and Atala (Atala 2000) are recommended.

Table 2.1 Tissue Engineering: State of the Art

Organ	Employed cell type	*In vitro* studies	Animal models	Clinical studies	Ref.
Bladder / ureter / urethra	Urothelial cells, smooth muscle cells		X		(Atala 2002)
Blood vessels	Vascular fibroblasts, endothelial cells, smooth muscle cells		X		(Niklason et al. 1999)
Bone	Osteoblastes			X	(Boyan et al. 1999; Niklason 2000)
Cartilage (joint, ear)	Chondrocytes, mesenchymal stem cells			X	(Perka et al. 2000; Temenoff and Mikos 2000)
Connective tissue (ligament, tendon)	Fibroblasts		X		(Woo et al. 1999)
Eye (cornea)	Corneal epithelial cells, corneal stromal cells, corneal endothelial cells	X			(Griffith et al. 1999; Schneider et al. 1999)
Genitals	Fibroblasts, smooth muscle cells, endothelial cells		X		(Park et al. 1999; Yoo et al. 2000)
Heart muscle	Cardiomyocytes, skeletal myoblasts		X		(Eschenhagen et al. 1997; Kofidis et al. 2002)
Heart valve	Vascular fibroblasts, endothelial cells, smooth muscle cells		X		(Sodian et al. 2000; Steinhoff et al. 2000)
Intestine	Enterocytes, Endocrinal cells		X		(Kaihara et al. 2000; Kim et al. 1999)
Kidney	Epithelial cells of the renal tubulus		X		(Yoo et al. 1998)
Liver	Hepatozytes			X	(Torok et al. 2001; Tzanakakis et al. 2000)
Meniscus	Mesenchymal stem cells		X		(Walsh et al. 1999)
Neuronal tissues, spinal cord	Neurons, oligodendrocytes		X		(Chamberlain et al. 2000a; Chamberlain et al. 2000b)
Oral mucosa	Mucosal epithelium		X		(Mizuno et al. 1999)
Pancreas	β cells		X		(Papas et al. 1999)
Salivary gland	Salivary epithelial cells		X		(Baum et al. 1999)
Skin	Skin fibroblasts, Epidermal cells			X	(Machens et al. 2000)
Trachea	Chondrocytes, smooth muscle cells, fibroblasts		X		(Spangenberg et al. 1998)

2.3
Cell Sources

In principle, differentiated cells can be useful for purposes of regenerative medicine. The term "somatic differentiated cell" designates all cells within our body, that have already reached a state needed to fulfil a specific function within a certain tissue. However, most somatic differentiated cells display only very limited proliferative potential. This frequently involves irretrievable loss of telomerase reverse transcriptase activity. Other cells discontinue proliferation due to changes in cell cycle regulation. So far, most somatic cell types cannot be expanded at all *in vitro*. For these reasons stem and progenitor cells are probably better suited for evolving applications of regenerative medicine. Given the right conditions, many of these cell types can be grown *in vitro* and sometimes indefinitely. At this point, it is important to comment on the terms "progenitor cell" and "stem cell". Stem cells have the ability to self-renew: if a stem cell divides, at least one of the daughter cells displays the same characteristics as the mother cell. The other one can be a stem cell too, or has alternatively undergone a certain degree of commitment, representing a progenitor or even a differentiated somatic cell. In contrast, progenitor cells do not express self-renewing capacity: if a progenitor cell divides, two cells with higher differentiation status arise.

2.3.1
Somatic Differentiated Cells

Cell lines corresponding to many somatic cell types that can proliferate more or less indefinitely have been isolated or created by 'immortalisation'. This can occur spontaneously, but it frequently involves transformation with a viral or cellular oncogene such as SV40 T antigen, vMyc or EBV. However, these cell lines have usually lost many features of the corresponding primary cells including the property of contact inhibition. Furthermore, these cells have already taken one or more steps towards a tumor cell (or are already tumor cells) and are therefore not appropriate for clinical applications. In contrast, normal somatic cells have lost the ability for unlimited proliferation. Many differentiated cell types can no longer divide at all; others show a very limited proliferative capacity (see table 2.2). So far, methods of isolation, culture and expansion have been developed for only a very limited number of somatic cell types. Cells that can be expanded to a significant extent include fibroblasts, endothelial cells, smooth muscle cells, chondrocytes, mesangial cells and urothelial cells. However, during prolonged cultivation the phenotype of these cells usually changes dramatically. Telomerase shortening can lead to genomic instability and it is not unlikely that mutations will accumulate (although this may apply to extended stem cell culture, too). In addition, isolation of significant numbers of the desired cell type is often impracticable, especially if those cells have to be collected within the affected part of the body.

A feasible strategy to improve proliferative capacity of somatic cells may be concepts of "transient immortalisation". Temporary expression of telomerase reverse transcriptase can result in unlimited or prolonged cellular proliferation in some cell types. Loss of Telomerase activity seems to limit the ability to divide in fibroblasts, epithelial cells (Bodnar et al. 1998) and partially in endothelial cells, adrenocortical cells, and hepatocytes (Rudolph et al. 2000; Sedivy 1998; Thomas et al. 2000; Yang et al. 1999). In other cell types, additional factors are responsible for cell cycle exit. Unfortunately, there is only very limited knowledge on cell cycle regulation in different somatic cells. Future detailed knowledge in this field may allow prolonged proliferation of somatic cells or even cell cycle re-entry of already resting cells for therapeutic purposes (Kobayashi et al. 2000).

Table 2.2 Isolation, Culture and Potential for *In Vitro* Expansion of Different Somatic Cells

Organ / tissue	Characteristical cell types	Identification / Isolation	Culture	*In vitro* expansion
Brain / neuronal tissues	Neurons	+	+	(+)
	Astrocytes	+	+	(+)
	Oligodendrocytes	+	(+)	-
Skeletal muscle	Muscle tubes	+	+ (myoblasts)	(+) (myoblasts)
Heart	Cardiomyocytes	+	(+)	-
Blood vessels	Vascular fibroblasts	+	+	+
	Endothelial cells	+	+	+
	Smooth muscle cells	+	+	+
Lung	Pneumocytes	+	(+)	-
Liver	Hepatozytes	+	+	-
Kidney	Epithelial cells of the renal tubuli	+	(+)	-
	Mesangial cells	+	+	+
Bladder / ureter	Urothelial cells	+	+	+
	Smooth muscle cells	+	+	+
Pancreas	β cells	+	+	-
Bowel	Enterocytes	(+)	+ (int. stem cells)	+ (int. stem cells)
	Endocrinal cells	(+)	+ (int. stem cells)	+ (int. stem cells)
Skin / hair	Skin fibroblastes	+	+	+
	Epidermal cells	+	+ (epid. stem cells)	+ (epid. stem cells)
Bone	Osteoblastes	+	+	+
Cartilage	Chondrocytes	+	(+)	+
Connective tissues	Fibroblasts	+	+	+

Abbreviations: int. = intestinal; epid. = epidermal

On the other hand, it might be easier and more realistic to generate sufficient numbers of differentiated cells using appropriate stem or progenitor cells. In conclusion, somatic differentiated cells are likely to be of very limited use for clinical applications and stem or progenitor cells will be the better choice for most cell transplantation or tissue engineering purposes.

2.3.2
Adult Stem and Progenitor Cells

In general, research on most adult stem cell types is just at the beginning. Most data has come from small animal models and very little has come from large animals including primate and humans. Despite some exceptions, it is currently unknown whether differentiated somatic cells obtained from stem cells *in vitro* are functionally comparable to normal differentiated cells within the body. Indeed, perhaps the latter can never be obtained from adult cells. Whereas (ES) cells have the potential to form all types of differentiated and functional cells within the body, as shown *in vivo* through the production of chimeric mice, currently we do not know whether this will hold true for any adult stem cell type. Although recent data especially on bone marrow derived mesenchymal stem cells are encouraging, much research has still to be done to define the true potential of adult stem cells. Nevertheless, even with a restricted potential, adult stem cells may work in some applications, especially for those tissues that have intrinsic regeneration potential *in vivo*. For those which do not regenerate *in vivo*, it will be less likely to achieve functional differentiation of appropriate cell types *in vitro*. Genetic manipulation, for example with certain transcription factors, is one approach that might help.

In contrast to ES cells, tissue-specific stem cells in the adult, e.g. in bone marrow, skin or neuronal tissues, are significantly committed towards certain cell types. Many adult stem cells, such as those in the gut and skin, fulfil important functions during continuous tissue-regeneration or wound healing. So far, adult stem or precursor cells have been identified and isolated for only a limited number of cell types. Moreover, it is not known for most adult stem cells whether their potential for proliferation is sufficient for therapeutic applications. Table 2.3 gives an overview on different stem cells types including data on differentiation and potential clinical applications.

2.3.3
Hematopoietic Stem and Precursor Cells

Hematopoietic stem and precursor cells have been used successfully to treat autoimmune diseases and leukemias for many years (Appelbaum and Buckner 1986; Slavin and Kedar 1988). They are isolated via bone marrow puncture or by leukapheresis after cytokine-induced mobilisation (Reik et al. 1997). For successful bone marrow transplantation it is crucial to match the histocompatibility antigens of donor and recipient. In this case, the use of autologous bone marrow stem cells is not possible since precisely these cells are affected by the disease to be treated. The search for a perfectly matched bone marrow graft is often difficult due to the limited number of potential donors (Beatty et al. 2000; Beatty et al. 1995) and has led to worldwide databases of registered donors. Frequently, patients die before a suit-

Table 2.3 Overview on plasticity and potential for proliferation of stem cells and terminally differentiated somatic cells, on future therapeutical applications of these cells and on tissue types, which might be engineered in the future. **ASC**: adult progenitor- / stem cells, **DSC**: differentiated somatic cells, **ESC/PGC**: embryonic stem cells / primordial germ cells, **EC/FC**: embryonic / fetal cells, **Diff.**: differentiation, **Exp.**: expansion, (+): so far, data not confirmed on human cells

Organ, Tissue	Disease /Failure	DSC	ASC		ESC/PGC		Transplantations	
		Exp.	Diff.	Exp.	Diff.	Exp.	Animal experiments	clinical
Bone marrow	Autoimmune diseases, leukemias	Not suitable	+	Limited	(+)	+	+ (ASC)	Clinical practice
Blood vessels, vascularized tissue etc.	Thromboses, valve defects, cardiac infarction, etc.	+	+	+	+	+	+ (ASC)	-
Brain, spinal cord	M. Parkinson, Alzheimer disease, multiple sclerosis, injuries of the spinal cord	-	+	Limited	+	+	+ (ASC, ESC)	Clinical studies (EC/FC)
Retina	blindness	-	+	(+)	?	+	-	-
Skeletal muscle	Injuries, myopathies	-	+	+	(+)	+	+ (ASC)	-
Cardiac muscle	Cardiac infarction, atrioventricular block, congenital malformations	-	(+?)	+	+	+	+ (ASC, ESC)	-
Bone	Bone fractures, degenerative diseases	-	+	+	(+)	+	+ (ASC)	Clinical studies
Cartilage	Attrition, arthritis, injuries	+	+	+	(+)	+	+ (ASC)	Clinical practice
Skin	Burn injuries	-	+	+	(+)	+	+ (ASC)	Clinical practice
Liver	Cirrhosis etc.	-	(+)	?	(+)	+	+ (DSC)	-
Pancreas	Diabetes	-	+	?	(+)	+	+ (ASC)	-
Intestine, stomach	tumours	-	+	(+)	(+)	+	-	-
Kidney	Insufficiency	Partially	?	?	?	+	-	-
Ureter, bladder	Congenital malformations, tumours	+	?	?	?	+	+ (DSC)	Clinical studies

able donor has been identified. One major limitation for clinical use of hematopoietic stem cells remains their low proliferation potential *in vitro*. Whereas detailed protocols are available to isolate these cells under GMP-conditions, it has so far not been possible to expand their number to a significant extent. Very recent data suggest that this situation may change as it has been reported that balanced expression of HOXB4 can result in a dramatically improved *ex vivo* expansion of these cells (Antonchuk et al. 2002).

Hematopoietic stem cells are probably the most extensively investigated adult stem cell type. $CD34^+$ hematopoietic stem cells reside within the bone marrow and progenitor cells derived from these stem cells migrate into peripheral blood and tissue. During and after migration to other sites of the body, bone marrow derived stem and precursor cells undergo a progressive commitment towards certain lineages. Whether they will develop towards one or another specific lineage is determined by a complex network of factors. This includes interaction with soluble cytokines, with other cell types and matrix substances. Besides these classical $CD34^+$ stem cells, other stem and progenitor cells exist. These include earlier stages, for example, the so-called "side population cells", which are defined by their ability to exclude certain dyes based on specific transmembrane transporter molecules (Zhou et al. 2001). Meanwhile, ABCG2-transporter-specific antibodies allow to detect SP-cells in histological sections, also.

The classical $CD34^+$ hematopoietic stem cells can differentiate into all hematopoietic lineages including T- and B-lymphocytes, NK-cells, neutrophils, eosinophils, basophils and mast cells, monocytes, macrophages, erythrocytes and thrombocytes. More recently, it has been shown that bone marrow stem cells of the "hematopoietic" lineage display a remarkable plasticity. It is now doubtless that different stages of "hematopoietic" stem and progenitor cells including $CD34^+$ cells found in the peripheral blood, can also be designated as endothelial progenitor cells (EPC), as they are able to form endothelial cells and contribute to vessel repair and neovascularisation (Asahara et al. 1997; Hirschi and Goodell 2001; Kaushal et al. 2001; Kocher et al. 2001; Lin et al. 2000; Rafii 2000). In contrast, data on their differentiation towards liver cells (Krause et al. 2001; Petersen et al. 1999), kidney cells (Forbes et al. 2002), neuronal cells (Meletis and Frisen 2001; Zhao et al. 2002), skeletal muscle including satellite cells and myoblasts (Seale and Rudnicki 2000; Ferrari et al. 1998; LaBarge and Blau 2002) or heart muscle (Jackson et al. 2001; Orlic et al. 2001a; Orlic et al. 2001b) are more preliminary. Many doubts have been expressed on another report on cardiac differentiation of blood derived endothelial progenitor cells (Badorff et al. 2003). Several studies detected very low numbers of recipient-derived differentiated cell types such as cardiomyocytes in biopsies of transplanted organs (e.g. Bayes-Genis et al. 2002; Laflamme et al. 2002; Muller et al. 2002). Whether hematopoetic or mesenchymal stem cells contributed to these cells and whether differentiation or cell fusion is the underlying mechanism is so far not known.

As shown by very recent studies, many of the results on adult stem cell plasticity might be explained by cell fusion events, rather than by stem cell differentiation (Terada et al. 2002; Vassilopoulos et al. 2003; Wang et al. 2003; Ying et al. 2002). In particular, all results on stem cell plasticity obtained by coculture experiments and all results on *in vivo* differentiation have to be judged very carefully. Accurate methods allowing discrimination between differentiation and fusion will have to be developed.

Indeed, very recent work suggests little evidence for plasticity (Castro et al. 2002; Vassilopoulos et al. 2003; Wagers et al. 2002; Wang et al. 2003). If there is transdifferentiation at all, the proportion of cells undergoing such transdifferentiation events seems to be extremely low. Furthermore, most of these studies have been performed in small animal models and it is unknown whether human cells will have similar properties. It is also not known whether the cells that have been generated are functionally comparable to the appropriate somatic differentiated cell, since in most cases only expression of one or few cell type specific markers has been detected (e.g. Jackson et al. 2001; Orlic et al. 2001a; Orlic et al. 2001b). More critical investigation of tissue samples obtained from patients that have received bone marrow transplants may help to answer the question whether human bone marrow stem cells are able to transdifferentiate and to contribute to ectopic locations such as brain, heart and liver.

2.3.4
Mesenchymal Stem Cells

Besides "hematopoietic" stem cells, another stem cell type resides within the bone marrow: The so-called mesenchymal stem cells (MSC) are part of the bone marrow stroma. These cells have been designated as bone marrow stromal cells and are still poorly characterised. Pioneering investigations on these cells have been done in the laboratories of Osiris Therapeutics, Cleveland, Ohio. Some experiments indicated that MSCs might be a cell population clearly demarcated from the fibroblast-like bone marrow stroma (Majumdar et al. 1998). Other data suggest that the morphologically fibroblast-like stromal cells represent an uniform cell population with stem cell potential (Seshi et al. 2000). Now, it seems that not even MSC themselves are a uniform cell population. Some years ago, MSC were regarded as such because they were defined by the expression or absence of a variety of different CD-markers (e.g. $CD45^-$, $CD34^-$, $CD90^+$, $CD105^+$) (Pittenger et al. 1999), but recent data revealed that MSC might comprise different cell populations (Colter et al. 2001). Moreover, another group has succeeded in isolating another adherent bone marrow stem cell type, which has been called MAPC (Reyes et al. 2001; Reyes and Verfaillie 2001). Whereas some groups isolate MSCs simply by means of density gradient centrifugation and adherent culture, others use CD-specific sorting procedures to isolate specific bone marrow stroma derived stem cells. In addition, the culture conditions, with respect to cell density and medium composition, vary considerably between laboratories. It is currently not known whether these sublineages of adherent bone marrow stem cells already exist in the bone marrow or whether different stages of stem cells arise in culture due to different isolation procedures or culture conditions. In fact, there are significant differences between these cell populations with respect to their potential for proliferation and differentiation.

For potential clinical applications, mesenchymal stem cells have major advantages. Autologous MSCs can be isolated relatively easily from bone marrow. The cells can be expanded to large numbers and seem to display an initially unexpected plasticity (see below). Furthermore, very recent reports provide some evidence to suggest they have low immunogenicity. Reasons for this are not known and it will be important to confirm these results. One disadvantage of MSCs is that their isolation and expansion needs at least several weeks, which precludes the use of autologous MSC for acute indications.

Mesenchymal stem cell populations can also be found in adipose tissue and show enormous plasticity (Zuk et al. 2001). First evidence that those cells can differentiate towards fibroblasts and mesenchymal cells were already available in the 1970s (Friedenstein et al. 1970; Friedenstein et al. 1974a; Friedenstein et al. 1974b; Friedenstein et al. 1976).

It has been supposed that differentiation of MSC towards cardiac myocytes might allow regeneration of injured myocardial areas after infarction. Clinical experience shows that such injuries do not repair themselves *in vivo*. This can be explained by the fact that the cell cycle is irreversibly arrested in cardiac myocytes immediately after birth (Chien 1995; Karsner et al. 1925). Although there are some data on remaining cell cycle activity (Beltrami et al. 2001), these are controversial and the broad opinion is that there is no significant proliferation of myocytes in the adult heart. Moreover, there are currently no data definitely demonstrating significant generation of new cardiomyocytes from immigrant or resident stem cells in these injured areas.

A few studies have been published demonstrating cardiac differentiation of rodent MSC (Makino et al. 1999; Tomita et al. 1999; Wang et al. 2000). Two studies reported cardiac differentiation of immortalised MSC-clones after 5-azacytidine treatment (Hakuno et al. 2002; Makino et al. 1999). 5-azacytidine is a hypomethylating agent, resulting in a completely altered gene expression pattern. Although the resulting cells were reported to contract spontaneously and to express a variety of myogenic markers and functional adrenergic and muscarinergic receptors (Hakuno et al. 2002), it cannot be excluded that the generated cells are rather skeletal than cardiac myocytes. Moreover, it is unlikely that the 5-azacytidine treatment of the reported immortalised MSC clones has generated cells comparable to normal cardiomyocytes. Finally, other investigators will have to confirm the data and further studies will have to proof the identity of those cells. All other data on cardiac differentiation of stem cells (e.g. Toma et al. 2002; Tomita et al. 1999; Tomita et al. 2002; Wang et al. 2000), similar to the transdifferentiation of endothelial cells towards cardiomyocytes (Condorelli et al. 2001), are based on transplantation or co-cultivation experiments and demonstrated expression of single cardiac markers and no contractions. These reports should be carefully re-evaluated, especially to exclude misinterpretations due to cell fusion events. In conclusion, it is still unknown, whether human MSC may be able to differentiate towards functional cardiomyocytes, but much more research is required.

It has also been shown that 5-azacytidine treatment induces differentiation of MSC towards skeletal myoblasts (Wakitani et al. 1995). These cells express several muscle specific markers and resemble skeletal muscle cells morphologically. Whether functional muscle tubes can be generated from human MSC is still unknown. Although recent *in vivo* data show such events in mice (Ferrari et al. 1998) it was not clearly demonstrated whether the generated myotubes were derived from MSC or hematopoietic precursors.

The first evidence that MSCs may be able to form bone or cartilage was described as long ago as 1980 (Ashton et al. 1980). Ashton et al. demonstrated *in vitro* formation of bone- and cartilage-like structures from rabbit stromal cells. Experiments from different research groups confirmed these results over the following years (Friedenstein et al. 1987; Grigoriadis et al. 1988). In contrast to the suggestion that MSCs may hold myogenic potential, it is not surprising that MSCs can form osteoblasts, in view of their localisation in bone marrow and of the natural potential for regeneration of bone and cartilage. Their chondrogenic potential may also be explained by the process by which bone fractures are healed. This involves the production of thin layers of cartilage that are subsequently replaced by bone. In 1992, it was shown that human MSCs express similar osteogenic potential (Haynesworth et al. 1992). Preclinical investigations on MSC-dependent formation of bone and cartilage then followed (Wakitani et al. 1994). In the meantime further preclinical data show that the use of MSCs in ceramic implants results in significantly improved bone formation (Bruder et al. 1998a; Bruder et al. 1998b) and the first clinical studies on children with inherited bone defects have already started (Prockop 1997). A phase 1 clinical study on transplantation of MSC-derived bone has been finished in 2002. Clinical studies on cartilage formation have been announced. (Osiris Therapeutics Inc.).

Other cells that can be formed by MSCs are adipocytes. The ability to generate fat cells has been well analysed (Grigoriadis et al. 1988), although it appears to be of very limited benefit for direct medical applications. They could, however, be genetically manipulated to produce critical blood-borne factors.

Verfaille et al. were able to demonstrate that the so called MAPC are able to differentiate *in vitro* towards endothelial cells (Reyes et al. 2002; Reyes et al. 2001). Moreover hepatocyte-like cells (Schwartz et al. 2002), astrocytes-like, oligodendrocytes-like, neuron-like and epithelial cells have been generated *in vitro* (Jiang et al. 2002). Injection of 10–12 cell clones into blastocysts resulted in contribution to most somatic tissues including brain, retina, lung, myocardium, skeletal muscle, liver, intestine, kidney, spleen, bone marrow, blood and skin in the resulting mice. Transplantation of MAPCs into SCID mice resulted in engraftment in hematopoetic tissues and epithelium of liver, intestine and lung. No contribution was seen to skeletal or cardiac muscle (Jiang et al. 2002). It is not known whether the injected or transplanted MAPCs gave rise to normal functional cell types or whether the detected cells evolved due to differentiation or cell fusion.

2.3.5
Cord Blood Stem Cells

Cord blood represents a stem cell source which is easily accessible from the umbilical cord directly after birth. These cord blood stem cells are somehow in a state between fetal and adult cells, but in addition they can be isolated, purified, and stored in liquid nitrogen for many years. This may allow autologous-stem-cell-based treatment of injuries and disorders during the lifetime of the individual from when they were obtained. However, autologous transplantation is not necessarily meaningful in cases of genetically based disease, since the underlying defect or predisposition is also present in the autologous cord blood cells. There are already cord blood cell collections in different nations, which have organised a worldwide network to supply MHC-matched cord blood stem cells similar to adult bone marrow stem cells. Moreover, a variety of companies in many industrialised nations offer the commercial isolation and storage of individual cord blood samples. Cord blood stem cells are taken from a source that has already served its useful function and is to be discarded. Some scientists, including those of the Vatican, have therefore tried to promote them on ethically non-contentions. But this notion depends critically on their potential for clinical use. Moreover, there is some concern over this commercial cell-banking with some ethics experts being very critical, since parents are pestered in a more or less subtle manner to finance long term storage of their child's cord blood. Moreover, it can not be foreseen whether the company responsible for the storage of your personal autologous stem cells will still exist if and when you need them. It also remains unclear whether the mode of purification, freezing and storage will result in viable cells able to differentiate appropriately.

Cord blood contains hematopoietic stem cells at different stages and some mesenchymal stem cell-like cells. In view of recent reports, the plasticity of hematopoietic cord blood stem cells seem to be very similar, maybe somewhat more extensive, than the plasticity of adult hematopoietic stem cells. Despite the fact that the number of stem cells in the small volume of cord blood is very limited, considerable amounts of stem cells can be prepared due to the higher proliferation capacity compared to adult hematopoietic stem cells (Cairo and Wagner 1997). Nevertheless, the number of HSCS available in each sample is generally not sufficient to treat an adult.

There are only preliminary data regarding the potential for proliferation and differentiation of cord blood derived MSCs (Wernet et al. 2002), but these demonstrate a high proliferative capacity and differentiation towards a variety of cell types.

2.3.6
Tissue-resident Adult Stem and Progenitor Cells

Other adult stem and progenitor cells are located in the tissues they give rise to. Some of these cell types might be derived from adult stem cells located elsewhere. Two examples of tissue-resident progenitor cells that can probably be generated from other adult stem cell sources are oval cells of the liver and skeletal satellite cells (Petersen et al. 1999), both of which are thought to be generated from bone marrow stem cells (Ferrari et al. 1998; Wakitani et al. 1995). Recently SP-cells have been detected in different tissue types including liver (Uchida et al. 2001), lung (Summer et al. 2003), kidney, intestine, skeletal muscle (Asakura and Rudnicki 2002) and heart (Oh et al. 2003). Whether bone marrow is the origin of all of these cells is currently not known. However, it has been supposed that most, if not all tissue-resident SP-cells are bone marrow derived and that differences are based on the local microenvironment. Some other examples for tissue-resident stem cells that have been investigated intensively are discussed below.

2.3.7
Intestinal Stem Cells

Over the last few years intestinal stem cells have moved more and more into the focus of stem cell research. Intestinal stem cells are located at the basis of the intestinal crypts, and are required to continuously regenerate the intestinal epithelium.

Recent data indicate that they show proliferation rates comparable to embryonic stem cells. This is not surprising considering the enormous regeneration rate of the intestinal epithelium. Similar to epidermal stem cells of the skin, intestinal stem cells form all cell types of the tissue concerned, in this case of the intestinal wall, with the exception of immune cells and cells of the enteric nervous system (Wright 1997). Recently, isolation and culture of mouse intestinal stem cells from the small bowel has been reported (Booth et al. 1999). Preliminary data reveal an up to now unexpected plasticity (Rolletschek et al. 2003). Markers of all three germ layers have been found on cells derived from intestinal stem cells. Considering therapeutic applications, the high potential for proliferation of these cells is especially encouraging. However, whether these cells may be a source for cell types other than those of the intestine has still to be explored.

2.3.8
Epidermal Stem Cells

The outer coating of mammalian skin consists of a multilayered epithelium, the epidermis, which is punctuated by hair follicles and sebaceous glands. Epidermal stem cells are, comparable to the intestinal stem cells in the gut crypts, located in the basal layer (Lavker and Sun 1982) and in the deep rete ridges. Hair follicles contain two zones of stem cell-like cells. The hair shaft and its surrounding sheaths are produced by matrix cells located on a basement membrane overlying the protruding piece of dermis known as the dermal papilla. Longer-lived stem cells reside more apical within the 'bulge' region (Cotsarelis et al. 1989; Rochat et al. 1994). Cell marking and follicle transplantation experiments have shown that the bulge contains stem cells that repopulate the matrix region, form new sebaceous glands, and replace epidermal stem cells (Oshima et al. 2001; Taylor et al. 2000). Epidermal stem cells are the basis for skin renewal and the healing of skin injuries. Besides hematopoietic and intestinal stem cells, they display the highest proliferation rates of all cell types within our body. Epidermal stem cells have been fairly well characterised (Niemann and Watt 2002; Watt 2001; Watt 2002), with the identification of specific molecular markers, including β1-Integrins (Watt 1998; Watt 2000), and the development of technologies to highly purify them (Watt 2000). Epidermal stem cells give rise to all the differentiated lineages of the interfollicular epidermis, hair follicles, and sebaceous glands. Skin cell cultures containing stem cells are already used to produce commercial bioartificial skin products (see below).

2.3.9
Neuronal Stem Cells

Studies in the late 1980's and early 1990's reported the existence of neuronal stem cells in the embryonic central (Cattaneo and McKay 1990; Cattaneo and McKay 1991; Kilpatrick and Bartlett 1993; Reynolds et al. 1992; Temple 1989) and peripheral (Stemple and Anderson 1992) nervous system. Later on, this was confirmed for the adult mouse brain (Lois and Alvarez-Buylla 1993; Reynolds and Weiss 1992). According to current knowledge (Steindler and Pincus 2002), adult neuronal stem cells are located in different parts of the brain, at least in the hippocampus, subventricular zone, the olfactory bulb and probably in the cortex. Others can be found in the spinal cord (Gage 2000; McKay 1997; Rao 1999) and the retina (Ahmad et al. 2000; Perron and Harris 2000; Tropepe et al. 2000). Recently, markers for neuronal stem cells (e.g. nestin and Sox2) have been identified, allowing the isolation and purification of these cells (Kawaguchi et al. 2001; Rietze et al. 2001; Uchida et al. 2000) (Billon et al. 2002; Zappone et al. 2000). Neuronal stem cells can be maintained as adherent cultures or within floating cell clumps, so called neurospheres. Differentiation towards different neurons and glia cell types can be observed within these cultures (Gage 2000; McKay 1997; Rao 1999). There is also recent data providing some evidence for transdifferentiation of neuronal stem cells. Spinal cord derived stem cells, which do not normally generate neurons, have been demonstrated to form interneurons if transplanted into the hippocampus. Adult hippocampus-derived stem cells can make olfactory neurons after transplantation into the subventricular zone. Other groups reported that adult murine neuronal stem cells are incorporated into different tissues and organs such as heart, blood and skeletal muscle after transplantation into early mouse or chick embryos (Bjornson et al. 1999; Clarke et al. 2000). Mueller et al. reported broad tissue contribution of murine neural stem cells after epigenetic modification (Kirchhof et al. 1993). As for other stem cell types, it is assumed that reprogramming and transdifferentiation, if not induced artificially, will depend on unknown factors present in the local microenvironment. But again, care has to be taken over the interpretation of some of the published data, especially over the issue of cell fusion.

The availability of sufficient numbers of neuronal stem cells might have implications for the treatment of a variety of diseases and trauma of the CNS including Parkinson's disease, Huntington's disease, spinal cord injury, stroke and multiple sclerosis. Most realistic is treatment of those diseases that are characterised by the loss of a special cell type producing important metabolites, such as in Parkinson's disease. Unfortunately, with the exception of those from the mouse, adult neuronal stem cells tend to senesce in culture and display only very limited proliferative potential. Another major disadvantage of adult neuronal brain stem cells is the limited access, especially for autologous stem cells. Allogeneic stem cells could only be collected from donors after brain death and it is not known whether significant numbers of viable stem cells could be isolated in this way. The collection of autologous stem cells may be even more complicated since such stem cells may be equally affected by the disease to be treated or even missing altogether, and surgical intervention in the brain of these patients may not be useful. Therefore, future aims in view of clinical applications are to develop efficient methods of *in vitro* differentiation and technologies to achieve *in vitro* or *in vivo* expansion of adult neuronal stem cells. Moreover, approaches to avoid rejection of allogeneic stem cells after transplantation into the brain have to be established. At present, the mechanical destruction of the blood-brain barrier during such cell injections may be the critical step leading to rejection of neuronal allotransplants within this immunological privileged organ. Consistent with this, it seems that immunosuppressive drugs need only be given for a relatively short period until the wound has healed.

2.3.10
Oval Cells of the Liver

In contrast to other internal organs, such as heart, lung and kidney, the adult human liver is able to regenerate within a relative short space of time. There is convincing evidence for hepatocyte-precursor cells, called oval cells according to the shape of their nucleus, to participate in liver regeneration. Oval cells are located around the bile ducts (Brill et al. 1993; Travis 1993). Besides differentiation towards hepatocytes, generation of bile-epithelium from oval cells also appears to be possible. The relative extent to which liver regeneration is based on immigration of bone marrow stem cells, on oval cells or on the remaining proliferative potential of adult hepatocytes (Krause et al. 2001; Petersen et al. 1999; Theise et al. 2000) is not known. In view of the possibility of fusion of bone marrow stem cells with hepatocytes recent results have to be re-evaluated carefully (Vassilopoulos et al. 2003; Wang et al. 2003).

2.3.11
Pancreatic Stem Cells

Dysfunction of the pancreas, especially of pancreatic β-cells is responsible for insulin-dependent diabetes mellitus. Transplantation of functional β-cells or even a bioartificial pancreas may allow treatment of diabetes mellitus. The pancreas is composed of at least 3 different tissue types: the islets with 4 different cell types, the exocrine zymogen-containing acini, and the ductal tree with centroacinar cells, ductules and ducts (Bouwens 1998). All pancreatic cell types appear to have a common precursor during embryogenesis. Investigations after experimental pancreas injury demonstrated the regeneration of new islets including insulin-producing β-cells. During the last few years it has been discussed albeit controversially, whether the observed pancreatic regeneration is due to tissue specific stem cells or due to transdifferentiation of fully differentiated cells (Bonner-Weir 2000; Bouwens 1998; Ramiya et al. 2000). Moreover, it has not been proven that insulin-producing β-cells evolve (Bonner-Weir 2000; Rao et al. 1990). Recent results indicate existence of a small portion of stem cells in the pancreatic ducts of adult rats and mice (Ramiya et al. 2000; Zulewski et al. 2001).

2.3.12
Transdifferentiation

There are increasing amounts of data purporting to show that stem or progenitor cells, or even differentiated cells, thought to be already committed to a certain lineage, may be capable of crossing lineage boundaries and to develop towards other lineages. As examples, transdifferentiation of umbilical vein endothelial cells towards cells expressing cardiac markers (Condorelli et al. 2001), transdifferentiation of neural stem cells (Kirchhof et al. 1993) or epidermal stem cells (Liang and Bickenbach 2002; Toma et al. 2001) or differentiation of liver derived stem cells towards cells with certain myocyte or pancreatic markers (Malouf et al. 2001; Yang et al. 2002) can be mentioned. Whether these transdifferentiation events are somewhat rudimentary or whether they can lead to functional cells is discussed controversially. There are several clear examples of transdifferentiation (Eguchi and Kodama 1993; Tosh and Slack 2002), such as within the eye of lower vertebrates, where cellular reprogramming occurs naturally. So we should not be too surprised if it occurs in cell culture or when isolated stem cells are returned to an ectopic site. However, it is essential that the techniques used are rigorous in being able to follow the fate of individual cells, ideally so that they can be seen to change from one cell type to another.

2.3.13
Previous Clinical Applications of Adult Stem Cells

So far, clinical applications of adult stem cells are rare with a few notable exceptions. As mentioned above, HLA-matched hematopoietic stem cell transplantation has been used successful for many years to treat autoimmune diseases and leukemias. Although very successful, there are still major problems in bone marrow transplantation. These include the unsatisfactory availability of matched bone marrow samples due to low readiness to donate and due to the fact that only minor parts of the population have been typed with regard to HLA. Other problems are of an immunological nature, e.g. the problem of contaminating T-cells resulting in the so-called graft versus host reaction. Moreover, it would be very advantageous if it were possible to significantly expand hematopoietic stem cells *in vitro* prior to use. Currently, the first clinical trials to treat liver failure, to reduce myocardial injuries after infarction (Stamm et al. 2003; Strauer et al. 2001), or to revascularize tissue areas after thrombosis or after ablative surgery are ongoing with HSCs.

Besides hematopoietic stem cells, mesenchymal stem cells have also begun to be applied in clinical studies. The first studies on transplantation of mesenchymal stem cells for bone repair have been performed and clinical trials on children with inherited bone defects have been started (Prockop 1997). Extensive clinical trials on bone and cartilage repair have been announced (Osiris Therapeutics Inc.).

The first clinical (and commercial) application of tissue engineering has been the transplantation of bioartificial skin substitutes, generated with epidermal stem cell containing cell preparations. The first use of autologous cells took place in the early 1980's (O'Connor and Mulliken 1981). Due to their location, these cells can be isolated together with skin fibroblasts relatively easily from skin biopsies and give rise in culture to an enormous number of cells. Such crude epidermal cell isolations can be used for *in vitro* engineering of bioartificial skin. A major problem in producing autologous bioartificial skin is the necessary culture period of 3–4 weeks. Since treatment of burn patients requires immediate wound covering to avoid severe infections, allogeneic skin substitutes are also applied. It has been proposed that rejection of allogeneic constructs occurs in a delayed fashion, since MHC class II is not expressed on keratinocytes, or at least at a very low level, and because antigen presenting Langerhans' cells do not exist in the cultures (Thivolet et al. 1986). Over time, the allogeneic cells seem to become replaced *in vivo* by autologous cells (Gielen et al. 1987). Allogeneic as well as autologous skin substitutes are now produced commercially. These products are still far away from natural skin, they lack sweat glands and hair, however they are useful to rescue patients with severe burn injuries.

In conclusion, the major hurdles for clinical use of adult stem cells are the unsolved questions of stem cell identification, isolation, expansion, differentiation and functional integration. In principle, this applies to all adult stem cells, in part even for hematopoietic stem cells, which have been already used clinically for many years. It is assumed that proliferation, programming and differentiation of stem cells are at least partially dependent on unknown factors of the local microenvironment. Identification of such factors as well as isolation of critical transcription factors essential for certain lineages, may result in the future development of cell culture systems allowing expansion and differentiation of a range of stem cell types for a variety of clinical applications.

2.4
Fetal Cells

In general, fetal cells show a higher proliferation rate compared to adult stem cells and especially to adult somatic cells. Due to the more immature state of most fetal cells, engraftment of these cells after transplantation is expected to be more efficient. Besides, fetal stem cells may display a broader potential for differentiation. Unfortunately, access to fetal cells is quite limited, since such cells can only be isolated after legal abortion. Moreover, comparable to the debate on embryonic stem cells, the clinical use of fetal cells is thought by some to be ethically controversial. It is also impossible to obtain autologous cells in this way. Therefore, pharmacological immuno-suppression will always be required to prevent regeneration of the fetal cells. Nevertheless, clinical applications of fetal cells and tissue have been reported in many cases. Fetal mesencephalic tissue or fetal dopaminergic neurons have been used with varying success with regard to treatment of Parkinson's disease (e.g. Freed et al. 1992; Michejda 1987; Price et al. 1995; Spencer et al. 1992). Fetal striatal tissue has been proposed for treatment of Huntington disease (Pundt et al. 1996) and transplantation of neuroretinal cells has been suggested as a way to overcome blindness (Little et al. 1998). Moreover, fetal pancreatic islets have been transplanted into diabetes patients (Farkas and Karacsonyi 1985; Schwedes et al. 1983), fetal liver stem cells have been used to treat adult patients with fulminant hepatic failure (Habibullah et al. 1994) and fetuses with immunodeficiencies or thalassemias (Touraine 1992).

It seems likely that clinical transplantation of fetal materials will never be performed on a large scale. This is in particular due to the need for an informed consent of the mother resulting in a limited availability of aborted fetuses and due to the low numbers of cells that can be isolated from one fetus. The last point is critical, as exemplified in the treatment of Parkinson´s disease, since several fetuses are usually needed to treat one patient.

2.5
Embryonic Stem Cells

During the first differentiation step of the very early embryo, trophectoderm and inner cell mass are formed. The trophectoderm gives rise to most of the placenta, whereas the inner cell mass forms other extra embryonic membranes and the embryo. Cells of the very early cleavage stage embryo may be totipotent and are able to contribute to all cell types. Although it has been demonstrated that human embryonic stem cells are still able to differentiate into trophoblasts (Xu et al. 2002), ES cells derived from inner cell mass are currently not considered totipotent, as single cells are unable to form a whole embryo.

Teratocarcinoma stem cells of gonadal tumors were the first cells to be recognized as pluripotent (Martin and Evans 1974). These largely aneuploid tumor cells would seem inappropriate for therapeutic purposes, although one trial in the U.S.A. has been reported which used NTera2 human teratocarcinoma cell-derived neurons to treat stroke (Nelson et al. 2002). Pluripotent murine embryonic stem cells (ES cells) were isolated over 20 years ago (Evans and Kaufman 1981). The generation of chimaeras by injecting ES cells into blastocysts (a routine procedure to produce mice carrying mutations in specific genes) demonstrated that ES cells contribute to all cell types of the body (Bradley et al. 1984), but, as mentioned above, ES cells alone are not able to form a whole embryo.

So far, ES-like cells have been isolated from many species (Cherny et al. 1994; Notarianni et al. 1991; Stranzinger 1996). However, most of these ES-like cell lines have not been characterised in a very detailed manner. It was often difficult to maintain these cells in an undifferentiated state and their potential for unlimited proliferation was not demonstrated. The proof of pluripotency by generation of ES cell chimeras has only been successful in mice. In humans, such proof is not possible for ethical reasons. *In vitro* studies may address this issue (Geber et al. 1995) but cannot finally establish pluripotency. In 1995, the isolation of ES cells from rhesus monkeys was reported (Thomson et al. 1995). The pluripotency of ES cells of two other non-human primate species has been demonstrated (Suemori et al. 2001; Thomson et al. 1996). Primate ES cell lines will be essential, especially for preclinical transplantation studies.

Human embryonic stem cells were first isolated in 1998 by Thomson and his collaborators. Up until now, probably more than 100 human embryonic cell lines have been isolated world wide, 78 of which have been included in the NIH registry. Some of these have been made available to public institutions. Only a minority of these cell lines have been characterised in any detail, and recent investigations revealed that there are major differences regarding culture properties and differentiation potential. Those cell lines that are well characterised show typical stem cell markers, express high levels of telomerase activity and have now been cultivated for months to years (Thomson et al. 1998a). ES cell clones showed normal karyotype. Telomeres had normal length and high telomerase activity was detected (Amit et al. 2000). However very recent reports demonstrated a certain level of aneuploidy independent of the passage number (Carpenter et al. 2003). Other researchers reported the evolvement of karyotypic changes including chromosomal translocations during prolonged culture (Draper et al. 2003).

Many of the existing human ES cell lines may not be suitable for clinical applications due to their low expansion potential, difficulties of culture and insufficient differentiation potential. Moreover, there is a potential risk of zoonotic infection, as almost all existing lines were derived on layers of mouse fibroblasts or feeder cells.

2.5.1
Isolation and Culture of Embryonic Stem Cells

Embryonic stem cells are derived from the ICM of blastocysts. As a rule, human ES cells are isolated from pre-implantation embryos generated by *in vitro* fertilisation (IVF). Up to 10 oocytes are collected after hormonal stimulation. After IVF, 3 developing embryos are generally transferred into the uterus. The transfer of 3 embryos represents a compromise to achieve a high rate of pregnancy on the one hand, and to avoid multiple pregnancies on the other. The remaining embryos can be stored in liquid nitrogen to generate subsequent pregnancies. The majority of human ES cell lines currently existing worldwide has been produced from surplus embryos from IVF cycles donated by the parents. In Germany it is prohibited to generate and transfer more than three embryos within one cycle, however pronuclear stage (the stage of oocytes after penetration of the sperm, but before dissolution of the nuclear membrane) can be stored for further use in later cycles. In Germany, more than 100,000 surplus pronuclear stage embryos are in storage, which could be used to derive ES cells. This is currently not allowed according to the law passed in April 2002 (see Chapter 3).

Immunosurgery is routinely used to isolate the ICM prior to derivation of human ES cells although this is not essential. Trophoblast-binding antibodies enable complement-mediated lysis of the trophoectoderm cells, while these antibodies do not have access to the ICM (Solter and Knowles 1975). Subsequently, the remaining ICM is transferred to a culture dish and after a few days passaged onto a feeder layer of mitotically inactive fibroblasts. Undifferentiated murine ES cells can be maintained and expanded indefinitely in the presence of feeder cells or cytokine 'LIF' (leukemia inhibitory factor). In the absence of feeder cells or after withdrawal of LIF and especially after addition of certain factors (e.g. retinoic acid derivates), ES cells start to form differentiated cells and tissues.

A recent study demonstrated that humane ES-cell cultures cannot be considered as pure populations of pluripotent ES-cells (Sathananthan et al. 2002). Instead, it is much likely that only a minority of the cells within an ES-cell colony represents "true" ES-cells. A majority of cells show signs of differentiation into different lineages and displays epithelial morphology. It seems to be crucial to optimise ES-cell culture conditions in order to achieve higher proportions of pluripotent undifferentiated ES-cells within the cultures.

Human ES cells have so far not been derived without feeder cells, although this has been done with mouse embryos in the presence of LIF. However, no effect of LIF has been demonstrated for human ES cells. It is not known whether additional factors are necessary to keep these cells in an undifferentiated state or whether LIF is not effective at all. For this reason, all human ES cell lines reported have been isolated and maintained on murine feeder cells with the exception of a very few lines that used human embryonic feeder cells (Richards et al. 2002; Itskovitz-Eldor, personal communication). Once established, human ES cells have been maintained in feeder-free culture systems with feeder-cell-preconditioned medium (Xu et al. 2001). Many groups have now also examined the use of serum-free media, not only to provide more uniform conditions for growth and differentiation, but also to help comply with GMP protocols for eventual clinical use. Although all human ES cell lines are derived from pre-implantation embryos of a comparable state, there are remarkable differences in differentiation and proliferation capacity. For example, culture of the H9 cell line seems to be relatively unproblematic whereas the H1 cell line proliferates much more slowly and is more difficult to maintain undifferentiated (N. Benvenisty, personal communication).

2.5.2
Differentiation of Embryonic Stem Cells

Differentiation of mouse ES cells has been investigated for many years (Doetschman et al. 1985). As mentioned above, embryonic stem cells are pluripotent and contribute to all cell types of the adult mouse. They can also give rise to many cell types *in vitro*. For example, murine ES cells can form hematopoietic cells (Burkert et al. 1991; Chen 1992; Doetschman et al. 1985; Nakano et al. 1994; Wiles and Keller 1991). Clones of hematopoietic precursor cells have been isolated and their specific gene expression patterns have been analysed (Keller et al. 1993; Schmitt et al. 1991). More recent results point to the existence of an early progenitor of hematopoietic cells and endothelial cells, the hemangioblast (Asahara et al. 1999a; Asahara et al. 1999b; Shi et al. 1998). In view of this relationship, it is not surprising that the formation of endothelial cells from ES cells has also been reported (Risau et al. 1988). Within embryoid bodies, differentiating aggregates of ES cells in suspension culture, ES cells are able to develop into spontaneously contracting cardiomyocytes (Doetschman et al. 1985; Klug et al. 1996; Muller et al. 2000; Sauer et al. 1999; Wobus et al. 1995). Although these cells appear to be in a fetal rather than an adult state, their gene expression patterns, contractile and electrophysiological properties are very similar to cardiomyocytes formed *in vivo*. Cells resembling pacemaker cells, atrial and ventricular cells, have all been identified and even purified from differentiating ES cells in culture (Muller et al. 2000). It has been shown that transplanted ES cell-derived cardiomyocytes survive in infarcted mouse myocardium (Klug et al. 1996; Roell et al. 2002), but proof that they bring about functional improvements is still pending. Murine ES cells are also able to form skeletal myoblasts (Braun and Arnold 1994; Rohwedel et al. 1998) and smooth muscle cells (Drab et al. 1997). Morphologically and on a molecular level, neuronal cells generated from mouse ES cells are very similar to neurons and glia cells of the CNS (Gottlieb and Huettner 1999; Lee et al. 2000). Up to now, only few transplantation studies have demonstrated successful integration. Some cells differentiated into dopaminergic cells, others formed serotonergic cells (Deacon et al. 1998). Brüstle was able to show differentiation of transplanted progenitors to neurons, astrocytes, and oligodendrocytes (Brüstle and McKay 1996). Generation of hepatocytes from murine ES cells has also been demonstrated (Yamada et al. 2002). Soria et al. and Lumelsky et al. generated insulin-secreting islet cells from ES cell cultures (Lumelsky et al. 2001; Soria et al. 2000) and very recent data demonstrate the efficient formation of insulin-producing β-cells from pdx1 and PAX4-transgenic ES cells (Blyszczuk et al. 2003). Other reports demonstrated the differentiation of murine ES cells to adipocytes (Dani et al. 1997) and keratinocytes (Bagutti et al. 1996).

Since 1995 data on the properties of nonhuman primate and human ES cells have begun to accumulate. This has enabled a comparison with mouse ES cells for which there is much more extensive data. Although there are significant differences regarding phenotype, proliferation potential, culture conditions and potential for differentiation, many properties of murine cells have been confirmed in human cells. Transplantation of human ES cells into SCID-mice resulted in the development of teratomas. Histological analysis revealed elements of all three germ layers (Amit et al. 2000; Thomson et al. 1998a). Benvenisty et al. were able to generate embryoid bodies similar to those obtained with mouse ES cells (Itskovitz-Eldor et al. 2000; Schuldiner et al. 2000). Within the embryoid bodies, expression of marker genes for all three germ layers and for different cell types including heart, kidney and liver was detected (Schuldiner et al. 2000). The effect of different growth factors on differentiation of huES cells has been analysed, again giving similar results to those obtained with the mouse (Schuldiner et al. 2000). Meanwhile, more detailed analysis on the differentiation of primate ES cells has become available. Several studies revealed evidence for neuronal (Reubinoff et al. 2001; Schuldiner et al. 2001; Thomson et al. 1998b; Zhang et al. 2001), hematopoietic (Li et al. 2001; Lu et al. 2002) and cardiac differentiation (Kehat et al. 2001). Human ES-cell derived cardiomyocytes have already been compared electrophysiologically with "normal" human cardiomyocytes and revealed high similarity with fetal ventricular myocytes (Mummery et al. 2003).

2.5.3
Genetic Manipulation of Human Embryonic Stem Cells

Genetic manipulation of embryonic stem cells will be essential for the detailed molecular analysis of differentiation events and may well be useful for future clinical applications. For example, highly purified differentiated stem cells can be prepared using cells expressing selection markers under the control of tissue specific promoters. cDNA isolated from such highly purified cells could be used to derive gene expression profiles. Similarly, undifferentiated ES cells will have to be eliminated by FACS-sorting or by expressing suicide-genes under the control of ES cell-specific promoters. This is necessary to prevent generation of teratomas after clinical transplantation of the differentiated cell population. Moreover, expression of certain transcription factors may be essential to induce cell-type specific differentiation (e.g. Blyszcuk et al. 2002).

The genome of murine embryonic stem cells can be manipulated relatively easily and efficiently by means of DNA electroporation. In contrast, genetic manipulation of human ES cells has proved to be much more difficult. Initial attempts using electroporation or liposome-related strategies failed to achieve significant transfection rates. Recently, a non-viral method was published that produced up to 5–10% transfected cells (Eiges et al. 2001), which is still a poor efficiency rate. Much better efficiency rates of up to 100% transduction combined with stable long-term expression have been achieved by using modern lentiviral vectors (Gropp et al. 2003; Ma et al. 2003; Pfeifer et al. 2002). The proposed use of lentiviral vectors for clinical applications is controversial due to the fact that these vectors have been derived from HIV-1, and stable integration into the host cell genome may result in malignant transformation events. A very recent report demonstrated efficient transformation of human ES-cells by a modified electroporation protocol. This protocol enabled first gene targeting in human ES-cells (Zwaka and Thomson 2003). One can thus anticipate that human ES cells will be also used in modified gene therapy protocols. It should be possible to control the integration site of the gene insertion *in vitro* before the cells are actually injected into a patient. This would avoid most of the problems associated with random targeting using modified viral vectors (including retrovirally induced tumors – see 1.2.2).

2.5.4
Approaches to Enable Clinical Transplantation of ES Cell-derived Cells without Immunosuppression

Human whole organ allografts have been transplanted for more than 30 years (Barnard 1967). Allotransplantation is frequently the last therapeutic option after terminal organ failure. Although immunosuppressive drugs, necessary to prevent rejection of the allografts, have been considerably improved, there are still major side effects. Pharmacological immunosuppression often leads to severe infections with usually apathogenic fungi, bacteria or viruses. Moreover, the incidence of malignancies, especially certain skin cancers and lymphomas, is drastically increased (2–8%). In view of the severe side effects of pharmacological immuno-suppression, such treatments would probably not be acceptable for the setting of stem cell based therapies.

2.5.5
World Wide Embryonic Cell Culture Collections

Clinical bone marrow transplantation is only possible due to the establishment of worldwide databases on major and minor histocompatibility antigens of potential donors. Although it is still difficult to find perfectly matched donors for rare combinations of histocompatibility antigens, the majority of patients could receive the bone marrow of a few thousand donors. In principle, it should be possible to organize a worldwide collection of human ES cell lines. In many countries, there are enough spare embryos available. The practically unlimited expansion potential of ES cells could guarantee large amounts of cells for all therapeutic purposes. Similar to bone marrow transplantation, such a collection should ensure the treatment of the majority of patients. For patients with more infrequent antigen patterns which cannot be matched perfectly, low levels of immunosuppression may be sufficient to avoid rejection. Beside the costs for such banks, ethical considerations on the use of such large numbers of embryos could be used to argue against worldwide ES cell collections. On the other hand, far fewer embryos would be sacrificed for the establishment of such a cell bank than for the large scale production of autologous ES cells by means of therapeutic cloning (see 1.5.4.4). The UK Medical Research Council is to establish a Stem Cell Bank, which will include human ES cell lines appropriate for both research and for potential clinical use (the later done under GMP conditions).

The probability of finding a match depends on several factors, including the size of the donor pool, the rarity of the recipient's HLA haplotype and the accepted degree of mismatch. Thus, assuming that a match is only required for HLA-A, HLA-B and HLA-DR antigens and that limited (one or two) mismatches are allowed, an individual is very likely to find a match even among a few thousand donors. A match of that type would nevertheless require immunosuppressive treatment for the rest of the recipient's life. It is unclear whether ES-cell based cell and tissue replacement therapies could be based on partial mismatches combined with immunosuppression. As some of the therapeutic targets (juvenile diabetes, for example) imply lifelong successful engraftment (with the possibility of repeated treatment), it may be advantageous to reduce to a minimum or eliminate completely the mismatch between the recipient and the donor ES cells. Such a requirement would significantly diminish the probability for any given individual to find a match even with a very large ES cell bank (Beatty et al. 2000; Beatty et al. 1995).

2.5.6
Genetic Engineering of a Universal ES Cell Line

Specific immunological elimination of transplanted tissues and cells depends on certain molecules necessary for T-cell activation. Cytotoxic T-cells, as well as helper T-cells, involved in the activation of antibody-producing B-lymphocytes, are activated via the binding of their T-cell receptors to major histocompatibility complex (MHC) molecules, expressed on antigen-presenting cells and target cells. A certain specific T-cell clone can only be activated if its T-cell receptor binds to a foreign MHC protein or to the self MHC protein expressed on another cell and loaded with a foreign peptide. Since all T-cell clones reacting with self MHC alone or loaded with self peptides are eliminated immediately after their differentiation from bone marrow stem cells, this system guarantees that reactivity is restricted to foreign structures. Target cells can evade such immune responses if they do not express so-called bystander molecules or if they express apoptosis-inducing surface molecules such as FASL. In these cases, attacking T-cells become anergic or apoptotic and die. Moreover, up-regulation of anti-apoptotic genes may help to survive antibody or T-cell mediated cytotoxicity.

Genetic modification of the surface structures involved in T-cell activation and artificial over-expression of anti-apoptotic genes has been proposed to allow graft survival without immunosuppression. To achieve this goal, inactivation of the natural MHC-molecules e.g. by means of RNA interference, ribozymes or gene targeting is likely to be necessary. However, to avoid natural killer cell mediated cytotoxicity, introduction of a MHC gene expressed by the recipient may be essential. Furthermore, down-regulation of bystander molecules such as B7 (Sharpe and Freeman 2002) and over-expression of FASL, TRAIL (Green and Ferguson 2001; O'Connell 2002) or anti-apoptotic genes such as Bcl-2 (Deveraux et al. 2001) might be necessary. Although many strategies to modify human ES cells have been proposed, it is currently unknown whether such genetic modifications will finally result in an universal non-immunogenic cell line (Bradley et al. 2002).

2.5.7
Tolerance Induction

Tolerance is a natural mechanism to exclude reactivity against self-structures. In contrast to immunosuppression, tolerance is antigen-specific. After achievement of tolerance, no or at least no significant external manipulation is necessary to maintain this state. Two different types of tolerance have to be distinguished: In central tolerance, all arising donor-reactive cells are eliminated in the thymus, in peripheral tolerance those cells can still exist in a non-reactive, so-called anergic state.

Pioneering work in this field was done in 1945 by Owen, showing that bovine heterozygote twins connected via placental anastomoses remain blood cell chimeras for their whole life, and by Medawar in 1953 demonstrating the feasibility of tolerance induction in newborn mice (Billingham et al. 1953). About 40 years ago, protocols were developed to achieve immunological tolerance to allografts in adult mice (Schaller and Stevenson 1965). Even in pigs and non-human primates tolerance induction was successful, although this was achieved by protocols not clinically applicable (Kimikawa et al. 1997). At the present time, novel knowledge and the availability of antibodies specific for human T-cells give rise to the hope that tolerance induction might become clinical reality in the future (Waldmann 1999). If clinically applicable protocols could be developed, all kinds of allogenic cells, tissues and organs could be transplanted without any immunosuppression.

Since one major component of current protocols on tolerance induction is the transplantation of hematopoietic bone marrow stem cells, and since there exist data that it may indeed be possible to generate functional blood cells from embryonic stem cells (Li et al. 2001; Lu et al. 2002), it seems feasible that tolerance towards ES cell derived cells and tissue will be able to be achieved. One open question is whether transplanted allogeneic human embryonic stem cells can survive long enough to allow establishment of tolerance in the recipient. It appears not impossible that transplanted ES-cells are able to survive for certain time in an allogenic host as at least undifferentiated (and unstimulated) cells express no MHC class II, and low levels of MHC class I. On the other hand it seems likely that after transplantation at least MHC class I will be upregulated after immunological activation (e.g. by IFN γ) or after differentiation. So far, the survival of fully mismatched murine or primate ES cells into non-immunosuppressed recipients have not been systematically investigated. A recent report demonstrated the induction of central tolerance in rats after intraportal injection of fully MHC-mismatched rat ES-like cells (Fandrich et al. 2002). According to the authors, transplanted ES-like cells survived and gave rise to stable thymus-dependent hematolymphoid chimerism. One explanation for the initial survival of the injected cells may be the observed low MHC class I expression as well as the expression of FasL and the lack of expression of MHC class II and costimulatory molecules B7.1 and B7.2. Although these data have to be confirmed independently, induction of central tolerance might be a promising strategy to achieve long-term survival of mismatched allogeneic ES-derived cells. Future investigations will have to show whether human ES cells indeed have low antigenic properties. Nonhuman primate models may be most appropriate to address this issue in a preclinical state. In the meantime, alternative strategies allowing transplantation of ES-derived cells without immunosuppression should be investigated. These include the above-mentioned cell banks, the production of universal donors as well as the derivation of autologous ES cells.

2.5.8
Production of Autologous Embryonic Stem Cells by Means of "Therapeutic Cloning"

Despite some promising data (Fandrich et al. 2002) it is at present not known whether it will be possible to transplant allogeneic huES cells without any immunosuppression. In contrast, transplantation of autologous ES cell-derived cells would definitely be possible without any immunosuppression.

Technologies have recently been developed to clone animals from somatic cells. Meanwhile sheep (Campbell et al. 1996), cattle (Lanza et al. 2001), pigs (Betthauser et al. 2000; Onishi et al. 2000; Polejaeva et al. 2000), goats (Keefer et al. 2001), mice (Wakayama et al. 1998), cats (Shin et al. 2002), rabbits (Chesne et al. 2002) and zebrafish (Lee et al. 2002) have been cloned. First cloned rhesus monkey embryos have been obtained using fetal fibroblasts (Mitalipov et al. 2002). A very recent paper reported cloning of human embryos by transfer of human somatic into rabbit oocytes and subsequent derivation of human ES-like cells (Chen et al. 2003). Although cloning efficiency rates are quite different in these species, these results indicate that cloning should be possible in all species including humans. As demonstrated in the studies mentioned above, it is possible to generate viable offspring using nuclear transfer technologies.

Unfortunately, somatic nuclear transfer is hampered by so far unsolved problems, mostly of epigenetic origin, resulting in very low efficiency rates, high mortality during pregnancy and malformations in the offspring. These problems are due to the fact that the somatic nucleus has to be converted into an embryonic state. This is usually done by fusion of the somatic cell (e.g. a skin fibroblast) to an enucleated oocyte (figure 2.2). The influence of unknown factors within the oocyte cytoplasm leads to reprogramming of the somatic nucleus, finally resulting in an early embryo which can be cultured until blastocyst stage prior to transfer into the uterus. Since reprogramming of the somatic nucleus obviously does not result in a completely normal embryo, the expression pattern of these embryos can be remarkably different from that of normal embryos (Humpherys et al. 2001; Rideout et al. 2001). One of the genes whose transcription level seems to have a major impact on the development of a "normal" embryo is Oct-4 (e.g. Boiani et al. 2002). A feasible strategy to overcome epigenetic problems of therapeutic cloning may be the use of nuclei derived from adult stem cells with at least some similarities to ES cells (e.g. intestinal stem cells).

Whereas these epigenetic problems are critical for reproductive cloning, they may be less significant for therapeutic cloning. The term "therapeutic cloning" designates the use of somatic nuclear transfer technologies not to produce viable offspring, but to generate autologous ES cells for therapeutic purposes (figure 2.2). In this case, the inner cell mass of blastocysts produced after transfer of a somatic cell nucleus from the patient into an enucleated oocyte is used to isolate autologous ES cells. These would then be differentiated into a variety of cell types for cell transplantation or tissue engineering applications.

Another potential use for therapeutic cloning is to derive human ES cells from individuals carrying rare genetic diseases or from individuals with specific combinations of genes that makes them predisposed to develop genetic or environmentally-induced disease. These ES cell lines, and differentiated cell types obtained from them, could potentially provide a valuable source of material and a means to study such diseases in vitro. They could be used to check for the effects of environmental agents, which at present can only be done using epidemiological techniques or animal models. Moreover, they could be used in screens to search for molecules that could alleviate the disease or to screen potential pharmaceuticals for toxicity. More broadly, different people respond to pharmaceuticals and other agents in different ways, and this will often depend on their genetic makeup, and panels of ES cells derived form a diverse range of individuals may be useful in research on this.

As development of a whole organism is much more complex than differentiation of a single cell type to be used e.g. for cell transplantation, certain epigenetic modifications which cause death or malformations in the setting of reproductive cloning might not have any detrimental consequences for therapeutic applications. It has been proven in mice that autologous ES cells produced by therapeutic cloning are suitable tools to correct certain genetic defects in the recipient (Rideout et al. 2002). In spite of the fact that the method seems suitable from a technical point of view for clinical purposes, some have argued against therapeutic cloning.

Major arguments of the opponents are:

1. If therapeutic cloning becomes widely accepted, then this will encourage reproductive cloning.
2. Somatic nuclear transfer results in a viable embryo, produced for the sole purpose of sacrificing it for ES cell production.
3. There is an insufficient supply of donated eggs.
4. The cost of such an individual treatment will be prohibitive for the majority of patients.

To bypass these arguments, different alternative technologies have been discussed or trials to realize such technologies are already ongoing. One possibility may be the use of animal oocytes instead of human oocytes. Although there is already a first report on therapeutic cloning using human nuclei and rabbit oocytes (Chen et al. 2003), it is currently not clear whether fusion of such oocytes with human somatic cells will result in cells that are functionally comparable to "normal" human cells, and while major concerns about the potential development of xenozoonoses have been expressed, it is a controversial point whether the resulting blastocysts would be human embryos at all.

Another solution which has been proposed is the generation of autologous ES cells from parthenotes. Parthenogenesis is the process by which an egg can develop into an embryo without fertilisation. It has been known for many years that this development can be triggered artificially in mammals using electric or chemical stimulation. A recent report demonstrated the derivation of ES cells from nonhuman primate parthenotes (Cibelli et al. 2002). Future analysis will have to prove whether these cells are comparable to normal ES cells and whether parthenogenetic stem cells can also be derived from human oocytes.

An additional alternative technology has recently been suggested: ES cells could be used to reprogram the nucleus of a somatic cell. Up to now this was not possible, since the amount of cytoplasm in an ES cell is too small to allow efficient nuclear reprogramming after fusion with the somatic cell. Moreover, enucleation of an ES cell is much more difficult than enucleation of a larger oocyte. A very recent report demonstrated multi-fusion of ES cells followed by a centrifugation-based enucle-ation step (Pralong et al. 2002). Fusion of these cytoplasts with ES cells resulted in viable cells. Fusion of such cells with a somatic cell may allow reprogramming of the somatic nucleus not to a totipotent but to a pluripotent state. The result might be an autologous embryonic stem cell, not able to form a whole embryo.

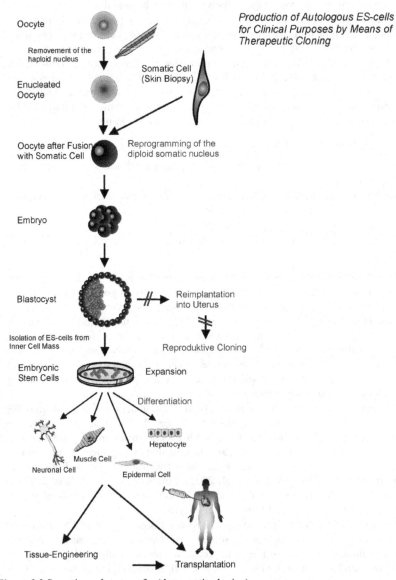

Figure 2.2 Somatic nuclear transfer (therapeutic cloning)

2.6
Embryonic Germ Cells

Pluripotent embryonic germ cells (EG-cells) are derived from primordial germ cells (PGC), the embryonic progenitors of oocytes and sperms. EG-cells express a high potential for proliferation and differentiation similar to that of ES cells. Studies on the development of the germline in mammals have been facilitated by the ability to trace germ cells in the embryo using the marker "tissue non-specific alkaline phospatase" (TNAP). PGCs and EG-cells are positive for alkaline phosphatase similar to ES cells. Murine PGCs have been isolated and cultured for many years now (Donovan et al. 1986). Isolation and culture of pluripotent murine EG-cells were reported in 1991 (Matsui et al. 1991; Matsui et al. 1992; Resnick et al. 1992). In contrast, the generation and culture of pluripotent human EG cells were not reported until 1998, (Shamblott et al. 1998). First data revealed that these cells are comparable to human ES cells as regards culture conditions and proliferation rates.

EG cells are very similar to ES cells in many respects, although studies on murine cells suggest differences in the regulation of gene expression and developmental potential. Both cell types are capable of forming embryoid bodies, they proliferate extensively and can differentiate into cells expressing markers of all three germ layers. Murine EG cells have been shown to differentiate *in vitro* into a variety of cell types including cardiomyocytes, skeletal myoblasts, and neuronal cells (Rohwedel et al. 1996). So far, there have been no reports on the functional analysis of murine EG-cells differentiated *in vitro*.

Many differences between murine ES- and EG-cells may be explained by "imprinting" effects. In the early embryo, genes necessary for embryonic development are activated, "adult" genes are inactivated. In contrast, in adult tissue "early genes" and some tissue-specific genes are repressed, whereas "adult" genes are active. Unlike ES cells, PEGs and EG-cells are in a hypo-methylated "reset" state. In the course of gametogenesis reprogramming occurs, resulting in oocytes and sperm capable of expressing genes that need to be activated in the early embryo. This different hypomethylated state may account for developmental defects observed in chimeric mice produced by injection of EG-cells into blastocysts (Kato et al. 1999).

Although both cell types have been available since 1998, there are significantly less data on the differentiation of human EG-cells than on the differentiation of huES cells. The initial study of Shamblot et al. demonstrated that human EG cells formed embryoid bodies (Shamblott et al. 1998). Within these embryoid bodies, markers of all three germ layers have been detected. Muscular, mesenchymal, hematopoietic-endothelial, neuroepithelial, and endodermal structures have been observed. (Shamblott et al. 1998). More detailed expression analyses revealed not only neuronal markers, but also muscular, hematopoietic, vascular and endodermal markers (Shamblott et al. 2001). All cell samples analysed showed a varied expression pattern. Specific differentiation towards certain lineages could not be proven (Shamblott et al. 2001). Initial transplantation studies have suggested that functional integration following the transplantation of differentiated cells derived from human EG cells is possible (John Gearhart, personal communication).

Primordial germ cells can be isolated from embryos aborted between weeks 5 and 9 post fertilisation (Shamblott et al. 1998). These cells can be used to establish pluripotent EG-cell cultures. For technical reasons, existing human EG cell lines have been isolated from embryos after planned abortion. However, the isolation of EG cells is technically complicated due to autolytic processes in aborted embryos and because of the necessity for precise timing of the abortion. Similar to the isolation of ES cells from early embryos, no autologous cells for a particular patient can be isolated from abortions.

2.7
Potential Risks of Upcoming Applications of Regenerative Medicine

As in most new technologies, all approaches of regenerative medicine are expected not only to bring advantages, but may also involve certain risks. This has recently become apparent in the case of gene therapy after the first death due to injection of adenoviral vectors (Lehrman 1999). So far, all gene transfer technologies shown to be able to efficiently deliver larger nucleic acid fragments *in vivo* are of viral nature. All of these viral vector sytems mainly involve two risks: Preparations of viral vectors may contain replication-deficient virus, or replication-deficient virus may be generated by recombination with other exogenous or endogenous viral elements. Such recombination events could result in wild-type type virus, which may be pathogenic per se, or, alternatively, new replication competent variants with different properties could arise. Besides, viruses, especially the ones displaying long-term expression and typically integrating into the genome such as retroviruses, lentiviruses, or adeno-associated viruses, can result in malignancies. This is due to the fact that such viral elements can inactivate (or activate) oncogenes (Li et al. 2002). The most striking example of this was recently reported. Following successful gene therapy for severe combined immune deficiency (Hacein-Bey-Abina et al. 2003), leukaemia developed in two out of eleven patients treated.

Similarly, it cannot be excluded that extended *in vitro* expansion of cells may result in the formation of immortalised tumor-like cells. In spite of the fact that spontaneous generation of immortalised cells is a rare event which depends on the accumulation of mutations in several oncogenes and protooncogenes, such events have to be taken into consideration. Detailed studies on the formation of cancer-like cells in cultivated somatic cells, adult or embryonic stem cells have not yet been carried out. A recent study suggests that there are remarkable differences in mutation frequency and type between embryonic stem cells and somatic cells (Cervantes et al. 2002). Nevertheless, aneuploidies and chromosomal alterations can be detected in cultured human embryonic stem cells, also (Carpenter et al. 2003; Draper et al. 2003).

Especially in the case of embryonic stem cells, the possibility of teratoma formation has to be mentioned. Undifferentiated ES cells tend to form tumors, so called teratomas, when transplanted into a recipient. This has not only been shown in murine ES cells, but also in human ES cells (Thomson et al. 1998a). Such teratomas usually contain cells differentiated towards cell types of all three germ layers. The remaining undifferentiated cells seem to proliferate indefinitely, causing a tumor-like growth. To avoid teratoma formation after ES cell transplantation, only cells pre-differentiated *in vitro* can be transplanted. Before clinical application, the remaining undifferentiated cells have to be removed. This can be done by FACS-sorting or, in the case of trans-genic lines, by the activation of a suicide gene. It has yet to be proved whether these methods are effective enough to guarantee complete depletion of undifferentiated cells.

Potential clinical use of one of the existing human ES cell lines poses a further risk. Up to now, almost all human embryonic stem cell lines have been cultivated on murine feeder cells. In considering xenotransplantation for use in clinical practice, it has been feared that porcine endogenous retroviruses might be transmitted from porcine xenografts to human recipients, potentially resulting in a new pandemic pos-sibly comparable to AIDS (Weiss 1998). Future clinical use of human embryonic stem cells and their isolation using mouse feeder layer has to be considered. Similar to pig cells, mouse cells contain functional humantropic endogenous retroviruses (e.g. Gardner 1978; Levy 1978; Morse and Hartley 1982; Stoye and Coffin 1987). These include B-type retroviruses of the "Mouse Mammary Tumour Virus" group as well as C-type retroviruses of the "Murine Leukemia Virus" group. Both groups have been shown to induce malignancies or neuropathological disorders. It is currently being analysed whether transmission of such viruses from murine feeder cells to the exist-ing human embryonic stem cell lines has already occurred or whether such transmis-sion is at all possible. It is likely that, due to endogenous murine retroviruses and other (as yet unknown) murine pathogens, none of the existing human ES cell lines will be suitable for clinical applications. However, this problem will not be permanent. Some groups have already developed human feeder cells and the first human ES cell lines have been derived and cultured without the use of murine cells (Richards et al. 2002; Itskovitz-Eldor, personal communication). Furthermore, it is likely that factors allow-ing feeder-free ES cell culture will be identified in the near future.

2.8
Clinical Approval

As with any new medical technology that offers chances and more or less unknown risks, clinical approval will mainly depend on the assessment of the clinical benefit for the intended patient cohort versus the risk for these patients. Since the risks con-nected with ES cell transplantation, i.e. the formation of teratomas or infection with unknown murine pathogens, are likely to be higher than those connected with the transplantation of autologous adult stem cells, only patients with acute life-threat-ening diseases will be candidates for first clinical studies on ES cell transplantation. Similar to ES cells, forthcoming studies on the application of adult stem cells will require risk-benefit-analysis. In each case it has to be decided whether it is likely that the intended stem cell therapy will provide results that are equal or superior to those obtained with conventional treatments.

2.9
Time Frame for Future Clinical Applications

It is difficult, if not impossible, to predict how long it will take to develop various stem cell-based therapeutic methods. At present, it remains unknown whether *in vitro* differentiation of stem cells will result in functional cells that can regenerate lost tissue structures. Even less foreseeable is whether future research will enable the production of vascularised tissues *in vitro,* enabling replacement of lost organs. Single clinical applications of adult stem cells, cell transplantation approaches and tissue engineering are already in the phase of early clinical studies. This includes the use of mesenchymal stem cells for bone and cartilage repair (see 2.3.4). Hematopoietic stem cells have been used successfully for many years to treat leukemias and lymphomas. Furthermore, the application of primitive bioartificial skin has been proven useful in certain burn patients. The time frame for future clinical use of certain stem cells depends on the cell type needed and on the question whether the cells to be transplanted have merely to supply certain metabolites or whether functional or even mechanical integration is required. Probably the most sophisticated procedure will be the generation of complex tissues and organs *in vitro.* Based on these considerations, the following time frame for future clinical applications of stem cells can be expected:

1. Transplantation of cells producing metabolites, hormones, cytokines etc. (e.g. pancreas β-cells)
2. Tissue engineering of simple types of tissue (cartilage, simple skin replacement)
3. Transplantation of cells with structural function (e.g. neurons, cardiomyocytes)
4. Tissue engineering of complex 3D (vascularised) tissues

The broad application of stem cells will not become reality within the next few years. It has been estimated that at least 5 to 10 years will pass until significant numbers of stem cell based therapies will become clinical reality. Whether ES cells or EG cells will be the first candidates for use in patients remains entirely speculative at this point.

3 The Regulation of Embryo Research in Europe: Situation and Prospects

3.1
Introduction

At the present time regulation of embryo research differs from country to country. Some countries prohibit embryo research altogether, some countries permit embryo research to varying degrees, while other countries do not have any regulation in this area at all, though institutions in which research can be carried out do exist or could easily be established.

In view of the numerous new prospects in reproductive and regenerative medicine considered in Chapters 1 and 2, and taking into account that many research projects are carried out in multi-centred studies crossing national borders, we have to consider whether regulation of research on human embryos should be kept on a mere national level or whether it should be regulated on an international or at least a European level. In an article previously published by the Europäische Akademie (Beyleveld and Pattinson 2001) it was argued that moral, social, and economic factors and pressures that exist render non-harmonisation of the law the optimal situation in relation to the interests (broadly conceived) of both those countries that currently prohibit research on embryos and those that permit it. Nevertheless, it was suggested that harmonisation that permits embryo research is likely to be the regulatory outcome in all the EU countries in the medium or long-term.

In this chapter, we revisit this issue, and consider the effect that the expansion of the EU will have on the situation. In order to do so, we first describe the current regulation of embryo research in the European Union (EU) and the Newly Associated States (NAS). Then, in order to gain a deeper understanding of the factors that affect the possibility of harmonisation, we look in some depth at the history and details of regulation of embryo research in the United Kingdom, Spain and Germany. These three countries were chosen because the UK and Germany are towards the opposite ends of the spectrum in relation to the restrictiveness of their laws in the EU, while Spain adopts a "middling" position. In order to further inform our analysis, we have also gathered information on laws in related fields (such as *in vitro* fertilisation, pre-implantation genetic diagnosis (PGD), and abortion) in not only the EU and NAS, but also in a number of other countries whose regulatory positions might assist with our analysis of the prospective regulatory situation in Europe. In the main, these are either countries that are economically powerful (such as China, Japan, and the USA) or countries that have a federal structure, which implies that they will be, or will have been be posed with, the issue of harmonisation themselves (such as Australia, Canada, and the USA). All of this additional information is not presented in our text, but in an Appendix (Appendix 4), to which reference is made when appropriate.

As science and the political and ethical debate progresses, legislation of embryo research is subject to constant scrutiny and revision. Any discussion of these issues necessary be work in progress. While this analysis focuses on the situation as it presented itself in 2002, when the bulk of our data was compiled,[1] we will also mention important later developments at end of this chapter.

Although much of our analysis must, perforce, be speculative, the conclusion we arrive at is that, taking into account the expansion of the EU that will be affected when the NAS join, we find no reason to disagree with the general view expressed in Beyleveld and Pattinson 2001.

3.2
Current Regulation in the EU and the NAS

3.2.1
The European Union

3.2.1.1
Regulatory Stances

Denmark,[2] Finland,[3] Spain,[4] Sweden,[5] and the UK[6] are the only countries in the EU that unequivocally have legislation permitting embryo research. However, Belgium,[7] Italy,[8] Luxembourg,[9] and the Netherlands[10] all have proposed legislation permitting embryo research. In Portugal, legislation permitting embryo research[11] was vetoed by the President.[12] In consequence, embryo research is permitted by default, though we understand that, to date, no research has been undertaken.[13] Greece also has no legislation. However, embryo research is permitted by a regulation of the Central Council of Health.[14] On the other hand, France[15] and Ger-

---

[1]   In some cases we were unable to obtain reliable information on the legal position. At times we were unable to obtain translation of potentially relevant sources or the advice of a legal expert for that country. Large gaps in the information collected by bodies such as the European Commission and the Council of Europe attest to the difficulties that exist in obtaining reliable information from a number of countries in this difficult and fast-moving area.

[2]   Law No. 460 of 10 June 1997.

[3]   Medical Research Act 1999.

[4]   Law 35 of November 1988.

[5]   Law 115 of 14 November 1991.

[6]   Human Fertilisation and Embryology Act 1990 ("HFE Act").

[7]   There are several proposals being considered. These can be found at http://www.senate.be under "Dossiers legislatifs-RECHERCHE" and the cue "embryon".

[8]   Draft Law s.2433 of 2000.

[9]   Law No. 4567 of 29 April 1999.

[10] Bill containing rules relating to the use of gametes and embryos ("Embryos Bill") 2000.

[11] Law on Assisted Procreation 1998.

[12] See European Parliament Directorate General for Research Scientific and Technical Options Assessment. Embryos, Scientific Research and European Legislation. Briefing Note No. 14/2001. Available at http://www.europarl.eu.int/stoa/publi/pdf/briefings/14_en.pdf.

[13] The statement made on page 62 of Beyleveld and Pattinson 2001 that embryo research has been carried out in the absence of legislation was made in error.

[14] Pattinson 2002, Appendix 3.

[15] Law 94–654 of 29th July 1994 and Decree No. 97–613 of 27th May 1997, which form part of the Public Health Code (Code de la Santé Publique).

many[16] both have legislation that may be regarded either as permitting embryo research under highly limited circumstances or as prohibiting it except under highly restrictive circumstances. This is because of the requirement that no investigation may be undertaken on an embryo that is not intended to benefit it (especially in respect of successful implantation) or unless it is directed to the improvement of medically assisted reproduction techniques.[17]([also 18]) That said, legislation is proposed in France that would allow the use of supernumerary embryos.[19] Both Austria and Ireland prohibit embryo research by implication. In Austria this is because no examination may occur unless this is for the purpose of trying to effect a pregnancy.[20] In Ireland, this is because the Constitution grants the unborn a right to life equal to that of the mother.[21]

3.2.1.2
Definition of "Embryo"

Those countries that have enacted legislation or propose to do so (permissive or not) do not agree on the definition of "an embryo". Indeed, some of them do not define "an embryo" at all, and of the nine different proposals for legislation in Belgium, only some contain definitions.

Finland (bearing in mind that the legislation in question governs the research use of embryos only) defines "an embryo" as "a living group of cells resulting from fertilisation not implanted in a woman's body".[22] Germany defines it as a fertilised human egg capable of development from the moment of fusion of the nuclei (and any totipotent cell removed from an embryo defined in the previous manner).[23] The proposed legislation in the Netherlands defines it as a "cell or complex of cells with the capacity to develop into a human being".[24] The UK legislation makes "an embryo" cover both "an egg in the process of fertilisation" and "a live human embryo where fertilisation is complete", completion being at "the appearance of a two-cell zygote".[25]

While the Spanish legislation does not actually define the term, it refers to general practice being to regard a "pre-embryo" as the fertilised egg up until 14 days or implantation, and the "embryo" as existing from 14 days to two and a half months.[26]

---

[16] Embryonenschutzgesetz of 13 December 1990 ("ESchG").

[17] Public Health Code Art. R. 152–8–1 to 12 (from Pattinson 2002, Appendix 3.; European Parliament (opcit. fn.12 ibid.); Code de la Santé Publique Art. 2141–8–5

[18] European Parliament – Fact Sheet on the Temporary Committee on Human Genetics and Other New Technologies of Modern Medicine. Available at http://www.europarl.eu.int/comparl/tempcom/genetics/fact_sheet_en.pdf.

[19] Art. 19 of the Draft Bioethics Act which would insert a new Article L2151 into the Code de la Santé Publique. The Draft was passed by the French National Assembly on 22nd January 2002 and is now under consideration by the Senate.

[20] Reproductive Medicine Act of 4 June 1992, s. 9 para. 1.

[21] Art. 40.3.3 of the Eighth Amendment to the Constitution.

[22] s.2(2) of the Medical Research Act 1999.

[23] Paraphrase of s.8(1) of the Embryonenschutzgesetz 1990.

[24] Embryos Bill Art.1c.

[25] HFE Act s.1(1)(a)&(b).

[26] Paraphrase from the English version of the Official Bulletin of the State No. 282, November 1988, Part II, 5.

None of the other countries provides a definition, though one may possibly be inferred in the Austrian legislation, which covers "developable cells", defined as inseminated ova and cells developed from them.[27] The Danish Act simply covers fertilised eggs, pre-embryos and gametes,[28] while the Swedish legislation refers to "fertilised oocytes of human origin".[29] The Irish, French, Greek, Italian and Portuguese laws or proposed laws provide no guidance at all.

3.2.1.3
Sources of Embryos for Research (Where Permitted)

In general, only embryos produced for IVF may be used for research. The one exception is the UK, which permits embryos to be created specifically for research.[30] That said, while Article 24a of the proposed legislation in the Netherlands prohibits embryos being created specifically for research, Article 32 of the proposed legislation provides for this prohibition to be withdrawn by Royal Decree five years after Article 24 comes into force in favour of Article 11, which permits the creation of embryos for research under limited conditions.[31]

3.2.1.4
Time Limit on Use for Research (Where Permitted)

The limit is generally a *maximum* of 14 days after fertilisation.[32] The exceptions are France (7 days)[33] and Germany (which does not state a limit). However, research cannot be carried out in Germany without the intention to implant (there is an implied limit of 5–6 days).[34]

Conditions

These may be divided into procedural conditions (specifically: need for consent; need for licences; and need for independent ethics committee approval) and substantive conditions (specifically: the purposes for which research may be carried out, amongst which we have singled out the benefit of the embryo for separate attention).

---

[27] European Group on Ethics opinion n° 12 23/11/1998 – Ethical aspects of research involving the use of human embryo in the context of the 5th framework programme. http://europa.eu.int/comm/european_group_ethics/docs/avis12_en.pdf; pg.6.

[28] Section 4, 7 para. 1, 9, 12, 15 para. 1, 16, 17 para. 1 & 2, 25 para. 2, 26 (fertilised eggs, 25 para 1 (gametes). The Danish Ministry of Interior and Health has stated that it is against the use of the term "embryo". However, the EC's Group on Ethics in Science and New Technologies states that the term is generally used (opcit fn.27 there pg. 13).

[29] Law 115, s.1.

[30] HFE Act Schedule 2, para. 3(1).

[31] Where the research is reasonably likely to yield new insights into the causes of infertility, in the field of artificial reproduction techniques, in the field of congenital diseases, or in the field of transplant medicine, *but only* where the research cannot be done using supernumerary embryos.

[32] For example, s.1(3)(a) & (4) of the HFE Act in the UK specifies that embryos may only be used for research up to the appearance of the primitive streak or 14 days, whichever is the earliest.

[33] opcit.at fn.18, p.16–17.

[34] This also applies to France, despite the specification of 7 days.

3.2.1.5
Consent

Belgium (proposed legislation),[35] Finland,[36] France,[37] Greece[38], Luxembourg, (proposed legislation)[39], Spain[40] and Sweden,[41] all require the consent of the gamete donors. In Germany, because research and PGD are prohibited, it is implicit that there cannot be a requirement for consent of the gamete donors. We have not been able to ascertain the position as regards this condition in Denmark, Italy and Portugal.

3.2.1.6
Licences

Finland,[42] and the UK[43] require licences to be obtained in order to carry out embryo research. In Germany, embryo research is prohibited. However, after the Stem Cell Act came into force, there is a formal procedure for obtaining permission to import embryonic stem cells for use in research, which involves getting a positive opinion from the Zentrale Ethikkommission für Stammzellforschung (Central Ethics Committee for Stem Cell Research, sec. 6 para 1 Stammzellgesetz (StZG)). However, the decision of this Committee cannot be regarded as a licence because the „zuständige Behörde" (the administration in charge of permitting imports) does not have to follow the opinion of the Committee. We have not been able to ascertain the positions in Belgium, Denmark, Greece, Italy, Luxembourg, the Netherlands, Portugal, Spain or Sweden.

3.2.1.7
Ethics Committee Approval

We have not been able to ascertain the position in Portugal. With two exceptions, all of the other countries require the approval of an independent ethics committee in their legislation or proposed legislation. The exceptions are Italy and Sweden. However, in Italy, approval is required in the default position that pertains by the National Bioethics Committee.[44] And, while the Swedish law does not require this approval, it is in practice often obtained. (Pattinson 2002 Appendix 4, citing Elisabeth Rynning).

---

[35] Doc 50-0429/001: Art. 3 No. 5; Documents legislatifs nos. 2-87/1: art. 5, 2-92/1: art. 8, 2-114/1: art. 3 sec. 2 & art. 7 sec. 2, 2-321/1: art. 6, 2-686/1: art. 3 no. 5, 2-695/1: art. 8, 2-716/1: art. 5 sec. 2, 2-726/1: art. 8.
[36] Medical Research Act, s.12.
[37] Art. L 2141-8-4 of the Public Health Code New Part.
[38] opcit.at fn.12, p.4.
[39] opcit.at fn.18, p.17.
[40] Law 35, s.15(1)(a).
[41] Law 115, s.1.
[42] Medical Research Act, s.11.
[43] HFE Act, s.3.
[44] opcit. at fn.12,p.5.

3.2.1.8
Benefit to the Embryo

Benefit to the embryo is not generally a requirement, except in France,[45] Germany,[46] in the proposed legislation in Luxembourg,[47] Italy (Pattinson 2002 Appendix 3 citing Roberto Mordacci) and in Spain (in relation to viable embryos).[48]

3.2.1.9
Other Purposes

Finland has, arguably, the most permissive position on this front, as research is only prohibited for genetic modification, and even then it is permitted if the aim is to prevent or cure serious hereditary disease.[49] In the UK, generally thought to be the most liberal, Schedule 2, paragraph 3(2) of the HFE Act originally limited research to where "necessary or desirable" to improve the treatment of infertility, increase knowledge of the causes of congenital disease or miscarriage, develop more effective contraceptive methods, or to develop methods for detecting gene or chromosome abnormalities of embryos before implantation. However, as permitted by the same provision of the Act, these purposes have now been extended by the Human Fertilisation and Embryology (Research Purposes) Regulations 2001[50] to include the purposes of increasing knowledge about the development of embryos, increasing knowledge of serious disease, and to enable the application of such knowledge to the treatment of serious disease.[51] While Spain restricts research on viable embryos to diagnostic, therapeutic or prophylactic purposes (on condition that "non-pathological genetic patrimony is not modified"),[52] in relation to non-viable embryos, research is permitted for 10 specified purposes[53] (which basically amount to the purposes originally permitted by the UK's HFEA. Where research is not for the benefit of the embryo, the proposed legislation in the Netherlands restricts research to the purposes of identification of new insights in the field of medical science. More restrictively, Sweden and Denmark only allow research for assisted reproduction by IVF.[54]

While France (like Germany, and the proposed legislation in Italy and Luxembourg) restricts research for the benefit of the embryo, there is a proposed extension to the legislation that would permit research on supernumerary embryos.[55] We have not been able to ascertain the positions in Belgium (re the proposed legislation), Greece or Portugal.

---

[45] Code de la Santé Publique Art. L2141–8–5.
[46] ESchG s.2. (Compare however the comments made in the German table at sections 1 (b) and (e)).
[47] opcit. at fn.12,p.5.
[48] Law 35, s.15(2).
[49] Medical Research Act 1999, s.15.
[50] SI 2001 No.188.
[51] This controversial development is discussed in detail in the UK case-study below.
[52] s.15(2) of Law 35 of November 1988.
[53] s.16 of Law 35 of November 1988. These purposes are extendable to any other purpose approved by regulation or the Multidisciplinary National Committee.
[54] Law 115 of March 1991 and Law 499 of June 1966 respectively.
[55] Article 19 of the Draft Bioethics Act which would insert a new Article 2151 into the Code de la Santé Publique.

3.2.1.10
Storage

In Finland, the period for which embryos may be stored for research (as permitted) is 15 years.[56] In France,[57] Spain,[58] Sweden[59] and the UK,[60] the period is five years. The Netherlands (proposed legislation) specifies no fixed limit, but makes the limit dependent on the consent of the gamete donors. More restrictively, Denmark only allows embryos to be stored for 2 years,[61] Greece[62] for one year (ibid.) and most restrictive of all, the proposed legislation in Italy would prohibit storage (Beyleveld and Pattinson 2001, 66 note 62, citing Roberto Mordacci). In Belgium, storage is not regulated by legislation and we are unaware of any other regulation. We have also not been able to ascertain the positions in Luxembourg (proposed legislation) or Portugal.

3.2.2
The Newly Associated States

This resume relies heavily on information gathered by the Council of Europe in 1998.[63] We have attempted to update this information, to clarify it and to elaborate it. However, with the exception of Malta, Hungary and the Czech Republic we have generally had very limited success, the most notable thing about the NAS being the difficulty of obtaining information on the regulatory position in these countries.

3.2.2.1
Regulatory Stances

According to the Council of Europe[64] embryo research is not regulated in Bulgaria, Cyprus, Latvia, Lithuania, or Romania. However, at the time that the Council of Europe gathered its information, legislation was planned in Lithuania (ibid., 135). Although Slovenia does not have legislation, the Council of Europe reports that it is proposed,[65] though we do not know whether it has been passed or whether it permits or prohibits embryo research. In the meantime, embryo research is currently permitted if authorised by the National Medical Ethics Committee (ibid., 144). In Esto-

---

[56] s.13 of the Medical Research Act 1999.

[57] opcit. at fn.12, p.5.

[58] Law 35 of 24 November 1988, s.11(3).

[59] Law 115 s.5 (Pursuant to an amendment in 1998. See European Parliament 2001b 4).

[60] HFE Act s.14(4). This is extendable by regulations (see s.14(5)), but none have, as yet, been passed. However, the Human Fertilisation and Embryology (Statutory Storage Period for Embryos) Regulation 1996 SI 1996/375 has extended the period of storage for treatment to a maximum of 10 years.

[61] Beyleveld and Pattinson 2001 stated that the period is 1 year. This was, in fact, correct prior to 1997 when Law No. 460 of 10 June 1997 entered into force in 1 October 1997.

[62] Statement of the General Council for Health 1988.

[63] Council of Europe. 1998. Medically Assisted Procreation and the Protection of the Human Embryo: Comparative Study of the Situation in 39 States/Cloning: Comparative study of the situation in 44 States. Council of Europe: Strasbourg ; Available at http://www.legal.coe.int/bioethics/gb/pdf/txtprep2b.pdf.

[64] opcit at fn.63 pp 144 and 173.

[65] Bill on the Treatment of Infertility and Fertilisation with Biomedical Assistance 1997.

nia, there is no legislation requiring embryo research to be authorised before it can be carried out. Nevertheless, the Council of Europe states that research may only be carried out under certain conditions (ibid.). However it does not indicate whether they are required by law, and we have been unable to obtain further information. While embryo research is not regulated in the Czech Republic (ibid.), the Bill of Fundamental Rights and Freedoms 1992 states that "Everybody has the right to life. Human life is worthy of protection even before birth" and Article 6(2) states, "Nobody may be deprived of life".[66] While this does not establish that embryo research is prohibited, because it might be held to be compatible with a proportional moral status of the embryo view, it does cast doubt upon the legal position.

In Hungary[67] and Slovakia,[68] there is legislation permitting embryo research.[69] In Poland the Council of Europe states that embryo research is "permitted for hereditary diseases and serious risks", but does not indicate whether this is a legislated position (ibid., 166). While there is no legislation in Malta, the Malta Bioethics Consultative Committee Report of 1999 recommended that embryo research be prohibited.[70]

3.2.2.2
Sources of Embryos (Where Permitted)

Whether embryo research is permitted by legislation or by default, it may be inferred that only supernumerary embryos may be used in all those countries that have signed and ratified the Convention on Human Rights and Biomedicine, Article 18(1) of which prohibits the creation of embryos specifically for research. This applies to all the countries except Bulgaria, Latvia and Lithuania, which, though they have all signed the Convention, have not ratified it. That only supernumerary embryos may be used is confirmed by the Council of Europe in relation to Poland[71] and Slovenia[72], and stated by Judit Sandor (1998, 8) in relation to Hungary.

3.2.2.3
Time Limit for Research (Where Permitted)

According to the Council of Europe, no time limit is imposed by regulation in Bulgaria, Cyprus, the Czech Republic, Lithuania, Poland, or Romania[73]. On the other hand, like most EU countries, there is a time limit of 14 days in Estonia (ibid.), Hungary (Sandor 1998, 8), and Slovenia.[74] The Council of Europe was unable to specify the position in Latvia and Slovakia[75] and we have not fared any better.

---

[66] Information provided by Lukas Prudil.
[67] Law No. 154 of 15 December 1997 on Public Health.
[68] (opcit at fn.63 pp 144 and 160.
[69] The Council of Europe does not specify what the Slovakian legislation is, but does state that embryo research requires authorisation by the Ministry of Health (opcit. at fn 56, p145). The legislation covers therapeutic research only (ibid. 160).
[70] Recommendation 8(2).
[71] opcit. at fn.63 p.182.
[72] opcit. at fn.63 p.164 and 182.
[73] opcit. at fn.63 p. 180.
[74] opcit. at fn.63 p.164 and 180.
[75] opcit. at fn.63 p.180.

Conditions

3.2.2.4
Consent

We have been unable to ascertain what the position is regarding the consent of gamete donors in any of the NAS.

3.2.2.5
Licences

Similarly, we have been unable to ascertain whether licences are required in any of the NAS.

3.2.2.6
Ethics Committee Approval

Ethics committee approval is required only in Estonia,[76] Hungary (Sandor 1988, 10),[77] Slovakia[78] and Slovenia.[79] The Council of Europe declares that approval is not regulated in Bulgaria, Cyprus, Latvia, or Romania, and that it is not required in Lithuania or Poland.[80]

3.2.2.7
Benefit to the Embryo

As far as we know, benefit to the embryo is not a requirement in Hungary, and not an informal requirement anywhere else.[81]

3.2.2.8
Other Purposes

We have no information on this. Hungary does specify permissible purposes, but we do not know what they are.

---

[76] opcit. at fn.63 p.149.

[77] The Law on Public Health (s. 186) requires medical research on embryos to be authorised by a competent scientific body set up under a different Act.

[78] opcit. at fn.63 p.149.

[79] opcit. at fn.63 p.149; approval of the National Medical Ethics Committee is required (opcit. fn.63, p.149.) The State Committee for Fertilisation with Biomedical Assistance monitors all activities that concern medically assisted procreation (ibid. 150). Follow up of research projects is left to Institutional Review Boards that are independent of the National Medical Ethics Committee (ibid., 157).

[80] We are unsure what the difference is supposed to be. The Index to the Council of Europe's document does not help when it says ,"When we were certain as to whether regulations or legislation existed, NR (not regulated) was the response given" (opcit. fn.63, pg 16). This should surely read, "When we were certain that no regulations or legislation existed, ...". However, this would not different it from a "No" response. So, perhaps, what is intended is, "When we were uncertain as to whether regulations or legislation existed, NR (not regulated) was the response given."

[81] As regards Poland; Slovenia and Slovakia, the Council of Europe (opcit. fn.63, p.160) states that therapeutic research on embryos in vitro is lawful. However, since those that permit embryo research at least by default will surely also permit therapeutic embryo research nothing may be inferred from this unless they prohibit non-therapeutic research.

3.2.2.9
Storage

The Council of Europe[82] reports that the draft law in Estonia imposes a limit of three years. The Council states that Slovenia imposes a limit of six years for medically assisted procreation, after which embryos created for that purpose may be used for research with the approval of the National Ethics Committee. This, however, leaves it unclear whether embryos created for assisted procreation may be used for research before the end of the six year period. There is no maximum period in Slovakia. The Czech Republic, Latvia, Lithuania, Poland and Romania do not regulate storage, while neither we nor the Council were able to obtain information about the position in Bulgaria, Cyprus and Hungary.

3.3
Case Study 1: Regulation in the UK

The Human Fertilisation and Embryology Act 1990 ("the HFE Act")[83] is the legislative foundation upon which embryo research is regulated. For the purposes of the Act an "embryo" is "a live human embryo where fertilisation is complete", "references to an embryo include an egg in the process of fertilisation" and "fertilisation is not complete until the appearance of a two cell zygote".[84]

Sources of embryos for research may either be surplus embryos[85] produced for IVF or embryos specifically created for research,[86] though the House of Lords Stem Cell Committee (HLSCC) has recommended that embryos should not be created for research unless there is a "demonstrable and exceptional" need that cannot be met by the use of surplus embryos.[87]

Embryo research is ultimately dependent on a "consent trigger".[88] For consent to be valid the donors must have received counselling[89] and have given their written approval[90] for the specific research. They may stipulate conditions under which research may be undertaken.[91] The purposes for which research may be conducted are broad and not restricted by a requirement that the embryo must benefit. They are:

i)   promoting advances in the treatment of infertility;
ii)   increasing knowledge about the causes of congenital disease;

---

[82] the following all: opcit fn.63 pg.89.
[83] Human Fertilisation and Embryology Act 1990 c.37. All UK Public Acts passed since 1988 and statutory instruments passed since 1987 are available at http://www.legislation.hmso.gov.uk.
[84] s.1(1)(a) & (b).
[85] Embryos left over from assisted reproduction procedures. See the Human Fertilisation and Embryology Authority Code of Practice (HFEACP) [Fifth Edition] para. 6.8b.
[86] Sch. 2, para. 3(1).
[87] Report of the House of Lords Select Committee on Stem Cell Research 13 February 2002, Conclusions and Recommendations ix, which is available at http://www.parliament.the-stationery-office.co.uk/pa/ld200102/ldselect/ldstem/83/8301.htm.
[88] Per Sch. 3, para. 6(3), an embryo cannot be used for any purpose without the consent of the donors whose gametes led to its creation.
[89] Sch. 3, para. 3(1)(a), (2) & para. 4.
[90] Sch. 3, para. 1.
[91] Sch. 3, para. 2(1)(c) & (2), (3) & (4).

iii) increasing knowledge about the causes of miscarriage;

iv) developing more effective techniques of contraception;

v) developing methods for detecting the presence of gene or chromosome abnormalities in embryos before implantation;

vi) to increase knowledge about the development of embryos;

vii) to increase knowledge about serious disease; or

viii) to enable the application of such knowledge to the treatment of serious disease.

The first five purposes were included in schedule 2 of the HFE Act,[92] while the latter three were added by regulations passed in 2001,[93] to take advantage of new scientific developments and, in particular, the prospect of stem cell therapies.

The Human Fertilisation and Embryology Authority (HFEA)[94] administers the Act. Research may not be undertaken unless licensed by the HFEA.[95] This is on a project specific basis,[96] and subject to independent ethics committee approval.[97] Research is permitted up to the appearance of the primitive streak or fourteen days, whichever is earlier.[98] Embryos for research may be stored for a maximum of five years,[99] after which they must be allowed to perish.[100]

3.3.1
Challenges to the Legislation

In litigation undertaken by Bruno Quintavalle on behalf of the Pro-life Alliance,[101] the Claimant sought a declaration[102] that human embryos created by cell nuclear replacement (CNR) or, to use the more common term, by cloning (for details of the biological details compare Chapter 1 and 2 of this book) are not created by "fertilisation", and hence not within the definition of an embryo[103] for the purposes of the Act. In response, the Defendant Secretary of State argued that section 1(1) of the Act should be given a purposive interpretation, according to which the essential

---

[92] Sch. 2 para. 3(2).

[93] The Human Fertilisation and Embryology (Research Purposes) Regulations 2001 SI 2001 No.188.

[94] Established by s.5.

[95] s.3 & s.11(1)(c).

[96] Sch. 2, para. 4(2)(b).

[97] HFEACP (Fifth Edition) part 11.6.

[98] s.3(3)(a) & (4).

[99] s.14(4) & Sch. 2 para. 2(3). For treatment purposes embryos may now be stored for a maximum ten years. See Human Fertilisation and Embryology (Statutory Storage Period for Embryos) Regulation 1996 SI 1996/375.

[100] s.14(1)(c).

[101] R v. Secretary of State for Health, ex parte Bruno Quintavalle (on behalf of Pro-life Alliance) [2001] EWHC Admin 918.

[102] The Claimant also sought a declaration in the alternative that, if CNR fell under the Act, it was unlawful as s.3(3)(d) prohibits "replacing a nucleus of a cell of an embryo with a nucleus taken from a cell of any person, embryo or subsequent development of an embryo". As he ruled in favour of the Claimant on the first declaration, Crane J. held that the issue raised in the alternative did not arise for determination though if it did he would not have agreed with the Claimant.

[103] " ... a live human embryo where fertilisation is complete, ... and includes an egg in the process of fertilisation, and, for this purpose, fertilisation is not complete until the appearance of a two cell zygote" (s.1(1)(a) & (b)).

subject of the Act is a live human embryo, not how it was created. In effect, the sub-section should be read as if the words defining an embryo were "a live human embryo where [if it is produced by fertilisation] fertilisation is complete ...", the clear intention of parliament being to control human reproduction by prohibition or licensing[104] by the HFEA. Despite acknowledging that the argument for the purpo-sive interpretation was a powerful one, in light of the history of the HFEA stem-ming from the Warnock (1984) Report and the White Paper[105] that preceded the HFEA, Crane J. held that such an interpretation would involve an "impermissible rewriting and extension of the definition".[106]

The Secretary of State for Health successfully appealed the decision of Crane J.[107] Lord Phillips MR, giving the main judgment of the Court of Appeal, held that "the legislation is largely about the treatment of embryos". Thus, although "the embryos expressly contemplated by the legislation are embryos created by fertilisa-tion", the crucial question is whether embryos created by CNR fall into the same genus as those created by fertilisation. On the basis that both types of embryos are essentially identical in being capable of developing into an adult of the relevant species, it was held that they clearly do.[108] On the question of legislative policy, Lord Phillips ruled that Parliament's intention was to "control the creation and use of human organisms" and, hence, it was essential to bring embryos created by CNR within the regulatory scheme created by the Act.[109] Furthermore, the Master of the Rolls considered that there are no countervailing considerations that militate against this purposive construction. The fact that some of the legislative provisions cannot be applied to embryos created by CNR, does not negate the importance of giving effect to parliamentary intent. Such inconsistencies[110] are not momentous because, in some cases, they can be remedied by regulations made by the HFEA, and in other cases they are not of such practical significance that embryos produced by CNR should be excluded from the Act.[111]

After the Court of Appeal refused leave to appeal against its judgment, the Pro-life Alliance successfully petitioned the House of Lords. However, when it heard

---

[104] Actually, a more accurate statement would be that Parliament intended to regulate human reproduction resulting from the creation, keeping or using of the embryo outside the human body (s.1(2) & (3) & s.3(1)).

[105] Human Fertilisation and Embryology: A Framework for Legislation Cm 259.

[106] R v. Secretary of State for Health, ex parte Bruno Quintavalle (on behalf of Pro-life Alliance) [2001], EWHC Admin 918, para. 62.

[107] See R (Quintavalle) v. Secretary of State for Health [2002] EWCA Civ 29.

[108] Ibid., paras. 28-35.

[109] Ibid., paras 36-42.

[110] For example, (1) s.3(4), which limits research to the time preceding the appearance of the prim-itive streak with reference to the time of the mixing of gametes cannot operate in respect of a CNR embryo in respect of gamete mixing, but is still relevant in respect of the appearance of the primitive streak; and (2) lack of provision in the Act for the donor of the nucleus in CNR to consent.

[111] See R (Quintavalle) v. Secretary of State for Health [2002] EWCA Civ 29, paras. 40, 41, 44-48. The Master of the Rolls disposed peremptorily of the alternative case advanced by the Pro-Life Alliance, by holding that there is no basis to treat an unfertilised egg prior to the insertion of a nucleus during a CNR process as an embryo (see note 19 supra). Consequently, the HFEA is perfectly at liberty to license CNR procedures, subject, of course, to the prohibition against placing any embryo so created in a woman (see Human Reproductive Cloning Act 2001 c.23).

the appeal, the House of Lords unanimously upheld the decision and, in essence, the reasoning of the Court of Appeal.[112]

3.3.2
The Broad Position of UK Regulation

The UK regulatory regime is neither a "prohibitive" one (which would not permit use of human embryos for any purposes other than their own benefit), nor a "liberal" one (which would permit use of human embryos for any legitimate purposes, subject only to the consent of their parents). Instead, it permits the use of embryos for the benefit of others alone, but makes this subject to specified purposes (benefits of others), the stage of development of the embryo, and regulatory scrutiny, in addition to parental consent. Insofar as this also characterises regulation in some other EU countries,[113] it is fair to say, however, that the UK's regime of what we will call "regulatory scrutiny" is, in general, less restrictive.

Accounting for the UK position is not an entirely straightforward matter. Granted, the UK position is in line with philosophical theories that accord a proportional status to the embryo, and this is not merely a coincidence, because ideas that the embryo as a potential born human being has a degree of moral status and that the embryo (and later the foetus) acquires greater moral status as it develops, figured quite strongly in the thinking of the Warnock Committee and are stated in its report.[114] However, it does not follow that the legislative position simply indicates that a majority of Parliament accepted proportionality thinking, let alone that if so Parliament was reflecting the general cultural opinion. The history of the Act (see Appendix 1) suggests a more complicated picture (though an at least widespread acceptance of proportionality thinking undoubtedly played a significant part).

In theory, the middle ground represented by regulatory scrutiny might simply be a consequence of the dynamics of power. For example, suppose that none of the three broad views on the moral status of the embryo (pro-life; proportionality; libertarian) has majority support, but a majority can be formed by an alliance between the proportionality camp and either of the other two. In such a case, we might expect regulatory scrutiny to be the outcome simply because the liberal camp is likely to be able to compromise in the direction of regulatory scrutiny, even if the pro-life camp is unwilling to compromise its ideals. While the libertarian camp might very much prefer a liberal regime, restricting embryo research in

---

[112] R (Quintavalle) v Secretary of State for Health [2003] UKHL 13.

[113] See the previous section.

[114] See Warnock (1984) paragraph 11.17 where it is stated "Although ... the law provides a measure of protection for the human embryo in vivo it is clear that the human embryo ... is not under the present law in the UK accorded the same status as a living child or an adult, nor do we wish it to be accorded that same status. Nevertheless we are agreed that the embryo of the human species ought to have a special status and that no one should undertake research on human embryos the purposes of which could be achieved by the use of other animals or in some other way." A similar gradualist view on the attainment and possession of moral status is evidenced in the Kennedy Report on Xenotransplantation where it concludes that though it is ethically unacceptable to use primates as a source of organs due to their close affinities with humans, it is ethically acceptable to use pigs. See Department of Health (1997 p.65).

the kind of way that the proportionality camp would want might be viewed as better (especially if the libertarian position is combined with utilitarianism, which is quite likely to be the case in the UK) than allowing an alliance between the pro-life and proportionality camps. On the other hand, the pro-life camp might very well consider that no compromise is possible – especially if this stance is rooted in religious belief, and there are strong indications that the pro-life camp, at least as an active political force in the UK, is strongly rooted in the Roman Catholic Church. Indeed, given that far less than 10% of the UK population are committed Roman Catholics,[115] and there is little evidence to suggest that other religious groups have anything like the same commitment to a strict pro-life position[116] what might, at first sight, seem surprising is how close run a thing the victory of regulatory scrutiny over prohibition has been at various points in the legislative and judicial history of the Act.[117] However, there are a number of factors that might account for this. For example, the pro-life position might be more prevalent (at least among Members of Parliament, including the House of Lords, and/or the judiciary) than is recognised. Related to this, Roman Catholics might be more heavily represented in key legislative and judicial positions than in the country as a whole. However, even if neither of these is the case, a disproportionate influence of the pro-life position might be due to those who adopt this position being more highly motivated or commanding more resources to participate actively in the legal arena (or both) where embryo research is concerned than those who adopt the alternative positions. Given the link between the pro-life position and Roman Catholicism in the UK, the resources of the Roman Catholic Church, and the willingness of the Roman Catholic Church to intervene in and support litigation in matters touching on the moral status of not only the unborn, but whenever questions of the right to

---

[115] 10% is the estimate given by the United States State Department of the percentage of the population that "identifies" with the Roman Catholic Church and 45% the estimate of those who identify with the Anglican Church. However as it estimates that only 8.7% of the population attends a Christian church on a regular basis, then patently the percentage of committed Catholics relative to the population must be significantly less than 10% unless Catholics are the only committed church goers. The report notes that official statistics of religious beliefs and church membership are not kept (which itself may be an indication of the state of the religious landscape). See United Kingdom: International Religious Freedom Report, Bureau of Democracy, Human Rights, and Labour, 26 October, 2001, available at http://www.state.gov/g/drl/rls/irf/2001/5716.htm.

[116] 1) Catholic/Orthodox Christians/Some Protestants-Full moral status commences at conception. Some Protestants argue that the early embryo should be given the benefit of any doubt. 2) Other Protestants, Jews, Muslims – The moral status of the formed embryo is greater than that of the unformed embryo, as "ensoulment" occurs in the formed embryo. See HLSCC, (opcit.fn.84) Appendix 4 and Sophie Boukhari, UNESCO Courier journalist, "Religion, Genetics and the Embryo" available at http://www.unesco.org/courier/1999_09/uk/dossier/ txt04.htm. Apart from the Catholic position the others stated are what are thought to be the views held by a substantial number of the particular group. The Catholic Church is unique in having one central arbiter of doctrine, the Pope. The other faiths each do not have one position held by all adherents.

[117] See, for example, the near passage of the Unborn Children (Protection) Bill which would have prohibited embryo research (discussed overleaf) and the "success" of the Pro-life Alliance in the High Court in the matter of R v. Secretary of State for Health, ex parte Bruno Quintavalle (on behalf of Pro-life Alliance) [2001], EWHC Admin 918, though this victory was subsequently reversed in the Court of Appeal (confirmed by the House of Lords). See R (Quintavalle) v. Secretary of State for Health [2002] EWCA Civ 29, and [2003] UKHL 13.

life arise,[118] there can be little doubt that such factors are significant in understanding the UK situation.

Related to this, the processes by which legislation is enacted in the UK can have an effect on what is enacted. For example, political parties can penalise their members for not taking the party line, through the application is what is known as the "party whip". Also, legislation that is promulgated near the end of a parliament is particularly vulnerable to being frustrated, because if it is not enacted before Parliament is disbanded, the whole process has to be gone through again, even if the Bill has popular support. The pressures of parliamentary business and political manoeuvring can mean that if the Bill is not enacted, it will be abandoned altogether. This was precisely the fate of the Unborn Children (Protection) Bill which was piloted through successive parliaments by Enoch Powell M.P. in 1984–1985 and Ken Hargreaves M.P.[119] in 1985–1986. (Lee & Morgan 2001, 57).[120] This Bill aimed *inter alia* to prohibit all embryo research. On first reading, the Powell Bill was passed by 238 to 66 (ibid.) and was only defeated on June 7, 1985, because of intense lobbying and parliamentary manoeuvring that caused it to run out of time.[121] The Bill was re-introduced by Hargreaves on October 21, 1986 within a fortnight of the end of the Parliamentary Session, and could therefore not practically have been passed in that session. The purpose of its introduction was to show that there was popular support for such a Bill and thereby convince the Government to introduce a similar Bill in the next session. At its first reading the Bill passed by 229 to 129 votes.[122] However, the Government, being in favour of embryo research, never introduced a prohibiting Bill (Kirejczyk 1999 891–892).

It is also possible on the other hand, particularly with private members Bills, to slip legislation through opportunistically by having it voted on late at night (as happened with s.156 of the Criminal Justice and Public Order Act 1994, which amended the HFE Act by incorporating a new section 3A prohibiting the use of germ cells derived from an embryo or foetus in fertility treatments).[123] Therefore,

---

[118] See, for example, Airedale NHS Trust v. Bland [1993] 1 ALL E R 821, *Re A* (Children: conjoined twins surgical separation) [2000] 4 ALL E R 961, The Queen on the Application of Dianne Pretty v. Director of Public Prosecutions and Secretary of State for the Home Department [2001] UKHL 61 as well as R (Quintavalle) v. Secretary of State for Health [2002] EWCA Civ 29.

[119] Both members of the Conservative Party which formed the government of the day.

[120] see also Repromed (Centre for Reproductive Medicine, University of Bristol) website at http://www.repromed.org.uk/history/20th_uk.htm.

[121] See the Repromed (Centre for Reproductive Medicine, University of Bristol) website at http://www.repromed.org.uk/history/20th_uk.htm and the Progress Educational Trust website at http://www.progress.org.uk/About/history.html.

[122] See Repromed (Centre for Reproductive Medicine, University of Bristol) website at http://www.repromed.org.uk/history/20th_uk.htm.

[123] The sponsor of this clause was Dame Jill Knight M.P. The members of the house who spoke on the issue were generally in favour of the ban but many, for instance, Secretary of State Bottomley M.P. and David Blunkett M.P. were concerned that the clause was being added before the House had the benefit of the completion of the HFEA consultation on the issue. Dame Knight declined to withdraw the clause and it was added to the Bill shortly before 12:45 a.m. on the 12th April 1994. See the UK Parliament website at http://www.parliament.the-stationery-office.co.uk/pa/cm/cmvo246.htm. The HFEA (1994) consultation ended and its report was published in July 1994. The report recommended that foetal ovarian tissue should not be used in infertility treatment, but could be used for embryo research subject to the need to obtain informed consent prior to the woman's death or undergoing abortion.

though it is customary for Members of Parliament to be allowed "a free vote" (i.e., not to be "whipped") on matters of "moral conscience", these are some of the factors which might have skewed the eventual outcome of the regulatory scrutiny position.

There is another possibility, however, and this is that regulatory scrutiny is a function of rational moral bargaining. As David Gauthier (1986) has brilliantly analysed, it is only rational for persons to enter into a compact that will bind them to the outcome of collective bargaining if they will not be left worse off by doing so than by remaining outside the bargaining arrangement. This can apply where the costs/benefits involved are moral values, provided only that any compromise arrived at does not involve the sacrifice of an absolute value and there is something of positive moral value to be gained (all things considered) by supporting the compromise.[124] With this in mind, a proponent of the pro-life position might argue that it is better to subject embryo research to regulatory scrutiny than not to regulate it at all, not on the grounds that it is better to restrict embryo research (as involving the killing of embryos) in at least some situations than in no situations (for that presupposes the proportionality view), but on the grounds that a position of regulatory oversight involves active recognition of the inherent value of having decisions made in a transparent and accountable manner (because it reflects the values of autonomy and reason giving as between individuals or groups acting in good faith who happen to fundamentally disagree on a moral issue).[125] In addition, a regime of regulatory scrutiny keeps embryo research in the public arena. As such, it facilitates ongoing debate about the matter, which is important, because, in a democratic society, it is always possible to review and alter previous decisions, and if the activities that the pro-life proponent considers morally abhorrent are kept in the public gaze then others might more readily come to see that embryo research is wrong. This point will be taken up again below (see Chapter 5).

Now, while we have no evidence to suggest that such thinking played a role in the victory of the regulatory scrutiny position in the UK, it is surely not impossible that it did.[126]

Alternatively, regulatory scrutiny might simply be a function of the pragmatic style – that it is almost trite to say – characterises the conduct of British public life (including its common law legal system). By a pragmatic style, we mean a commitment to look at matters very much on a case by case basis, weighing together all relevant factors (rather than one that works from first principle) which, by its nature, is likely to gravitate to the middle-ground whenever required to produce a general stance. Such a style, which might simply be a function of "a British character" to seek the middle ground in all things, can be linked to utilitarianism (though, whether as cause or effect we cannot say), which, as we have noted already is very much part of the UK's philosophical-cultural landscape.

---

[124] See the Chapter 5.

[125] Some commentators have argued that the law is the vehicle used to engineer democratic compromise in pluralistic societies. See for example Clothier (1986, 1–2) and Charlesworth (1990, 149).

[126] It may, in any event, explain Margot Brazier's (1999, 187–188) view that having lost the battle to absolutely prohibit embryo experimentation persons opposed to any embryo experimentation (as she is) should approve of attempts to agree some intermediate status for embryos and controls on research.

Two other factors that should also be taken into account are (1) that the Warnock Committee, whose proposals led to the HFE Act was chaired by a highly respected moral philosopher, Mary Warnock (now Baroness Warnock), who favoured the proportionality view; and (2) that the fact that the HFE Act was preceded by a "trial period" in which there was the Interim Licensing Authority during which nothing disastrous appeared to happen might have assuaged some persons fears about the "knock-on" or "slippery-slope" consequences of permitting research on embryos at all.[127]

Finally, it is possible when dealing with moral issues that some persons, because they find such issues so difficult, or simply because of their felt ignorance, will opt to follow the lead of experienced persons whom they respect. Thus, it is important to note that the position taken by Baroness Warnock appears to have been crucial in the passing of the Regulations on Embryo Research.[128]

3.4
Case Study 2: Regulation in Spain

3.4.1
Main Regulations on Embryo Research

Embryo research in Spain is essentially regulated by the Assisted Reproduction Techniques Act (Law 35/1988 sobre Técnicas de Reproducción Asistida) of 1988.[129] It only permits certain research activities on human embryos.[130] Further regulations are to be found in the law regulating the donation and use of human embryos and foetuses or the cells, tissues or organs therefrom (Law 42/1988 of 28 December, de donacíon y utilización de embriones y fetos humanos o de sus celulas, tejidos u órganos).[131] These two are complemented by Chapter V of the second part of the Penal Code entitled „Offences related to Genetic Manipulation".[132]

The Law 35/1988 uses the term pre-embryo (preembrión) without giving a legal definition about it, the other term also used is pre-implantation embryo (embrión preimplantatorio) to describe a group of cells resulting from progressive division of the fertilised ovum until approximately fourteen days later. The term „embryo" is used for the period following implantation when the organogenesis takes place,[133] a

---

[127] We are grateful to Prof. Sheila McLean and Baroness Warnock, respectively, for suggesting these points.

[128] The Human Fertilisation and Embryology (Research Purposes) Regulations 2001, SI 2001 No.188.

[129] Boletín Oficial del Estado (BOE, State Gazette) No. 282 of 24 November 1988, correction of errors in BOE No. 284 of 26 November 1988.

[130] The following text bases on an inofficial translation provided by Professor Dr. Carlos Maria Romeo Casabona.

[131] BOE No. 314 of 31. December 1988. Only non-viable or dead embryos can be made as a donation or be used, art. 2 e; abortions to obtain embryos for donation or use are not permitted, art. 3.2.

[132] See also Lema Añón 2000a, p. 47, Romeo Casabona 2002, p. 331.

[133] The Statement of Purpose of the Law 42/1988 refers to the moment when they are implanted stable in the uterus. See Revista de Derecho y Genoma Humano (Law and the Human Genome Review), No. 8, 1998, p. 119, 120.

time of about two and a half months (postimplantation embryo – embrión posim-
plantatorio). Here a definition is given in Statement of Law 42/1988: Embryos and
foetuses are considered as such from the moment when they are implanted stable in
the uterus, establishing a direct, dependent and vital relation to the pregnant
woman.[134] The First Final Provision of the Law states that donation and use of
developing fertilised ova up to the fourteenth day after fertilisation are regulated by
the Law 35/1988. This stipulation concurs with the criteria to keep the pre-embryo
not longer than fourteen days, art. 15.1 b Ley 35/1988.[135] Besides the lack of a more
or less descriptive definition of what the embryo is, Spanish law does not even provide a definition of its legal status. Therefore information about the real status of an
embryo derives from different regulations, mostly Law No. 35, and jurisdiction on
this matter and can be obtained while focussing on these.[136]

The (postimplantation) embryo is protected by a special law, regulating the donation and use of human embryos and foetuses or the cells, tissues or organs therefrom (Law 42/1988 of 28 December, de donacíon y utilización de embriones y fetos
humanos o de sus celulas, tejidos u órganos).[137] That law was necessary to fill the
legal gap which existed, because the Transplantation Act (Law 79/1979 of 27 October, de extracción y trasplante de órganos ...) was not applicable to embryos and
foetuses.[138]

For practical reasons and to avoid reiteration the Donation and Use of Human
Embryos and Foetus Act does not refer to gametes, in-vitro fertilised ova and preimplantation embryos, because they are subject to the regulation of the Law
35/1988.[139] Therefore, in Spain, for the field in which embryo researchers are
interested, there is actually only one law that gives the legal framework for all
activities.[140]

Generally research or experimentation, whether of diagnostic or general character on live pre-embryos shall only be authorised if it meets the following requirements:

– The concerned parties and, if such be the case including the donors, must have
  given their written consent after having received a detailed explanation of the aims
  of the research and its implications. The pre-embryos are not permitted to be developed beyond fourteen days after the fertilisation of the ovum (deducting the time of
  cryopreservation). Furthermore the research has to be carried out in Health Centres and by legally recognized and authorised teams, art. 15.1 c Law 35/1988.

---

[134] See Romeo Casabona (ed.), Codigo de Leyes sobre Genetica, § 1.2, p. 48.
[135] See Romeo Casabona (ed.), Codigo de Leyes sobre Genetica, § 1.1 II, p. 23.
[136] Lema Añón 2000a, p. 50 sees that the law regulates the legal status of the embryo whereas
Casabona, 1996, p. 361 states that there is a need for a new legal status of the pre-embryo, also
Javier Gafo Fernández, Second Report of CNRHA, Dissenting vote, p. 50.
[137] BOE No. 314, of 31. December 1988. Only non-viable or dead embryos can be made as a
donation or be used, art. 2 e; abortions to obtain embryos for donation or use are not permitted,
art. 3. 2.
[138] The first Final Provision Primera of the Law 42/1988; see also Romeo Casabona (ed.), Codigo
de Leyes sobre Genetica, § 1.2, p. 4; Spanish Constitutional Court No. 212/1996 of 19 December 1996 (English version); Revista de Derecho y Genoma Humano (Law and the Human
Genome Review), No. 8, 1998; p. 119, 121.
[139] See Romeo Casabona (ed.), Codigo de Leyes sobre Genetica, § 1.2, p. 48.
[140] See Romeo Casabona (ed.), Codigo de Leyes sobre Genetica, § 1.2, p. 48.

- Research on viable in vitro pre-embryos shall only be authorised for diagnostic, therapeutic or preventive purposes, art. 15.2 a Law 35/1988, and if the non-pathological genetic patrimony is not modified, art. 15.2 b Law 35/1988, that means that the embryo must benefit.[141] This intention of the law is enforced by art. 12.1 which demands that every intervention on the live pre-embryo in vitro with diagnostic purposes shall have no other aim but the assessment of its viability or non-viability or the detection of hereditary diseases, in order to treat them if possible, or to advise against its transfer for purposes of procreation.
- Research on pre-embryos not for verification of viability or diagnosis is only permitted on non-viable pre-embryos after the scientific demonstration that it cannot be carried out in the animal model, art. 15.3 b Law 35/1988. In every case a license by the state authority is needed, art. 15.3 c Law 35/1988. Aborted pre-embryos shall be considered non-viable, art. 17.1 Law 35/1988.
- Fertilisation of human ova is prohibited for other aims than human procreation, art. 3 Law 35/1988, a prohibition which is reinforced by art. 161.1 Penal Code. Anyone who fertilises human ova for other reasons than those of procreation can be punished with prison from one year up to five years and a ban from office or profession from six to ten years. Thus in-vitro fertilisation cannot be a primary source to obtain embryos for research. The second section of this article (161.2) penalizes the creation of human beings by cloning or other procedures directed towards race selection.[142]

3.4.2
The Ruling of the Spanish Constitutional Court of 19. December 1996[143]

In 1989 about 80 members of the Spanish parliament who were members of the Partido Popular (Popular Party, a conservative party) filed a challenge on the grounds of unconstitutionality against Law 42/1988. The claimants considered that this law had breached the mandatory constitutional protection of human life guaranteed by art. 15 of the Spanish Constitution and alleged that the law contravened several of the statutory requirements governing the obligatory use of a law to regulate certain issues as principles of penal legality. The dignity of the human person was also mentioned on various occasions by the appellants.[144]

In its judgement the Spanish Constitutional Court recalled the purpose of Law 42/1988: It should regulate the donation and use of human embryos and foetuses, cells, tissues or organs therefrom, because it was an issue which was not dealt with by the existing law. The court continued by stating that art. 15 of the Spanish Constitution indeed recognises, as a fundamental principle, the right to life of all naturally born persons but it is not extendible to nascituri: "the argument ... cannot be considered to have provided grounds to affirm that the nasciturus has the right to

---

[141] See also summary on the Spanish situation in Romeo Casabona 2001, p. 127.
[142] For a critical explanation of these offences see Romeo Casabona 2002, p. 331.
[143] Sentencia del Tribunal Constitucional (STC-Spanish Constitutional Court Ruling) 212/1996 – Supplement to the BOE No. 019 of the 22 January 1997, p. 00032. B 16. The decision is published in English in: Revista de Derecho y Genoma Humano (Law and the Human Genome Review), No. 8, 1998, p. 119.
[144] Ibid., p. 120, 121.

life."[145] There is no fundamental right, as such, but rather a constitutionally protected interest which derives from the normative content of art. 15 of the Constitution.[146] Therefore the State is obliged to protect the natural process of gestation and to put in place a legal system including penal provisions as the ultimate guarantee for the protection of life.[147] The art. 15 of the Spanish Constitution implies a protection also to nascituri. But the Law 42/1988 deals with a situation in which by definition the human embryos and foetuses cannot be viewed as being nascituri. The term non-viable means, that they will never be born in the sense, that they will never have a life independent from their mother. And the Court concludes: "...no nasciturus is involved."[148]

The criticism that the law entailed viewing human beings, irrespective of their degree of development, as elements of property, was considered by the Court to be unfounded, because the Law expressly prohibits in art. 2 d: "Donation and subsequent use shall never be for commercial purposes." Therefore the Constitutional Court dismissed the challenge in all aspects mentioned above.[149]

3.4.3
The Spanish Constitutional Court Ruling of 17. June 1999[150]

Also in 1989, 73 members of the Spanish Parliament and members of the Popular Party, as well, lodged a challenge on the grounds of unconstitutionality against Law 35/1988 on Assisted Reproduction Techniques, which the party considered to be unconstitutional.[151] The first reproach to the law of being unconstitutional was made on the statement of purposes, when it stated that „the moment of implantation is necessarily of biological valuation, before it, the development of the embryo is moving in uncertainty, and with it, the gestation begins and biological reality inside the embryo can be proved"[152]. The Court referred to its own constant doctrine, that the statement of purpose has no normative scope and hence, it cannot be object of a challenge on the grounds of unconstitutionality.[153]

Concerning the violation of the right to life, art. 15 of the Spanish Constitution, alleged to be possible by the criticised law, the court revised its ruling STC 212/1996 that the legal protection does not cover the degree of development of the pre-embryo.

---

[145] STC 53/1985, Point of law No. 7.
[146] This argument was made in the ruling 53/1985 concerning a challenge on the grounds of unconstitutionality of a law modifying the regulations concerning abortion of the Penal Code. Here one can see the strong relation between legislation on abortion and embryo research.
[147] That means for instance, that the value of life and integrity of the embryos demand the reduction of the number of supernumerary embryos as a criteria of in-vitro fertilisation, see Lema Añón 2000b, p. 106.
[148] Revista de Derecho y Genoma Humano (Law and the Human Genome Review), No. 8, 1998, p. 119, 124: Aborted embryos means that all expectations of viability have been frustrated; an aborted embryo merely a cellular structure with no possibility of subsequent development.
[149] Ibid., p. 133.
[150] STC 116/1999.
[151] Revista de Derecho y Genoma Humano (Law and the Human Genome Review), No. 11, 1999, p. 97.
[152] See Romeo Casabona (ed.), Codigo de Leyes sobre Genetica, § 1.1 II, p. 24.
[153] Revista de Derecho y Genoma Humano (Law and the Human Genome Review), No. 11, 1999, p. 98.

The court reads article 15, which concerns research and experimentation on embryos, as follows: It is evident that the law in no case permits experiments on viable pre-embryos just as little it permits more investigation on this than for diagnostic, therapeutic or preventive reasons.[154] Other treatment on embryos the law only permits, if it is done to non-viable pre-embryos, that means (here the court is citing its ruling 212/96 concerning the donation and the use of embryos and foetuses) that it is incapable of developing to become a human being, a "person" in the fundamental sense of art. 10.1 of the Constitution.[155] This distinction between the time before the implementation and thereafter is strictly followed by the court. If the applicants alleged that art. 2.4 of the Law would permit the woman to terminate the pregnancy whenever she wanted, because she shall be able to request to have their implementation suspended at *any moment* during their (techniques of assisted reproduction) implementation, it is just a misinterpretation of the regulation. The Court does not see any indication in the law and the relating materials that constitute a further permission for abortion.[156]

In this sense the court sees no violation of the constitution, if the law permits in art. 12.1 after the detection of a hereditary decease which cannot be treated, to advise against the transfer of the pre-embryo, because the legal protection guaranteed by the constitution is less for the pre-embryo than for the embryo in uterus.[157]

The concept of cryopreservation was alleged to be unconstitutional too. The court argued that at first the limits of technical standards are the cause for surplus embryos. But to cryopreserve them avoids further fertilisations. Therefore this preservation is not an attack on human dignity but on the contrary a method of a better use of already existing embryos.[158] In consequence of these considerations the Spanish Constitutional Court dismissed the challenge on the grounds of unconstitutionality concerning the above mentioned aspects.[159]

3.4.4
The National Committee on Human Assisted Reproduction (Comisión Nacional de Reproducción Humana Asistida – CNRHA)

The Law 35/1988 obliged the Spanish Government to establish by Royal Decree a permanent National Committee which should offer guidance in the use of reproduction techniques, art. 21.1. This Committee has to collaborate with the administration with respect to the collection and updating of scientific and technological knowledge, establishing criteria for the running of the Centres or services which use these techniques and to improve them. The committee shall have delegated functions and be able to authorise scientific, diagnostic and therapeutic projects of experimentation or research, art. 21.2. The Committee in the first report made it clear that it has no executive powers.

---

[154] Ibid., p. 103.
[155] Ibid., p. 103, 104.
[156] Ibid., p. 104.
[157] Ibid., p. 106.
[158] Ibid., p. 105.
[159] Ibid., p. 113.

The committee was appointed on 21 March, 1997 after the Royal Decree[160] was released and constituted on 11 November, 1997, nine years after the law had come into force.[161] The members are representatives of the government, the administration and also of enterprises involved with human fertility and related techniques, furthermore there has to be a council, representing a broad social spectrum.

3.4.4.1
The First Report of the CNRHA[162]

The Committee's first report, published in 1999, which together with the ruling of the Constitutional Court is said to be marking a change in the development of legislation that had become inactive, is nevertheless fixing the limitations of its examinations: It is only the first report, which only contains the most important questions (Lema Añón 2000a, p. 48).

In No. 9 of the report, the Committee explains that there is an increasing number of cryopreserved embryos of about 25.000 in the year of the report. 15% of these are already beyond the time of permitted storage of five years. The law does not say what is to be done with these embryos after this time has passed.[163] The Committee gives only one recommendation: To establish a deadline for storage of embryos relating to the average period of reproduction of the mother, which corresponds to a period of about 50 years.

Being aware that an increased time of conservation produces more stored embryos, the committee mentions two solutions: destruction or use for research. But the committee made clear that the second solution was not sufficiently discussed by the members of CNRHA and therefore reserved its opinion for later after further analysis of the problem.[164]

After the discussion of different methods of cloning the committee considers that the cloning of human beings by transferring the nucleus is not recommendable.

After reporting that there are different opinions among the different members concerning the status of the embryo and the beginning of human life, the committee gives preference to the use of adult stem cells rather than to embryonic stem cells.

Finally a concrete recommendation is made to the legislature: Article 161.2 of the Penal code shall be modified to prevent doubtful interpretations which can suggest that cloning should be prohibited only in cases of race selection.[165]

---

[160] RD 415/97.

[161] The Law 42/1988 also provided for an adequate committee in the First Additional Provision lit. f.

[162] A Spanish summary is published in: First Report of CNRHA, Revista de Derecho y Genoma Humano (Law and the Human Genome Review), No. 10, 1999, p. 246. A list of all committee members is to be found in this summary.

[163] Ibid., p. 248; in Catalonia the number of cryopreserved embryos increased by 13% between 1995 and 1998, those of which have passed the time limit of five years increased by 100% at the same time: see Lema Añón, 2000b p. 109.

[164] First Report of CNRHA, Revista de Derecho y Genoma Humano (Law and the Human Genome Review), No. 10, 1999, p. 246, 249.

[165] Ibid., p. 250.

3.4.4.2
The Second Report of CNRHA[166]

The second report of the committee was dedicated exclusively to the problem of research on surplus embryos. In this report the committee recommends the use of supernumerary embryos for research.[167] This use is preferred as an alternative to the destruction of cryopreserved embryos at the end of the period of cryopreservation. The Committee demands the informed consent and an institutional control, as already well developed in other fields, and the establishment of guarantees of protection according to the Convention on Human Rights and Biomedicine of Oviedo in 1997. The committee advises therefore to modify the regulations especially Law 35/1988.

One of the dissenting votes argued that surplus embryos have to be considered non-viable because at the moment of the decision to use them for research the only alternative would be destruction. The term viability used by the law is not a pure biological one. Furthermore the law itself considers in art. 17.1 aborted pre-embryos as dead or not viable.[168]

The three other dissenting votes were opposed to the use of surplus embryos for research, focussing several aspects of the problem: One vote would prefer the destruction of embryos rather than using them as a means, because it could be the "lesser evil", whereas keeping them alive without a real chance to continue its development would be a "painful possibility".[169] The second dissenting vote argued that permission to use surplus embryos could provoke an increasing number of surplus embryos, therefore the recommendation failed the aim as set in the first report to reduce the number of surplus embryos.[170] The third dissenting vote proposes implementing surplus embryos in other women rather than the biological mother and using adult stem cells for research.[171]

3.4.5
Practical Aspects and Problems

Like in Germany, or better more like in the UK, there was almost no case brought to the courts except one in which debate took place on the selection of sex of the descendants in 1990 (Lema Añón 2000a, p. 47, 51).[172] On the other hand, there are two very important decisions of the Constitutional Court, the later one, which converted the provisional legal situation – the danger that the law would be declared unconstitutional – disappeared ten years after entering into force (Lema Añón 2000a, p. 57).

---

[166] Second Report of CNRHA Madrid, May 2000.
[167] Second Report of CNRHA, p. 38. The report contains four dissenting votes of members of the committee, one concerning the way of recommending research on surplus pre-embryos, three voted against the use itself.
[168] Vote of Manuel Atienza and Antonio Garciá Paredes, Second Report of CNRHA, p. 46.
[169] Vote of Javier Gafo Fernández, Second Report of CNRHA, p. 51.
[170] Vote of Conzalo Herranz, Second Report of CNRHA, p. 52, 53.
[171] Vote of Maria Dolores Vilacoro, Second Report of CNRHA, p. 55, 56.
[172] For further information about this case see footnote 5 of that source.

The most important problem is considered to be the increasing number of cryopreserved embryos, which have no perspective for the future. The CNRHA, now in its second report, recommended the possibility of using surplus embryos for research. The problem is that the Spanish legislation only permits research on viable embryos for diagnostic, therapeutic or preventive reasons, art. 15.2 Law 35/1988. Further research is only permitted on non-viable embryos but the cryopreserved embryos are viable. A change of legislation is considered to be difficult remembering the content of the ruling of the Constitutional Court No. 116/1999 on assisted procreation: The prohibition of research on pre-embryos meets the requirement of protection made by the constitution (Lema Añón 2000b, p. 113).

To complete the overview on Spanish regulation it has to be mentioned how the country is representing its position in the international discussion. As far as the cloning of human beings is concerned, the General Director of the Spanish Institute of Health Carlos III. has stated at the United Nations General Assembly's Ad Hoc Committee on an International Convention against the Reproductive Cloning of Human Beings, that the Spanish Government considers that all human cloning, including the so-called therapeutic cloning, is contrary to human dignity. Therefore the Spanish Government supported with enthusiasm the French-German proposals for a convention against reproductive cloning by extending the object thereof, that means a convention against the so-called therapeutic cloning too.[173] This position was reiterated in September 2002 when the Spanish delegation declared that, taking into account the success of regenerative medicine, only adult stem cells should be used for research and all human cloning shall be banned.[174]

3.5
Case Study 3: Regulation in Germany

3.5.1
Protection of Embryos Act and Stem Cell Act

The primary set of regulations governing the treatment of human embryos and therefore also the research on and with embryos in the Federal Republic of Germany is the ESchG (Embryonenschutzgesetz – Embryo Protection Act)[175] from 13 December 1990, which came into force on 1 January 1991 and has not yet been changed. Its title already clearly expresses the intention of the legislature having explicitly denoted the law as a protective law. In addition, the structural approach within the law is unequivocal: Section 1, paragraph 1 of the ESchG carries a potential penalty of up to three years' imprisonment, thereby demonstrating the overall protectionist attitude of the legislature, supplemented by criminal law as the most severe means at the disposal of the state. Thus, the leading commentary on the ESchG points out: "The Protection of Embryos Act is a criminal law."[176] In consequence of the compe-

---

[173] Statement by the Director General, New York, 26 February 2002, p. 3.

[174] See Statement by the Director General, New York, 23 September 2002, p. 3 & 4; also Lilie 2003, p. 729, 730.

[175] Bundesgesetzblatt (Federal Law Gazette) I, p. 2746.

[176] Keller/Günther/Kaiser, Embryonenschutzgesetz Kommentar (ESchG-Kommentar), B II. 1 Rn. 1.

tence of law-making in the federal system, the German Parliament, the Bundestag, could only use the possibility to pass a criminal law, Art. 74 I No. 1 of the constitution, Grundgesetz. Only in 1994 the Constitution was amended by a further competence of the federation, the Bund, Art. 74 I No. 26, to regulate artificial procreation, gene technology, transplantation of organs.[177] It is said that at least a coalition of pro-life groups within the conservatives and the greens pushed the passing of this law.[178]

As well as the ESchG, the "Act to safeguard the protection of embryos in conjunction with the importation and use of human embryonic stem cells" (Stem Cell Act – StZG)[179] stipulates the permitted treatment of human stem cells since 1 July 2002. In this law the term research in connection with embryos appears for the first time. The penal provisions, however, are no less severe even though they are only stipulated in the last section of the law. Admittedly, this law only contains penal provisions in section 13, i.e. almost at the end, and it also carries a maximum sentence of three years' imprisonment. Nevertheless, the StZG, allowing the importation of stem cells, appears to be simply an exceptional regulation in a tangle of rules, giving highest priority to the protection of the embryo as a starting point. Basically, the StZG permits very little. On the whole, the protection of the embryo is not granted without contradictions, as will be shown later on.[180]

This becomes clear when one compares the protection of the embryo in vitro with the regulations governing abortion. Regarding how embryo research is legally stipulated, the provisions of the Criminal Code, sections 218ff governing the termination of pregnancy are relevant.

By means of the StZG – although still a protective law in its terms – the legislature comprehensively takes position on embryo research for any purposes other than their own benefit and even this only applies to components of embryos having been extracted abroad. After all, it is already possible to establish at this point that Germany has regulated embryo research in a restrictive manner. Even the explicit exceptional rule for existing stem cell lines has not brought about any changes in this respect.

3.5.2
Relevant Legislation

By passing the ESchG, only a limited number of the range of problems in view of reproduction has been covered by statute. A comprehensive law governing reproductive medicine and a law on human genetics are yet to be expected.[181] Until the ESchG came into force, embryos that had been created extracorporally, i.e. by in

---

[177] Rosenau, 2002, p. 3.; Neidert 2002, p. 467, 469.

[178] Compare the statements from the main political parties, which also show similarities between the attitudes of Greens and conservatives in this chapter.

[179] Bundesgesetzblatt I, p. 2277.

[180] Robert Goebbels at the European Conference on stem cells: "My criticism of German legislation applies essentially to an aspect that to me seems illogical: the embryo is protected until the time it is aborted." European Commission, Stem Cells Therapies for the Future?; 2002, p. 13.; a criticism that Deutsch 1991 p. 721, 724 already foresaw.

[181] Keller/Günther/Kaiser, ESchG-Kommentar, Vorwort, p. V. For comparison see regulations in Austria, Switzerland and France. Neidert 2002, p. 467, 470 demands such a code but doubts that there will be a sufficient majority in the present parliament.

vitro fertilisation, were not protected by criminal law before implantation.[182] In addition, this protection did not exist until the point at which nidation was completed. As regards the criminal law being in effect when the ESchG came into force, actions before the nidation were not considered as termination of pregnancy pursuant to section 219 d of the Strafgesetzbuch (StGB, Criminal Code).[183] Also the statutory penalties meant to protect life and health did not provide protection for the embryo, as in the case of those meant to protect property.[184] The Genetic Engineering Act of 1990[185] does not contain any regulations on human genetics either.

A decade of legal political discussion had passed before the ESchG was adopted.[186] Positions taken in this discussion are still advocated in similar ways even today. From a legal perspective, it has to be mentioned that the argument centred on the question of whether a law on the treatment of embryos should include criminal sanctions.[187] It was also pointed out, however, that comprehensive embryo protection could result in a contradictory evaluation with regard to the regulation of the termination of pregnancy.[188] In this case, the legislature has opted for a protection of human life from the time of fertilisation, although in its decision in 1975 the Bundesverfassungsgericht (Federal Constitutional Court) did not answer the question of whether at the time of nidation human life within the meaning of Art. 2 para. 2 s. 1 of the Grundgesetz (Federal Constitution) had already emerged. Concerning the termination of pregnancy, this protection has a lower priority as an exception because of the particular conflict of interests purported not to exist in the same way if embryos have been created in vitro. In its second important decision concerning abortion the Bundesverfassungsgericht explicitly confines itself to the time after nidation in affirming the right to protection on the part of the embryo in relation to its mother.[189] The intensive discussion on the StZG began when a German researcher wanted to research stem cell lines and to import them for this purpose.[190] All of a sudden it was realised that the regulations of the ESchG might no longer

---

[182] See also v. Bülow 2001, margin nos. 303–306.

[183] Keller/Günther/Kaiser, ESchG-Kommentar, V. I. 1 margin no. 2f; compare also Jähnke, in: Leipziger Kommentar zum Strafgesetzbuch (Commentary on the Criminal Code), 10th edition, 1989, s. 219 d margin no. 1f; The legislature has introduced this regulation on the termination of pregnancy with the same wording into the Penal Code, s. 218 para. 1 s. 2, currently being in effect, see below on this.

[184] Keller/Günther/Kaiser, ESchG-Kommentar, B.I.2. margin nos. 4-6; In respect of property, the aspect of human dignity has been pointed to particularly for which it would be forbidden to regard the early embryo as a chattel. This is not precise insofar as it would have to be established beforehand who may invoke this dignity at all. Zuck 2002, p. 869 pointed to this in view of the current discussion. It is probably more important that the proprietor – whoever this may be – should not be granted the exclusive power of Provision pursuant to the BGB (Civil Code), s. 903.

[185] The law governing issues of genetic engineering (Genetic Engineering Act – Gentechnikgesetz-GenTG) from 20 June 1990 (BGBl., I, p. 1080) came into force on 1 July 1990.

[186] Keller/Günther/Kaiser, ESchG-Kommentar, B. II, III margin nos. 1–36 provides an overview on commissions having worked on this subject and on legislative initiatives, amongst other things also at Federal State level.

[187] Critic from the Austrian point of view e.g. Schick 1991, p. 15.

[188] Keller/Günther/Kaiser, ESchG-Kommentar, B. IV, 4g, margin no. 24, 26.

[189] See BVerfGE (official collection of the decisions of the Federal Constitutional Court) 39, 1, 37; and BVerfGE 88, 203, 276.

[190] See Frankfurter Allgemeine Zeitung (daily broadsheet) from 28 June 2001, p. 1.

suffice to guarantee comprehensive embryo protection. That is because embryos have to be consumed in order to create stem cells. Pursuant to the ESchG, it is not permitted to create stem cells in Germany because this would mean that embryos would be required for purposes other than for their own preservation. Also this discussion clearly demonstrates the prohibitive approach of the German legal position: the main emphasis is not on research, which may conceivably have to be restricted, but rather on the protection of the embryo, which has to be rendered less stringent, which in turn, however, can lead to contradictions. A possible way to circumvent the protection intended by the ESchG was to go abroad. This is because pursuant to sections 2 and 6 of the ESchG, a German researcher collaboratively working on the extraction of stem cells abroad does not incur criminal liability if the action is not punishable there, section 7, para. 2 of the Penal Code.[191] To order stem cell lines abroad does not result in prosecution according to current German criminal law either: Incitement to commit an offence would merely come into consideration if the cells had only been created upon the order of the German researchers. This does not apply regularly, however. On the contrary, the researchers want to resort to cell lines already existing. Criminal liability for aiding and abetting by psychological means, which is possible in conjunction with a principal offence not punishable abroad pursuant to section 9, para 2 s. 2 of the Penal Code, cannot be incurred here either since it would have to be proved that the Germans, by their order, exerted an influence on the production.[192] In addition, the import of embryonic stem cells that were not totipotential had not been punishable before the StZG came into force.[193]

3.5.3
Current Decisions

Court decisions applying the ESchG as a criminal law have not been rendered as yet.[194] In those cases of infringements of the law, the ordinary courts of law would be competent, i.e. those criminal courts being competent in respect of all other infringements of criminal laws. Thus the effectiveness of the ESchG can be seen as extremely comprehensive. It has now been in force for twelve years and so far no infringements of its regulations have come to light, neither has there been the least shadow of suspicion that would have warranted investigations. There are no known statistics regarding unconfirmed cases. On the contrary, when researchers demand that the law should be amended or when they threaten to go abroad, this is a demonstration of conformity to the law. In short: the law is well accepted.

Nevertheless three decisions could be found referring to the EschG as part of the legal system. In one case the Bundesverfassungsgericht dismissed an appeal against the decision of a civil court which had not accepted the claim against private health

---

[191] Lilie/Albrecht 2001, p. 2775; Schroth, 2002, p. 171.

[192] Lilie/Albrecht 2001, p. 2776.

[193] Lilie/Albrecht, ibid., 2776; accordingly Schroth 2002, p. 172, who points out that the importer must not have exerted any influence on the production of the stem cells.

[194] In Great Britain the decision R v. Secretary of State for Health, ex parte Bruno Quintavalle (on behalf of Pro-life Alliance) did also not refer to an actual case as a basis but the argument dealt with the lawfulness of the Act, compare case 1: Regulation in the UK, Challenges to Legislation.

insurance to be reimbursed for expenses spent for PGD.[195] Another case concerned the claim against an insurance company which refused to pay for an embryo transfer of an embryo created by using the egg of a woman other than the mother. The Bundessozialgericht denied the claim, arguing that an insurance company cannot be obliged to pay for medical treatment which is prohibited by the ESchG, section 1 para. 1 no. 2 (Prohibition of embryo donation).[196] In a third case a couple wanted to reduce its tax debt by that amount of money it had spent on surrogate motherhood in the USA. The Munich Tribunal dealing with tax decided that it is not possible to reduce the tax by acting against German laws, in this case section 1 para. 1 no. 2 & 7 ESchG, even if it is legal abroad.[197]

3.5.3.1
Statutory Definition of the Embryo and Prohibition of Cloning

The ESchG defines the embryo in section 8. This states:

> The fertilised human egg cell capable of development is regarded as an embryo from the time of conjugation; furthermore, this applies to every totipotential cell extracted from an embryo which may divide and develop into an individual human being once the necessary additional conditions have been met.

The prohibition of cloning human beings has been formulated by the legislature as follows:

> A term of imprisonment of up to five years shall be imposed on anyone who artificially causes a human embryo to emerge using genetic information identical to that of a different embryo, a foetus, a human being or a deceased person. (section 6, para. 1 EschG)

3.5.3.2
Criticism of the Statutory Definition of the Embryo and of the Prohibition of Cloning

By defining the embryo, the law lays down normatively – as is shown by the term "is regarded" – that the protection by criminal law already begins at the early stage of the development of the embryo.[198] This definition has come in for criticism in view of the transfer of the nucleus of a somatic cell into a denucleated cell.[199] By this process, a totipotential cell emerges which has almost identical genetic information compared with that of the transferred somatic cell. However, section 6 of the ESchG imposes a prohibition of cloning only referring to an embryo with *identical* genetic information. The transfer of the cell only causes an embryo with *almost identical* genetic information to emerge as the rest of the genetic information originates from the so-called mitochondrial material of the carrier egg cell (see Chapter 1 and 2). Therefore it has to be asked if an embryo with almost identical genetic

---

[195] BVerfG: File-No: 1 BvR 1764/01.
[196] BSG, File-No: B 1 KR 33/00 R, published in NJW 2002, p. 1517.
[197] Finanzgericht München, File-No 16 V 5568/99.
[198] Keller/Günther/Kaiser, ESchG-Kommentar, s. 8, margin nos. 1, 4.
[199] Gutmann 2001, p. 355.

information may be considered as an embryo with identical genetic information. Article 103 para. 2 of the GG (Grundgesetz, Federal Constitution) and section 1 of the Penal Code prohibit any analogous conclusion to the detriment of the person to be charged. According to one view, this prohibition is not violated because the law would not require identical genetic information but merely almost identical.[200] The opposing school of thought, according to which "identical" is not "similar", regards this as sophistic.[201] Especially, because it would remain unclear when almost identical information would become not identical anymore. Thus, the provision would become too uncertain. This dispute would doubtless be irrelevant when using optimised genetic material, because in this case, a being would be created with enhanced genetic substance and section 6 ESchG would not be applicable.[202]

Regarding the element of the offence "causing the emergence of an embryo" in section 6 ESchG, the definition of the embryo in section 8 ESchG becomes difficult. Embryo within the meaning of the ESchG is the human egg cell already fertilised and capable of development from the time of conjugation. When a somatic cell is transferred into a denucleated human cell, no nuclei fuse together. If one understood "already" as "also",[203] an embryo within the meaning of section 6 ESchG would have emerged. But according to one view, it would make more sense to construe "already" as temporal.[204] Consequently, this would mean for an embryo to be protected by the ESchG it would have to have emerged through fertilisation. Cloning by the so-called "Dolly-technique" would not be punishable pursuant to the German ESchG, critics argue.

Nevertheless, according to general opinion, this genomic copying is viewed as punishable. That is because it is still regarded as cloning, i.e. a procedure prohibited by the ESchG.[205] Gutmann points out that the legislature knew about this second technique of cloning[206], although at the time it had not yet successfully been applied. Gutmann concludes from this that the legislature either wanted to redefine the term "embryo" as soon as fertilisation would become technically feasible or did not want to cover this form of cloning at all.

Another problem is seen to lie in section 6 ESchG. It is said to indirectly impose an obligation to kill[207] because if the implantation of an embryo having been produced illegally is punishable, then such an embryo can only be killed. That is why

---

[200] Accordingly v. Bülow 2001, margin nos. 303–306; Recommendations by the Deutsche Forschungsgemeinschaft (German Research Foundation) on the research on human stem cells from 3 May 2001, available at http://www.dfg.de/aktuell/stellungnahmen/lebenswissenschaften/empfehlungen_stammzellen_03_05_01 html; Hilgendorf 2001, p. 1160.

[201] Schroth 2001, p. 170; Gutmann 2001, p. 353; see also Taupitz 2001, p. 3434.

[202] So Rosenau 2002, p. 5; Hilgendorf 2001, p. 1160; von Bülow 1997, A 721; Neidert 2002, p. 470.

[203] Accordingly a theoretical line of reasons by Schroth 2001, p. 170, 172.

[204] Schroth 2001, p. 172 by referring to Gutmann 2001, p. 353.

[205] BT-Drs. 13/11263, p 13; Rosenau 2002, p. 3; Hilgendorf, 2001, p. 1160.

[206] By referring to: Keller/Günther/Kaiser, ESchG-Kommentar, s. 6 margin no. 2.

[207] Accordingly already Keller/Günther/Kaiser, ESchG-Kommentar, s. 6 margin no. 11; Gutmann 2001, p. 356; Schroth 2001, p. 172, which is said to be absurd as it regards a protective law. One could not protect the embryo by obliging the doctors not to transfer it. However, the legislature attempts this by prohibiting the transfer concerning the cloned embryo. Also Neidert 2002, p. 470 who sees this problem also caused by the prohibition of implantation of an embryo against the will of the woman who previously accepted this treatment.

section 6 ESchG is said to have completely failed to properly establish the elements of a criminal offence and to require revision if a prohibition of cloning is actually intended. Gutmann considers section 6 ESchG as too uncertain for this reason and consequently as unconstitutional, Art. 102 para. 2 GG.[208]

Because the critics argued that cloning by transferring the nucleus of a cell would be permissible pursuant to the current ESchG, a modification of the wording of the prohibition of cloning was proposed: In future, criminal liability should be incurred by anyone who "artificially causes a human embryo to emerge in any other way than through fertilisation by a human sperm".[209]

Beyond this, the report recommended the Federal Government to extend the definition of the objects protected by section 8 ESchG. The totipotential tissue aggregation and the totipotential cells separated from totipotential cell aggregations should also be protected by the ESchG in future.[210] At all events, it can be seen that a prohibitive law, intended to be totally comprehensive and comprising penal norms is very soon likely to become contradictory and ineffective when applied to a matter that, as a result of scientific advance, is subject to relatively rapid change. A far more general criticism denounces the relationship between the protection of embryos by the ESchG and the protection of unborn children as granted by the regulations on the termination of pregnancy. Accordingly, a pregnancy may be terminated without stating the reasons after a mandatory consultation up to the twelfth week, section 218 a para. 1 Penal Code. The law states that the termination does not constitute a punishable offence ("no punishable offence has been committed")[211]. Furthermore, the abortion is not illegal when, in consideration of the present and future living conditions of the pregnant person, it is deemed medically indicated in order to prevent risk to the life of that person or to avert the danger of the grave impairment of her physical or mental health and when the risk cannot be averted in any other way acceptable to the said person, section 218 a para. 2 Penal Code. Furthermore: In the last reform of abortion law in 1995 embryopathic indication was deleted from the Act out of consideration for the organisations of the disabled.[212] Consequently, the provision on the termination of pregnancy laid down in the Code now merely contains a medical-social indication in section 218 a, para. 2 Penal Code.[213] This unequal treatment of human life, however it is defined – absolute pro-

---

[208] Gutmann 2001, p. 356; not clear insofar Taupitz 2001, p. 3434, who declares the transfer of the nucleus illegal because it infringes the prohibition of cloning but himself does not decide whether identical genetic information is involved or not. Rosenau, 2002, p. 5 also regards the reform of this regulation as necessary. In the same way, Schroth, 2001, p. 172, considers § 6 ESchG to be a completely ineffective paragraph and therefore in need of reform.

[209] Report of the Federal Government on the question of whether the Protection of Embryos Act was required to be modified, BT-Drs. 13/263, p. 18; see also Art. 15 of the French draft of an Act on Bioethics, Country Report France, Appendix 4.

[210] Report of the Federal Government on the question if the ESchG was required to be modified, BT-Drs. 13/263, p 23, 24.

[211] Tröndle in: Tröndle/Fischer, Kommentar zum Strafgesetzbuch, 49. edition, s. 218 a, margin no. 3 makes the observation that it is a dogmatic novelty in penal law that cannot be integrated and that it arises from inconsistencies in the constitutional law and creates a mass of contradictions and hardly resolvable problems.

[212] See Tröndle in: Tröndle/Fischer, ibid., s. 218 a, margin no. 9, 9a.

[213] Tröndle in: Tröndle/Fischer, ibid., s. 218 a margin no. 9a, which establishes that in this case voluntary manslaughter is medically indicated and recognised as justified merely for the handicap.

tection in vitro, on the one hand, and accessibility for the pregnant person, on the other hand, leads to extensive legal, political discussion: One of the arguments states: If embryos, according to the will of the pregnant person, may be destroyed prior to birth, there is no argument to support not being able to use less developed embryos, anyway not intended to produce human life, for research purposes and thereby possibly help others. Either one must define the right to terminate pregnancy more strictly or loosen up the ESchG.[214] Thus it is clear that the discussion concerning embryo research is constantly conducted from the point of view of embryo protection and with the inclusion of the aspect of abortion.

3.5.3.3
The Discussion on the Status of the Embryo

The regulations discussed up until this point do not allow conclusions to be drawn on the status of the embryo at all under the German legal system. Section 219, para. 1 Penal Code states that in view of the objective of counselling pregnant women in conflict situations that the woman must be aware that the unborn children have their own right to live also in relation to her at *every* stage of her being pregnant and a termination of pregnancy would come into consideration only in exceptional situations. Section 218, para. 1 s. 2 Penal Code explicitly excludes actions before the embedding of the fertilised egg in the womb from being regarded as termination of pregnancy within the meaning of the Criminal Code. As a consequence, a pregnancy within the meaning of the Criminal Code only starts with nidation. Therefore the embryo does not have any protection until embedding under the provisions of the Criminal Code. For those embryos having emerged from in vitro fertilisation, which are not protected, the ESchG applies, but not in view of the destruction of surplus embryos (feticide in cases of multiple pregnancy and destruction of cryopreserved embryos).

The fundamental problem is already to be found in the constitutional law. The question is whether the protection granted by the ESchG (albeit not without contradictions) is demanded by the Grundgesetz, i.e. the German constitution. It is a matter of dispute whether the embryo is to be accorded human dignity and to what extent it should be protected. [215] There appears to be a growing number of those who see an embryo in vitro as not being endowed with human dignity and thus not possessing the right to protection.[216]

3.5.4
Embryo Research

3.5.4.1 Admissibility of Embryo Research

To sum up what has been established up until now, research on embryos is prohibite Germany pursuant to section 2, para. 1 ESchG, as far as it does not directly serve the individual embryo's preservation. This means the Act prohibits any use of the

---

[214] Faßbender 2001, p. 2753; Dederer 2002, p. 24; see also Schroth, 2001, p. 173; Mildenberger 2002, p. 293.

[215] Instructive Kloepfer 2002, p. 420 with many references. See also Höfling 2001, Merkel 2002.

[216] Taupitz 2002, p. 6; Schlink 2002, p. 19; Heun 2002, p. 523; Ipsen 2001, p. 991.

embryo having become available extracorporally for a different purpose than its own benefit.[217] Thus, the Act also prohibits the use of surplus embryos. However, no rule prescribes what has to happen with these embryos. Therefore it is worth considering assigning these surplus embryos to consuming research.[218]

3.5.4.2
Extraction of Embryos

To sum up the above still further, embryos must not be produced for purposes of research. Beyond that, surplus embryos must not be used for consuming research as it does not serve the individual embryo's preservation, section 2, para. 1 ESchG. Research is supposed to be possible on the impregnated egg cell until the fusion of the pronucleis into the zygote with a diploid chromosome number for the following reason: section 1, para. 2 ESchG prohibits the triggering of such pronucleis phases but their use is not sanctioned by section 2 ESchG.[219] As a consequence, the legislature does not protect the egg cell as much as the embryo.[220]

3.5.4.3
Admission of the Importation of Embryonic Stem Cells

The Stem Cell Act, StZG, followed the compromise draft subsequently to two further drafts that had been more restrictive or liberal respectively. The restrictive draft[221] had demanded the Federal Government to guarantee that the import of stem cells having been extracted from human embryos would not be permitted corresponding to the spirit of the ESchG. The liberal draft[222] planned on permitting the import of embryonic stem cell lines. Subjects should have been unused embryos having emerged from IVF that had been donated out of altruism for research purposes.[223]

The German Bundestag (Federal Parliament) opted for a middle course in a well-publicised vote. In section 1 StZG the legislature states the purpose of the Act. It is supposed to respect and protect human dignity and the right to live, as well as to guarantee the freedom of research.[224] The Act is limited to a narrow field of application; pursuant to section 2 StZG it covers the import and use of embryonic stem cells.[225] The StZG again shows how restrictive the German regulations are. Pur-

---

[217] Keller/Günther/Kaiser, ESchG-Kommentar, s. 2 margin no. 1; Neidert 2002, p. 469.

[218] Taupitz 2002, p. 6; Schlink, 2002, p. 19; Heun 2002, p. 523 does not see constitutional objections; Ipsen 2001, p. 991; Dederer 2002, p. 24.

[219] Keller/Günther/Kaiser, ESchG-Kommentar, s. 2 margin no. 10. It also does not matter if these egg cells were produced contrary to the prohibition laid down in s. 1. Taupitz 2002, p. 3434 submits a summary of what is allowed and what is forbidden when using embryos in the field of research.

[220] This is remarkable in so far as also at this point the ensuing human being is already genetically determined, which is pointed out by Faßbender, 2001, p. 2748.

[221] BT-Drs. 14/8101. Available in German at http://dip.bundestag.de/parfors/parfors.htm.

[222] BT-Drs. 14/8103. Available in German at http://dip.bundestag.de/parfors/parfors.htm.

[223] A commission not mentioned by the draft in detail was supposed to have given prior consent. Artificial insemination merely for research purposes was still intended to remain illegal.

[224] The freedom of research enjoys explicit constitutional protection by the Grundgesetz of the Federal Republic of Germany (Constitution), Art. 5 para. 3, 1.

[225] It follows from the StZG, s. 3 para. 1, that these are embryonic stem cells.

suant to section 4 para. 1 StZG the import and use of embryonic stem cells are prohibited on principle. Exceptions to this prohibition exist under certain procedural and substantial conditions: Embryonic stem cells may be imported and used for research purposes if:

- they were extracted from surplus embryos from in vitro fertilisations in the country of origin before 1 January 2002,
- the persons entitled to disposal under the law of the country of origin have properly consented to the extraction of stem cells,
- no remuneration or benefit in kind has been granted,
- no other regulations, especially those of the ESchG, are violated.[226]

Research projects dealing with embryonic stem cells are only permitted if they

> serve high-ranking research objectives in association with the progress of scientific understanding within the framework of basic research or with the extension of medical knowledge when developing diagnostic, prophylactic or therapeutic techniques to be applied on human beings".[227]

Furthermore, research activities may only be carried out if the questions posed have already been provisionally answered as far as possible by using animal cells or animal embryos and if no equivalent results can be expected from research on anything other than embryonic stem cells.[228]

The import and use of embryonic stem cells have to be approved by the competent authority, section 6, para. 1 StZG. There is a legal entitlement to this approval ("The approval *has* to be granted...")[229] if the conditions set by sections 4 and 5 have been met and if the statement of the Zentrale Ethikkommission für Stammzellforschung (Central Ethics Commission for Research on Stem Cells, in the following short: Ethics Commission) has been presented after consulting the competent authority. The competent authority is determined through a legal regulation of the Bundesgesundheitsministerium (Federal Ministry of Health).[230]

The accumulation of the conditions is interesting at this point: The project has to be ethically sustainable and the statement of the Ethics Commission has to be presented. In the case that the authority does not want to agree with the assessment of the Ethics Commission, it has to provide the reasons for this decision in writing. Pursuant to the Act it is therefore possible for the Ethics Commission to deem a research project ethically unsustainable although the competent authority does not have any objections under ethical aspects. The Ethics Commission could just as well opt for the project but the competent authority could refuse its approval since it deemed the project ethically unsustainable. The legislature has thereby laid down two rules: The Ethics Commission only serves as advisory body. However, its func-

---

[226] s. 4 para. 2, StZG, providing a statutory definition in nos. 1 and 2.
[227] s. 5 no. 1 StZG.
[228] s. 5 no. 2 StZG.
[229] Compare also the statement of reasons for the Act, p. 10; a critical look at this legal construction is to be found in Beckmann 2003, p. 593.
[230] StZG, s. 7; In the draft law the Robert Koch/Paul Ehrlich Institute in Berlin had already been determined as the competent authority. The Regulation was passed on 10 July 2002 and designates the Robert-Koch-Institut as the competent authority. Thus changing the competent authority does not require an amendment of the law by Parliament.

tion is so important that the authority making the eventual decision has to state the reasons for its dissenting opinion.[231]

The StZG also stipulates the composition of the Ethics Commission. It consists of nine members who are appointed by the Federal Government for three years and elect a chairman from their midst. Four of its members come from the specialist fields of theology and ethics and five come from the specialist fields of biology and medicine (section 8 para. 1 and 2 StZG). The members of the Ethics Commission are independent and not subject to instructions (section 8 para. 3 StZG).

This law provides a safe legal basis for the research on certain embryonic stem cells in Germany[232]. It confirms the ESchG by pointing out in section 4 para. 2 no. 4 that the research project must not violate provisions of the ESchG.

The provision of the section 4 para. 3 StZG is remarkable. Approval has to be rejected pursuant to it if the extraction of the stem cells has obviously taken place contrary to fundamental principles of German law. The Act does not mention any examples. It thereby implies, on the other hand, that the protection of embryos or the protection of emerging life as the respective background (whatever the definition of emerging may be) is not particularly a fundamental principle of German law. Therefore it would be hard to argue against those reproaching the responsible parties as having applied double standards which have been glossed over by utilitarian thinking or even on the grounds of short-sighted nationalism – "no German embryos for research purposes".[233] Nevertheless, this regulation also shows that fears in less developed countries that talk of CLONEialism are not unfounded. Of course, the StZG prohibits the import if the techniques used for the extraction of stem cells obviously contravene German principles. It is impossible to foretell how a German authority will check if these conditions have been met.[234]

On top of this, the regulations of the StZG appear dubious more on the basis of practical arguments from the perspective of research. Only last year researchers in Singapore managed to culture human stem cells surviving without mouse cells and extracts from mouse cells.[235] This means these stem cells are interesting for therapeutic purposes because they have not been contaminated with mouse viruses.

---

[231] Statement of reasons for the Act, BT-Drs. 14/8394, p. 10.

[232] The Federal Chancellor declared that no new legal situation had been created but only legal security had been provided for. See http://www.bundesregierung.de/documente/Schwerpunkte/Gentechnik/Bundestag, p 2; Comments on the preceding resolution are given by Kloepfer 2002, p. 418; regarding the Law, see Rosenau 2002, p. 20.

[233] Deutsch 1991, p. 725, in an initial discussion on the ESchG, citing from a statement made by Johannes Gross, had already posed the question as to whether this kind of division of work in the modern world was to mean that: some nations did the research whilst others took care of the corresponding moral issues. Also Neidert 2002, 467, 470 who points out that the StZG may be politically correct but less so ethically – a notorious compromise (p. 470).

[234] A further error in the rationale of the law was pointed out by Rosenau 2002, p. 22: Though no new stem-cell lines have to be produced for German stem-cell imports, if the Germans buy up those produced by the deadline, new ones have to be created for further interested buyers. Thus the spotlessness of the German solution is only on the surface.

[235] Nature Biotechnology, publication online, 5 August 2002, doi 10.1038/nbt726 available at http://www.nature.com/cgi.taf/DynaPage.

Nevertheless, German researchers must not use them because they were not produced before the deadline laid down in the StZG.[236]

3.5.4.4
Preimplantation Genetic Diagnosis (PGD)

Preimplantation diagnosis is closely connected with questions regarding embryo research and provides examples by which not only problems concerning the legal position but also of the entire discussion may be illustrated.[237]

So far PGD has not been practiced in Germany. Some sensation was aroused by the case in which a couple was denied PGD in Germany, but was able to receive a healthy child after PGD in Belgium.[238] The legalisation of preimplantation genetic diagnosis is now under intensive discussion. The main point of the argument held by the critics of PGD is that it abandons the commitment to protect unborn human life and thus opens the way to a process of selection being applied to human embryos.

Selection procedures involving human life immediately remind us of the terrible experiences suffered in Germany during the time of the Third Reich. One of the most imminent problems is that physicians are pressing for permission to fertilise and implant human embryos, while reserving the right to abort these embryos afterwards if certain pathological results are diagnosed.

Critics argue that the new possibilities to choose between wanted and unwanted human beings will pave the way to a programme of undesirable embryo research and a kind of new eugenics.[239] The validity of such arguments, usually called 'slippery slope arguments', will be discussed below in Chapter 5.

Since our culture is devoted to the protection of human life, this opinion, which opposes every form of PDG, would appear to be cogent. It seems to be founded on the ESchG. Whereby, it is claimed, according to the spirit of this law, a fertilised, potentially human egg cell from the very point of fertilisation as well as every totipotent cell extracted from an embryo and able to develop into an individual are to be defined as „embryos".[240] Therefore, extracting a totipotent cell from an embryo and using it for PGD is not permissible by law[241]. At present, especially the

---

[236] See also Frankfurter Allgemeine Zeitung from 5 August 2002, No. 179, p. 30.

[237] See also Schneider 2000, p. 360; Bartram et. al. 2000; Herdegen 2001, p. 776; Renzikowski 2001, p. 2753; Schroth 2002, p. 172; Mildenberger 2002; Dederer 2002; Schlink 2002; Benda 2001, p. 2147; Sendler 2001, p. 2148; Ipsen 2001, p. 995; Laufs 1998, p. 1753; Laufs 1999, p. 1762; Laufs 2000, p. 1765; Herzog 2001, p. 339; Graumann 2001, p. 39 (the same theses of Graumann with older statistics can be found in: Zur Problematik der Präimplantationsdiagnostik, Politik und Zeitgeschichte, Das Parlament, 29. 06. 2001, p. 17).

[238] Renzikowski 2001, p. 2754; Schneider 2002, p. 1. All in all it may be assumed that worldwide not more than 500 children have been born after PID, see the information supplied by Faßbender 2001, p. 2747 (200 children by the year 2000) and Graumann, p. 40 (276 children by the year 2001). In a survey by the Allensbacher Institut in October 2001 66% of the Germans questioned that they had so far never heard of, Allensbacher Jahrbuch 2001, p. 884.

[239] Graumann 2001, p. 44 ff; so also Herzog 2001, p. 396, who, however, rejects a probationary sentence.

[240] See "the definition of the embryo".

[241] Keller/Günther/Kaiser, ESchG-Kommentar, s. 2, margin no. 54; Renzikowski, ibid., p. 2754; Schroth 2002, p. 172, (III.1); Mildenberger 2002, p. 296; Faßbender 2001, p. 2747; Laufs 1999, p. 1762; Laufs 2000, p. 1765 f; See also draft discussion paper of the Federal Medical Council, DÄBl. 2000, A-525.

further fate of the embryo must not depend on the results of the diagnosis. Proponents of this point of view regard such a procedure as destruction of the still totipotent cell. Accordingly, they assert, the ESchG would be violated by this method of diagnosis because it does not aim at the preservation of the embryo[242].

Proponents of another view use the argument that PGD serves to avoid the possible need for a later termination of pregnancy for reasons of the so-called embryopathic indication or – as it is now dealt with in Germany – as a maternal indication.[243] The ethical legitimacy of PGD is therefore derived from the being in a position to avoid a later termination of pregnancy. The extraction of a totipotent cell for biopsy also violates section 6 paragraph 1 ESchG[244]. By extracting a totipotent cell, which is defined as an embryo[245], a new embryo comes into being showing genetic information identical to the remaining, original embryo. The current discussion in Germany also concerns the question of whether the extraction of blastomeres during the eight-cell-stage is permitted by law. In accordance with the present level of knowledge these cells are not totipotent any more, so that the restrictions contained in the ESchG would not be applicable[246]. Nevertheless it is under consideration as to whether these tests with blastomeres during the eight-cell-stage violate section 2 paragraph 1 ESchG, i.e. whether the tests in question are, or are not, designed for the preservation of the embryo.[247]

Nevertheless the Bundesärztekammer (German Medical Association) prohibits the diagnosis before the transfer of the embryo unless to exclude severe sex specific diseases within the meaning of sec. 3 ESchG[248] and the Deutsche Ärztetag (German Medical Congress) has just recently supported a prohibition of preimplantation genetic diagnosis.[249]

3.5.5
Current Discussion – Statements from the Main Political Parties[250]

If one applies the principle of democracy, more precisely the system of representative democracy predominant in western democracies, one is forced to concede that it is necessary to heed the attitudes and views of the population, because it is the people that elects those that decide in Parliament on what is allowed and what is forbidden.[251]

---

[242] Laufs 1999, p. 55.

[243] Schlink 2002, p. 20; Ipsen 2001, p. 995.

[244] See also Günther in: Keller/Günther/Kaiser, ESchG-Kommentar, s. 2 margin no. 54.

[245] S. 8 para. 1 ESchG.

[246] See also Neidert 1998, p. 353.

[247] This procedure is nevertheless regarded as incompatible with the ESchG by Renzikowski 2001, p. 2753, because it is not driven by the required intention to effect implantation; a position held also by Faßbender 2001, p. 2748; Mildenberger 2002, p 297; and supported by Kloepfer 2002, p. 424. In contrast Schroth, 2002 p. 172, Schreiber 2000, A-1136; Neidert 2002, p. 470 says that the intention to perform PGD is not to eliminate pathological embryos but the birth of a not disabled child.

[248] Musterberufsordnung (MBO-Professional Rules) D IV. No. 14.

[249] Compare http://www.bundesaerztekammer.de/ -BÄK-INTERN.

[250] On this see the comprehensive analysis for Western Europe by Rafaelo Pardo, in this publication.

[251] This foundation, in view of the examples given by Beyleveld and Fraser, in this publication, on the tactics applied when passing laws, is certainly not sufficient on its own. Therefore the premise shall remain valid that in a country, in the long run, no legal position can continue to exist that is conspicuously in violation of the views of the predominant majority of its legal subjects.

The attitude of the population in Germany as well as in other European countries will be presented and discussed below in Chapter 4. Within the political parties, basic assessments are discernible that lead one to expect a certain direction in future legislation. On the whole, the political parties express their positions rather cautiously. The CDU/CSU (Christian Democratic Union/Christian Social Union) has declared in its government programme for the years 2002 to 2006 that it does not want to be captivated by scientific delusions of feasibility. Utilitarian arguments should not be put before the protection of all human life. That is why it advocates adherence to the strict principles of the German ESchG.[252]

The SPD (Social Democratic Party) is more cautious (being less explicit and leaving room for interpretation) in its government programme. It wants to continuously determine the chances and limits of research in the field of genetic engineering from scratch. In its view, research serving the people respects the moral and ethical boundary. Furthermore, the SPD welcomes the debate on scientific and ethical questions and guarantees to promote them in the future as well.[253]

By contrast, the GRÜNEN (the Greens) disapprove of research on embryonic stem cells in their guiding paper. They say embryos would have to be consumed for this, i.e. life would have to be annihilated. Those who want this research set off life against life and it is not convincing to construe the concept of human dignity differently by trivialising the issue.[254]

The FDP (Liberals) called for a careful relaxation of the provisions of the ESchG already in 2001 in order to put surplus embryos from in vitro fertilisation at the disposal of research.[255]

The Federal Election in September 2002 was not insignificant for future legal policy in the field of embryo research in so far as the two parties that came together to form a coalition hold contrary points of view on the decisive issues. The more liberal SPD is shoulder to shoulder with the GRÜNEN with a far more protectionist fundamental attitude.[256]

How big the pressure is from each side, the life protectionists on the one side and the researchers on the other, can be seen from their respective public statements: In 2001 the DFG (Deutsche Forschungsgemeinschaft, German Research Foundation) had not only welcomed the import of stem cells in a statement but also demanded their production in Germany.[257]

---

[252] Government programme of the CDU/CSU, http://www.cdu.de/regierungsprogramm.

[253] Government programme of the SPD, pp. 33, 34. http://regierungsprogramm.spd.de/servlet/PB/show/1076396/spd-regierungsprogramm.pdf.

[254] Bündnis 90/The Greens, Politik in der Verantwortung – Eckpunkte für eine Gentechnikpolitik der Bundestagsfraktion, p. 4, available in German at http://www.gruene-fraktion.de/themen/bildung/01051eckpunkte-gentechnik.pdf.

[255] Gerhardt/Schmidt-Jortzig/Flach/Parr, Gegen Fundamentalismus in der Gen- und Biotechnologie, press release from 31 May 2001, p. 2, available at http://mdb.liberale.de/fraktion/aktuelles.php?id=32423.

[256] See also Frankfurter Allgemeine Zeitung 20 September 2002, p. 4; interestingly the conflict within a coalition between the CDU/CSU and the FDP would be similar, except that then the smaller coalition party would be representing the more liberal views.

[257] Recommendations by the DFG on the Research on Human Stem Cells from 3 May 2001, no. 10, available at http://www.dfg.de/aktuell/stellungnahmen/lebenswissenschaften/empfehlungen_stammzellen_03_05_01 html. The DFG is the central self-governing institution of science for the promotion of research at universities and publicly funded research institutes in Germany.

Both main denominations had however turned to the MPs with the joint plea for a clear decision in favour of the protection of the human being right from the beginning before the vote on stem cells.[258]

3.5.6
Commission of Inquiry and National Ethics Council – an Attempt at a Solution

The degree to which the political community is intent on finding a ruling capable of consensus in questions concerning research and the protection of life can be seen by the means they are applying. Thus, for example, at the political level two ad-hoc bodies convened simultaneously, albeit arriving at different proposals as a result.

The National Ethics Council was set up in June 2001.[259] In May the Federal Government had passed a corresponding resolution on this and appointed 25 members from business, the denominations and the organisations of the disabled. The respective idea came from the Federal Chancellor who for the first time in February 2001 wanted to create a body primarily providing information for and spreading it to the population. His vote was also not supposed to be binding on Parliament. On 20 December the Ethics Council presented its opinion. It argued in favour of a strictly controlled import of embryonic stem cells.[260] Among other reasons, the establishment of the Ethics Council came in for criticism because a Parliamentary Commission, the Commission of Inquiry "Law and Ethics in Modern Medicine", with 13 MPs and scientists respectively worked simultaneously.[261]

It presented its results in an extensive final report in May 2002.[262] In a second intermediate report on the research on stem cells from 12 November 2001,[263] the Commission of Inquiry had cast a majority vote against the import of human embryonic stem cells. It had deemed the use of human embryos even abroad not ethically sustainable and had regarded the scientific reasons as insufficient.[264] The minority opinion had favoured admitting a controlled import of already existing stem cell lines which became law, i.e. the StZG, later on.

What could have been assessed in the event of a consensus of votes in favour of the predominance of the political discourse now shows something different: the conflict over decisions on matters of conscience cannot be resolved by majority decisions. There exist opposing positions and, as has been the case since the first ruling of the Bundesverfassungsgericht (Federal Constitutional Court) on the question of abortion,[265] a legal ruling will always bear the character of compromise and thereby accommodate the relative positions of both sides.

---

[258] Letter from 14 January 2002, available at http://dbk.de/presse/pm2002/pm200211701.html.
[259] http://www.ethikrat.org/ueber_uns/einrichtungserlass.html.
[260] Available at http://ethikrat.org/publikationen/stell_stammzellen/inhal.html.
[261] Established by Resolution of the German Bundestag from 24 March 2000 – BT-Drs. 14/3011.
[262] Deutscher Bundestag (editor), Enquete-Kommission Recht und Ethik in der modernen Medizin (Commission of Inquiry. Law and Ethics in Modern Medicine), Berlin 2002, available at http://dip.bundestag.de/btd/14/090/1409020.
[263] BT-Drs. 14/7546, available at http://www.bundestag.de/gremien/medi/2zwischen_engl.pdf.
[264] BT-Drs. 14/7546, p. 100.
[265] BVerfGE 39, 1.

3.5.7
Possible Solutions for the German Legal Situation

Since the Stem Cell Act has been adopted, consuming research on embryos or research for any purpose other than the embryo's own benefit remains illegal in Germany. The Stem Cell Act has so far only clarified a partial area: Merely pluripotential stem cells, i.e. cells which can no longer develop into an individual human being, may be assigned to research purposes. The necessary destruction of embryos has been explicitly accepted as a compromise. In order to prevent supply from emerging through demand a definite deadline was set. It is only permitted to import stem cell lines, which have already been produced before this deadline. A practical problem in conjunction with this has already been mentioned: The legislature will have to make a new decision if stem cell lines already in existence turn out not to be feasible. There is another obvious starting point for criticism: That which is prevented by rigid legislation stipulating the protection of embryos because certain ethical or moral boundaries have been established is circumvented under the approval of the state as the unpopular conduct may occur and be exploited abroad. Another contradiction has evolved from the last Act of the legislature and adds to the existing contradiction between the levels of protection for the embryo in vivo and for the one in vitro. However, that which has been established regarding the termination of pregnancy may apply again: The decision follows from weighing up the advantages and disadvantages for the interests of different parties against each other.

The currently existing rulings in Germany show that questions of reproduction and therefore of the creation of human life are discussed as fundamental ethical issues. On sober reflection, in view of the dictions in the rulings existing so far, it may be recognised that irrationality driven by emotions will also have some influence on future decisions. A rational approach[266] to this issue will have to take this irrationality into consideration. How this could possibly be done, will be discussed below in Chapter 5.4. The Need and Limits of Procedural Solutions to Problems of Substantive Rationality.

Changes of the regulation on embryos have to take into consideration that the conflict of the present situation is based on the contradiction between a quite liberal abortion law – which permits the termination of pregnancy almost until the birth of the child especially when there are handicaps – and the very strict ESchG – which prohibits dealing with embryos for other purposes than their own benefit. Therefore the legislation comprises that the same legally protected right is differently protected by the ESchG and by the Penal Code as the latter permits to abort the embryo.

Therefore a fundamental change in German legislation is needed. The most comprehensive solution would be reached by making a Reproduction Law, which contains regulations of all fields related to this matter. For this reason it has to be considered to reduce the very strict protection of embryos by the ESchG approaching it to the level of protection of embryos provided by the Penal Code. We prefer a gen-

---

[266] This is attempted for example by Gethmann/Thiele 2001, p. 36 and 45.

eral legal regulation by the parliament to a less comprehensive solution which could consist of single decisions, cases and suggestions of professional associations like the Bundesärztekammer.

The disadvantage of a fundamental solution by a parliamental law could be that such a law is less flexible to be changed if it is required by scientific progress. But regarding the values in question we could not advice other solutions.

Of course, it will take a while until such a law is prepared. Thus interim-solutions need to be considered which might obviate the necessity of "preimplantation-tourism" into other European countries. In view of the various national legislation, as presented in the appendix, the problem that couples are seeking preimplantation diagnosis in Belgium or the United Kingdom is currently under intensive discussion. Such effects are neither to be encouraged nor are they desirable. To prevent them it is necessary to subject the ESchG to a fundamental analysis and to reconsider its possible applications. Bearing in mind the risk that preimplantation diagnosis might be used for morally contentious reasons, such as the pre-determination of sex, all possible solutions need to consider that this method shall not be available for general use. On the other hand, it needs to be taken into account that a severe contradiction of value judgements occurs if preimplantation diagnosis is prohibited when the existence of a genetic defect is highly probable, thereby obliging the mother to accept the implantation of a presumably defective embryo. All things considered, in view of the possibility of subsequent abortion, which would then be unpunishable, a couple cannot be expected to take on responsibility for such a pregnancy.[267] In a value analysis of the legal system as a whole, the regulations governing abortion must be drawn into consideration, comparing their suitability to solve conflict. In the course of a teleological reduction[268], s. 2 para. 1 ESchG would have to be excluded in those cases where abortion would be medically indicated in accordance with s. 218 of the Penal Code. Similar arguments are used by the majority of the National Ethics Council (Nationaler Ethikrat).[269] It is claimed that it would not be morally supportable to start out by bringing about or worsening a concrete situation of conflict only to subsequently legitimise a certain decision aimed at solving

---

[267] For a comprehensive rationale, compare the report of the Bioethik Kommission Rheinland Pfalz (Bio-Ethics Commission of the Rhineland Palatinate), dated 20/6/1999, Thesis II 9.

[268] According to the teleological theory of interpretation, the teleology of the law is examined in regard to the immanent sense of a particular legal institution or set of regulations applying to a certain field (Fikentscher 1976, p. 676). In the process of arriving at legal judgements, preference is given to the criterion of teleological interpretation. In particular the judgements made by the criminal divisions of the Federal Court of Justice are reached in consideration of the sense and the aims of a norm, which is derived from the pertinent circumstances applicable to each case and in the overall context governing the construction of the norm. In this process, the objective teleological criteria are not merely targeted on interpretation but are guided by the aim to achieve a just and reasonable interpretation of the law (Larenz 1995, p. 337). The scope of objective teleological criteria also includes legal ethics, thus creating the benchmarks against which justice, as a central criterion of interpretation, is measured (Fikentscher, ibid., p .679).

[269] The question as to whether an unpunishable blunder is implicated, as is sometimes held, (Neidert 2000, p. A-3484) appears to be somewhat problematic. On the basis of the treatment contract one should at least assume the appointment of a guarantor. It remains difficult to answer the question as to whether the requirement of equal treatment, according to section 13 Penal Code is met when, after completing the relevant diagnosis, the physician, instead of implanting an embryo, allows it to perish in the Petri dish.

the very conflict that was created thereby in the first place.[270] In view of this, a teleological reduction of the scope of application of s. 2. para. 1 ESchG needs to be considered.

3.6
Prospects

3.6.1
The European Union

In Beyleveld and Pattinson 2001 it was argued that the moral, social, and economic factors and pressures that exist render non-harmonisation the optimal situation in relation to the interests (broadly conceived) of both those countries that currently prohibit research on embryos and those that permit it. Nevertheless, it was suggested that harmonisation that permits embryo research is likely to be the regulatory outcome in all the EU countries in the medium or long-term. The analysis may be summarised as follows.

In countries that prohibit embryo research (apart from that which offers direct benefit to the embryo itself), moral or religious arguments to the effect that human life is sacred from the moment of conception appear to be the main reason for prohibition. However, the potential for embryo research to be used for "eugenic" purposes, and an association of this with Nazi ideology, which is to be avoided at all costs, also plays a part, particularly in Germany and Austria. Nevertheless, in prohibition countries, there are factors that argue for a permissive stance. In the main, these are, first, the loss of scientific prestige that embryo research might confer that follows prohibition (which might have economically disadvantageous effects); second, a perception that it is hypocritical to use the products of embryo research (such as IVF technologies) if one is not prepared to carry them out oneself; and thirdly (related to this) a perception that it, in the final analysis, the moral prohibitions are not that overriding, for, if they were, prohibition countries should surely not be prepared to associate with permissive countries, as (according to a strict pro-life stance) these license murder.

In permissive countries, the moral position tends to be that the embryo has only a proportional moral status, which immediately renders it possible to weigh the life of the embryo against benefits (particularly health and life benefits) that embryo research might produce for others. Consequently, it was suggested that the only considerations (apart from moral or religious conversion, or a huge scandal involving embryo research clearly motivated by and carried out in accordance with "Nazi" type ideals and methods) that might cause these countries to prohibit embryo research are economic or political factors. If this is so, what we must imagine is that the permissive countries regard the European Union as an overriding value (politically/economically), and that the prohibition countries are not prepared

---

[270] Statement of the National Ethics Council, Genetic Diagnosis before and during pregnancy, January 2003, vote in favour of a limited approval of PID, p. 59 f (This vote was signed by 15 members of the council; 9 members voted to keep and define more precisely the prohibition of PGD stipulated in the ESchG, pp. 42 and 58).

to associate closely with countries that do not prohibit embryo research. However, both premises are counterfactual!

However, if such considerations in balance argue for harmonisation in a permissive direction, there are also factors that argue for retention of the non-harmonised status quo: viz. cultural sovereignty and, in the eyes of permissive countries, the economic and prestige advantage they gain over prohibition countries (with which they do not need to compete); as well as, in the eyes of prohibition countries, the thought that because the "natural forces" (i.e., economic and social pressures) are in the direction of harmonisation for permission, non-harmonisation might be the best that can be done to preserve their own principles. On this basis, it was contended that, provided that permissive countries are not wedded to a liberal position (as against a regulatory one) as a matter of overriding principle (and there is no evidence that they are), non-harmonisation is optimal for permissive and prohibition countries.

Nonetheless, despite this being the case, because the natural forces are in the direction of harmonisation for permission, positive action is required if non-harmonisation is to be maintained. However, this would require the competitive disadvantage of prohibition countries to be reduced.[271] But, because the constitution of the EU makes it difficult for EC legislation to levy taxes on Member States (as they must be portrayed as single market measures), and funding strategies that might be adopted[272] will not prevent researchers obtaining funds elsewhere, it has to be concluded that harmonisation in the direction or permission is the likely direction in which the EU will move.

We do not feel able to offer a definite prediction, but will limit our view to presenting an hypothesis. This is that (ignoring for the moment the impact of the NAS joining the EU) the EU situation will be driven by what happens in the UK (which will, in turn, be driven by the US). What has been happening in Germany recently seems to us to show that there is intense economic pressure on prohibition countries (as long as they are not willing to maintain an absolute stance equivalent at least to that of Ireland) to compromise in order not to suffer economic/scientific prestige disadvantage (if not to actually gain such advantage), and we suspect that these developments are not completely independent of what has been happening in the UK. If this is so, then had the UK House of Lords overturned the Court of Appeal in the *Quintavalle* case *and* were Parliament then subsequently to legislate for prohibition, then not only would a move to harmonisation for permission have been halted, but a move to harmonisation for prohibition might even have been set in train. However, we doubt that even this would be sufficient, unless the US were to become a

---

[271] For a more detailed discussion see Beyleveld and Pattinson 2001.

[272] In Beyleveld and Pattinson 2001 it was stated that the EC's Group on Ethics in Science and New Technologies recommended in its report, National Regulations in the European Union Regarding Research on Human Embryos that the European Commission withhold funding for research involving embryonic stem cells. This was a mistake. The Group (as Anne McClaren kindly pointed out to us) was merely considering the possibility of making such a recommendation, which had been raised by one of its members. However, it is worth noting that the European Commission has in fact made it its policy not to fund research using embryonic stem cells in the 6th Framework Programme. This policy is currently being reviewed and we anticipate that the embargo on funding such research might well be lifted.

prohibition country. For, while the current Bush administration seems to have the ear of the pro-life lobby in the US, it would take a major constitutional upheaval for this to be translated into legislation for prohibition in the US, where the liberal ideology translates into government policy leading to withholding of Federal funding rather than legislation. This, however, is pure speculation, given that the House of Lords upheld the decision of the Court of Appeal. Consequently, we hypothesise that an economically driven move to harmonisation to permission will occur.

Of course, it might even be argued that, whatever the UK does in the future, the US will exert pressure on the EU to prompt various countries to take a permissive stance. However, while this is, indeed, possible, it is the UK, surely, that (because of its closer affinity to the libertarian and pragmatic thinking of the US) is most susceptible to US pressures, and this makes it difficult to see why, if the UK could come to adopt a prohibition stance, other countries more inclined to do so morally should be affected by US pressures so as to become permissive.

3.6.2
The Impact of the Newly Associated States

However, this is not now the whole story. The impact of the accession of the NAS must now be taken into account. Four things, it seems to us, stand out about the NAS.

(1) There is generally a lack of regulation of embryo research, coupled with a lack of debate on this issue.
(2) Most of the countries have signed and ratified the Convention on Human Rights and Biomedicine, which contains a prohibition on the creation of embryos for research.
(3) There have been debates in the NAS on abortion and IVF, which have generally evinced a "pro-life" attitude.
(4) The NAS are not strong economically.

Putting (1), (2) and (3) together suggests that, morally, the NAS might (in the main) favour prohibition. However, even if this is so, it is doubtful that the NAS would have the economic "clout" to influence the EU as a whole. That said, this lack of economic power will not be accompanied by an equal lack of legislative influence. However, it is arguable that this lack of economic power might even make the NAS susceptible to adopting a permissive stance in order to attract research investment.

3.7
Conclusion

We "predict" or hypothesise, then, that the EU (including in its expanded form) will continue to move towards harmonisation for permission. But before we finally conclude this, it is, perhaps, useful to consider the Australian case, where it seems that the individual States have managed to regulate in rather different ways. If lack of harmonisation is possible in Australia, why ought it not to be possible in the EU? Well, of course, it is possible in the EU (as it in fact is at present). However, there is a difference between the EU and Australia that limits what can be inferred about the former from the latter. This is simply that Australia is a single country, which means that the Federal Government has a responsibility for all its citizens that the EU as a whole and individual Member States do not have for citizens of other member States. This means that decisions taken by individual States in Australia do not necessary impact on their citizens' advantage or disadvantage to the extent that this is so in the EU. In short, if some States in Australia choose to be prohibition jurisdictions and this, in principle, carries an economic disadvantage, then it is arguable that such disadvantage will not necessarily be felt by citizens of the State (or felt by them to the same extent) as long as other States of Australia are permissive. Consequently, the Australian experience is not a reliable model for making predictions about the dynamics of the European Union in its current constitutional form.

3.8
Recent Developments

We anticipate that it will be objected to this analysis that we have overlooked that the EU could simply decide to legislate against embryo research generally or research to produce ES therapies or ES therapies themselves. Indeed, in April 2003, the European Parliament amended a proposal produced by the European Commission for a directive on setting standards of quality and safety for donation, procurement, testing, processing, storage and distribution of human tissues and cells (COM(2002)319-C5-0302/2002-2002/0128 (COD)), by introducing provisions to ban not only research on human cloning for reproductive purposes but also the creation of human embryos for research or to supply stem cells, "including by means of the transfer of somatic cell nuclei" (Amendment 30) and the transplantation of ES cells and embryonic tissues produced by cloning (Amendment 51 and Amendment 68). The basis that was given for these amendments was Article 152(4) of the EC Treaty, which gives the EU competence to legislate for human health, following an opinion expressed in the Draft Report (2002/0128(COD) of the Committee of the Environment, Public Health and Public Policy on the Commission's proposal for a Directive of 4 February 2003. The Draft Report (see Section VI paragraph 7) claims that competence to regulate research on embryos derives from Article 152(4) of the EC Treaty because the obtaining of egg cells from which embryos will be produced is hazardous for the health of the donor woman, (2) "protection of the donor also includes the human embryo", and, above all, there is considerable risk to the recipient from transplanted cells produced by cloning becoming cancerous.

The Commission, however, rejected the above mentioned amendments (and many other "ethics" amendments proposed by the Parliament) as being beyond the scope of Article 152(4) "that provides for public health protection and not the implementation of ethical objectives as such" (C1 of the Explanatory Memorandum of /*COM/2003/0340 final – COD 2002/0128*/), and took the same view on in vitro research (Recital 6 of the Commissions revised proposal making it clear that research in vitro is not covered by the Directive).

The Commission's position on this is surely correct. In particular, the three arguments advanced for holding that the EU has competence to regulate embryo research in vitro are spurious. As far as argument (1) is concerned, this has equal force against in vitro fertilisation and no specific force against the use of embryos for research. Argument (2), on the other hand only seems to have any application where the embryos are to be implanted, which is not the case where they are to be used for in vitro research. While argument (3) might be a consideration when considering therapies, it does not follow that this militates against research into those therapies. All medicines are hazardous. So, if argument (3) is sound, all medicines research should be banned.[273] Indeed, if safety is the paramount consideration then this argues not against research but for research to assess the risks and attempt to deal with them.

Testifying to the dynamism of the area, a number of legislative changes are on the horizon as this chapter goes into print: In *France* a revision of the Bioethics Law of 1994 was approved by the Senate in January 2003 and was discussed by the Parliament in the first semester of 2003. It proposes to allow for research on supernumerary human embryos including the procurement of human ES cells for 5 years under certain conditions.[274] *Belgium* looks likely to adopt an even more liberal approach: a bill on research on human embryos in vitro was approved by the Belgian Senate and Parliament and now awaits signature. It not only proposes to authorise the procurement of embryonic stem cells from supernumerary embryos under certain conditions but also allows for the creation of human embryos for research purposes including by means of somatic cell nuclear transfer.[275] These developments would seem in line with our conclusions above, that predict a move towards more permissive regulation.

---

[273] In June 2003, the Parliament, Commission and Council reached a common position, which in effect, means that the Commission's revised proposal will become the Directive.

[274] http://www.assemblee-nationale.fr/12/projets/pl0593.asp

[275] http://www1.dekamer.be/FLWB/pdf/49/1122/49K1122001.pdf

4 Attitudes toward Embryo Experimentation in Europe

4.1
Perceptions of Scientific Advances and Public Policies

The measurement and analysis of public perceptions of science and technology have gradually gained in importance as an input to the design of public policies and the regulation of scientific and technological advances. Although science and technology had traditionally remained closed to the formal intervention of the lay public, the last two decades have brought about an increase of public participation in various forms, ranging from consensus conferences (Joss and Durant 1995, Einsiedel et al. 2001) to direct consultation of the public (referenda, see Bonfadelli et al. 2002), and sometimes simply policy-makers' sensitivity to the public's views and preferences, as gleaned from studies of public opinion.

Since the early eighties, some analysts have predicted that science and technology would also eventually be opened up to democratic participation, as had previously occurred with other, less complex and, for ordinary citizens, more familiar areas (Prewitt 1983, Miller 1983). Therefore, in the view of these authors, it was crucial to increase the entire population's scientific literacy so as to avoid decisions that would be unsupported by sound scientific knowledge. Other authors have gone further, arguing that, in this as in other complex areas, there is no democratic reason whatsoever to require scientific literacy as a prerequisite for participation in public policies involving science and technology: conscience should take precedence over competence (Lévy-Leblond 1992). Yet another group of analysts – mostly scientists inquiring into the status of science versus other forms of knowledge in the last third of the century – accepts the idea that society might decide on some aspects and goals of science (for instance, on the areas that should get more public funding), but has advocated for a clear dividing line between science and the public. They argue, in other words, that the main dimensions of science should be kept under the exclusive control of the relatively small group of individuals who are trained as scientists, a professional elite – based on merit, not hereditary privilege – whose "entry ticket" is a rigorous and highly demanding training process (Shamos 1995, Levitt 1999).

In the current spectrum of viewpoints on the science-public relationship, the dominant perspective of some influential public bodies (House of Lords 2001, EU's Directorate of Science) is a mixed one, combining the principle of the scientific community's independence in purely cognitive questions (those "internal" to scientific practice) with external regulation by parliaments, governments and public agencies in areas with ethical consequences or with significant impacts on the environment or other key social issues (for instance, privacy, safety, or diverse hazards). This procedure may at times be informed by reports and recommendations from ethics or interdisciplinary committees, or by other forms of technology assessment. More tentatively, and in some countries more than others, the public has been given the chance to participate in major decisions on science and technology policies. In this context, knowing the public's views of scientific issues, especially in the areas of biomedicine and biotechnology, has become an increasingly important component of public policies. The long tradition of measuring public knowledge and attitudes toward science and technology in the United States (National Science Board 2002), has been emulated by the European Commission with a large number of surveys and studies on science and technology in general and biotechnology in particular (INRA Europe 1993, Gaskell et al. 1997, 1999).

Knowing how the public views certain key scientific advances of the turn of the century is a useful input for regulators, the scientific community and other agents wishing to have their voices heard or to influence decisions about the knowledge and manipulation of basic biological processes. This chapter does not seek to make recommendations so much as to provide systematic evidence, for the first time, on perceptions of a broad array of life science issues in Europe, from animal and human cloning and IVF techniques all the way to experiments on human embryos and PGD (preimplantation genetic diagnosis). As we will show, the nine European societies under study have important reservations about some recent scientific-technological developments in biotechnology, which vary according to their goals and applications. These commonalities by no means add up to a uniform standpoint across all of Europe. In this area, as in other basic spheres of culture and values in Europe, shared ground coexists with marked differences between countries which, in the issues that concern us, derive principally from the dominant religious beliefs of each society (concretely, from differences between Catholic and Protestant countries, and within each country between believers and nonbelievers). This plural reality is not only evident in comparing countries, but also when we observe what happens inside each country. As we will show, perceptions of embryo experimentation are strongly polarised in most European societies, since this issue seems to touch on the deeply held beliefs and values of the great majority of people. All this suggests that the most reasonable course of action with regard to human embryo experimentation is to deal with it by regulations that are based on not just one ideological principle or belief, but that take on board varying criteria and reservations, in consonance with the plural and at times deeply opposing views that co-exist in today's Europe.

Before examining the public's perceptions and assessments of animal and human cloning, IVF, embryo experimentation and PGD, it is worth giving a brief outline of attitudes toward science in advanced societies at the beginning of the twenty-first century, which can serve as a frame of reference for interpreting the public's specific attitudes toward the life sciences.

4.2
Science in a Crisis of Legitimacy?

4.2.1
Perceptions of Science in Late Modernity: Environmental Awareness, "Zero Risk" Culture and the "Postmodern Condition"

The cultural transformations of modern society that are most likely to have affected perceptions of science are the rise of environmental awareness, the emergence of a "zero risk culture" in the most advanced societies, and the change in sensibilities that some authors have called the "postmodern condition". Cultural historian Leo Marx has observed that the belief in progress which characterised modern Euro-American culture has eroded over the last three decades. He argues that the main factor in its decline has been the growing pessimism about the role of human beings in nature, that is, the awareness that our industrial production system and modernity in general, sustained by science and technology, are having serious adverse effects on the global ecosystem (Marx 1998). This judgment is based on ample and converging evidence: the public in late modern societies is aware of some aspects of environmental degradation such as urban and industrial pollution, global warming, greenhouse effect, loss of plant and animal biodiversity (Worcester 1993; Mertig and Dunlap 1995). At the same time, it takes for granted and is apparently unwilling to sacrifice the many advances in living standards provided by techno-scientific developments and their social dissemination through the productive system. These two opposing vectors are in large part responsible for the public's ambivalence towards science at the turn of the century.

Some analysts have labelled the society of the late twentieth century "the risk society" (Beck 1992, 1999). The hard-wiring of science and technology into a wide array of systems that are fundamental for the late modern way of life – such as transportation, communication, food and energy production – has generated systems so tightly coupled that local problems or accidents can spread rapidly and affect numerous system elements, thereby magnifying risks and complicating their management (Perrow 1999). At the same time, greater knowledge of many processes that were previously opaque to science has contributed to an increased awareness of risk factors. As the ever lengthening list of potential hazards has become part of the cognitive schemas of lay people, it has led to a "culture of fear" and "zero tolerance" of risk. The latter attitude has flourished especially when the risk is involuntary, causes long-term effects, remains invisible to the non-scientific eye, and might contribute to diseases that are dreaded in our culture, as well as other properties that lie outside expert definition and assessment, as has been pointed out in the "psychometric paradigm" investigation of risk perceptions (Slovic 1987). The increase of interdependencies and systemic couplings, brought about by science and technology, between what were earlier only loosely connected components and public risk perceptions reinforce each other and lead to the distorted image of advanced societies as more at risk than those of the past and to the idea that science and technology are largely to blame for this alleged increase in risk.

In their turn, studies of postmodernity have identified a major change in the culture of the late 20th century which is of interest here, even without any endorsement of their general viewpoint and theoretical premises. Jean-François Lyotard has referred to this change as the decline of the "grand narratives" or "metanarratives" (Lyotard 1984) that had given meaning to the organization of society in the modern age as well as to the values and lives of individuals, and which had legitimised science itself as a kind of discourse. The metadiscourse structured around the advance of reason and the continued and unlimited creation of wealth has accompanied science, and justified its role, since its emergence in the first stage of modernity. In the late 20th century, these conceptual schemas gave way to what this French philosopher has called "incredulity toward metanarratives" or, to put it another way, the proliferation of "microvisions", value frameworks that are more fragmented and incoherent, or even overtly or latently self-contradictory. These microvisions manifest themselves in the culture as well as in the cognitive schemas of individuals, and correspond to the proliferation of different lifestyles in advanced societies as well as extremely fragmented and rapidly evolving individual lifestyles. Science and the values associated with it, which had displaced more traditional forms of culture in early and high modern modernity, have paradoxically lost cultural ground in the highly science-dependent period of late modernity. The legitimating claims of science (basically, its cognitive superiority, but also some of its material effects and benefits) have been called into question, in the wake of the re-emergence of more traditional cultural forms and the emergence of new "narratives" or postmodern discourses, from art, architecture and literature all the way to politics.

4.2.2
The Cultural Appropriation of Science at the Turn of the Century

The major cases of resistance to science and technology during the earliest modernization processes focused on their threat to the way of life and subsistence of certain occupational groups. One such manifestation of social resistance, the Luddite rebellion, has become a metaphor and symbol for all episodes and expressions of non-acceptance of scientific and technological advances, and has been overquoted in the literature on the social acceptance of these advances. But this image of opposition by groups of craft workers displaced by the advent of new technologies (the new textile machinery of the first decade of the 19th century) bears the unmistakable stamp of the industrial era (or early modernity) and lacks sufficient power to capture the phenomena of ambivalence, opposition or resistance to some technoscientific developments characteristic of the post-industrial age. Since the 1950s, recurrent protest and controversy have related more to the (observed or alleged) undesirable effects of certain subsets of science on the natural environment, on the core values of our culture (religious ones among them), on conceptions of human identity, the boundaries between species and lay perceptions of risk (Pardo 2001). This resistance is associated with the preservation of more universal interests (the whole of humanity, future generations), public goods (sustainability of the natural environment beyond one's own surroundings, i.e. not so much NIMBY ["not in my back yard"] as NIABY ["not in anyone's back yard"]), activism on behalf of other species (animal rights groups), religion-based worldviews, or human self-conceptions. The figure of Victor Frankenstein, fruit of Mary Shelley's literary imagination in about the same period as the Luddite phenomenon, has demonstrated its much greater relevance for describing reactions to *potential* problems that arise when science ventures into areas supposedly reserved for "natural" law or the designs of the Creator (depending on whether the version of natural order is secular or religious).

Despite the public's more latent than explicit anxieties, one can say that the basic terms of the implicit founding contract for scientific autonomy were not seriously questioned until the 1980s. Opposition phenomena were until then sporadic and focussed on particular technological applications (nuclear power, recombinant DNA, pesticides) (Nelkin 1992). It is only in recent years that they have taken on a broader conceptual perspective (ecology, postmodernism, "democratisation" of science). Various forces and currents that had been gathering since the mid-sixties converged in the final years of the twentieth century with other developments such as scientific policies in knowledge-intensive societies, and has led to a change of context for scientific practice.

This emerging framework is characterised by *conditional consent* to science, greater *regulatory pressure* and *ambivalent beliefs about progress* in broad segments of society. The scientific community is gradually being drawn into a force field where the rules of the game (the "social contract for science") that had structured its relationship with society or the public are being redefined (Guston and Keniston 1994). Public officials call for useful research (not simply the most advanced "blue-sky research") that contributes directly to economic growth, national security or the satisfaction of other social needs (health, environmental protection), while it tries to minimise potential risks and side effects, and keeps scientific developments compatible with the beliefs and moral values of most members of society (or even of influential minorities that are particularly "vocal"). The public itself – in particular, associations and interest groups like environmental, consumer and religious organizations and, more recently, single-issue groups calling for the "democratic control of science" – is demanding a "voice" in public policies about science and technology and their commercial development, and also in matters pertaining to their impact on health or the environment, or even on the moral sphere. What we are witnessing, then, is an increased number of experiments with the institutional forms of public involvement in defining these policies and regulations. These changes, in turn, generate resistance from the scientific community, which generally advocates for maintaining the distinction between science and public that took definitive shape at the end of the nineteenth century, with the institutionalisation and professionalisation of scientific activity (Pardo and Calvo 2002).

Some postmodernist analyses suggest that societal perceptions have undergone a radical transformation from unrestricted scientific-technological optimism to a new perspective of no less unrestricted scepticism and distrust vis-à-vis science. But this description does not really match the available evidence, which suggests, first of all, that we are up against a more general phenomenon, a fragmentation of values and culture, which explains why no principle or orientation, however universal (such as science), can act as a central cultural axis in more than one or a few domains. More specifically, the data point toward the coexistence of a general belief in progress through scientific and technological advances with a simultaneous and growing anxiety over certain sets of technologies (more specifically still, certain elements of these sets), whose applications and, in some cases, experimentation and research procedures affect images, beliefs and moral or religious values that are deeply rooted in western culture.

4.3
Ambivalence toward Science at the Turn of the Century: The Case of the Life Sciences

Most areas of science and their application to social needs present no problem to the majority of people, and many are seen as clearly beneficial. Typically, technological and scientific advances still take their place silently in the background of the complex system of collective satisfaction of needs at the end of the century and, more weakly, in individuals' cognitive schemas for interpreting the world and organising the realm of everyday experience. In general, only little and temporary attention is paid to these advances outside the scientific community; or, to put it another way, nowadays scientific themes have to vie for the interest of a public that already has access to a range of information channels and subject areas far beyond their cognitive abilities and available time.

The positive assessment of science in general and of a vast range of technological developments in advanced societies is compatible with ambivalent and critical attitudes toward certain subsets of science or, even more precisely, toward certain applications of techno-scientific advances. To couch this in the language of attitudes, favourable general views of science in the abstract and of scientists as a professional group do not automatically imply equally favourable attitudes toward specific scientific areas. Hence the proposal to replace the study of general attitudes (with greater universality but low predictive power) with that of predispositions toward defined families of techno-scientific developments (or relatively homogeneous subsets of science and technology that give rise to relatively uniform valuations) (Daamen, Van der Lans and Midden 1990). A more balanced picture that fits the data better, is that *general* images of science and the scientific community are explanatory factors for *specific* attitudes, but they are not strong predictors, when taken in isolation, of people's attitudes to areas of science that impinge on core values and beliefs (Pardo, Midden, and Miller 2002).

One assumption of many studies on public perceptions of science and technology is that people hold either positive or negative attitudes toward them. Frequently, however, individuals have positive expectations (a "promises of science schema"), while at the same time they maintain significant reservations in other respects (a "reservations about science schema") (see Miller and Pardo 2000). For some individuals, usually those with low or medium-low educational background and/or no particular interest in an area, this duality is due to the lack of structure in their attitudes (i.e. due to poorly formed attitudes) (Converse 1964, 1970). For other, more sophisticated individuals with a high educational level, mixed positions are mainly due to their ability to distinguish between different aspects or dimensions of a specific scientific development and its potential impacts (Evans and Durant 1995). A comparative look at attitude patterns in different societies shows varying degrees of this ambivalence, due to the simultaneous presence of positive and negative views of science. In some societies (particularly the US), science and technology have penetrated the cultural, social, and institutional architecture so deeply that most people hold science in high esteem and see mainly the "promises of science", with very low "reservations" about it (i.e. an individual with a high score on an indicator measuring "promises of science" typically scores low on another indicator measuring "reservations about science"). This configuration can lead, at a societal level, to a combination of small minorities who are very vocal in their opposition to scientific advances with a very large majority who consents to advances and tries to benefit from them as quickly as possible. In these countries the "promises of science" schema clearly predominates. In many other societies (e.g. most European ones), more people score almost equally high on both schemas or indicators, which leads to greater ambivalence and longer time lags in the acceptance of those technological applications which affect deeply rooted cultural values, beliefs and images (Miller and Pardo 2000).

From time to time, some subsets of science come to the forefront of public opinion and turn into objects of controversy – particularly those with potentially undesirable impacts on the environment, on the range of risks and hazards, or on central images and moral values of our culture (such as the identity of and boundaries between species, beliefs about the beginning of life and the status and protection of the human embryo, the individuality or uniqueness of human life). Biotechnology is perceived today, by broad segments of the public, as one of the problematic techno-scientific areas. Some of its sub-areas have replaced nuclear energy technologies as the main target of criticism and fear. One of the fundamental roots of current reservations about biotechnology are changes in the perception of nature and the "natural" in the last third of the twentieth century. More than three decades of intensive media coverage of major threats to the environment by human activities, and the emergence of powerful single-issue organisations for the preservation of the environment, have had a major effect in shaping public perceptions of nature, in which awareness and pessimism are particularly salient. Moral and religious reservations and fears regarding the manipulation of animal and particularly human life are perhaps the other most important vectors of concern and anxiety about some subsets of science and technology today.

The public seems increasingly concerned about the consequences of some recent advances in biotechnology. Opposition to genetically modified (GM) foods and crops has reached a high level in many European countries, and the birth of Dolly the sheep, which attracted a very high level of media coverage, was received with perplexity and concern. Cloning was perceived as a monumental and "direct challenge to the sacredness of the biological individual" (Horning Priest 2001). The possible cloning of humans has caused alarm among the public at large and the regulators, and reopened a heated debate on the limits and controls of scientific research in the area of life sciences. For its part, the scientific community worries about what it sees as the progressive crystallisation of negative attitudes without a proper understanding of the scientific issues at stake.

In this context, recent research on human embryonic stem cells has elicited considerable debate in most countries, becoming the most salient scientific area of confrontation between the different parties involved, and certainly one of the most difficult ones to regulate. The current situation and dilemma have been aptly summarised by a European Commission Working Paper as follows:

> as the life sciences and biotechnology develop, they contribute considerably to securing welfare on the personal and societal levels as well as to creating new opportunities for our economies. At the same time, the public is increasingly concerned about

the social and ethical consequences of these advances in knowledge and techniques as well as about the conditions forming the choices made in these fields (Report on Human Embryonic Stem Cell Research, Commission Staff Working Paper, Commission of the European Communities, Brussels, 3.4.2003, SEC (2003) 441).

The Commission Staff Working Paper endorses the line of action proposed by the European Group on Ethics (EGE): "a continuing dialogue and education to promote the participation of citizens, including patients, in scientific governance, namely in social choices created by new scientific developments". This working paper claims that "the ethical and social debate should be a natural part of the research and development process, involving society as much as possible".

The prevalent position advocated by the scientific community is the traditional one, i.e. autonomy, freedom of research and self-regulation, whose combined effect has often proved instrumental in advancing research and solving practical needs in the past, without compromising critical moral criteria and standards. For their part, religious groups (particularly Catholic groups), on the basis of their particular beliefs, would like to ban once and for all any research involving human embryos, while groups of patients suffering from serious diseases that could be treated in the future if current research proves successful, would like not only to remove any restriction on research oriented to clinical goals, but also to encourage and strengthen public support for research in this area. Interest groups and religious institutions are able to have their voices heard, not infrequently speaking on behalf of the public at large or the whole of society, but currently very little is known about the public's perceptions and attitudes. This chapter tries to offer a first, exploratory picture of the public's concerns and expectations about new developments in the biomedical sciences. We will make extensive use of data obtained in a recent survey carried out in nine European countries, covering a wide range of general attitudes to science combined with a special focus on the natural environment and perceptions of biotechnology. This survey is, to the best of our knowledge, the first systematic study carried out in a large number of European countries of perceptions and attitudes toward many different aspects of cloning, embryo experimentation and PGD.[1]

I would like to thank Professor Robert McGinn, Stanford University, for his insightful comments on a draft for this chapter. Mariana Szmulewicz and Javiera Barandiaran, of the BBVA Foundation Department for Public Opinion Studies offered also valuable comments on an earlier draft of this chapter.

[1]   Description of the survey. *Population*: National coverage in nine countries. *Universe*: Population aged 18 or over. *Sample Distribution:* Random sampling based on a distribution by population and community size. Final selection of the individual varied in each country, using both sex and age quotas as well as other random selection processes. *Data Collection Method:* Personal face to face interview, with a structured and pre-codified questionnaire. Denmark and the Netherlands used a CAPI system. *Sample Size:* National samples in each country of 1,500 cases. Each country has followed its own weighting system based on region, sex and age. Spain used a self-weighted sample representative at the national level which does not require further weighting. *Fieldwork Period:* Fieldwork began in November 2002, finishing in most cases by February 2003. Reference for this survey: Rafael Pardo, European Study on Biotechnology and Embryo Experimentation. Bilbao: BBVA Foundation, 2003.

4.4
General Attitudes to Science and Technology in Europe

It is well known that attitudes to complex objects may be influenced by a number of different value orientations and worldviews, and that some of these concurrent viewpoints and values may be in opposition (see, in the context of political objects, Campbell, Converse et al. 1960). One "worldview" or "orienting disposition" to be taken into account for understanding people's attitudes toward biotechnology is public perceptions of science at large. From the perspective of the scientific community, it is only logical that the public's attitudes toward biotech would follow from the more general class of attitudes to which, from a formal viewpoint, they pertain: predispositions toward science and technology in general. But it is clear that they may also be related to attitudes towards the natural environment, to religious and moral beliefs, economic competitiveness, health, risk perceptions and several other sets of goals and values (Pardo, Midden and Miller 2002). On the basis of findings reported in the literature on attitudes toward science, we should expect that *general* perceptions of science would have only a moderate role in explaining the public's attitudes toward *specific* areas of science such as biotechnology (Daamen, Van der Lans and Midden 1990), but it is also important to bear in mind that their influence is far from negligible. At any rate, it is important to assess whether current reservations about embryo experimentation (to be documented later) are just the result of a more general detachment from science in this period of late modernity, as the scientific community not infrequently claims, or whether it has more specific roots. As we will see below, general expectations regarding science and confidence in the scientific community are high in all the countries studied, and this suggests that the basis for reservations and opposition to biotechnology and embryo experimentation must be sought elsewhere.

4.4.1
Interest in Science

A canonical indicator of people's attitudes to science is their (declared) level of interest in it. In a period in which the number of interest areas and information channels has dramatically expanded, interest in scientific issues fares rather well in the nine countries studied, compared with other areas, and scores in the medium to high range, just below medicine and the environment. In most countries, these two areas, along with science in general, make up the cluster of issues that commands the highest level of sustained attention. On a scale from 0 to 10, the mean and median values of interest in science in each country (given in descending order) are as follows: 1. France 6.5 (median 7), 2. Italy 6.3 (median 6), 3. Denmark and Netherlands, both 6.2 (median 7), 4. Germany 5.7 (median 6), 5. Austria 5.5 (median 5), 6. Spain 5.3 (median 5), 7. Great Britain 4.5 (median 5), 8. Poland 4.4 (median 5).

4.4.2
Expectations about Science and Technology

A second and more direct indicator of people's general predispositions toward science is the expectations they have about the net social impact of future scientific and technological advances, particularly in some of the most salient areas. The question asked in a number of surveys, including our own, reads "I will read a list of new technologies to you. Please tell me, in each case, if you think this technology will improve our quality of life in the next 25 years, if it will have no effect, or if it will worsen the quality of life". The list of technologies includes solar energy, computers, biotechnology, telecommunications, new materials, genetic engineering, space exploration, animal cloning, the Internet, and nuclear energy. The areas perceived in a favourable light are, in all the countries, telecommunications, computers, solar energy, new materials and the Internet, and in some countries biotechnology also (Denmark with 70% responding that it will improve the quality of life, France with 61%, Spain 59%, Italy 56%, Germany 53%, and Austria 52%). The most problematic areas in most countries are (from most to least problematic) animal cloning, nuclear energy and genetic engineering. It is interesting to note that in the majority of countries animal cloning has replaced nuclear energy as the area with the most negative connotations. While the label "biotechnology" seems to elicit a positive response, the reverse happens with "genetic engineering": in Austria there are 33 percentage points of difference, in the Netherlands 20, in France, Germany and Great Britain 16 points, in Denmark both areas are perceived favourably but with 15 percentage points of difference, in Italy 12, in Poland 11; only in Spain is there virtually no difference in the positive appreciation of both.

To summarise, there is neither a general acceptance of current techno-scientific advances nor a general rejection, but different evaluations depending on the specific area. The prevalent response is clearly positive, and the cluster of telecommunications, computers, the Internet, new materials, and solar energy is seen without exception in a favourable light. At the opposite extreme are nuclear energy and cloning, which trigger a markedly negative perception. Finally, in seven countries out of nine, genetic engineering is perceived negatively, in contrast with biotechnology, which in seven countries receives approval. Although we lack precise data to ascertain the differences seen by the public between biotechnology and genetic engineering, it is plausible that the explicit reference, in the latter expression, to the manipulation of life at the gene level triggers deep-seated fears in an ample segment of the public.

4.4.3
Confidence in the Scientific Community

Finally, strong indicators of the public's general attitudes toward science and technology are confidence in the scientific community and the image of the scientist. In recent years, it has been proposed that trust in scientists as a professional group should, if not replace scientific literacy as a predictor of attitudes toward science, then at least be introduced as a critical variable in the regression models used to explain them (Gaskell et al. 1997), and current research on social capital has reinforced the importance of this variable. The cognitive resources and time for dealing with complex objects are extremely limited for most individuals and, as the cognitive science and social cognition literature has shown, people tend to rely on "cues" and reference groups in their decisions. The differences in credibility and positive attributes among various groups who make competing claims on a complex issue are critical for people's responses to these multifaceted and conflict-prone situations. Given the relative novelty and importance of this variable, it was measured through four indicators reflecting different aspects of trust in and prestige of the scientific community, taken generally and in the context of specific scenarios.

The first indicator is the perceived contribution of a number of professional groups, including scientists, to the improvement of living conditions and human progress, on a scale from 0 to 10. In all countries, the scientific community is perceived as one of the groups with the highest contribution to human progress, along with medical doctors and, in virtually all countries (with the only exception of the Netherlands), environmentalists (or ecologists). Medical doctors occupy the first position in Austria (mean value of 7.7), Denmark (7.8), the UK (7.3), France (7.7), Italy (7.5), the Netherlands (7.9), and Spain (7.3), followed closely by scientists. In Germany and Poland, the scientific community ranks in the first position (with a mean value of 7.4 and 7.1, respectively). Politicians and the clergy are at the opposite extreme.

A second indicator of general confidence in different groups was introduced at the end of the questionnaire (at considerable distance from the first indicator so as to avoid contamination) with a direct and general question on trust: "How much trust do you have in what each of the following groups of professionals do or say?" This indicator produced an almost identical ranking, with the sole difference that environmentalists took the second position in most countries, below medical doctors, and immediately above scientists. Once again, in Germany and Poland the scientific community ranked first. The two remaining indicators tap confidence in the information about environmental and biotechnology issues provided by different groups, and the pattern here is similar to the one just described. Confidence in the scientific community and environmental organizations is high in both cases, contrasting with low confidence in the national government and even lower confidence in companies (particularly multinational corporations).

Confidence in the scientific community is high both in absolute terms and relative to other professional groups and institutions, and scientists are respected by most people as reliable sources of information on sensitive issues. But it has to be noted that environmental groups are also perceived as reliable and trustworthy, and this is an important novelty of the late modern period, opening the way for a possible conflict of legitimacy in cases in which the scientific assessment and the environmental position collide.

Science, therefore, is just about the only modern institution that still enjoys high legitimacy. Most individuals have high expectations regarding the social effects of many of the most dynamic techno-scientific areas. This level of public appreciation of science is compatible with important reservations regarding the technologies of gene manipulation and animal cloning, either because of concern about their impact on the natural environment and "natural" processes (both seen in a radically different light than in the high-modern period), or due to moral and religious beliefs, an issue we will deal with in the following sections.

4.5
Perceptions of Animal and Human Cloning in Europe

The announcement of the cloning of Dolly the Sheep by Roslin Institute researchers in 1996 was greeted with surprise and concern by public opinion. It was covered widely by the media and led to generally critical or at least cautious expressions of opinion among different groups and institutions. There were two prime causes of concern, both of them still present years later. The first was that scientific research had helped breach a new and basic limit in the human manipulation of the natural world, opening the door to the mass production of animals. The second – hotly debated at the start of the twenty-first century – emerged from the dynamic known in ethics as the "slippery slope"; that is, the risk that cloning animals would be a first step to cloning human beings, a possibility that has long been the subject of science fiction literature (Heise 2003) and has typically provoked "images of fear" (to use the expression proposed by Weart with reference to nuclear power) (Weart 1989) and the "horror of monsters" (Turney 1998).

Our data cover a broader range of perceptions of cloning than most studies conducted to date (including the Eurobarometers on biotechnology): they cover both acceptance/rejection of human and animal reproductive cloning in a general and abstract sense, and the way in which perceptions may be modified in specific cases or situations. This broader evidence shows that in virtually all countries (with the sole exception of Spain), animal cloning is rejected by most people, both for stock-breeding or food production, and even more importantly to obtain future health benefits for humans. Opposition is stronger in the Netherlands, Germany, France and Austria, and is also the prevalent position in all countries except Spain (see figure 4.1).

In contrast with the high level of confidence in the scientific community at large, the public does not give a "blank check" to researchers on cloning. Trust that "scientists will use research on cloning or genetic copying of animals in a responsible manner" is low in all countries, particularly in Germany with a mean value of just

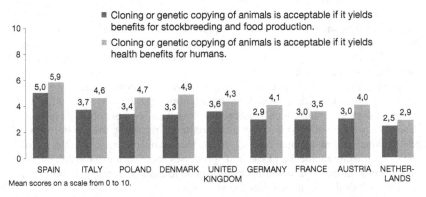

Fig. 4.1 Attitudes to animal cloning (on a scale from 0 to 10, where 0 means disagree completely and 10 means agree completely)

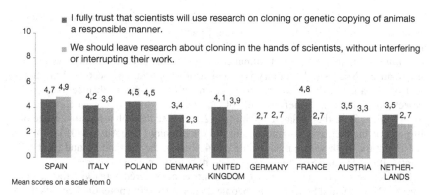

Fig. 4.2 Attitudes to researchers on cloning (on a scale from 0 to 10, where 0 means disagree completely and 10 means agree completely.)

2.7 (on a scale from 0 to 10), Denmark (3.4), Austria (3.5), and the Netherlands (3.5). Accordingly, the European public is not willing to leave research into animal cloning "in the hands of scientists, without interfering or interrupting their work". The willingness to allow researchers to develop their work without public input is lower in Denmark, France, Germany, and the Netherlands (see figure 4.2).

In order to capture more variability and to secure more robust measures of attitudes toward animal cloning, two summated scales or aggregated indicators were built, using eleven items tapping different aspects of evaluations of cloning. The eleven items were subjected to principal component analysis with varimax rotation, and two components emerged in the expected direction (positive, negative), accounting for about 60% of the total variance. Seven items loaded on the first component and the remaining four on the second component. The first component or indicator, reflecting negative predispositions, was labelled "reservations about cloning", and the other, capturing positive aspects, was labelled "promises of cloning". Before collapsing each group of items in the corresponding indicator ("reservations", "promises"), they were checked for internal reliability using Cronbach's alpha formula for additive scales, with very good results[2]. The "reservations" scale has a range of 70 (0–70) and the "promises of animal cloning" scale a range of 40 (0–40).

Reservations about animal cloning are very high, particularly in the group of the most advanced countries: the Netherlands, Germany, France and Austria attain a mean value of 45, ten points above the "neutral" point (i.e. 35), followed by Denmark (43), Italy (42), the UK and Poland (both with 40). Only Spain departs from the general pattern, with a mean value of 34, consistent with recent findings in

---

[2] The Cronbach's Alpha values for the "reservations" scale were: 1. UK 0.92, 2. Denmark 0.89, 3. Italy 0.89, 4. Austria 0.88, 5. Poland 0.88, 6. Germany 0.87, 7. France 0.86, 8. Spain 0.86, 9. Netherlands 0.85. The values for the "promises" scale were: 1. Italy 0.88, 2. Austria 0.85, 3. Netherlands 0.84, 4. UK 0.84, 5. Spain 0.82, 6. Denmark 0.81, 7. Poland 0.81, 8. Germany 0.79, and 9. France 0.77.

other surveys on biotechnology and the life sciences showing that concerns about these new scientific developments among Spaniards are low both comparatively and in absolute terms (Gaskell et al. 1997, 2003).

Promises – that is, support of animal cloning – reach very modest mean values in all countries. Not a single country is enthusiastic about the prospects and the conduct of research in this area and, once again, in all the most advanced societies expectations about the beneficial side of cloning are extremely low, at a considerable distance from the "balance" point (i.e. 20, on a scale with a possible range of 40 points): the Netherlands (mean value 11), Germany (12), Austria (13), France and Denmark (both with a mean value of 14 points), the UK and Poland (both 15 points), Italy (16) and Spain (19, i.e. one point below the neutral score).

It is clear that in most of the European countries considered here (with the exceptions of Denmark and Spain), many people do not yet see the potential benefits that animal cloning may make possible, whereas they do see its risks, its impact on the natural environment, and even the "slippery slope" opened up by genetic copying of animals. Most people seem to agree with the strong statement "animal cloning serves no useful purpose": Austria reaches a mean value of 6.6, followed by the Netherlands with 6.4, Germany and France both with a mean value of 6, Italy 5.5, the UK 5.4, Denmark 4.9 and Spain 4.3. There are several reasons for the lack of approval for animal cloning, but certainly the cost-benefit analysis many individuals use when assessing a particular technological development must wait until the range of possible benefits has been understood by the public. Most people seem willing, in the current scenario, to go as far as banning research on animal cloning in all EU countries, foregoing possible future benefits (an inclination that is particularly marked in Austria, with a mean value of 6.8, followed by France with a mean score of 6.8, and the Netherlands with 6.7).

A review of attitudes to animal cloning in specific scenarios or cases provides some clues as to the causes of public resistance. Genetic copying of animals for a widely accepted goal such as the preservation of highly endangered animal species softens opposition to cloning a little bit, but not to the extent of moving the mean value to the approval zone (with the exceptions of Spain, with a mean value of 5.4, and Poland with a mean score of 5.2).[3] However, it is revealing that the rejection level increases dramatically when the goal is to replicate specimens (bulls or horses) of exceptional quality: in no country does the mean score rise above 3.5, and in most it remains below 2.5 (Germany 2.4, France 2.4, Netherlands 2.5, Austria 2.5). Apparently mass production of exceptional specimens evokes strong negative feelings: the "direct challenge to the sacredness of the biological individual" that Hornig Priest referred to in the general context of genetic engineering may be at work here (Hornig Priest 2001).

There is still little empirical evidence on how people see the human cloning issue, and even fewer analytical studies. The report published by The Wellcome

---

[3]   The statement offered to the interviewees reads as follows: "If a particular animal species were so endangered that only a few specimens were left, to what extent would you find it acceptable or unacceptable to clone some of those animals so as to try to save the species? Please use a scale from 0 to 10, where 0 means you find this absolutely unacceptable, and 10 means you find it completely acceptable".

Trust in 1998, *Public Perspectives on Human Cloning* (The Wellcome Trust 1998), is a useful resource for assessing how the public has received this new development, although the methodology used in this study – a variant of "focus groups" –, means that results cannot be formally extrapolated to society as a whole (in this case, British society). Nevertheless, some of its main conclusions are of interest here. The prospect of research into human reproductive cloning and related research for medical applications, two years after the cloning of Dolly was announced, still met with serious reservations among all discussion group participants. These reservations were not attenuated by an increase of information, or by the level of knowledge about the techno-scientific side of cloning. Even some segments who would potentially benefit from advances in cloning appear to have significant doubts about its application. Another relevant finding of the *Wellcome Trust* study is the low level of trust in both the scientific community working on cloning and the effectiveness of regulations: focus group members were convinced that scientists will, sooner or later, start doing research into human cloning, whatever existing regulations say. The most striking and instructive conclusion of the *Wellcome Trust* report is the scant weight of two canonical variables – the level of *knowledge* on cloning, and the *practical interest* of potential beneficiary groups of this line of research[4] – in shaping attitudes toward reproductive cloning, suggesting that the strongly negative perception of cloning derives from moral values, images, stereotypes, fears and emotions in the culture of advanced societies at the turn of the century.

A major factor shaping people's negative reactions to animal cloning today may be the presumed link between cloning in animals and humans: since many people seem to share the belief in a "slippery slope" dynamic, it is only logical that they oppose animal cloning as a way to prevent research from moving into the realm of human cloning. We have seen that many people have serious doubts about the self-control of scientists working in this field, and so they fear that "if animal cloning is allowed, human cloning will be allowed sooner or later". According to our data, this belief is particularly strong in Austria (with a mean score of 7.4), the UK (7.2), and in France, Poland and Germany (7.0). Even in Denmark, the country in which the inevitability of this linkage is perceived as weaker, the prevalent position assumes a scenario in which animal cloning is followed by the reproductive cloning of humans.

The cloning of Dolly and other animals, as covered in the press, has apparently convinced most people that the scientific and technical knowledge for cloning humans will be in place a few years hence. Once cloning moved out of the realm of science fiction and appeared to the public as a reality in the case of animals, and as not-too-technologically-remote a prospect in the case of humans, the moral debate has gained particular intensity. Few prospects provoke such universal moral rejection in European societies as reproductive human cloning. On a scale of 0 to 10, the mean score in answer to the question "to what extent do you think it is morally acceptable that human beings might be cloned in the future?" runs no higher than 2 in any coun-

---

[4]   Note, however, that the main benefits posed were in the field of reproductive cloning, not in possible medical applications for diseases like Parkinson's or Alzheimer's. It is reasonable to think that people from communities affected by such diseases would be favourably disposed toward research into non-reproductive cloning.

try, and in most (countries) stands at around 1.5. This shows the extremely strong level of moral rejection that human cloning triggers in the public imagination. Not even a series of different scenarios to which human reproductive cloning could hypothetically be applied succeeds in improving the score; indeed, in some cases, the rejection becomes even stronger than in the standard scenario.

The case of "a couple with infertility problems whose only chance of having a child is through human cloning" is rejected categorically as a pro-cloning argument in all the European societies analysed. In the two least critical countries, the UK and Spain, the mean scores, on a scale from 0 to 10, are 2.6 and 2.5 respectively. The remaining countries respond in the range between 1.5 (France) and 1.7 (Austria). Other cases in which reproductive cloning might be applied, selected from those that were debated in the press after Dolly's birth, met with even stronger opposition than that of the couple with fertility problems. Especially the scenarios of cloning "to replace a loved one after his/her death", "to replace a young child who had died", or "to clone people considered to be geniuses who have made great contributions to society", were met with extremely pronounced opposition in all societies.

The "social representation" of cloning, as it now stands, conjures up Frankenstein-type visions, and evokes deep-seated fears – evident in countless novels and films – that are basic to our culture, like the annihilation of human individuality through mass production. As two science and technology analysts put it, "appearing to promise both amazing new control over nature and terrifying dehumanisation, cloning has gripped the popular imagination [...] Dolly has become far more than a biological entity; she is a cultural icon, a symbol, a way to define the meaning of personhood, to describe social issues, and to express concerns about the forces shaping our lives. She provides a window on popular beliefs about human nature and the social order, on public fears of science and its power on society, and on concerns about the human future in the biotechnology age" (Nelkin and Lindee 1998).

4.6
Attitudes to Reproduction Technologies: IVF and Semen Bank Donors

Before examining current European attitudes to embryonic experimentation and PGD techniques, it is worth taking a glance back at the initial controversy surrounding human reproduction options which are now more or less accepted by the broad public, despite the Catholic church's doctrinal opposition since 1987. Specifically, this section looks briefly at the issues of IVF techniques and, in particular, the use of semen bank donors. IVF techniques also have an obvious tie-in with the issue of embryonic experimentation to obtain stem cells, given that one of the main sources of stem cells are the frozen embryos left over from in-vitro fertilisation treatment cycles.

After Louise Brown was born in 1978 through *in vitro* fertilisation (IVF), infertility joined the ranks of biomedically treatable problems. The technique rapidly gained social acceptance, and is now an expensive but more or less routine option, depending on the country. It is reckoned that since 1978 approximately one million

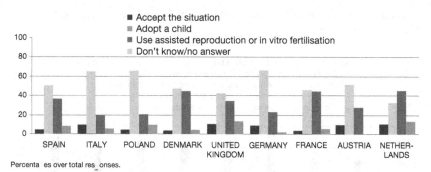

Percenta es over total res onses.

Fig. 4.3 From the following alternatives, which one would you recommend to a couple who wish to have children, but are unable to do so?

children have been brought into the world thanks to IVF techniques. We should remember, though, that twenty years back this medical breakthrough met with a mixed reaction. Debate about the ethics of the issue, with strong opposition from the Catholic church, was accompanied by gloomy (though contradictory) predictions about animal-human monsters and eugenics. A few years later, the fears had practically vanished, although a low-key debate around moral issues continued, mainly inspired by the Catholic church and pro-life groups. However, the recent debate about the use of spare embryos to obtain stem cells has brought a new moral angle to bear on the issue, though without any significant influence on social perceptions as of early 2003.

Although IVF techniques do not provoke the same moral reservations and fears now as when they first emerged over two decades ago, there is variation in social responses to the infertility problem. The preferred alternative among the three main options for a couple who wish to have children, but are unable to do so – to accept the situation, to adopt a child, or use assisted reproduction or *in vitro* fertilisation techniques – varies widely between countries (see figure 4.3). The passive option of accepting the situation finds scant support, with no more than 10% of the population advocating this viewpoint. The adoption of a child, as against the use of IVF techniques, is the alternative mainly chosen (in descending order) in Catholic countries or those with a large Catholic population: Germany (66% vs. 22% respectively), Poland (66% vs. 21%), Italy (65% vs. 20%), Austria (52% vs. 28%), Spain (51% vs. 37%) and the UK (42% vs. 34%). At the opposite extreme, the use of IVF is the prevalent position in the Netherlands (45%, vs. 33% opting for adopting a child), while in Denmark (47% vs. 45%) and France (46% vs. 44%), the two main options are virtually on an equal footing.

Recourse to semen bank donors is an IVF option for couples where the man has sterility problems, but it can also provide a means to have children in other social situations (this was one of the reasons, though not the main one, given by the Catholic Church in 1987 in its doctrinal opposition to IVF techniques). The use of semen bank donors also theoretically opens the door to the selection of children's physical characteristics. According to the empirical evidence obtained in our survey, European societies exhibit strong commonalities in judging the ethical status (acceptance or rejection, as the case may be) of the use of semen bank donors for a series of social scenarios.

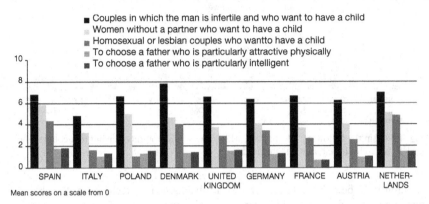

■ Couples in which the man is infertile and who want to have a child
■ Women without a partner who want to have a child
■ Homosexual or lesbian couples who wantto have a child
■ To choose a father who is particularly attractive physically
■ To choose a father who is particularly intelligent

Mean scores on a scale from 0

Fig. 4.4 Attitudes to use of semen bank donors (IVF treatments) (on a scale from 0 to 10, where 0 means disagree completely and 10 means agree completely.)

High response variability is not so much due to general ethical or religious stances that are applied uniformly, as it is to the highly specific civic and moral judgments made in each country about the conception of children outside the couple (not necessarily marriage), about the gender and sexual orientation of the mothers or fathers, and about the possible selection of the child's physical characteristics. European public opinion on this issue can be summed up as follows: a) there is extremely strong rejection of the hypothetical use of semen bank donors to make an a priori or "designer" selection of a baby's main physical characteristics (the median value on a scale from 0 to 10 is 0 in all countries, and the highest mean values are 1.8 in Spain and 1.6 in the UK); b) at the opposite extreme, we find a high or very high degree of acceptance in the case of couples where the man is sterile (three countries – Germany, Austria and the UK – have median values of 7, four countries – France, Spain, Poland and the Netherlands – have a median of 8, Denmark, with a median score of 9, has the most favourable position, and only Italy shows a strong division of opinions with a median of 5 and a mean value of 4.8); c) the use of semen bank donors for women without a partner meets with mixed responses in most countries, with rejection tending to win out (majority acceptance is only found in Spain with a mean value of 5.9, the Netherlands with a mean of 5.4, Poland with a mean of 5.0 and Denmark with a mean of 4.7); d) the case of homosexual and lesbian couples who want to have a child meets with rejection in all countries except the Netherlands (with a mean value of 5.2 and a median of 6).

These results demonstrate two points. The first – and formal – point is that the degree of acceptance of a given techno-scientific advance (in this case, IVF) may depend not so much on the technique *per se*, but rather on its most salient and visible applications and their perceived acceptability. The second and substantive point is the outright rejection, in all countries, of the use of genetic manipulation techniques for the selection of individuals' genetic traits. Eugenics and the notion of "designer babies" are received today with almost universal repulsion in Europe.

4.7
Stem Cells, Embryo Experimentation and the Status of the Embryo

As noted earlier, the great majority of scientific and technological advances are integrated silently, and many are only noticed by the public when products or services into which the new discoveries are "embedded" appear on the market. Generally, the basic research that enables process and product innovations remains hidden except to experts in the field and the limited segment of the "attentive public" that tries to keep track of developments in a given area (Miller 1983). In contrast, in some cases research attracts the interest of the mass media, some interest groups and organizations, and even of the general public before it has left the laboratory. Stem cell research is one of the most striking examples of this process. No other area of research in recent years has achieved greater salience or controversy, and drawn so much regulatory attention from governments. The controversy in this case does not revolve around possible biomedical applications, which are generally assessed positively, but around the research itself, specifically the use of human embryos as a source of stem cells.

4.7.1
Awareness and Information about Stem Cells

Despite the intensity of the current debate about embryo experimentation, specific issues in stem cell research are covered by the mass media with widely differing levels of salience and continuity between one country and another. It may be that a significant part of these differences in awareness and information about stem cells depend on how much they featured in national communications media in the period just prior to data collection; we know, for instance, that media coverage and debate on stem cell research kicked off one year earlier in the UK than in other societies like Spain, where it has recently become a hot topic, while media interest has declined in the UK.

We can classify the nine European societies surveyed into three large groups in terms of the public's level of declared awareness about stem cell research:[5] 1) *Medium-high level:* Italy (68%), Germany (66%), Denmark (60%), France (56%), Spain (54%), 2) *Medium level:* UK (48%) and the Netherlands (47%), 3) *Low level:* Austria (38%) and, at a considerable distance, Poland (28%). Awareness is a limited but interesting indicator: it suggests that, despite the debate raging among regulators, churches and other interest groups, and despite the issue's high media profile, almost half the adult population in five countries has never even heard about it (which doesn't mean that, given the right stimulus such as a mere mention of "embryo experimentation", it would not draw defined reactions from this broad subset of society).

---

[5] The question serving as a basis for this classification was: "Recently information about a type of cell called 'stem cells' has appeared. Have you read or heard any news about this kind of cell?".

But the "awareness" score may result from unequal coverage – in intensity and temporal distance from the survey – in national media of the research, its possible applications and its ethical implications. In addition, it does not measure people's knowledge of the topic. It must therefore be supplemented by an indicator combining awareness of stem cells with a minimum of information about them.[6] This combined indicator sets two conditions for considering a person minimally informed about stem cells. First, they must declare having heard, seen or read information on the subject, and second, they must have responded "probably false" or "absolutely false" to the statement "it is possible to extract stem cells from human embryos without destroying the embryos". The idea was to "filter out" the subjects who declared they knew nothing about stem cells before the interview, but guessed the right answer to the question on stem cell extraction and the destruction of the embryo. According to this indicator (which does not require a high level of scientific literacy), even in the best-informed country, Germany, less than a third of the adult population had reached even an elementary threshold of familiarity with the stem cell issue by the first quarter of 2003. Estimates of the minimally informed population segment in the surveyed countries are (in descending order): 1. Germany 28.4%, 2. Italy 21.3%, 3. Denmark 20.8%, 4. Netherlands 20.4%, 5. Spain 16.5%, 6. UK 15.9%, 6. France 13.6%, 7. Austria 11.6%, 9. Poland 8.2%. To judge by these results, for a large majority of Europe's population, attitudes and responses to embryo experimentation for the purpose of obtaining stem cells are only weakly related to knowledge and strongly to images, fears, stereotypes and moral beliefs about embryos, their status and their rights. Our findings also indicate the size of the task awaiting the scientific community in explaining the science of this issue to the lay population in years to come.

4.7.2
The Status of the Embryo

Acceptance or rejection of current research on embryonic stem cells depends on a wide range of variables, which can be grouped into the following categories: degree of familiarity with the scientific dimension of the research, perception of possible biomedical (and, secondarily, economic) benefits, general worldviews (prominently, perceptions of and trust in science, images of nature and of what is "natural", general religious beliefs), risk perceptions and, evidently, the perception of the moral status of the embryo. Given people's deficient knowledge about stem cell research, as identified in our survey, and their scant awareness of its possible medical benefits, we can assume that the vast majority base their value judgements on worldviews, risk perceptions, and especially – as a variable related more directly to the attitude object – their perception of the status of the embryo. Obviously this perception is not sufficient to explain the variability of attitudes towards embryo experimentation, but it does function as a powerful predictor. It is also likely that perceptions of the status of the embryo are significantly (although by no means exclusively) tied to individuals' religious beliefs or, from a broader perspective, to the dominant religion in each society;

---

6   As a rudimentary and, certainly, rather indirect measure of knowledge on stem cells, respondents were offered the following statement: "It is possible to extract stem cells from human embryos without destroying the embryos", preceded by the introduction: "Please tell me, for each of the following sentences, the extent to which you believe they are true or false".

the Catholic church especially has taken an actively critical stance that is markedly different from the positions of the Anglican and Lutheran Protestant churches. Even more importantly, the degree of belief in a religion (Roman Catholic, Protestant or other) vs. no belief in any religion is a significant predictor of attitudes to embryonic stem cells, with the non-religious segment more in favour that the religious one.

The definition of the moral embryo's status is therefore central to our analysis. In order to secure the widest possible variability in the empirical data on perceptions of this status, the survey questionnaire offered four options that appear under different guises in current debates among the better informed population segments and stakeholders: 1. "a human embryo that is a few days old is a mere cluster of cells, and it makes no sense to discuss its moral condition"; 2. "a human embryo that is a few days old has a moral condition halfway between that of a cluster of cells and that of a human being"; 3. "a human embryo that is a few days old is closer in its moral condition to a human being than to a mere cluster of cells"; 4. "a human embryo that is a few days old has the same moral condition as a human being".

The first step is to estimate the percentage of the population choosing either of the two extremes, that is, positions 1 ("cluster of cells") and 4 ("same moral condition as human being"), versus those taking more moderate views. Other things being equal, debate on the issue is likely to be more ideological, or inversely more piecemeal – open to dialogue and shades of opinion – depending on the degree of polarisation in each society. The size of the group choosing the intermediate option (i.e. "a human embryo that is a few days old has a moral condition halfway between that of a cluster of cells and that of a human being") may also indicate whether, given the right circumstances, there is a social base for a rational debate that is neither holistic nor ideologised. Furthermore, the degree of acceptance of position 1, plus that of the intermediate position, can be taken as a measure of each society's openness to human embryo experimentation for biomedical ends. Table 4.1 sets out the response distribution for each country surveyed.

Table 4.1 Perceptions of the Status of the Human Embryo (%)

Country	Cluster of Cells	Halfway cluster cells-human being	Closer to a human being	Same moral status as a human being	Don't know
Austria	6.4	22.3	24.3	32.6	10.3
Denmark	36.6	23.5	11.6	22.2	5.2
France	22.8	22.6	16.5	28.3	6.9
Germany	8.8	21.8	25.9	28.9	10.7
Italy	20.3	15.7	13.7	37.3	10.4
Netherlands	19.5	34.5	16.5	21.5	7.1
Poland	13.8	15.4	9.8	38.2	18.2
Spain	21.5	18.8	12.5	29.8	15.5
UK	25.5	16.8	11.7	23.6	9.6

Several dimensions that are condensed in this table stand out. One is that Europe (represented here by nine countries) certainly offers a plurality of perceptions regarding the status of the human embryo, as evidenced by the wide variability of responses within each country, and even more so between countries. Secondly, in all the countries surveyed, albeit with marked differences, there is significant support for the intermediate ("halfway") and the other nuanced position ("closer to a human being"). Obviously, the larger the segment opting for the intermediate position, the broader the social base will be for regulations that balance different viewpoints and moral sentiments. Such regulations are also facilitated if the "neutral" ("halfway") segment is flanked by approximately equal weights at the two extremes.

In six of the nine countries, the position that finds the greatest approval (although it falls well short of absolute majorities), is the one that ascribes to a human embryo a few days old the same moral condition as to a human being. These are countries with very large Catholic majorities, where the Roman Catholic Church plays an active cultural role (the exception being Germany, where Catholic and Protestant religions are comparable in number, so that there is a sizeable Catholic population, but well below the other societies in the group). The six countries are, in descending order: 1. Poland 38%, 2. Italy 37%, 3. Austria 32%, 4. Spain 30%, 5. Germany 29%, 6. France 28%. Since estimates give a very high percentage of Catholics in all these societies, it is clear that people are either not well aware of the Catholic church's opposition to embryo experimentation, or the official doctrine is not shared by broad segments of this religion.[7]

Only in two of the nine countries surveyed is the majority perception (again, relative majority) that "a human embryo that is a few days old is a mere cluster of cells, and it makes no sense to discuss its moral condition". These are Denmark, with 37% and, at a significant distance, the UK with 25% (in this case, almost on an equal footing with the 24% who believe that a human embryo has the same moral condition as a human being). In Denmark, most of the population is Evangelical Lutheran (around 95%) while in the UK a large majority belongs to the Anglican church. Whatever factors may be at work, clearly neither of these societies has a widespread organized religious faith that strongly opposes distinguishing between an embryo and a human being, at least in the context of embryo experimentation. Finally, in one country, the Netherlands, the intermediate option attracted the greatest support, with 35%. The majority of the Dutch are "unaffiliated" with any religion, with Catholics and Protestants weighing in at around 31% and 21%, respectively. If we add that 19.5% of respondents perceive a human embryo of a few days of age as "a cluster of cells", it seems in principle likely that this society would offer more space for the development of embryonic stem cell research.

As far as the ratio between the two extreme options is concerned, i.e. the ratio between the "embryo as human being" option and the "embryo as a cluster of cells" option, large differences among European societies emerge. In one group of countries, for each individual who sees the embryo as a cluster of cells, there are from

---

7　Estimates of these countries' Catholic populations in 2002 are: Poland 95% (75% practising), Italy (97%), Austria (78%), Germany (34%), Spain (94%) and France (83–88%), although according to our own data, based on the survey used in this chapter, those estimates are on the high side, i.e. actual figures may be lower for most countries.

just under two to just over five who give it the same status as a human being. These are, as we can see from table 4.2, Austria, Germany, Poland and Italy. In three other societies, there is greater equilibrium: from greatest to smallest, the Netherlands, France and, at greater distance, Spain. Finally, the ratio is reversed – i.e. more people see the embryo as a cluster of cells than as a human being – in only two countries: the UK, where the difference is minimal, and Denmark, where the difference is marked.

Table 4.2 Ratio, by Country, between the Embryo as Human Being and as Cluster of Cells

Country	Ratio
Austria	5.1
Germany	3.3
Poland	2.8
Italy	1.8
Spain	1.4
France	1.2
Netherlands	1.1
UK	0.9
Denmark	0.6

Finally, perceptions of the human embryo in Europe can be analysed in terms of a size comparison between three segments in each country: the *first* one comprising those who see the embryo as a mere cluster of cells, a *second* one including those who see it as halfway between a cluster of cells and a human being, and a *third* one that aggregates the semantically and evaluatively close options of the human embryo as a human being and the embryo as closer in its moral condition to a human being than to a mere cluster of cells (see table 4.3). To the extent that the regulation of embryo experimentation might be a function of each society's perception of the status of the embryo, one would expect a tendency toward restrictive rules in Austria, Germany, Italy and Poland. Conversely, a more permissive regulatory stance (assuming always that perception of the embryo's status is a fundamental, though by no means exclusive variable) could be expected (in descending order) in Denmark, the UK, and the Netherlands. The middle ground, with more uncertain positions susceptible to being swayed one way or another depending on other variables and factors (mainly the "voices" of various organised stakeholders from religious groups and institutions to political parties, patient asociations and scientific and medical institutions), is represented by Spain and France.

Table 4.3 Perceptions of the Status of the Human Embryo (%)

Country	Cluster of Cells	Halfway cluster cells- human being	Closer to a human being + Human being	Don't know
Austria	6.4	22.3	56.9 (1)	10.3
Denmark	36.6	23.5	33.8 (9)	5.2
France	22.8	22.6	44.8 (5)	6.9
Germany	8.8	21.8	54.8 (2)	10.7
Italy	20.3	15.7	51.0 (3)	10.4
Netherlands	19.5	34.5	38.0 (7)	7.1
Poland	13.8	15.4	48.0 (4)	18.2
Spain	21.5	18.8	42.3 (6)	15.5
UK	25.5	16.8	35.3 (8)	9.6

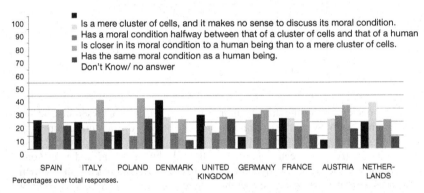

Fig. 4.5 Perceptions of the status of the embryo. "A human embryo that is a few days old...."

4.7.3
Beliefs about the Beginning of Individual Human Life

The debate about the *status of the embryo* is a relatively new issue for public opinion as a whole (though not for certain minorities like pro-life associations or pro-choice associations). A more familiar "frame of reference" is that of differing views on the *beginning of individual human life,* which, as we know, has influenced perspectives and regulations on abortion. The two issues are related from a logical standpoint but, depending on the level of the debate or the decision (individual, collective or regulatory), one or other frame can be "activated" (or retrieved) from memory or the repertoire of arguments and used as a conceptual framework. The standard views about the beginnings of human life which have crystallised in the current map of societal and individual beliefs are: 1. At the moment when the egg and the sperm unite, 2. Two weeks after conception, when different tissues can be distinguished, 3. Three months after conception, when growth of the foetus begins, and 4. At the time of birth. The response distribution in the nine European countries can be seen in table 4.4.

Table 4.4 Beliefs about when Individual Human Life Begins (%)

Country	Egg-sperm	2 weeks	3 months	Birth
Austria	33.2 (6)	20.8	23.7	6.5
Denmark	30.9 (8)	11.4	39.4	15.3
France	34.6 (5)	17.1	30.6	14.5
Germany	38.3 (4)	15.7	31.5	9.0
Italy	52.0 (2)	10.9	18.4	7.3
Netherlands	25.9 (9)	20.0	41.3	8.9
Poland	52.1 (1)	10.3	12.2	11.3
Spain	44.3 (3)	11.9	20.8	12.3
UK	31.6 (7)	14.0	24.8	14.9

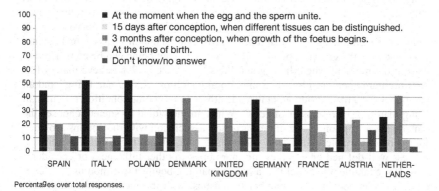

Percentages over total responses.

Fig. 4.6 Beliefs about when individual human life begins

A glance at the table reveals, as in the case of the status of the embryo, the wide variability existing in most European societies: the map of opinions on human life is highly differentiated and plural in the Europe of the turn of the century, which suggests that standardising regulations across countries (despite the latent aspiration in Europe to converge on multiple planes, including core beliefs and values) would be a very difficult task at this point in time. Moreover, in some countries opinions are so evenly divided between the two dominant options ("when the egg and the sperm unite" and "3 months after conception") that regulating matters like embryo experimentation would demand extensive dialogue if one wanted to achieve a high level of consensus and legitimacy in a question affecting deep moral values and beliefs.

To say that views on the status of the human embryo and on the beginning of human life have obvious logical connections is not to imply that these linkages are obvious or automatically activated for all individuals. In some cases, as in the assessment of alternative views on abortion, it is likely that the frame of reference activated will be that of the beginning of human life, a frame that is currently structured around the four positions stated above. The debate about embryo experimentation has yet to crystallise in the public opinion, and its frame of reference (for instance, how people visualise an embryo) is less consolidated than that of the beginning of human life. Research on social cognition has shown that individuals form and use attitudinal schemas

> which don't impose strong logical dependencies among issues, but are open to many kinds of linkages among them, from associations based on previous experience to the 'glue' provided by emotions, and social influences ranging from the immediate network of social relationships all the way to the culture of a particular society at a specific point in time. (Pardo, Midden, and Miller 2002)

In the case that concerns us here, though there may be no strict logical dependence between the responses to the two issues (status of the embryo, beliefs about the existence of a new human being), we do observe strong statistically significant associations with substantive significance (the non-independence of these two beliefs was measured through the Pearson *Chi*-square test for significance of the relationship between categorical variables, and found to be highly significant in all countries). For instance, in the case of Denmark, among those responding that a new human life doesn't start until birth, 68% believe that a human embryo is simply a cluster of cells and only 7% that it has the same moral condition as a human being. Conversely, of the group responding that "a new human being can be considered to exist at the moment when the egg and the sperm unite", 57% believe that "a few days old human embryo has the same moral condition as a human being" and an additional 15% that "a human embryo is closer in its moral condition to a human being than to a mere cluster of cells". When an individual or society chooses the "intermediate" option in one of the variables, association with the options provided for the other variable will show a wider dispersion, suggesting either that attitudes are less well-defined or that they are more open to other influences or considerations (for example, the belief that a new human life begins three months after conception is compatible, for a number of reasons, with the idea that the embryo is more than a mere cluster of cells and that the status of a-few-days-old embryo is halfway between that of a cluster of cells and that of a human being; conversely, there is a fairly low probability for this to occur with someone opting for an extreme position, like that life begins when the egg and the sperm unite). Evidently, then, there is a large degree of association between the two *schemas* about life (although this association does not occur in a deductive manner or with logical-formal implications), and the statistical distribution of both points to the major differences existing within each of a number of European societies and, more dramatically, between them.

4.7.4
Religious Beliefs, the Status of the Embryo and the Beginning of Individual Life

The beliefs about both the moral status of the embryo and the beginning of human life have been shaped in modern societies by many cultural traditions, but mainly by scientific advances in the Biomedical Sciences and, more importantly, by religion. There are other relevant secular moral criteria and beliefs operating in our culture, but they have gained influence on visions about matters of life and death only slowly and in opposition to mainstream religious principles and guidelines. Thus, although its role varies from country to country, the religion variable is of great interest for the differentiation of beliefs held in European societies about when individual life begins and about the moral condition of the embryo, beliefs which in turn influence attitudes toward embryo experimentation.

The main divide runs between individuals who believe in one of the major religions and those who declare themselves "unaffiliated" or as not having any particular religious belief. If we focus on the two extreme positions for each of the moral frames of interest here (i.e. the status of the embryo and the beginning of individual life), we observe a common pattern through all the European countries included in our study: religiously affiliated individuals (no matter if they are Roman Catholic, Lutheran Protestant, Anglican or other) strongly believe that "a new human being can be considered to exist at the moment when the egg and the sperm unite", and also that "a few days old human embryo has the same moral condition as a human being". In general, the "nonbelievers" tend to maintain a position closer to scientific facts and images of the embryo and the beginning of human life than the believers. According to our data, the percentage of non-religious people is high or even very high in most European societies analyzed here: in descending order, the Netherlands 54.0%, the UK 47%, France 41%, Germany 39%, Austria 34%, Spain 24%, Denmark 22%, Italy 11%, and Poland 3%.

If we look at the two extreme categories of beliefs about the beginning of individual human life (i.e., "at the moment when the egg and the sperm unite" vs. "at the time of birth"), the differences by country are very pronounced, particularly in Catholic countries (see table 4.5). The difference between the two extreme responses given by Catholics is particularly high in Spain, Italy, Poland, and Germany (above 40 points) and, although also strong, at least ten points lower in the Catholic populations of Austria, the Netherlands and France (with a difference between the two extreme choices of 30 points or less).

The same pattern with some variations is found in relation to beliefs about the moral status of the embryo. The largest differences between the two end categories, i.e. "a human embryo that is a few days old is a mere cluster of cells, and it makes no sense to discuss its moral condition" and "a human embryo that is a few days old has the same moral condition as a human being" are observed in the Catholic population of Italy, Poland, Germany, Austria, and France, and in the Protestants of Germany and the Netherlands (see table 4.6).

The most salient result is the demarcation line that seems to exist between believers and nonbelievers regarding their views about the beginning of individual life and the moral condition of the embryo. The size of the subset of non-religious believers in each country has grown over the last decades of the XXth century, and it could have a significant impact on attitudes toward embryo experimentation and other biomedical advances that touch on moral principles inspired by religious faiths. At any rate, policy-makers in countries with significant minorities of nonbelievers have to be careful not to impose regulations that strongly depend on religious principles, while at the same time taking into account the ample influence of moral criteria inspired by the major religious faiths.

In the short term, ruling elites and interest groups can take positions and make decisions that are far removed from the perceptions and feelings of public opinion. But further ahead, under democratic conditions, issues framed within powerful force fields that affect core values and beliefs, will tend to fall into a regu-

Table 4.5 Beliefs about when Individual Human Life Begins by Religious Beliefs (%)

		Egg-sperm	2 weeks	3 months	Birth	Don't know	Difference 'egg-sperm' – 'birth'
Spain	Catholics	49.4	12.4	18.6	9.1	10.6	40.3
	Nonbelievers	29.1	11.3	27.1	21.1	11.3	8.0
Italy	Catholics	54.8	11.2	17.6	5.8	10.5	49.0
	Nonbelievers	37.7	10.1	22.6	14.2	15.3	23.5
Poland	Catholics	53	10.6	11.6	11.2	13.5	41.8
	Nonbelievers	35.6	6.6	20.3	14.1	23.4	21.5
Denmark	Protestants	32.3	11.3	38.9	14.3	3.1	18.0
	Nonbelievers	21.4	10.7	45.4	20	2.6	1.4
United Kingdom	Anglican	33.3	14.9	29.6	14.2	8.1	19.1
	Other Christian/ Evangelical	36.3	16.9	17.8	15.8	13.2	20.5
	Nonbelievers	28.3	12.6	25.6	15.7	17.7	12.6
Germany	Catholics	48.3	17.2	28.4	5.3	0.8	43.0
	Protestants	43.9	11.2	27	7.1	10.7	36.8
	Nonbelievers	30.2	17.6	35.3	11.7	5.1	18.5
France	Catholics	40.9	18.7	26.2	10.5	3.8	30.4
	Nonbelievers	28.3	14.7	36.2	18.1	2.6	10.2
Austria	Catholics	35.6	22.3	21.7	5.1	15.3	30.5
	Nonbelievers	28.8	19.9	25.5	8.6	17.2	20.2
Netherlands	Catholics	28.7	20.2	40.7	6.8	3.7	21.9
	Protestants	37.3	24.6	30.4	4.3	3.4	33.0
	Nonbelievers	20.1	16.4	47.3	12.3	4	7.8

latory treatment aligned with the majority view of the general public. Otherwise, regulations can have no social authority or even efficacy, especially when their effects can be avoided by crossing national frontiers. Meantime, values and moral sensitivity may change if the consequences of adhering to them clash with widely accepted goals – if, for instance, they block or delay research aimed at finding cures for serious diseases. Therefore, in addition to the frames of refer-

Table 4.6 Beliefs about the Moral Condition of the Embryo by Religious Beliefs (%)

		A cluster of cells	Halfway cluster cells-human being	Closer to a human being	Same moral status as a human being	Don't know	Difference 'same moral status as a human being' – 'a cluster of cells'
Spain	Catholics	16.8	18.6	11.8	33.5	19.3	16.7
	Nonbelievers	35.4	18.8	14.3	19.6	11.8	-15.8
Italy	Catholics	17.7	15.6	14.5	40.2	11.8	22.5
	Nonbelievers	33.0	17.6	9.5	22.0	17.9	-11.0
Poland	Catholics	13.4	14.2	9.9	39.5	22.9	26.1
	Nonbelievers	19.7	26.2	9.1	19.9	25.1	0.2
Denmark	Protestants	35.6	23.2	13.1	22.9	5.3	-12.7
	Nonbelievers	44.2	25.1	8.7	14.6	7.4	-29.6
United Kingdom	Anglican	28.0	14.9	16.2	25.3	15.5	-2.7
	Other Christian/ Evangelical	20.6	23.5	13.3	22.7	19.8	2.1
	Nonbelievers	26.9	16.7	9.6	21.5	25.4	-5.4
Germany	Catholics	9.6	21.6	25.2	34.9	8.7	25.3
	Protestants	6.2	20.4	23.7	32.0	17.7	25.8
	Nonbelievers	9.4	23.1	26.5	24.7	16.3	15.3
France	Catholics	14.6	20.7	18.8	35.6	10.3	21.0
	Nonbelievers	32.1	23.7	14.8	20.1	9.3	-12.0
Austria	Catholics	4.9	21.1	23.4	33.8	16.8	28.9
	Nonbelievers	8.4	24.8	25.1	27.5	14.2	19.1
Netherlands	Catholics	19.3	29.0	19.0	24.3	8.3	5.0
	Protestants	9.4	25.4	21.6	35.1	8.5	25.7
	Nonbelievers	26.0	36.5	11.9	16.2	9.4	-9.8

ence for embryo status analysed in this section, we will complete our study of European public opinion with a profile of predispositions for accepting or rejecting embryo research for specific purposes and under given conditions and assumptions.

4.8
Attitudes toward Embryo Experimentation

For the great majority of the population, moral debates and dilemmas do not present themselves as abstract matters to which a uniform criterion or principle is applied (although this can be the case with some minorities). Rather, they arise in specific contexts, usually cases in which various core values or principles could apply. Frequently, these are competing values leading to different decisions or courses of action. In such cases, individuals and society as a whole have to seek a trade-off between all possible decisions by weighing up each set of values and principles against the others.

4.8.1
Spare Embryos and Embryos Created for Biomedical Research

The debate on the acceptibility of using human embryos as a source of human stem cells has revolved around two focal points or concrete scenarios: the use of "spare embryos" (embryos left over from fertility treatments, i.e. embryos that will never develop) and the use of embryos created specifically to further biomedical research. It was stressed in both cases that the ultimate goal of the research is to acquire knowledge that may in the future be applied to the treatment of serious diseases (diabetes, Alzheimer's, Parkinson's, cancer). These goals were carefully specified when gathering the data presented here, under the just-stated assumption that though some population segments of society may base their judgment on a universal, religious or secular moral principle, the majority will use several criteria, whose weight will vary according to specific situations or scenarios. The questions presented to the interviewees made clear the following points: a) first, that the idea was to judge the morality of experimenting with "human embryos that are *a few days old*", thereby establishing, as much as possible, a uniform frame of reference so responses would not based on different images of what an embryo is; b) second, that the purpose of the research was "to obtain stem cells so as to find *treatments for serious diseases*", thereby introducing the trade-off between the embryo's rights and a possible future cure for grave illnesses; c) finally, the two use scenarios, that of *spare embryos* and that of *embryos specifically created for conducting biomedical research*, were presented separately.

The mean values of the response distribution obtained in each country show that, in all of them, the two use scenarios are evaluated in clearly different ways: as one might expect, the spare embryo scenario scores higher. Whereas in seven out of the nine countries (all except Austria and Poland), research with spare embryos was judged acceptable, only in one case (Denmark) did respondents approve embryos being created for research (see table 4.7). The differences between the mean values for the two scenarios are highest in Germany (a difference of 2.3, from moderate approval in the first case to strong rejection in the second), followed by France (difference of 2.1, from the second highest level of acceptance of the nine to rejection by a fairly small margin), the Netherlands (difference of 2.0, from approval to rejection), the UK (difference of 1.1, from approval to non-approval, with neither opinion very emphatic) and Denmark (a difference of just 1.0, with approval clearly winning out in both scenarios).

Table 4.7 Attitudes to Embryo Experimentation: Mean and Median values

Country	Spare Embryos		Embryos created for research	
	Mean	**Median**	**Mean**	**Median**
Austria	3.6	3	2.7	2
Denmark	6.7	8	5.7	6
France	6.1	7	4.0	4
Germany	5.1	5	2.8	1
Italy	5.5	6	4.1	5
Netherlands	5.7	7	3.7	3
Poland	4.9	5	3.9	3
Spain	5.6	6	4.7	5
UK	5.5	6	4.4	5

These mean and median measures of the distribution of attitudes toward embryo experimentation tell us part of the story, but also conceal a fair amount of relevant information, from an analytical standpoint (insight into attitudes) as well as from a practical one (they exclude issues of interest to the regulator). The first conclusion suggested by the differing mean and median scores – markedly different in some countries – is that we are up against skewed distributions where the mean is being pulled toward the long tail, i.e. the extreme observations. The median, a more resistant measure of the centre of distribution, has to be taken into account in interpreting the data, and we also have to analyse the overall distribution by means of a graphic or frequency table. Another important issue is the number of missing cases in the calculation of these measures of center. As we can see from table 4.8, the countries included fall into two groups in terms of the percentage of missing cases. The first group, including Poland, Spain, Austria, the UK and Italy, stands out by the unusually high number of missing cases. Although interpretations of this number may vary, it suggests that a large subset of the population did not have defined attitudes toward embryo experimentation by the first quarter of 2003. The second group of countries, comprising the Netherlands, France, Germany and Denmark, all fall within the conventional range for missing cases, though the numbers for Denmark and Germany are unusually low. In principle, this would suggest a higher level of information and/or judgment capacity in the adult population on the issue of embryo research. Although the measures of center and percentages were calculated using "valid cases" only (i.e. with no missing values for the two variables of interest), it should be borne in mind that, particularly in the first group of countries, the percentages are higher than would have been the case if all sample cases had been counted in (cases with and without missing values).[8]

Table 4.8 Missing Cases on Attitudes toward Embryo Experimentation (%)

Country	Spare Embryos	Embryos created for research
Poland	27	28
Spain	22	22
Austria	21	20
UK	20	22
Italy	17	17
Netherlands	9	8
France	8	9
Germany	7	4
Denmark	3	3

There are important aspects of people's attitudes toward embryo experimentation which central tendency values do not express. One is the degree of *polarisation* of attitudes, most graphically (though of course not exclusively) reflected in the percentage of individuals giving extreme responses ("0", total rejection; "10", total approval) to each of the two scenarios (spare embryos, embryos created for research). Table 4.9 shows that there are great differences in the way each scenario is scored. In the case of spare embryos, at least half the countries record fairly similar percentages at the extreme ends of the scale. By contrast, we find radically different percentages at the two extremes in the case of embryos created for research. What we observe in the latter scenario are very broad segments vehemently opposed to this research option, and only a small minority energetic in its support. The highest levels of outright rejection to the creation of embryos for research were recorded in Germany and Austria and, at ten points' distance, Poland. The case of

---

[8] As an example, the degree of acceptability of experimenting with spare embryos was 0 in Poland in 24% of cases, if percentages are worked out not on the basis of total cases in the sample, but including only those individuals able and willing to give a response (what are technically known as "valid cases"). If, on the other hand, the percentage is calculated on the whole of the sample (whether or not respondents have a view on the issue presented), the result is 17.8%. For the purpose of this analysis, all calculations were made only on the basis of valid cases, those whom one can assume to have an attitude or at least a formed opinion. This procedure facilitates intra-country comparison of the percentages opting for one or the other response, as well as the prevalence of certain responses between different countries. Note, however, that in some areas and countries, we have learned little about the possible attitudes of a subset of the population that did not wish or felt unable to reply. It is plausible to postulate that most of the cases of non-response indicate the absence of an opinion (due to lack of information, the esoteric nature of the issue, etc.).

Table 4.9 Polarisation of Attitudes toward Embryo Experimentation (%)

Country	Spare Embryos		Embryos created for research	
	0 (Total rejection)	10 (Total approval)	0 (Total rejection)	10 (Total approval)
Austria	29.5	5.3	42.2	3.3
Denmark	10.3	28.7	15.2	21.3
France	12.1	19.3	29.5	9.5
Germany	17.5	13.7	43.9	4.3
Italy	15.9	14.0	28.0	7.8
Netherlands	14.1	10.9	27.3	4.8
Poland	24.4	20.4	33.4	14.9
Spain	15.2	16.9	20.6	12.4
UK	15.8	14.4	22.4	8.3

Denmark is an exception to the general pattern, with the segment totally in favour of this type of research outnumbering the segment totally against.

Another perspective on attitudes toward embryo experimentation that is of interest analytically and to the regulator, emerges from the response distribution across three subsets or segments. The first is made up by the population clearly rejecting embryo experimentation (represented by scores 0 to 3, both inclusive), the second those opting for relative neutrality or mild rejection/approval (scores 4 to 6, both inclusive), and the third those decidedly in favour of this type of research (scores 7 to 10, both inclusive). Table 4.10 shows the results of this partition.

Again we find that Poland and, above all, Austria take a rejection stance that diverges from that of other countries, which approve the use of spare embryos for biomedical research ends. The majority of respondents in seven European countries who accept experimentation with spare embryos clash with minorities ranging from 19% (in Denmark) to 33% (in Germany). All in all, however, the existence of a substantial neutral (or mild rejection/approval) opinion alongside the subsegment explicitly in favour suggests that these societies could adopt permissive and more or less homogeneous regulations on the use of embryos left over from IVF treatments for biomedical research. By contrast, the creation of embryos for the same purpose meets with widespread rejection, Denmark being the sole exception. Opposition is less pronounced in Spain and the UK, where opinion is quite clearly divided. We conclude that in this case (embryos created for research), at the time of writing, the course of action most in keeping with public perceptions and sensibilities would be specific national regulations with a restrictive orientation (wholly or partially, depending on the country, with the exception of Denmark).

Table 4.10 Attitudes toward Embryo Experimentation: Three Segments (Rejection, "Neutral", Approval) (%)

Country	Spare embryos			Embryos created for research		
	0–3 Rejection	4–6 "Neutral"	7–10 Approval	0–3 Rejection	4–6 "Neutral"	7–10 Approval
Austria	52.4	27.6	20.1	63.0	23.8	13.2
Denmark	19.1	17.5	63.4	28.8	21.9	49.8
France	22.1	23.4	54.4	48.1	23.1	28.8
Germany	33.1	26.3	40.7	63.7	19.0	17.1
Italy	25.2	30.7	44.2	44.4	27.6	27.9
Netherlands	24.8	25.0	50.2	50.9	25.7	23.2
Poland	39.7	21.6	38.7	50.9	20.7	28.3
Spain	27.4	27.6	44.9	37.9	29.8	32.9
UK	28.0	25.0	47.0	41.1	27.3	31.6

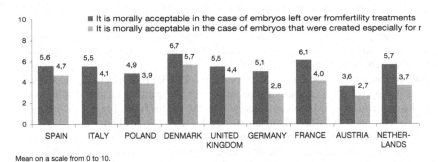

Mean on a scale from 0 to 10.

Fig. 4.7 Moral acceptability of human embryo experimentation (on a scale from 0 to 10 where 0 means it is absolutely unacceptable and 10 means it is completely acceptable.)

4.8.2
Medical Benefits vs. the Embryo's Rights

The acceptance (or rejection) of embryo experimentation for biomedical research can be measured more directly, and in a way that is more accessible to public perceptions, as a trade-off between the interests and rights of the embryo and the possible medical benefits to many people. The statement used to gauge the public's attitudes was: "The medical benefits for many human beings that can perhaps be obtained in the future thanks to research with embryos that are a few days old are much more important than the embryos' rights". Respondents were asked to state their agreement on a scale of 0 to 10, in which 0 signifies total disagreement and 10 total agreement.

The mean values of four countries indicated acceptance that the medical benefits of the many takes precedence over the rights of embryos, although in no case without reservations. These countries were, in decreasing order of approval: Denmark with a mean value of 5.5, Spain 5.3, UK 5.0, and France 5.0. Not far behind came the Netherlands with a mean value of 4.6. The highest levels of rejection for the primacy of medical benefits over embryos' rights were found in Germany (with a mean value of 3.9), Italy (4.0) and Austria (4.1). It is interesting to regroup these responses, as we did above, into three subsets of rejection, neutrality (or moderate rejection/approval) and firm approval. The results (see table 4.11) give a deeper insight into the earlier findings and confirm that the strongest approval is in Denmark, followed by more lukewarm approval in Spain, the UK, France and the Netherlands, and contrasting with the rejection verdicts of Germany, Italy, Poland and Austria.

These results clash with those of Eurobarometer 58.0 (2002), reported in *Europeans and Biotechnology in 2002* (Gaskell et al. 2003), which state that

> contrary to what might be expected, the term 'cloning' does not lead to automatic rejection. When cloning is employed in an application that is seen to be useful, people are prepared to discount the risks and affirm support (p. 13).

The question put in the Eurobarometer refers only indirectly to the stem cell issue and does not mention how they are obtained.[9] This means interviewees would not have recognised the dilemma between the search for medical benefits and the embryo's rights; consequently, the survey's findings cannot be read as support for embryo experimentation. Hence the mistaken conclusion from the Eurobarometer data that

> strong support is evident for medical uses of biotechnology, including the use of genetic testing to diagnose inherited diseases and, perhaps, surprisingly, for the 'cloning of human cells or tissues to replace a patient's diseased cells that are not functioning properly'. Positive attitudes to the latter suggest that the European public (including those in predominantly Catholic countries) may take a more liberal view than their legislators and policy-makers".[10]

The public may have more open views than legislators in cases such as the use of spare embryos, but as we have seen, respondents in most countries have strong reservations about creating embryos for the advancement of biomedical science.

These reservations are particularly evident in the low level of confidence in researchers working in this area, which contrasts with the high esteem and trust extended to the scientific community at large. The statement "I completely trust that scientists will use research with embryos that are a few days old in a responsible fashion" only gets a mean approval score in France (5.74), the Netherlands (5.4), and Poland (5.2), while the lowest levels of trust are observed in Germany (3.3) and Austria (3.6).

---

[9]   The question only refers to "cloning human cells: cloning human cells or tissues to replace a patient's diseased cells that are not functioning properly, for example, in Parkinson's disease or forms of diabetes or heart disease".

[10] The Wellcome Trust, "Europeans' views on biotechnology: the Eurobarometer report", http://www.wellcome.ac.uk/en/genome/geneticsandsociety/hg16n003.html.

Table 4.11 Medical Benefits for many Human Beings vs. the Embryo's Rights[11]: Three Segments (Rejection, Neutral, Approval) (%)

Country	Segments		
	0–3 Rejection	4–6 Neutral	7–10 Approval
Austria	**40.2**	**40.9**	18.9
Denmark	26.7	32.0	**41.3**
France	28.5	**39.6**	31.9
Germany	**44.4**	35.8	19.8
Italy	**40.3**	**40.7**	19.0
Netherlands	33.9	**40.4**	25.7
Poland	**40.3**	34.8	25.0
Spain	22.1	**48.2**	29.7
UK	27.6	**41.4**	31.0

Another, complementary approach to measuring attitudes on embryo experimentation is the use of summated scales which do not rely on direct answers to one or two questions, but assign a global score to each respondent on a latent dimension via a battery of questions, each touching on a different aspect of the attitude under study. This procedure gives measurements greater solidity by minimising or cancelling out the error component intrinsic to any isolated reading of a complex reality that is only indirectly accessible to the researcher (in the case that concerns us, through verbal stimuli in the social context of an "interview").

This aim of more precise measurements was addressed by using 10 items, each of them touching on a significant aspect of attitudes toward human embryo experimentation.[12] The analysis phase began with the standard procedure of identifying the main components or vectors underlying the different attitude aspects. First of

---

[11] The wording of the question is as follows: "The medical benefits for many human beings that can perhaps be obtained in the future thanks to research with embryos that are a few days old are much more important than the embryos' rights."

[12] The statements read literally as follows: "1. Research with human embryos that are a few days old is an unacceptable interference into the natural processes of life"; "2. Restricting research with embryos that are a few days old means opposing progress and welfare"; "3. It is profoundly wrong to damage the least developed forms of human life, such as embryos that are a few days old"; "4. The medical benefits for many human beings that can perhaps be obtained in the future thanks to research with embryos that are a few days old are much more important than the embryos' rights"; "5. We have already gone too far in the manipulation of life, and we should not continue along this path"; "6. I completely trust that scientists will use research with embryos that are a few days old in a responsible fashion"; "7. Allowing research with embryos that are a few days old in order to obtain stem cells for use in medicine will open the door to other morally reprehensible uses"; "8. Research with stem cells from embryos that are a few days old should be supported as a means of finding efficient cures for diseases such as Parkinson's, Alzheimer's, or diabetes as soon as possible"; "9. If research with stem cells obtained from embryos that are a few days old is permitted, we will end up creating monsters"; "10. Research with embryos that are a few days old is morally acceptable if it leads to medical benefits for many people in the future".

Table 4.12 Promises and Reservations of Embryo Experimentation

Country	Promises		Reservations	
	Mean	Median	Mean	Median
Austria	18.5 (9)	19	27.5	27
Denmark	27.6 (1)	29	23.3	22
France	27.3 (2)	28	26.8	26
Germany	20.4 (8)	20	27.7	27
Italy	23.1 (6)	24	27.6	28
Netherlands	24.4 (4)	26	24.4	24
Poland	20.6 (7)	20	26.5	26
Spain	24.1 (5)	25	23.8	24
UK	24.4 (3)	25	24.5	25

all, the 10 items were submitted to a principal components analysis with varimax rotation, which resulted in two components in all countries (except Denmark)[13] with an explained variance of 60–63%. The two components grouped the expected items together: one expressing a favourable view and positive expectations about embryo experimentation (so-called *"promises* of embryo experimentation"), and the other expressing moral reservations and fears about its impact and consequences (*"reservations* about embryo experimentation"). Each component served as the base for constructing a summated scale, whose internal reliability, as measured by Cronbach's Alpha Coefficient, stood between 0.75 and 0.88.[14] The summated scales thus obtained had a score range from 0 to 50. The mean scores on each of survey countries are set out in table 4.12. The countries scoring highest on the *promises* indicator were Denmark, France, the UK, and the Netherlands, while those scoring lowest were Austria, Germany, Poland and Italy. It was precisely this last group which recorded the highest scores on the *reservations* indicator (in decreasing order from more to fewer): Germany, Italy, and Austria. The lowest reservations of all were found in Denmark, Spain, the Netherlands and the UK.

The correlations between *promises* and *reservations* indicators pointed in the expected direction (negative correlations) and attained medium to high values, all of them statistically significant (see table 4.13). The correlation was high in Denmark and Germany, followed by the Netherlands, indicating a dominant pattern in these

---

[13] In the case of Denmark, only one component was obtained, with the 10 items recording high "loadings". The group of items making up the second component in the other countries appeared in Denmark's one, but with the opposite sign. This and other results for Denmark suggests that attitudes on this issue are more structured than elsewhere.

[14] The Cronbach Alpha values for the "reservations" scale were: 1. UK 0.84, 2. Italy 0.84, 3. Denmark 0.83, 4. Poland 0.83, 5. Netherlands 0.82, 6. Germany 0.82, 7. France 0.82, 8. Spain 0.79, 9. Austria 0.75. The values for the "promises" scale were: 1. Austria 0.88, 2. Germany 0.87, 3. Italy 0.87, 4. Poland 0.87, 5. Spain 0.87, 6. UK 0.86, 7. Denmark 0.85, 8. France 0.84, 9. Netherlands 0.84.

societies: the tendency for an individual who scores high on one indicator to score low on the other. The same pattern showed up in the other countries but less markedly so, which suggests that a high score on one scale (*promises*, for instance) is compatible with varying scores on the other for a greater number of individuals. This means predicting scores on the second scale from those observable on the first would give rise to a significantly higher rate of error than with the first group of countries, where attitudes to embryo experimentation are firmer and more structured.[15] The more stable attitudes of this group imply that a shift in one or the other direction would require the emergence and dissemination of new, extremely significant information on embryo experimentation, and would involve a clash with other fundamental values and an intense level of public debate. Any regulations at odds with the dominant perception in these societies would suffer strong opposition and a lack of legitimacy. In the countries where positive and negative expectations interact more freely, we could postulate a greater shift in attitudes among broad segments of the population (under the stimulus of more public information and debate), and a regulatory environment with fewer constraints from public opinion as a whole (though this is not to rule out the presence of vocal minorities with a greater or lesser degree of lobbying power, pushing for particular regulatory orientations). It should be stressed, in any case, that these observed structural differences are a question of degree, and in all countries we observe a medium-high level of attitude structure among the majority of the population.

Table 4.13 Correlations between 'Promises' and 'Reservations' to Embryo Experimentation

Country	Correlations
Denmark	-.66
Germany	-.60
Netherlands	-.51
France	-.46
Austria	-.39
Italy	-.38
Poland	-.36
UK	-.30
Spain	-.25

4.8.3
Religious Beliefs and Attitudes toward Embryo Experimentation

We have seen that attitudes toward embryo experimentation have a radically different pattern depending of the source of embryos: spare embryos resulting from IVF treatments, which are generally approved, vs. embryos specifically created for bio-

---

[15] For more on the structure of attitudes, see Philip Converse.

medical research, which at present meet with strong rejection in most societies. There is also variation in attitudes toward both categories of embryos by religious belief. People who believe in a religion and take into account moral criteria aligned with their religious credo have to balance those guidelines with other valuable goals (among others, the advancement of research, the future treatment of serious diseases), so it is no surprise that we find different positions in that subset of the population (see tables 4.14 and 4.15). But the main differences unfold along the axis marked at one end by no religious belief and at the other by religious belief (irrespective of whether it is Roman Catholic or Protestant). In accordance with the findings about differences in opinions about the beginning of individual human life

Table 4.14 Acceptability of Experimentation with Spare Embryos by Religion (Rejection, "Neutral", Approval) (%)

		Rejection (0–3)	"Neutral" (4–6)	Approval (7–10)	Don't know	Difference 'approval' – 'rejection'
Spain	Catholics	24,1	22,3	29,9	23,8	5,8
	Nonbelievers	14,1	20,1	49,7	16,1	35,6
Italy	Catholics	21,6	26,2	36,6	15,7	15,0
	Nonbelievers	16,9	19,1	38,9	25	22,0
Poland	Catholics	28,6	15,7	27,8	27,8	-0,8
	Nonbelievers	25,2	15,3	40	19,5	14,8
Denmark	Protestants	18,1	16,5	63,5	1,9	45,4
	Nonbelievers	16,1	16,1	65,1	2,8	49,0
United Kingdom	Anglican	22,5	26	37,2	14,2	14,7
	Other Christian/ Evangelical	32,5	11,7	36,8	19	4,3
	Nonbelievers	19,1	21,1	37,9	21,9	18,8
Germany	Catholics	28,6	26	39,7	5,7	11,1
	Protestants	34,5	25,1	35,2	5,2	0,7
	Nonbelievers	27,8	24,6	38,8	8,8	11,0
France	Catholics	24,9	21,8	46,6	6,8	21,7
	Nonbelievers	14,9	21,8	54,4	8,9	39,5
Austria	Catholics	39,9	21,7	12,9	25,5	-27,0
	Nonbelievers	40,5	22,8	19,6	17,2	-20,9
Netherlands	Catholics	26,4	22	46,2	5,4	19,8
	Protestants	33,3	22,3	37,9	6,8	4,6
	Nonbelievers	17,4	21,8	50,6	10,2	33,2

Table 4.15 Acceptability of Experimentation with Embryos Created for Research (Rejection, "Neutral", Approval) (%)

		Rejection (0–3)	"Neutral" (4–6)	Approval (7–10)	Don't know	Difference 'aproval' – 'rejection'
Spain	Catholics	31,9	22,3	22,2	23,6	-9,7
	Nonbelievers	22,6	25,4	34,9	17,1	12,3
Italy	Catholics	38	23	23,5	15,5	-14,5
	Nonbelievers	28,9	21,7	22,7	26,7	-2.2
Poland	Catholics	36,6	14,7	19,6	29,1	-17,0
	Nonbelievers	33,6	15,3	33,8	17,3	0,2
Denmark	Protestants	27,8	21,9	48,2	2	20,4
	Nonbelievers	25,1	19,2	53,5	2,3	28,4
United Kingdom	Anglican	34,3	25,8	25,2	14,7	-9,1
	Other Christian/ Evangelical	44,6	13,3	22,2	19,9	-22,4
	Nonbelievers	26,8	23,5	25,8	23,9	-1,0
Germany	Catholics	60,6	23,8	14,6	0,9	-46,0
	Protestants	68,4	13	12	6,6	-56,4
	Nonbelievers	57,4	18,4	19,5	4,8	-37,9
France	Catholics	49,6	18,7	23,9	7,8	-25,7
	Nonbelievers	36,7	23	30	10,2	-6,7
Austria	Catholics	48,9	18,6	9	23,5	-39,9
	Nonbelievers	48,8	20,6	13	17,5	-35,8
Netherlands	Catholics	51,1	24,7	17,9	6,3	-33,2
	Protestants	57,2	19,9	16,1	6,8	-41,1
	Nonbelievers	39,8	25	25,8	9,4	-14,0

and the moral condition of the embryo, the segment of nonbelievers has a more flexible view of embryo experimentation, and doesn't seem willing to grant the embryo the same level of protection as persons, particularly if medical benefits are at stake. Once again, this shows that in European societies the moral differences regarding the embryo are closely linked to the religiosity of individuals and social groups, and Europe is in this respect a pluralistic society, with significant numbers of nonbelievers in many societies.

4.9
Attitudes to PGD

Preimplantation Genetic Diagnosis (PGD) is another biomedical application that provides clear medical benefits and, at the same time, poses serious difficulties from an ethical standpoint. In fact, it is not permitted in several of the countries studied here. A core reason for the moral reservations is, as Patzig has remarked, the "slippery slope" argument – if couples are allowed access to Preimplantation Genetic Diagnosis and its use becomes widespread, it might trigger a dangerous dynamic leading to permissiveness in gender selection, and fueling the desire for "perfect children" who are completely free from any known hereditary diseases (Patzig 2001). This scenario, in turn, could alter the way in which our society regards certain minorities or patient groups in need of social support. The benefits of using PGD to avoid passing on to children serious hereditary diseases such as cystic fibrosis, thalassemia and others, would therefore be overshadowed by the adverse consequences to society as a whole.

Looking at this question from the standpoint of public perceptions, we find a total divergence between opinions on two contrasting applications. There is extremely high approval for the use of PGD as part of IVF processes, where couples suffering from genetic diseases or indeed any couple (without any known family history of genetic diseases) can find out whether their child will also suffer from any of the diseases in question and, if so, choose not to have the embryo implanted. In contrast, recourse to PGD in order to know the sex of a future child so a couple can decide whether or not to go ahead with the implant is strongly opposed in all countries (see table 4.16). Substantial medical benefits "deactivate" moral reservations, while the idea of pursuing other ends with the same technique "activates" moral reservations to an extreme degree. Therefore, from the standpoint of public perceptions and contrary to what is implied by the "slippery slope" argument, there is no automatic causal link between permitting and using PGD for one type of case (morally acceptable or neutral) and lending legitimacy to other scenarios that provoke almost universal rejection. Nevertheless, the doubt remains as to whether the uniformly positive verdict on access to PGD for couples suffering from hereditary diseases and couples not known to suffer any such illness might not be read as a latent rejection of "playing the genetic lottery". If this is indeed what underlies the general response (which we are not in a position to state with certainty), we can postulate that as the list of genetic diseases identifiable through PGD gets longer, the vast majority of people will have no problems accepting its use and discarding embryos that carry a defective gene. In any case, the distinction between "negative" and "positive" eugenics, that is, between using genetic techniques to eliminate or minimise "ills" (diseases) and using them to achieve a positive desired outcome (gender and other physical traits) is implicit in the different responses which the two scenarios trigger in the population (see our remarks in the section on acceptance of IVF techniques and access to semen bank donors).

Table 4.16 Acceptance of PGD Techniques in Different Cases

Country	For any couple		For couples which suffer from genetic diseases		For gender selection	
	Mean	Median	Mean	Median	Mean	Median
Austria	6.8	8	6.8	8	1.4	0
Denmark	7.3	8	7.7	9	1.2	0
France	8.0	9	8.2	9	1.2	0
Germany	6.8	8	7.2	8	1.3	0
Italy	6.9	7	7.0	8	1.8	0
Netherlands	6.3	7	6.7	7	1.5	1
Poland	7.1	8	7.2	8	2.3	1
Spain	7.4	8	7.4	8	2.7	2
UK	7.4	8	7.6	8	2.1	1

■ So that any couple can find out whether their future child suffers from a serious genetic disease a implantation.

  So that a couple which suffers from a genetic disease can find out whether their child will also suffer that case, choose not to continue with the implant.

■ So that a couple can find out the child's gender and, if it doesn't coincide with the gender they des with the implant.

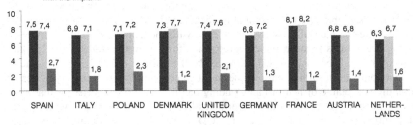

Fig. 4.8 Acceptance of PGD techniques in different cases (on a scale from 0 to 10 where 0 means it is absolutely unacceptable and 10 means it is completely acceptable.)

4.10
Regulation of Embryo Experimentation: EU vs. National Level

As we have observed throughout this chapter, turn-of-the-century Europe is home to contrasting attitudes on embryo experimentation for biomedical research ends, both within and, particularly, between countries. The desire to find effective treatments for some of the most devastating illnesses that afflict advanced societies clashes with deep-seated religious and moral values, beliefs about the point where individual human life begins and, in some cases, with the memory or realization of how eugenic principles were applied in the first half of the last century. Given the polarisation that exists, the diversity of opinions within and among countries, and simultaneously the importance of the possible medical benefits at stake, it is clear that regulators in this area face a hard challenge.

EU countries have arrived at different regulatory frameworks, inspired by their historical experiences and belief systems, which range from permissive and ambivalent to firmly prohibitive legislation. Things are complicated further by continuous research advances, modifying the data on which regulations are based, and the high researcher mobility that characterises our age. In this context, it is not just the content of regulations which matters. Even more important is their degree of supranational harmonisation and how a deeper convergence can be achieved. In the case of Europe, this means debating at which level embryo research should be regulated, that is, whether it should come under national or European Union legislation.

One variable affecting the decision of whether to opt for regulation at the local national or global European level could be the dominant preferences of European societies, particularly of their elites and the so-called "influentials" (Weimann

Table 4.17 Preferences for the Level of Regulation on Embryo Experimentation

Country	EU	National	Don't know
France	78.9	7.1	14.0
Germany	73.9	18.7	7.4
Netherlands	73.9	19.9	6.2
Italy	65.1	24.0	10.9
Austria	60.5	26.4	13.1
Denmark	59.6	35.0	5.4
Spain	59.2	24.3	16.5
UK	47.3	31.3	21.4
Poland	40.0	43.3	16.7

■ So that any couple can find out whether their future child suffers from a serious genetic disease a implantation.

So that a couple which suffers from a genetic disease can find out whether their child will also suffer that case, choose not to continue with the implant.

■ So that a couple can find out the child's gender and, if it doesn't coincide with the gender they des with the implant.

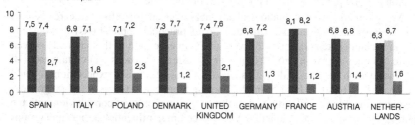

Fig. 4.9 Preferences for the level of regulation on embryo experimentation.

1994), but also of the population at large. Unfortunately, to the best of our knowledge, no significant information is available on the views not just of policy-makers but also of opinion leaders. What we do have, however, is a picture of European attitudes in the year 2003. In the survey discussed in this chapter, the following question was posed: "Today there are different kinds of regulation and legislation about embryo research in European Union countries. In your view, should embryo legislation be the same in all of the European Union, or do you think it's better if each country has its own legislation according to its beliefs?". A clear majority of the total number of respondents in the nine countries surveyed opted for a common regulation for the whole EU (see table 4.17). Only in Poland was opinion sharply divided, with those wanting national regulation just outnumbering those in favour of EU-wide regulation. In the cases of Denmark and, above all, the UK, large minorities supported national-level legislation (also, many individuals in the UK had or stated no preference). At the opposite extreme, the countries most strongly in favour of European legislation were France, Germany and the Netherlands. Note also that in five of the nine countries surveyed, numerous respondents could not formulate a clear position on the issue. This lack of opinion possibly resulted from the combination of a complex underlying issue (embryo experimentation) with a regulatory dimension that is remote from people's normal concerns (regulation is an esoteric area for the public except maybe in economic matters, which can have a significant effect on their living and welfare standards).

There is evidence here that the EU could take some steps towards the future harmonisation of regulations on embryo experimentation, perhaps starting with an analysis of the issue and the formulation of general guidelines by ethics committees and scientific associations with a European reach. However, the existence of marked differences within and between countries requires that this pre-regulatory stage be accompanied by a prolonged and rational debate.

4.11
Conclusions

Several of the chapters of this book advocate regulations of embryo experimentation that are based on state-of-the-art scientific knowledge and a rational debate between the different ethical viewpoints in turn-of-the-century Europe. If such regulations are to be capable of affecting core cultural values, while also facilitating possible biomedical benefits for some of the most serious diseases, it must take into account public perceptions and preferences on the issue.

Many scientific and technological advances have emerged from research centres and even entered the market (the domain of practical applications) without any broad social debate and, certainly, without the intervention of the public. However, there has been a growing conviction over the last two decades that certain areas of research which might collide with societal values, beliefs and expectations, should be opened up to some kind of public participation. Such access could range from just knowing what people think and value to various forms of consultation with the public on the part of the regulators. At the same time, influential authors and groups have started arguing for the institutionalisation of "voice" mechanisms or the extension of democracy into certain areas of science (though not, of course, into its strictly cognitive dimension).

Not infrequently, the scientific community and some regulatory agencies attribute public resistance to certain scientific advances to a single variable, the generally low level of "scientific literacy". The solution, from this standpoint, is equally simple – improved formal and informal education of the public about the scientific worldview. There is ample evidence for public's poor knowledge of the cognitive aspects of science on both sides of the Atlantic, but it is less clear that resistance or critical attitudes toward some scientific developments derive exclusively and directly from the level of scientific knowledge. Other authors postulate that a fundamental change has taken place in advanced societies over the last third of the twentieth century, from a positive assessment and total consent to science to no less universal criticism and suspicion. If the latter view were accurate, and if, as seems irrefutable, the global society of the turn of the century cannot do without science in everything from economic growth and leisure to healthcare, the only sensible option for policy-makers would be to support scientific progress, and pay little heed to sporadic episodes of public protest. But the situation is rather more complex.

The available evidence shows unequivocally that science – in its contents, its applications and its practice by the scientific community – is the modern institution that enjoys the highest levels of legitimacy and esteem in all advanced societies. Expectations about future scientific developments are globally positive, and the level of trust in what the scientific community does and says is unrivalled. There is no sign that the population at large resists science in general or most of its currently most dynamic areas. But people do have a growing ability to differentiate between areas and applications, and refuse to extend a blank cheque to any scientific research activity or goal. At the same time, pessimism about the outcomes of human intervention in the natural world has deepened and, precisely due to scientific progress, awareness of the risks associated with the current way of life has increased compared to a few decades ago. In keeping with these changes, ecological movements have taken their place alongside the scientific community as repositories of the public's trust, which signals a potential for conflict in cases where the positions of these two groups differ.

Within the framework of this generally positive image held of science, some subsets of the broad area of the life sciences and biomedicine provoke feelings of perplexity, concern and resistance. Whereas information and communication technologies, the Internet, new materials, and solar power research give rise to positive expectations, nuclear energy, cloning and genetic engineering belong to the group of disciplines and areas that are perceived more critically. The potential for ethical problems and risks is clearly different in areas that directly affect human and animal life than in those whose influence is more indirect and more subject to control. The gene stands as a cultural symbol or icon and its manipulation – even if it aims at biomedical benefits – triggers "images of fear". Non-therapeutic cloning and embryo experimentation are areas that can affect images of human self-identity and the boundaries between species, ethical values and religious beliefs. This potential, along with their newness and the public's lack of familiarity with the underlying science, makes them a seed bed for controversy and regulatory intervention.

Less than a decade ago animal cloning left the realm of science fiction to become a reality and the object of ample media coverage. Seven years later, the great majority of the European population does not perceive any clear benefits from the technique to set against its risks and the implications of this human intervention in the world of animal reproduction. Fears and moral reservations are, as we write, the most prevalent reaction. One of the aspects of cloning or genetic copying of animals that is most vigorously rejected by the population is the serial mass production of animals, even if the copies are based on individuals with exceptional qualities: it is "a direct challenge to the sacredness of the biological individual", in the words of Horning Priest. Equally obvious is the public's belief that animal cloning could lead to human cloning in the not too distant future. The idea that we have begun to go down the "slippery slope" is a fundamental aspect of the moral reservations and fears that the issue of animal cloning triggers today. Another indirect consequence of animal cloning has been a higher level of salience and public awareness with regard to the life sciences, as people are willing to pay greater attention to developments in this area.

When we move from the techno-scientific intervention into the animal world to human beings, we find two significant spectra of opinions and one common aspect. The common thread is that for a large majority of the European population, attitudes and responses to obtaining stem cells from embryos are based not on knowledge, but mostly on images, fears, stereotypes and beliefs about embryos, their status and their rights. As mentioned earlier, this finding suggests the enormous task that await the scientific community in explaining the science of this issue to the lay population.

The first important spectrum of opinions runs from strong reservations about any scientific or technological development that might open the door to "biological design" or the selection of human genetic traits, at one extreme, to applications such as IVF or semen bank donors that can help fight medical conditions and hereditary diseases, and apparently do not affect core moral values today, at the other. The second spectrum concerns ethical-religious reservations about embryo experimentation for obtaining stem cells, which are especially dominant in countries with large majorities of religious believers (Catholic or Protestant) and, particularly, countries in which the Roman Catholic Church has had an influential role in shaping moral criteria and in orienting debates about controversial issues, as compared to countries with ample religiously unaffiliated segments.

One of the fundamental predictors for attitudes toward embryo experimentation is a person's images and beliefs about the status of the embryo. Our evidence shows that Europe (represented here by nine countries) encompasses enormously divergent perceptions of the status of the human embryo, as evidenced by the wide variability of responses within each country, and even more so, between countries. Status responses ranged from seeing the embryo as 'a mere cluster of cells' to ascribing to it 'the same moral status as a human being. Although this issue frames opinions and decisions in favour of or against embryo experimentation, moral debates and dilemmas, for the great majority of people, are not abstract matters to which a uniform criterion or principle is applied (although this may be the case with some minorities). Rather, such questions arise in specific contexts, usually cases in which various fundamental and competing values or principles could apply. In such cases, individuals and society as a whole have to seek a trade-off between all possible decisions by weighing up each set of values and principles against the others. The issue of abortion illustrates how beliefs about the status of the embryo may be qualified or even trumped in the presence of other values or objectives.

The debate about embryo experimentation has been has been framed by two different assumptions: the use of "spare" embryos from IVF treatments, and the creation of embryos specifically for research. Responses to the two cases are radically different: the first meets with (lukewarm) approval, while the second is strongly rejected in every country except Denmark. Nevertheless, it is worth noting that most European societies are highly polarised between segments in favour and against, and that a broad population appears not to have formulated an opinion by 2003. Catholic countries or those with a large population adhering to this denomination tend to move in a more restrictive direction, while the other countries tend toward clear approval or at least toward more flexible and permissive viewpoints. The size of the population with no religious beliefs also has a profound impact on the approval of embryo experimentation. Spain stands somewhat apart from the group it should belong to in terms of religion, while German respondents were particularly sensitive to and cautious about anything hinting at the manipulation of genetic traits or the creation of embryos, due perhaps to the dramatic experiences during the Nazi regime.

Attitudes toward PGD exhibit a similar pattern: approval when the goal is to prevent transmission of hereditary diseases, and total rejection of its use for the selection of other characteristics (gender selection). No other moral reservations were identified in this case, not even in countries like Germany.

The foregoing analysis of citizens' views of several controversial innovations and related technical practices in the biological sciences and biotechnology has revealed some strong commonalities of perception and assessment across national boundaries. Such shared assessments are striking in a Europe that is remarkably pluralistic, both within and between national societies. However, these commonalities obscure a major and important tension that is latent in the total respondent population. On the one hand, we have seen that the dominant aspiration in most of the surveyed countries is to achieve a common transnational regulatory framework vis-à-vis the controversial innovations. On the other hand, the survey data reveal that there are deep differences in the perception and evaluation of these innovations and practices (and of the beings involved in them) from one country to another, whether the innovation or practice in question be that of animal or human cloning, IVF for human reproduction, embryo experimentation, or PGD. Given the considerable differences we have documented here, a common regulatory framework could probably only be constructed in a piecemeal fashion on the basis of a broad-ranging and rational moral debate, and the scientific education of policy-makers and the public at large.

5 Toward a Rational Debate on Embryo Research

Introduction

Over the past years embryo research has aroused an enormous interest in the public arena and in expert literature ranging from the life sciences to reflective disciplines such as philosophy and jurisprudence. It seems, however, that results of scientific evaluation of embryo research are not adequately taken up in public debate. It is generally admitted that scientific expertise is a necessary part of any rational debate on technological advances, but as a matter of fact public opinion is not strongly influenced by such considerations. There are two reasons for the persisting rift between scientific and popular opinion. On the one hand many scientists refrain from participating in controversial public debates, because they find such exchanges abusive, chaotic and inconclusive. On the other hand, the general public is reluctant to learn from scientists who in general aim to define their position carefully and precisely. As a result many current debates, especially those published in popular press, fail to meet serious standards of rationality. The purpose of this chapter is to review some of the causes of this failure and to recommend a remedy. The guiding principle of this approach is that workable solutions for moral conflicts should not be based on a set of incontestable moral norms. Rather a procedural approach should be developed that allows one to come to a conclusion in a discursive manner, even if consensus on the underlying moral principles or ethical theories has not and could not be achieved.

This chapter will first give a description of different policy positions taken with respect to the regulation of embryo research (5.1). These positions will be called for convenience "Prohibitive", "Permissive", and "Regulatory". They are put together from different arguments and opinions that can be found in public discussions, and though they make use of arguments that are also present in philosophical debates, they themselves are not in any way well defined philosophical positions. Additionally it might be important to notice that they are not directly related to any national regulation as might be suspected in view of our discussion of the regulations in the United Kingdom, Spain and Germany in Chapter 3 "The Regulations of Embryo Research in Europe: Situation and Prospects". Though these three countries have been chosen as subjects of legislative case studies and they can be seen as representing different regulative approaches, the considerations behind their regulations are not to be identified with the ideal type positions discussed in this chapter. In the second section (5.2) some frequently encountered arguments and argumentative strategies are discussed – such as slippery slope, naturalism, taboos, abusive rhetoric, etc. – and they are analysed with the use of methods of academic philosophy. In doing this we will try to make it clear that philosophical expertise can contribute to improvement of the quality of public debates. We will argue, however, that philosophical critique is not always sufficient to resolve moral conflicts about embryo research. One of the main reasons for this is that policy positions are often adopted in order to achieve political aims and not in order to produce a well-founded structure of philosophical arguments that might be revised in the light of strong counter-arguments. To deal with this situation the third section will outline a procedural way of resolving moral conflicts of embryo research (5.3). This approach combines respect for various moral standpoints towards embryo research with the insight that reaching a political compromise in a rational way is a necessary precondition of social cooperation and understanding.

5.1
Three Policy Positions: Prohibitive, Permissive and Regulatory

None of the three policy positions to be considered is uniform or homogeneous. The basic stance in each case results from various, and sometimes vague, influences: emotional associations, religious sentiments, scientific rationality or, occasionally, nearly fantastic imagery. They may be delivered as moral convictions in association with religious faith or deep-seated intuitions.

5.1.1
The Prohibitive Position

Opponents of embryo research will contend more often than other groups that embryo research involves interference with natural processes, is conducted by performing presumably repulsive acts, such as gene manipulation or the killing of immature or incipient human beings, or enhances human arrogance and irresponsibility. The prohibitive position tends to be associated with several of the following prescriptions or recommendations: Prenatal experiments must be prohibited, and even non-sentient embryos and fetuses should be treated as if they could suffer harm in order to enhance the value of human dignity in public imagination. Stem cells that are needed for medical use should be developed not from embryos but from bone marrow or muscles of individuals who need replacement organs. Cloning of entire human organisms, brainless or not, must be strictly forbidden to accentuate human individuality and uniqueness. Full human rights are to be accorded equally to adults, children, embryos and zygotes.

In order to support these claims advocates of the prohibitive positions variously use some of the following strategies: They form their opinions on embryo research relying on speculative theories of personhood; believe in moral authorities that are beyond human critique; defend traditional values and customs.

Widespread versions of the prohibitive position characteristically declare that human life begins at conception, because this is the moment when the soul enters the body. This is the position of the Catholic Church, for instance, though by no means was it always, or is now, espoused by all Christians. Following Aristotle's *Politic,* medieval Christian doctrine assumed that life begins with "quickening", or when the mother first feels the movements of the baby, which usually happens in the fourth month of pregnancy. This doctrine was only rejected by the Catholic Church in the 19th century[1] (Kovács1996: 223), and the rejection has been justified by the biological argument that at the moment of conception two gametes are fused together and become one biological organism.

The fusion of cells is undoubtedly a biological event. If it acquires an additional theological meaning, however, we must ask: What are the grounds for saying that at the moment of cell fusion God infuses a soul in the body? If by soul we mean life, as the term was literally understood in the *Bible*, then the claim is plausible but innocuous, as no one ever questioned the proposition that a zygote is a living entity, as of course are the gametes, the egg and sperm that precede and give rise to the zygotes. If, however, by soul is meant something like Aristotelian form or a self-subsisting form of Thomas Acquinas, then the belief needs to be supported by impartial evidence. It is by no means clear that at the moment of conception human senses, intelligence and moral qualities are implanted in the body but inactive. No *observable* facts confirm such an assumption, and the very concept of a theological fact, i.e. a fact proposed by theology but not detectable to science is rather dubious.

Theories of personhood claiming that a person exists at least from the fusion of an egg and sperm onwards have been heavily criticised in recent years.[2] However, the Catholic Church has reaffirmed its view that embryos have human rights from the moment of conception in the encyclical *Evangelium Vitae* (1995).[3] This position has important implications. If an embryo has full human rights, its life must be defended as resolutely as the life of a child or a mature human being. Consequently the Church

---

[1] "[T]he church had already accepted the doctrine of immediate animation in 1708 but did not apply it with full rigour to abortions prior to quickening. In 1854, Pius IXth reaffirmed that animation occurred at conception, and in 1869 the distinction between the ensouled and the non-ensouled embryo was officially removed from Canon-Law." By the same decision those who were found guilty of committing abortion were *ipso facto* excommunicated from the Church.

[2] It is untenable that a zygote consisting of several cells could be considered a human individual. As long as the cells are totipotent, they can develop into several complete organisms, moreover, some of the cells can develop into placental tissue. It seems most confusing to consider a blastocyst made up of totipotent cells an individual organism that can later on remain an individual organism or develop into several organisms or become part of tissue of an already existing organism. Ref. e.g. Donald Evans: "[T]he facts of totipotentiality and pluripotentiality of the cells of the early conceptus are obstacles to our identifying that conceptus as a human life. Whilst I may retrospectively link a given human life with an early conceptus I cannot know that any early conceptus will become only one life – or even a life at all as it might develop into a hydatidiform mole or, given that it contains too many chromosome it must perish." (Evans 1996: 34–35).

[3] "Though the Magisterum was not engaged directly in scientific discussions and philosophical statements, the Church always taught and still teaches that the fruit of human procreation from the first moment of its existence has an unconditional right to respect that is due to every human being in its integrity and in its corporeal and spiritual unity. Human being should be respected and treated as a person from the moment or its conception, and for this reason from the same moment it should be granted the rights of person, one of which is the inalienable right of every innocent human person to life." (*Evangelim vitae*, section 60, in Polish edition 1996: 922) Alongside this exhortation the encyclical specifically declares that techniques of artificial insemination (section 14) and embryo research (section 63) are incompatible with Catholic teaching. "Various techniques of artificial insemination that seem to serve life and are often practised with this intention, in reality create an opportunity of assault on life. They cannot be accepted from a moral viewpoint, because they separate procreation from a truly human context of the conjugal act, and moreover because those who a high percentage of failures is noted by those who perform these techniques. This refers not so much to the moment of insemination but to the initial phases of the development of an embryo that is

cannot authorise a request for prenatal screening, if it is assumed that an abortion will be sought by the pregnant woman if she discovers that the fetus has serious congenital defects. For similar reasons it is unacceptable to the Church to conduct research on embryos whenever such research results in their death. The position of the Catholic Church is founded on the belief that the Catholic 'magisterium' or *teaching authority* is ultimately the only legitimate source of moral orientation. The Pope is credited with a literally interpreted infallibility in religious and ethical matters, so his pronouncements *ex cathedra* cannot be doubted or questioned. Other Christian denominations, Judaism and Islam do not have a single teaching authority and seem more tolerant of abortion and embryo research[4] (see: Boukhari 1999).

Many adherents of the prohibitive position defend traditional value systems. They seem to be worried by a possible landslide of moral degradation when a society departs from traditional values. Thus they demand a total ban on human cloning not because they necessarily consider the procedure itself unacceptable, but because in their opinion it will tend to corrupt public morality. They also believe that the question of the acceptability of therapeutic cloning need not be evaluated on its own terms, because in their opinion therapeutic cloning is inseparable from reproductive cloning, and by permitting the former one gives implicit permission to the latter.

It would be difficult to find a single author who holds all these views in the form described above. Obvious specific differences distinguish advocates of the Prohibitive position from one other. But Leon R. Kass, the Chairman of the United States President's Council on Bioethics, would be a good example of a writer who believes that the latest advancements in biotechnology do more harm to the moral organisation of society than they do good to patients whom they as yet cannot offer much help. Kass holds that medical sciences should concentrate more on fortifying natural powers of healing that are present in all organisms and less on the development of

---

exposed to imminent death. Besides, in many cases more embryos are created than are needed for the transfer into the mother's womb, and subsequently the 'supernumerary embryos' are killed or used in scientific research." (John Paul II 1996: 858) "The moral assessment of termination of pregnancy is to be extended to new forms of procedures performed on human embryos, which, even in the case when they are conducted for worthy purposes, lead inevitably to embryo's death. This is particularly true of experiments on embryos that are ever more often performed within biomedical research and accepted by the law of some countries. And though one should consider permissible to perform procedures on a human embryo that respect the life and integrity of the embryo and do not expose it to an improportionally great danger but are conducted for therapeutic purposes in order to improve its health or save its endangered life, it must be emphasised that using human embryos and fetuses as an object of experiment is a criminal act violating dignity of human beings which have the same right to respect as new-born children and as every man." (John Paul II 1996: 926).

4   Other religions invite more open discussion. Jewish rabbis enjoy high authority and esteem, but they cannot formally make ultimate pronouncements about what is right and what is wrong. Also priests in the Orthodox Church and masters among the Buddhists are trusted as experienced teachers, but they do not pretend to make religious morality. Muslims too have different schools of jurisprudence and, in general, do not try to establish one moral orthodoxy: "While all the major religions generally believe human life and dignity should be respected, the Church of Rome is the only religion that considers the embryo 'as a human being from the moment of conception,' and it sticks firmly to this doctrine." (Ref. Boukhari 1999) One fragment in the Koran says explicitly that a fetus and a child are two different beings, a fetus is at some point made 'into another creation': The Koran says in Surat 23: "And certainly We created man of an extract of clay, then We made him a small seed in a firm resting-place, then We made the seed a clot, then We made the clot a lump of flesh, then We made the lump of flesh bones, then We clothed the bones with flesh, then We caused it to grow into another creation". (Ref. Boukhari 1999).

new sophisticated techniques, especially when the advanced medical procedures treat human individuals as helpless bodies rather than complete psycho-physical wholes.

> For example, should not the remarkable powers of self-healing, present in all living things, make us suspect that dumb nature in fact inclines purposively toward whole-ness and is not simply neutral between health and disease? [...] Can biochemistry and neurophysiology ever do justice to what we know first and best: our inward experience of ourselves as passionate, purposeful and thoughtful beings? (Kass 1985: 8)

Kass seems to be impressed by the fact that no matter how powerful new scientific discoveries prove to be, we will never be able to find a cure for every ailment. Some conditions will always remain incurable, and eventually all patients will have to die. Thus we should not pin too much hope on new medical techniques because they will fail us sooner or later. We should prepare ourselves for the worst and reconcile ourselves with the inevitable facts of infirmity and death. Scientists tend to promise too much. They show disregard for the traditional methods of coping with the vicissitudes of life and want to make every human being a perfect specimen of the human species.

> Regardless of the intent of scientists, the teachings of science, as they diffuse through the community, do not stay quietly and innocently on the scientific side of the divide. They challenge and embarrass the notions about man, nature, and the whole that lie at the heart of our traditional self-understanding and our moral and political teachings. The sciences not only fail to provide their own standards for human conduct; their findings cause us also to doubt the truth and the ground of those standards we have held and, more or less, still tacitly hold. (Kass 1985: 6)

It is remarkable, Kass points out, that we do not have rational expectations about our future, and tend to demand ever more from society that we live in and from its leaders. He find this overoptimistic and demanding attitude simplistic and irritating.

> How much more life do we want, assuming it to be healthy and vigorous? [...] The simple answer is that no limit should be set. Life is good and death is bad. Therefore, the more life the better, provided of course, that we remain fit and our friends do, too. (Kass 1985: 305)

The infantile expectation that our physical and mental condition can be continuously repaired makes us indifferent to and intolerant of those who are less able than ourselves. We want to eliminate all misery and disability from the surface of the earth forgetting that deprivation and discomfort have always been part of the human condition. Instead we should accustom ourselves to the presence of the weak and the handicapped. We should realise that even though they do not excel in anything that present society values most, on the other hand neither do they contribute to the worst of its evils.

> After all how many architects of the Vietnam War or the suppression of Solidarity suffered from Down's syndrome? Who uses up more of our irreplaceable resources and who produces more pollution: the inmates of an institution for the retarded or the graduates of Harvard College? And which of our genetic mutants display more vanity, self-indulgence, and the will-to-power, or less courage, reverence, and love of country than many of our so-called best and brightest? It seems indisputable that the

world suffers more from the morally and spiritually defective than from the genetically defective. (Kass 1985: 46)

5.1.2.
The Permissive Position

Proponents of unrestricted embryo research (whose position we have called "permissive") will argue more often than other groups that embryo research should be conducted freely and without external supervision, because no institution has the authority to impose its moral opinions on people already having their own, different moral
views. They hold that scientific progress is unstoppable. Whatever is technically feasible will be tried and implemented by somebody sooner or later irrespective of what
moralists say about such ventures. They also point out that all attempts to subject
social behaviour to traditional values is contrary to individual autonomy.

More specifically advocates of the permissive position would prescribe or recommend the following practices. Experiments on embryos and fetuses should proceed
without regulation. Embryos may be used to produce a reservoir of human organs.
Parents may pre-select their children and reject embryos (in the IVF procedure) or
fetuses (through abortion) that do not meet their specifications for healthy, normal
children. In moral terms, runs the argument, such decisions are not very different
from consciously choosing one's mate for the characteristics that may benefit one's
children. If congenital defects cannot be detected early, infanticide should be considered a rescue option for parents who are unwilling to raise defective babies. Taking a
cue from Peter Singer, who condones infanticide in cases of severe congenital
defects that were not discovered at the time when abortion was still a possibility
(compare e.g. "Should the Baby Live?" Kuhse and Singer 1985), a radical advocate
of the permissive position may tolerate infanticide in every case when the parents are
not satisfied with the qualities of their newborns. He may remind us, for instance,
that in Northern India some parents dispose of their new-born daughters to this day.[5]

Consequently advocates of the permissive position accept variously the following
beliefs: If parents decide to have babies with predetermined features, then it is permissible for practitioners in gene or embryo manipulation to help them have such
children. It is permissible to treat some organisms instrumentally if their sacrifice is
more than compensated by advantages to other organisms; technical and social
progress are more important than traditional values.

The permissive position grants persons full rights to decide how their gametes
will be used, even if such decisions clash with widespread public sentiments. The
most obvious potential conflict concerns the choice of babies and the permissibility
of repeated production of embryos that are later discarded until the couple concerned

---

[5] Persons who dispose of female fetuses or infants are not considered moral monsters in Northern
India. Philip Kitcher says: "Northern Indian uses of prenatal technology to reduce the number of
daughters are morally repellent if they are inspired by parental contempt for women. But it is not
necessary to conjure up moral monsters in the Punjab. Many of those who go to the new medical
centers may do so reluctantly, bowing to the combined pressures of the caste system and their
economic situation, or, perhaps, recognizing that the lot of women in their local society is so
hard that it would be better not to give birth to girls doomed to be brutalized." (Kitcher 1997:
216–217).

find an embryo that meets their expectations. To some people this is a concealed form of abortion performed for frivolous reasons. Yet an advocate of the permissive view will hold that an early embryo is little more than a lump of cellular material that can make no moral demands on us. This view does not usually find eloquent champions, but the position has been described by John Robertson (1994) in his book *Children of Choice*. Robertson discusses "procreative liberty", and outlines a proposal to have "dream children" that parents could be proud of. He seems to be aware that unrestrained permissivism may be self-defeating and harmful. If we make babies according to their parents' specifications, the practice can seriously damage life styles and personal plans of those who engage in the process. Some authors call this side-effect "commodification of children" and strongly disapprove of it (see: Peters 1994).[6] This discussion, however, is premature for biological reasons. As yet it is impossible to enhance some desirable properties in the offspring without at the same time eliminating some other of their desirable characteristics. (see Chapter 1. Theoretical and practical possibilities in human embryo research.)

The scope and effectiveness of future technologies is hard to assess. Advocates of the permissive position frequently hold that numerous chronic conditions known to man will be alleviated when new methods based on embryo research have been employed. These hopes find support from various scientific findings. Through experiments on embryos it was possible to develop measures that limit maternal-fetal HIV transmission, for instance. Fetal kidney tissue was used in research that led to the development of polio vaccines. Tissue derived from embryos is helpful in the treatment of diabetes and neurological ailments such as Parkinson's disease and Huntington's disease (as we have shown in Chapters 1 and 2). Embryo experiments are also important for research on preimplantation genetic diagnosis, for research on isolating pluripotent stem cell lines that can have a clinical use in transplantation, for research on nuclear transplantation needed to avoid disorders due to maternally inherited cytoplasmic defects, etc. (For further reading refer also to Brody 1998).

Advocates of the permissive position do not see a major difference between the sale of semen or blood and authorising the sale of embryos. In their opinion, all kinds of human tissue that are not endowed with sensitivity can be commodified and used as material to enhance the quality of life of sentient human beings. They do not feel the scruples that make some people uneasy when they find that a potential human being will have no chance of further development but will be turned into a stem cell line. They respect individual liberty and autonomy above all, and they accord every individual the right to plan his/her life in a way that promises fulfilment and satisfaction. In their opinion mature individuals can make choices that have no precedent because their lives may be very different from the formerly encountered patterns (Margaret Mead 1970).

An interesting champion of the permissive position is John Harris. A compassionate utilitarian, he is ready to adopt many unconventional schemes to maximise happiness of as many persons as possible. According to Harris our attempts to solve moral dilemmas that arise in medical practice are often guided by what he would call

---

[6] For further discussion see e.g. O'Neill (2002), Autonomy and Trust in Bioethics, Chapter 3, Cambridge UP; Stock, G./Campbell, J., Engineering the Human Germline, Oxford UP 2000; Stock, Redesigning Humans. Choosing our Children's Genes, Profile Books Ltd. London 2002.

moral prejudice and pay tribute to unexamined moral convictions that lack boldness of vision.

> Take a trait like intelligence. Most people [...] would not regard intelligence as a trait associated with disease at all. [...] Imagine two groups of mentally handicapped children, or children with severe learning difficulties. In one the disability is traceable to a specific disease state or injury. In the other it has no apparent cause. Suppose now that gene therapy held out the prospect of improving the intelligence of children in both groups so that they could lead normal lives. [Opponents of genetic therapy seem] to say that it would be unethical to attempt to help either group, but might possibly be interpreted to mean that we could help those whose condition is traceable to a disease but not the remainder. (Harris 1994: 221)

Such considerations make Harris an advocate of a bold program of biotechnological improvement of human qualities. He strongly disagrees with advocates of traditional values.

> "The possibility that some day parents may be able to 'order' children with particular characteristics such as the ability to do mathematics is not only abhorrent but also unlikely ever to be achieved." [...] For my part I do not find it abhorrent, and I wonder why, if it is legitimate to try to educate our children to acquire the ability to do mathematics, it is not legitimate genetically to engineer into them a like ability. Why, ethically, might it be wrong to attempt something via genetic engineering that it is not wrong to attempt via what used to be called 'social engineering'? (Harris 1994: 221)

It is obvious that in order to make such changes possible extensive embryo research is needed. Harris endorses the consequence. He realises that to many persons radical enhancement of personal talents and abilities through genetic engineering is unacceptable, because they view it as an attempt to alter individual properties by the use of highly intrusive artificial methods which totally alter individual identity. But Harris finds such opposition unfounded and obscurant. It perpetuates moral prejudices.

> [Opponents of genetic therapy feel charged] with such a heavy responsibility [...] as finding "an ethical position which would command acceptance in this country for the foreseeable future". They tend to believe that the best way of achieving this is to arrive at a position which already commands wide acceptance. This means that they almost inevitably follow popular prejudice, the views of those who have come to swift judgements without the benefit of the expertise, access to evidence and information and time for reflection available to [themselves]. (Harris 1994: 221)

5.1.3
The Regulatory Stance

Advocates of the regulatory position concerning embryo research will argue more often than other groups that embryo research should be allowed to continue, but it has to be supervised by agencies that enjoy public trust and are capable of enforcing socially acceptable policies. Advocates of this position believe in particular that debates concerning embryo research are unnecessarily exacerbated by emotional and abusive language overloaded with poor conceptual distinctions; that these conflicts probably will not be resolved by the endless struggle to identify the "one and only" correct moral standpoint on embryo research. Most importantly, proponents of the regulatory stance propose pragmatic regulations that enable social co-operation while respecting various moral standpoints.

They may recommend or prescribe the following policies: Invasive research shall be performed on embryos up to a specified age, but not on fetuses. Human tissue can be derived from experimental embryos, and produced in fairly large quantities to make it available to many patients, especially if the quantity and/or quality of pro-genitor cells of adult donors prove insufficient. Therapeutic cloning can be con-ducted by numerous licensed scientific laboratories. Even reproductive cloning is to be allowed in exceptional circumstances. If artificial prosthetic equipment cannot be used, or if its use is impractical, cultured tissue should be accepted without moral or religious qualms. Human rights are not associated with human genetic make-up, but with human-like sentience, behaviour and mental capacities.

In particular advocates of the regulatory position more readily than other groups will share some or all of the following convictions: To be a human organism or a proto-organism (a lump of undifferentiated cells) is not the same as to be a human person, and not all human organisms must be treated like persons. Embryos should be treated with special consideration but they can also be utilised and destroyed for the benefit of science and/or medical therapy if they are treated in a manner indi-cated by carefully designed procedures. It is advisable to distinguish between thera-peutic and reproductive cloning, and introduce such regulations that protect zygotes and embryos from commodification, while assigning embryos special status that makes them neither persons nor property.

Advocates of the regulatory position usually hold that human embryos – and by this term they mean not only pre-embryos but all embryos until they have become fetuses (which usually means by the end of the second week) – are not human per-sons but organisms belonging to the human species with a potential to become human persons. As such, they do not possess full human rights. For instance, they do not have the right to life, nor the right to be endowed with individuality. But they have a special status that guarantees them a degree of considerate treatment and grants them the privilege to be handled according to certain rules. The main question to be answered here is, How should "considerate treatment" or "the right to be han-dled according to certain rules" be defined?

Advocates of the regulatory position will argue that there is no reason to assume that a zygote acquires rights of its own as long as it remains insensitive and unable to reflect or decide. It is unclear in what sense one could say that a fertilised ovum "has been harmed," for instance, if only its own interests are taken into consideration. The problem is very different if the ovum is intended by its parents to become a child. Then it is possible, though wrong, to interfere with its potential to become a child. But if an embryo has been donated to scientists in order to be transformed into stem cells, it is unclear who it can be harmed in its potential to become a child, if it has never been intended to become one and if on is own it can have as yet no preferences at all. We may put it in more general terms by saying that incipient human beings command respect, which is to increase in proportion to the extent to which they develop behavioural and structural indicators of the ability to feel, perceive, etc. (see, e.g., Beyleveld and Brownsword 2001, Chapter 6; and below section 5.3). Advocates of the Regulatory position will probably all agree that harming an embryo or harming a fetus before it can feel pain, means nothing more than doing to it something that might constitute a damage to its well-being when and if it develops into a sensitive human individual. We do not want to harm embryos only insofar as we do not want to harm future babies, children and adults, argue advocates of the regulatory position.

This position seems rational to them, because they view it in the light of the belief that all rights arise from two sources, natural or conventional. Sensitive creatures possess a *prima facie* right not to be hurt, and it is their natural right, because they feel pain when mistreated. Children possess a conventional right to acquire heritage from their parents, because this is how accumulated wealth is usually disposed of. But embryos have neither a natural right nor a conventional right to become fetuses in the two first weeks of their lives, say advocates of this position, because they lack visible agency and/or a sufficiently determined course of life. Comparatively few embryos develop into full human beings (see Chapter 1) and this is especially true of embryos made *in vitro*, as their chance of further development into human beings is limited to the situation when they are put into a womb of a receptive woman, which in itself requires a clinical intervention and cannot be considered a universal right.

Advocates of the regulatory position will also point out that there are overriding reasons not to consider zygotes separate individuals either in biological or philosophical terms.[7]

Advocates of the regulatory position want embryos to be treated carefully and with respect. They disapprove of wasteful and careless treatment of embryos, for

---

[7] The question of individuality becomes highly speculative unless it is put in common sense terms. "[I]t must be emphasised that no particular concept of individuality can be applied uniformly and consistently throughout the living world." (Alex Mauron 1996: 60) One could argue that a specimen is more individual if it is haploid rather than diploid, because as haploid it has only one set of chromosomes, and as diploid it has two. Certain algae have alternating cycles of haploid and diploid development, and accordingly could be considered more or less individual in the two stages. On the other hand, however, the haploid stage produced by meiosis possesses separate female and male parts looking like long leaves, so perhaps it is two organisms or a colony of organisms. The diploid stage is produced by fertilization and is sexually undifferentiated, so in this respect it is more like one organism. "Generally speaking in the plant kingdom all the thinkable variations on this theme exist." (Alex Mauron 1996: 63). In philosophy one author found it useful to revive some ideas of Duns Scotus. For Scotus, the crucial problem was not how common nature is invested in a human individual, but rather how an individual is to be distinguished from the species in which it belongs. Scotus's answer was that any individual is what it is due to the individuation principle. "Scotus uses the term 'common nature' to describe what is common to both the group and the individual... . [W]hat horses share in common is indifferent to whether we are referring to a single horse or to all horses. For Scotus, then, the common nature needs something else, an individualizing principle, to constitute a particular horse. ... We can think of the preimplantation embryo as our common human nature for two reasons. First, even though this entity is genetically distinct from its parents and even genetically unique, it is not yet individualized. [These human cells] are morally privileged by being human cells, cells that manifest the human genome, and as such are an entity that represents the essence of human nature... . And because these cells are our common human nature and individualized human nature, I argue that cells from this entity may be used in research to obtain and develop stem cells for use in transplantation or to develop specific human tissue or perhaps even organs." (Shannon 1998) Thus, Thomas A. Shannon says that early embryos are more human than adult human individuals, because they are less individuated. They have qualities that belong to the species but not to a specific specimen. We are human before we are persons. This view makes it permissible to use zygotes as a source of tissue for any human organism, because we are obliged to protect human species and human nature rather than human individuals.

instance when embryos are mishandled by persons who engage in incompetent biological experiments. Advocates of this position urge that embryo research must be
conducted in well equipped and properly staffed scientific establishments whose
efforts have a high probability of success. Those demands notwithstanding, advocates of the regulatory position avoid talking about presumed rights that embryos
might possess. They claim that embryos should be treated with care and respect
because they have inherent value, either as potential human beings or as potential
stem cells. They also have a practical value, because they are difficult to obtain and
cannot be easily replaced. But as their future is not predetermined as to whether they
become persons, or tissue, or lost entities that have never been noticed and used, they
cannot have a specific right to become either persons or stem cells. At the same time,
however, as long as these two options, and perhaps some more, remain open,
embryos should not be treated as inanimate matter, because they are incipient human
organisms or incipient human tissue. It seems best, therefore, to give them the special status of "neither persons nor property" and emphasise the conditions which
imply that we have certain obligations when we handle them.

The special status of human embryos is a legal or theoretical construction that can
be developed in statutes and by-laws, but which cannot be adequately presented as a
list of natural or conventional rights. Thus different communities and different legal
systems may have their own regulations specifying conditions on which embryo
research and research on stem cell lines should be considered permissible.

Advocates of the regulatory position would generally say that embryos can be
used in experiments undertaken in order to develop new therapeutic procedures, but
should not be used (barring special cases) to create human individuals by cloning.
The mere technical ability to make a replica of an existing person is not a good reason to make a human clone. Creating a new human being is always connected with
great responsibilities and the prospective parents should be prepared to meet them.
Hence, we should be wary of the false analogy with irresponsible parents in this context. It is true that society can do very little to discourage reckless couples from having babies that they will not care for. This is no reason, however, to exonerate scientists who might engage in producing clones. No laboratory should be allowed to
make a cloned baby simply in order to prove their technical ability, and it is highly
recommended that objectives and methods used in embryo research are subjected to
rigorous professional and social control.

The regulatory position is not necessarily incompatible with any religion. It is
based on a system of rules and this aspect is usually favoured by religious institutions. It emphasises the importance of minimising harm and suffering, which again
is an aspect favoured by many religious institutions. It is supported by scientists and
patients, which makes it a potentially popular undertaking, and popularity is often
appealing to religious institutions, too. Thus even at present most religions seem to
espouse the regulatory position.

In general, it would be wrong to assume that advocates of the regulatory position
are motivated primarily by practicalities, such as research success, for instance. They
contend that all non-renewable parts of human body as well as zygotes and embryos
have special status and must be treated according to a code of behaviour. In other
words, we should not pattern our obligations towards embryos on the duties we have
to sentient human beings, because not all embryos are destined to become human

beings. On the other hand, it is also important that we do not reject all responsibility for the future of embryos, because embryos possess unparalleled potential of more than one kind. As this potential changes in time, it seems advisable to the advocates of the regulatory position that zygotes, embryos, genomes and genes are assigned special status that is separately defined for each of them. This view has been championed by Paul Lauritzen, who emphasises that embryos should be given special status defined by law (see: Lauritzen 2001).

There are analogies in law that elucidate the concept of "neither property nor a person". The classic locus is *Roe v Wade* where the status of implanted embryos was defined by pointing out that they were, strictly speaking, neither part of a woman's body, nor an independent organism, but something of a mixed character. The state has no interest in their well being up to a certain moment chosen more or less conventionally, but it will step in to defend their future after they have sufficiently matured. Similar to the concept of "neither property nor a person" is the construction of "pro-attitude" proposed by Donald Evans (Evans 1996b). Evans says, following Nowell-Smith, that before we consciously decide what our specific attitudes to an entity should be, we should read our feelings and behavioural dispositions in order to make sure that they agree with our rational convictions. The resulting pro-attitude is a composition made of basic emotional reactions and correlated opinions. The attitude cannot be supported by either pure beliefs or pure emotions. It has to combine and harmonise feelings and convictions.

There are several advocates of the regulatory position, both individual and institutional, and their approach is well exemplified in the writings of Donald Evans. He tries to offer impartial procedures to solve practical controversies that arise between advocates and opponents of embryo research. Taking his cue from some aspects of Ludwig Wittgenstein's philosophy Evans claims that we should resist the temptation to solve the debate by making new regulations or by proposing new linguistic or legal concepts. It is more promising, he holds, to make closer observations about our emotional reactions that underlie our moral beliefs. As we seem to be unable to decide in a rational manner at what point a human being begins to exist, we may want to inquire what it means emotionally to us to establish a relation with a member of our own species. His answer is that whenever we meet a human being we activate a repertoire of feelings and behaviours that are appropriate in our contacts with other human persons but not in different contexts. We activate them unreflectively and spontaneously. These reactions can be corrected, they can be suppressed and they can be artificially invoked, but even though they are not sufficiently rigid (or perhaps because they are pliable) they are more important and reliable than any theoretical considerations that we might use in the attempt to construct a new (or impose an old) authoritative definition of human person.

> [T]he language of persons is rooted in very general, shared reactions. These reactions are not eccentric though they need not be universal. Occasionally individuals may be cut off from them, and consequently probably cut off from whole areas of human discourse and understanding. Some of these deficits may have considerable moral significance. Nevertheless for those who are able to talk of human beings as living souls, the relation between bodily appearances and behaviours on the one hand and concepts on the other, are internal. One cannot separate one's concept of a human being from the appearances and behaviours which draw from us such reactions. (Evans 1996b: 38)

When we look at an adult human person, a child or a fetus we tend to identify with them and, rightly or wrongly, make assumptions about their feelings and other mental states. When we see an embryo under a microscope no such emotional reactions stir us. We may be told to invoke such reactions in our breast and even obey such an instruction. But usually we will not be induced to having such feelings simply when we are told that the concept of human being which includes embryos is correct, and one which excludes embryos is incorrect, or the other way round.

> [I]t is not the question of whether our concept of a human being is correct. This is not a question which it makes sense to ask. It is not the language which is correct or incorrect, but what is said in the language. Thus to assert that a given human being is reckless, brave, stubborn, diseased and so on, is to claim that something is true. The concept of a human being makes no claim. (Evans 1996b: 39)

If we can establish the same emotional relation to an embryo that we have when we get in a contact with a sensitive human, then the embryo makes a successful demand on our sensibilities and becomes one more human being. If not, we cannot admit honestly that the embryo represents a human being to us. It is then no more than a group of undifferentiated cells that deserves some respect but does not have human rights. Evans draws an important distinction between human life, and *a* human life. He admits that an embryo is human life, because it is living and because it is human. But Evans denies an embryo is a human life, i.e. that embryos are sufficiently individuated to be considered separate human individuals. He is ready to acknowledge that there are people who may feel differently, and may have emotional reactions that are appropriate in contacts with humans whenever they encounter a human embryo or even human tissue. But their position is not compelling to every decent and rational individual.

Evans appreciates that differences between basic reactions will not be easily overcome or neutralised. He proposes regulation as an optimum solution. A consensus can be created by minimising moral claims and highlighting points of agreement, he argues. Being particularly concerned with practical proposals, he suggests a procedural method for procuring more gametes needed in research and *in vitro* fertilisation. He sees the problem as essentially pragmatic and not ideological. It is important, he claims, that (1) more gametes are made available, (2) the differential treatment of male and female providers is discontinued, (3) the selling of gametes is submitted to regulations (Evans 1996c: 319–320). His scheme is built on the analogy with the practice of recruiting healthy volunteers for medical research. They are both remunerated and saved from exploitation, which otherwise they might themselves accept, by regulations and fixed monetary incentives. In this way harms and injustices are minimised.

5.2
Rationality and Persuasiveness of Specific Arguments

The three positions described above can be supported by a variety of arguments, some more reliable, some less. As we have mentioned, however, it is impossible to define these positions by specific sets of characteristic arguments. Advocates of the three policy standpoints may use whatever arguments they wish. We must therefore discuss these arguments as a separate problem, keeping in mind that the same policy defended with one set of arguments on one occasion, may be supported with another set of arguments on a different occasion.

5.2.1
Taboos

Contemporary society has not entirely purified itself of the sense of taboo. Certain forms of behaviour strike some observers as sacrilegious even though they are not so intended by those who act in the suspect way. It is often argued for instance that by undertaking embryo research one breaks a fundamental taboo of our culture that demands the highest respect for all that is essentially human. This claim is characteristically made by advocates of the prohibitive position.

Sigmund Freud believed that taboos arise from a clash of emotions over which reason is unable to arbitrate. When we are exposed to ambivalent feelings that are so strong as to terrify our minds, instead of trying to decide what is the best solution in such circumstances, we conclude that we face something awe-inspiring or holy that cannot be properly described. If a whole society shares such instinctive reactions, a system of elusive beliefs and defensive behaviours develops, and it constitutes taboo. It is remarkable that in many cultures taboos are connected with death. An encounter with a deceased person is shocking and confusing, and even if one would like to treat the corpse as if it were still alive, one knows it would be senseless and disrespectful to do so. Freud, who investigated the phenomenon of taboos in primitive cultures and compared it to psychic neurosis occurring in western societies, offers an explanation that starts from the ambivalence of feelings and ends with a sense of confusion (Freud 1938: 95).

When members of a tribe encounter a dead man, they can neither treat him as a friend nor as an enemy. He gets a special status that cannot be captured by assimilating him to other, well known objects. But still witnesses do not know what to do, so they resort to ritual reactions. In some tribes, the anxiety generated by these conflicting feelings is diffused by putting blame on the dead man's enemies, his friends, family or neighbours, on witches, strangers, infidels and so on. And if it is impossible to find culprits, blame is attributed to, and almost gladly accepted by, those who arrange the burial. Grief is transferred on the grave diggers who in turn become taboo, too.

There are some important parallels between a corpse and an embryo. Both are an extension of a live, fully developed human being. An embryo already lives but is not fully developed; a corpse is fully developed but no longer alive. We are easily confounded by such border line entities. The confused feelings are relieved if we resort to mysticism, magic, neurosis or bad logic. But in every culture what is incomprehensible inspired abomination, and abomination must be diffused by traumatic social reactions such as a ritual of social contempt, war, change of life style or neurotic withdrawal. Incomprehension always produces a sense of danger, and danger must be dispelled.

> [C]onsider beliefs about persons in a marginal state. These are people who are somehow left out of the patterning of society, who are placeless. They may be doing nothing morally wrong, but their status is indefinable. Take, for example, the unborn

child. Its present position is ambiguous, equally its future. For no one can say what sex it will have or whether it will survive hazards of infancy. It is often treated as both vulnerable and dangerous. The Lele regard the unborn child and its mother as in constant danger, but they also credit the unborn child with capricious ill-will which makes it dangerous to others. When pregnant, a Lele woman tries to be considerate about not approaching sick persons lest the proximity of the child in her womb caused coughing or fever to increase. (Douglas 1970: 115)

To some extent every culture grows on such mechanisms, which are not easy to identify, but which nevertheless organise all of social life. These attitudes are responsible for the collective identity of the people who respect them and for their collective life plans. Leszek Kołakowski calls such values "symbols", and holds that they always have a religious aura around them.

Anything can become a religious symbol: a man, a word, an individual biography, every object, a stone, a building, an animal, a relic.[8] But nothing can be turned into a religious symbol by a rational agreement, by a free contract or an arbitrary design. An object can acquire a symbolic power only if a congregation of the faithful has bestowed on it this particular status in a spontaneous and authentic concordance. A religious symbol can be no more imposed by an efficient method of persuasion than, for instance, a symbol incarnating national values of a given community. (Kołakowski 2000: 224)

But in contemporary society we cannot return to taboos and magical thinking. Our basic attitude to nature has substantially changed, because our knowledge of natural processes is much better than in primitive societies. Our basic world view is comprehensive, empirical, scientific and open to verification. A primitive society has a different attitude. They personalise their contacts with nature, they try to communicate with hidden forces, and try to find a detailed interpretation of every single fact that occurs in their lives. This attitude cannot be recreated, once it has been lost.

It is misleading to think of ideas such as destiny, witchcraft, mana, magic as part of philosophies, or as systematically thought out at all. They are not just linked to institutions [...] but they are institutions. They are all compounded part of belief and part of practice. They would not have been recorded in the ethnography if there were no practices attached to them. Like other institutions they are both resistant to change and sensitive to strong pressure. Individuals can change them by neglect or by taking interest. (Douglas 1970: 108)

The lesson to be learned from this is that in the primitive society institutions are the framework in which individual opinions emerge and develop. A rational society makes institutions conform to the opinions and attitudes that have been chosen by their members. The vision of a laboratory that produces chimeras is terrifying if we associate it with a witch's cave or a wizard's castle, if we suspect that the laboratory kills innocent embryos and produces miserable monsters. If we realise, however, that these laboratories develop substances that will have therapeutic effect, we may decide that they are also an integral part of our culture, as much as hospitals, univer-

---

[8] Another example for a new cultural symbol, perhaps even an icon, has been mentioned in Chapter 4 in discussing people's attitudes towards Dolly the sheep. See Chapter 4.6 Perceptions of animal and human cloning in Europe.

sities or pharmaceutical companies. Taboos cannot be revived in the critical, independent and questioning mind.

5.2.2
Redefinition

When someone engages in a practice that is questionable to other people, its ultimate acceptability may depend on the language that both parties will use to present it. Pre-selection of embryos is described by some people as a method of assisting parents to have healthy children, by others as an attack on innocent human life. In order to win social approval embellishing terms or misleading insinuations are used. Sometimes new words are invented. The concept of pre-embryos was adopted, for instance, to make sure that no criticism levelled at embryo research would apply to very young embryos. When Advanced Cell Technology started their experiments on embryos, the company's spokesman emphasised that the company was not trying to create a human being, just an embryo (Economist #8250). Taken literally, the claim was true, because the experiment was not continued beyond the 14th day. But the claim was also misleading (in its technical rather than moral aspect), because even though the scientists had committed themselves to interrupt it on the 14th day, the embryo did not begin to develop.

Public acceptance of new scientific ventures depends heavily on the language in which they are described. A project that is presented as an attempt to develop stem cell lines "in experiments conducted on human embryos who cease to exist in the process" will be evaluated very differently from the same project if it is advertised as an attempt to "extract a special kind of tissue from human embryos while making sure all the time that the tissue remains living and capable of multiplication". Emotional associations are more important to the general public than details of scientific procedures. We believe that it would be helpful to minimise the influence of emotional associations evoked by stylistically different though factually equivalent formulations. One possible approach to this problem is to present different ways of talking about the same things from the very beginning and show that they are all factually equivalent. This seems to be a rational procedure but it can create a lot of confusion as well. Another possible approach is to use, whenever possible, a neutral language that does not imply any normative conclusions at all. Instead of speaking of "human embryos" one might use such phrases as "proto-human organisms", "human proto-organisms", "biological carriers of future human persons", "human blueprints", etc. Such formulations are of course cumbersome and will not be easily accepted. What is more important, even if they were accepted, they would quickly lose their neutrality in a heated debate. We must remember that the term "human embryo" was also neutral some time ago but lost that status when champions of embryos' integrity undertook to endow embryos with human dignity.

The debate about admissible limits of biotechnology remains rational as long as the discussants are ready to examine all relevant arguments and agree to rephrase them in a different language. Whenever such a translation is not possible, it is evident that something is either wrong with the proposed rephrasing or with the arguments used. An open-minded, understanding approach is characterised by the fact that its arguments can be presented in different languages. For instance the Regulatory position can be expressed in the language of sociology, anthropology, law or philosophy, whereas some Prohibitive opinions, can only be expressed in the language of one religious denomination or one segment of social activists, such as Pro-Life groups, for instance, which as a rule insist that all human embryos are "potentially human individuals" and refuse to admit that they are "balls of cells without small legs and arms".

5.2.3
Emotions and the Rhetorical Language

The debate about embryo research is also vitiated by more straight forward rhetorical assaults and insults.

> It is not enough to say that the wicked deed has already been done, that the embryos have already been killed. The purpose of that killing was to obtain the stem cells. One ought not to implicate oneself in that process, not even for the noblest and the most beautiful ends.... We human beings very easily reason ourselves into taking positions that end up having the most tragic of consequences, positions of which we would never have approved had we seen those consequences at the time. For the fruit of the tree of knowledge over yonder appears to be very sweet, and we feel sure that if we eat of it, then happy endings [...] will result. Those endings have always turned to sulphur in our cheeks. (Novak 2001: 17)

Such proclamations do not make a serious entry in any rational discussion. They may captivate the hearer's mind but they are devoid of the logical power to convince. Rational discussion is made very difficult by authors who deliberately distort utterances made by their opponents or who impute evil motives to them. Suppose someone says that embryos in utero and embryos *in vitro* have different prospects. It is a matter of fact statement that cannot be denied. The former embryos will grow into babies or will abort (spontaneously or when induced), the latter will turn into babies or become frozen. Yet the distinction seems artificial to some polemicists.

> Senator Hatch (R.-Utah) suggested that he [the President] adopt the geographical theory of human life. Whether you are a human being or not depends on where you are, not who you are. An embryo in a test tube, Hatch argued, was not human life, even if the same embryo in a Fallopian tube was. According to this thinking, if someone built a large test tube and dropped Hatch into it, he would cease to be a human life. (Jefferey 2001: 6)

Though the remark might be intended only as a joke, it makes a serious discussion impossible. It does not constitute a proof that an embryo has the same rights no matter where it is, in a test tube or in a Fallopian tube. By the same logic one could say that a fertilised human egg has the same rights irrespective of whether it has left the woman's body or nested in her uterus. Rhetorical phrases may help to present certain issues in a vivid manner, but they complicate serious efforts to find consensus.

5.2.4
Potentiality

The potentiality argument says that all embryos are potential human beings (i.e., they possess such qualities that confer upon them full moral status even before they look and feel like humans), because in favourable circumstances they can develop into human beings. It is, therefore, obligatory, says the argument, that nothing is done to them that might preclude their becoming mature human beings (see for instance van Niekirk 1996). In a more technical formulation the argument runs:

1. X is a biological entity with potential for personhood.
2. The core harm in killing [...] is involved in any killing of an entity with potential for personhood.
3. This harm is of a sufficient magnitude to ground a prima facie right to life
   Therefore
4. X has a prima facie right to life... (Holm 1996: 216)

Those who make use of the potentiality argument emphasise that the principle of potentiality should be respected by IVF practitioners, for instance. If this demand is taken literally it imposes strict obligations that are impossible to observe in practice. The official policy of IVF services would have to contain three points: (i) No surplus embryos are created; (ii) All surplus embryos are cryopreserved until the time when they can be re-implanted; (iii) If they are not implanted and no longer viable, the surplus embryos should be allowed to die a peaceful death. This policy is motivated by humanitarian concerns. However, as discussed in Chapter One the first two conditions are regularily violated in the course of almost every IVF treatment. The third condition allows some surplus ova to die a peaceful death, so it is not accurate to claim that they remain under the protective influence of the potentiality principle. The potentiality principle requires that "nothing is done" to eliminate the possibility that the embryos might develop into mature human beings. So it effectively eliminates all IVF services and embryo research. The principle gives its support to the prohibitive position only, and renders the two remaining positions invalid.

Those who find it useful and permissible to conduct embryo research must reject the potentiality principle. They can do so in good conscience if they avail themselves of the argument offered by Michael Tooley.

> Suppose at some future time a chemical were to be discovered which when injected into the brain of a kitten would cause the kitten to develop into a cat possessing a brain of the sort possessed by humans, and consequently into a cat having all the psychological capabilities characteristic of adult humans. Such cats would be able to think, to use language, and so on. Now it would surely be morally indefensible in such a situation to ascribe a serious right to life to members of the species *Homo sapiens* without also ascribing it to cats that have undergone such a process of development: there would be no morally significant differences. (Tooley 1999: 31)

Tooley concludes that future kittens would have no right to a brain injection that turns them into miraculous kittens, that they would have no right to life other than real kittens have before they had the injection, but they would have to be protected like humans if someone turned them into sensitive, intelligent new-breed cats.

Tooley's thought-experiment also shows that no one possesses any rights at present by virtue of what one can become in the future, and no one should be ascribed several different sets of different rights now relative to various incompatible future statuses that one may possess. Otherwise the potentially miraculous kittens would have two different sets of rights, one set connected with their nearly human status, another set connected with their uninterrupted feline status. The conclusion to be drawn for human fetuses is similar. They may possess certain rights as already gestating children of the parents that want them as their own children. Then they will have even more rights when they are born, become babies and grow up. But they do not have the right now to become babies of their potential parents against the parents' wishes any more than the future kittens would have the right to become intelligent cats.

5.2.5
Naturalism

It is often objected that abortion, IVF, the use of stem cells in therapy and selection of desirable traits in children through prenatal screening are all unacceptable practices because they are not "natural". It is not clear, strictly speaking, what "natural" means in this context, but the general intention of such protests is obvious. Whenever "traditional" life styles clash with "modern life styles" traditionalists say that "modern life styles" are unnatural. The controversy does not arise in medicine only. In the last decade similar fears were expressed with respect to transgenic foods or excessive exposure to radiation from computer screens. (As has been shown in Chapter 4.4 "Ambivalence toward science at the turn of the century". In the case of life sciences, the situation is completely different today and the cluster of telecommunication, computers, the Internet etc. is seen without exception in a favourable light.) However, if biotechnology is unnatural, how is surgery any different?

> [T]he argument from unnaturalness seems to apply to many practices which have been traditionally considered quite acceptable. If genetic engineering is to be condemned because of its power to change individuals by technical means, then most medical interventions should be condemned as well. Surgical operations, for instance, often alter people by transforming them from fatally ill patients into perfectly healthy citizens. (Häyry 1994: 210)

Another meaning of "natural" is associated with a theological vision of the world. Thomas Aquinas believed, for instance, that God created two kinds of laws. One law was openly given in the Ten Commandments. The other was implanted in nature to be discovered by the curious human mind. On this view, those who engage in practices that were not included in the divine plan subvert God's intentions and try to impose their own laws, incompatible with divine laws. An obvious deficiency of this approach is that one cannot find any serious indication that God in fact had made the two kinds of laws and wished the tacit law to be discovered and observed by all people.

In general, the argument from the "natural" or the "unnatural" character of an activity is not conclusive, because anyone can use the term for his/her own purposes and no practically useful criteria can be selected to determine what is and what is not natural. Klaus Eder shows that European attitude to nature is very ambiguous. On the one hand, we try to identify with nature, consider ourselves part of it, and try to establish closer, more sympathetic relations with everything that lives. On the other hand, we consider ourselves masters of nature, and wish to use it for our own purposes. It is impossible to say why one of these attitudes should be considered better than the other.

> Without distorting the image too much, one can say that the history of western Europe is viewed as a history in which empathy with nature and equal respect for nature is increasing. Rights apply equally to everyone, and what is understood by "everyone" tends to expand to include "every living thing". This development runs parallel to scientific enlightenment, the decline of magic and the formation of secular moral notions. [...] The picture of increasing movement towards a moralization of nature is, of course, not as clear as these suggestions would lead one to believe. Not only are there exceptions to this image of an increasing sympathy with nature, these exceptions in fact fit together into an independent history of the exploitation of nature: nature as an object to be used, whose ability to resist can be controlled by knowledge. (Eder 1996: 144–145)

We also use an evaluative concept of nature sometimes, but apply it inconsistently, too. Sometimes we believe that nature itself evolves and produces more perfect, more diversified species, sometimes we think that everything in nature is good as it is, and either hoping that it might be better or trying to make it better is an "unnatural" wish.

In view of these shifting and highly differentiated opinions the argument from "naturalness" seems philosophically unsound. Neither can we agree on what is natural, nor are we willing to accept that the "natural" arrangement of things (whatever that might mean) is superior to any other arrangement – artificial, technological or cultural.

5.2.6
Dignity

The concept of dignity, like the concept of nature, enters in the discussion on embryo research in different ways. To the defenders of the research, human dignity is enhanced whenever new, more sophisticated and effective methods of treatment are available. To their opponents, dignity is offended whenever embryos are put to death in order to retrieve stem cells and use them to cure patients.

The question 'How to appreciate dignity of an embryo?' is not very different from 'How to show respect to embryos?' Advocates of the prohibitive position would say, by sustaining their life whenever possible. Advocates of the permissive position would say, that no specific signs of respect are necessary. Advocates of the regulatory position give the most elaborate answer. Michael J. Meyers and Lawrence J. Nelson from the Hastings Centre argue that it is possible to use embryos respectfully as a source of cellular material if we employ the method of ascertaining moral status elaborated by Mary Anne Warren.

> In an abbreviated form [the] seven principles are: (1) respect for life – living organisms may not be killed or harmed without good reasons, (2) anti-cruelty, (3) agent's equal rights, (4) human rights, (5) ecological importance, (6) interspecific communities, (7) transitivity of respect [meaning mutual respect]. (Meyers and Nelson 2001: 30)

With respect to embryo research these conditions are satisfied because (1) embryos are not killed wantonly but used to retrieve stem cells, (2) they are not made to suffer, (3) they are not agents, and as such they do not possess rights equal agent's rights, nor (4) any specifically human rights at all, (5) the impact of embryo research on the environment is minuscule, (6) embryo research affects no other species, (7) and embryos cannot be said to fail in showing us due respect. The wisdom of choosing these particular conditions may be debated, but Meyers and Nelson are right insofar as scientists show no kind of disrespect to embryos when they conduct research on live embryos in a competent manner. Their respect is manifested by adopting strict procedures. The code of behaviour employed in well organised medical laboratories is sufficient to make sure that embryos are not deprived of their dignity.

There may be other ways of interpreting dignity, but none seems to indicate conclusively that using embryos for research or selecting genes for future children offends dignity of the donors, the recipients or the researchers. Deryck Beyleveld and Roger Brownsword (2001) argue that human dignity resides essentially in the capacity of agents to set ends for themselves. Embryos do not display that capacity in the first 14 days of their lives to say the least. They may be considered agents who have this capacity only by disposition but cannot display it (they are "locked-in agents") or they may be considered no agents at all. Which of the two possibilities is true cannot be known with certainty, so it must be supposed that they have the dignity that all inactive agents possess. However, this possibility has no practical consequences, because, as Beyleveld and Brownsword have argued, it is not rational to appeal to the dignity of embryos, but only to their possible dignity, and the constraints that this imposes are minimal as they remain in proportion to the degree to which the embryo has or displays indicators of having the capacity to act (see Beyleveld and Brownsword 2001: 122–126). Embryos can indubitably acquire dignity later, when they come to behave like self-conscious agents. Consequently we can deprive of dignity those embryos who will develop into mature humans but whom we have treated in such a way that as mature humans they will have a limited capacity to set ends for themselves. If they do not transform into mature humans, however, they will never have a chance of exercising the capacity to act as agents, and then their dignity will not be diminished by instrumental treatment. We may conclude that objections to embryo research based on the assumption that the research deprives embryos of their dignity are groundless.

The concept of human dignity becomes particularly important in some countries, as becomes obvious in the German Constitution, whose first article claims that "Human dignity is inviolable". However there is considerable disagreement on the question of whether or not these words apply to embryos so that one can directly deduce their right to live from the German Constitution. Ulla Wessels, for example, believes that it is impossible.

> What the principle of human dignity forbids [...] is making man a mere means. Yet what does it mean to make someone a mere means? It means to ignore his ends; to treat his preferences as irrelevant. But embryos do not have preferences; *a fortiori* none that we could ignore: hence whatever we do with embryos, we do not thereby violate their dignity, for it cannot be violated. (Wessels 1994: 238)

Of course, other authors come to a different conclusion. Thus, although it seems morally advisable to handle embryos with special consideration, the manner in which they should be treated cannot be deduced from the concept of non-exploitative relationship, from the principle to always treating a person as an end or from the concept of dignity.

5.2.7
Rights

Rights enter the debate on embryo research in many ways. One can ask, e.g., if embryos have the right to survive, if their donors have the right to make decisions about their future, if researchers have the right to transform embryos into strains of stem cells and if anybody has the right to sell embryos or products derived from them. To some writers it is important to assert that cellular material must not become subject to commodification. If one chooses to use the language of rights, this belief can be expressed by saying that totipotent and pluripotent cells possesses the right not to be treated like mere tissue. Let us have a closer look at this proposal.

Commodification seems wrong for several reasons. One is that we would like to avoid the situation when people feel compelled to sell their body parts or products to pay their financial obligations or to meet demands of retributive justice. Though, from a different perspective, this might be better thought of as a problem of injustice, rather than one of rights.

Some authors assert (see: Resnik 1998) that special rights may arise in the field of embryo research because gametes and zygotes exist as patterns, permanent connections or established sets of relationships between physical objects, and cannot be lifted from their carriers without destroying the carriers. In a philosophical language, they are like universals, rather than individuals. It may be obligatory to protect them not for their own sake, but in order to ensure that individuals to whom they belong retain their physical uniqueness.

It is a matter for further discussion if genomes and genes should be considered intellectual property and/or if any of their parts or sequences can be patented. In general, no natural processes can be patented, but human inventions can be. The distinction becomes blurred in some applications. When a piece of DNA is included in a biological process that leads to the creation of biological material of considerable market value, e.g. antibodies, it is difficult to decide if this piece of DNA is a natural object or part of a technological invention. The European Parliament discussed in 1995 the scope of restrictions on the use of human genetic material. Its particular concern was to determine what restrictions should be applied in patent law. Some social activists were worried by the possibility that biotechnological companies may acquire patents that would make them sole possessors of technologies that can be used to make medical diagnoses or produce new biological materials. In this context the concept of "pirating human qualities" was created (see: Rogers 1998).

In general it is difficult to establish if someone possesses a right or not, when this right is questioned. As we have seen, it can be debated if embryos possess human rights, if researchers have the right to transform embryos into strains of stem cells, if anybody has the right to sell embryos or products derived from them or even if totipotent and pluripotent cells possess the right not to be treated like special tissue. We can conclude that the language of rights is effective only if the rights themselves are firmly established or if we have institutions that can solve hard cases. But even if we have such institutions they occasionally face formidable problems, especially when historically honoured obligations clash with conditions of justice. Ronald Dworkin comments on this problem in an interesting manner.

> Political rights are creatures of both history and morality: what an individual is enti-
> tled to have, in civil society, depends upon both the practice and the justice of its
> political institutions. So the supposed tension between judicial originality and insti-
> tutional history is dissolved: judges must make fresh judgements about the rights of
> the parties who come before them, but these political rights reflect, rather than
> oppose, political decisions of the past. When a judge chooses between the rule estab-
> lished in precedent and some new rule thought to be fairer, he does not choose
> between history and justice. He rather makes a judgement that requires some com-
> promise between considerations that originally combine in any calculation of politi-
> cal right, but here compete. (Dworkin 1978: 87)

Dworkin strongly believes in the continuity of law and in the validity of rights
confirmed by law. He says that in general politically established rights are just and
fair. But he admits that occasionally entrenched rights may become unfair and then
the judge must find a compromise that reconciles traditional entitlements with currently pursued goals. This proposal is very fair but not very effective. Traditionally
embryos enjoyed no rights because they could not be manipulated by scientists. Is it
fair to say they have no rights now, because they did not have them in the past? Advocates of the prohibitive position would certainly disagree. And wherever they are
successful in establishing their own legislation, they turn the argument around and
say that in the law they have introduced, and which by now has become history,
embryos possess some rights that cannot be violated by considerations of fairness.
One can imagine that it would be extremely difficult for any judge to find a compromise here. So we must conclude by saying that reference to rights is not helpful in
solving hard cases connected with embryo research.

5.2.8
The Slippery Slope Argument

It is often claimed that biotechnology initially introduces procedures that are helpful
and morally acceptable, but as further inventions are made or new goals are adopted
these procedures imperceptibly change into something more dangerous and morally
suspect.

> Genetic intervention has the potential to be eugenic, that is to "improve" inborn traits
> to the ultimate advantage of the race. The categorization of genetic engineering of
> resistance to infectious disease as an ethically permissible technique falling within an
> indistinct divide between negative and positive eugenics might promote the danger of
> the slippery slope into enhancement engineering, which is generally viewed with
> such apprehension as to be beyond consideration. (Wood-Harper 1994: 134)

Janice Wood-Harder essentially says that we must not give blanket approval to whatever will be done under the guise of an activity that has already been approved but that
may be conducted very differently in the future. This is a legitimate request. But it
remains debatable what additional conclusions can be deduced from such exhortations.
Shall we say that a legitimate activity should be stopped now, because in the future it
may evolve into an unacceptable activity? Or shall we say that we must be particularly
careful about what we endorse at present, because our acceptance may be extended to
something new in the future without our knowledge and approval? Or shall we say that
we must make wide-ranging decisions now, because in the future we will not have

enough will power or enough discrimination to distinguish right from wrong any more? All these presumably adverse consequences are pointed to as the effects produced by a slippery slope slide. It does not seem that they are all equally convincing.

It is obvious that the moral exhortations are sometimes marked by wisdom and foresight, and sometimes by fear and ungrounded predictions. Consider the following list of events: Knowledge of Reproductive Processes, Family Planning, Contraception, In-Vitro-Fertilisation, Embryo Research, Extraction of Stem Cells from Embryos, Destruction of Fetuses, Infanticide. The list is a modification of several enumerations made by Rosalind Hursthouse (Hursthouse 1987) who intended to contrast conservative and liberal attitudes in evaluation of procreative behaviour.

The list can have at least two interpretations, both of which illustrate the slippery slope argument. On one interpretation it presumably shows how some types of behaviour lead to new types of behaviour, and that what was initially acceptable can give rise to new practices that are no longer acceptable. This is a historical interpretation and it is very inaccurate. Although in general it is true that the entries listed earlier also happened earlier in time, they did not all happen in the order suggested and they did not all happen in the same order in every society. Some practices of birth control were used before sexual education was systematically taught, some forms of embryo research predated IVF practices. If the succession of events that we have seen so far does not occur commonly or universally in the same order, there is no reason to say that it will happen in any predetermined order.

On the second interpretation the list shows different types of behaviour, ordered from less objectionable to ever more immoral. But again, is it true that the extraction of stem cells from embryos is something worse than family planning? Two further understandings of Slippery Slope arguments need to be discussed: the predictive and the evaluative interpretation (See Lamb 1998). Let us review their validity. The predictive version of the Slippery Slope Argument says:

> What you want to do now is not bad in itself. But others will follow you. And then your actions, together with theirs, will produce unacceptable results. Consequently, we must prohibit you from doing what you propose to do.

The Predictive Version can easily be refuted. Suppose someone asserts:

> I want to do something that is not questionable in itself. I will take full responsibility for my acts, and I will act alone, taking no external assistants or external partners. But I will not accept that other people's faults will be placed on me. There is no reason why somebody else's unacceptable behaviour is lumped together with mine, if mine is acceptable.

This rebuttal is conclusive. Imagine a person or an organisation that has chosen to act alone, and clearly defined the scope of their activities. Suppose also that all their activities are legitimate and morally acceptable. It is possible that someone will imitate these procedures but will act incompetently, illegally or immorally. We cannot simply declare that because the imitator acts illegally, the legally operating institution should lose its license.

The Evaluative Version of the Slippery Slope Argument says:

> What you want to do now is not bad in itself. But the proposed action belongs in a class of acts not all of which are permissible. We can order them from innocuous to

> absolutely unacceptable, but we do not know, where the innocuous part ends and the unacceptable begins. Thus we feel it is best to say that all these acts are unacceptable.

The evaluative Version can be refuted as well.

> If it is indeed true that we do not know where to draw the line, it follows only that wherever we draw it, the line will be arbitrary.

It does not follow that the line must be placed at the beginning of the list. An opponent may as well say that it should be placed at the end, and all the enumerated acts are equally innocuous. This rebuttal is fully effective, too. No one can say in response to it that although processing stem cell lines is permissible, it should be considered impermissible, because infanticide is impermissible. Nothing in the world is bad by remote resemblance to something else, not even in the case when it is difficult to decide where to draw the dividing line.

We should note, however, that there is a related problem here that may give us more serious trouble. Sometimes one person's activities occurring together with another person's activities bring forth bad results, even though the two activities performed separately would not have such results. An obvious example would be smoking a cigarette and filling a tank with gasoline. But this case is not a "slippery slope" connection, but a bad coincidence. We should prevent such coincidences, of course, but it would be wrong to prevent them by making one or the other activity absolutely blameworthy. It is thus possible that embryo research may coincide with a general change of moral attitudes and a growing popularity of unwelcome attitudes. But this is not in itself enough to make embryo research illegitimate.

5.2.9
Caution

Exercising caution is always a rational policy. By exercising caution we can avoid unnecessary mistakes caused by haste, recklessness or negligence. Evidently, if embryo research is to be conducted at all, it must be conducted in a cautious and prudent manner. If it is not, the entire endeavour will be killed by justified criticism. Scientists should not make promises that they cannot keep, doctors must not recommend risky procedures, policy makers should not implement institutional arrangements that are impractical and unpopular.

In other words caution is mainly a practical issue. It is indispensable whenever we must solve a dilemma, but cannot solve it by using established, uncontroversial rules. For instance, for many infertile couples the problem arises, if they should use hormone therapy or IVF. They may want to have a review of probable consequences of either procedure and their rate of success. When Mandy Allwood had octuplets, a doctor said: "It is difficult to tailor the dose", intimating that the use of IVF would have been preferable (Economist, 8/17/96 #7979). For millions of infertile couples the fact that IVF allows them to control the number of children, while hormone therapy does not, may be a decisive reason to choose the former rather than the latter.

It should be clear that the principle of caution, if applied, as we have just indicated, to practical action rather than theoretical reflection, supports the regulatory position much better than any other position. The regulatory position is consistently cautious and pragmatic. It proposes to solve all socially important problems, even if it does not have clear rules on which the solutions could be based. It warns, however, against creating new problems, which might arise from championing unpopular proposals.

5.2.10
Injustice

Injustice is not a specific problem connected with embryo research and its potential applications. It is a general problem of inequality between nations, classes and individuals. On the global scale donors and recipients of sophisticated medical procedures usually belong to different social classes and/or different countries. For example, if a US hospital undertakes a kidney transplant in which an American and an Indian are involved, it is more probable that the Indian is the donor and the American the recipient than the other way round. Some social activists already use the term "clone-ialism" in the attempt to show that medically advanced countries will try to exploit less advanced ones and biotechnology will facilitate this trend.[9] Accordingly it is more likely that research institutes will prospect for egg donors and embryo donors in disadvantaged countries. Obviously the terms of trade are not equal. But is this fact a sufficient reason in itself to decide that embryo research should be stopped?

Some argue that those who pay for organs to inhabitants of poor countries perpetrate some form of exploitation. We are not going to enter this debate in depth since this debate would be beyond the scope of this book. However, with regard to the exploitation problem that might occur with regard to the obtaining of human embryos it is not necessary that such practices are to be considered illegitimate because of the mere fact that terms of trade are not equal. One could also argue that such a transaction may give a person a legitimate opportunity to acquire a sum of money, that will help his/her family to escape from permanent poverty. It is a demand of informed consent that even in such transactions procedures need to be developed that are able to safeguard that both parties engage in such a transaction act willingly and knowingly.

Taking another look at the arguments, which have been presented above, we have to ask ourselves, how valid these frequently used arguments are. And, perhaps even more important: What are their merits for solving the current controversies regarding embryo research? We do not believe that the arguments we have discussed above can be very helpful in further debates on embryo research, since they turn out to be rather inefficient if taken without additional presumptions.

A recourse to *taboos* will remain ineffectual even if it is true that some essential symbols are constitutive for every culture. These symbols, unlike taboos, must be openly discussed and consciously adopted. If they are not, they will be ignored and forgotten.

Redefinition is a skilful method to reach a common position but only on the assumption that the two initial stances were sufficiently close from the beginning. New lan-

---

[9]   See e.g. J. Rifkin, The Biotech Century: Harnessing the Gene and Remaking the World, Tarcher New York 1999; A. Buchanan, D. Brock, N. Daniels, D. Wikler, From Chance to Choice, Cambridge UP 2000.

guage may help discussants realise that their views were not very different. But it does not have the power to induce a person to abandon a view that she firmly trusts.

Emotions and Rhetorical Language may only be used to stop a discussion. This is a proper attitude to adopt in contacts with persons who are stubborn, incommunicative, motivated by various personal reasons, unwilling to reach consensus. It is interesting to note, however, that such characters are usually the first who employ abusive language.

Potentiality is an important mode of reasoning, because when critically used it shows that every living entity has different potentialities. One also discovers that there is nothing in the nature of the argument itself that points to one of the possible continuations of life than another. Embryos, children and adults have different potentialities and there is no reason to believe that only the most sublime potentiality must be pursued.

Naturalism is an attitude so many faceted and diversified that it can not serve as an argument in moral debates. It is not clear what nature expects of us, if it expects of us anything at all, if it is better to live in a natural way, and what it means to live in such a way. Thus by calling a practice unnatural one does not offer a useful critique. Perhaps one only points out that the practice has not been known before and as yet is not very popular.

Dignity need not be understood as a supreme principle. What it requires can be better formulated as a set of practical rules that should be observed with respect to entities that we value for practical, emotional or philosophical reasons.

Representatives of various positions agree that *rights* will not solve moral debates. The advocate of the prohibitive position will agree that rights always clash and are often blown out of proportion by defenders of narrowly designed social causes. Advocates of the permissive position find it difficult to accommodate all rights on the principle of reciprocity. The regulatory position prefers to use other terms than rights, such, e.g. as specific obligations towards entities that cannot defend their interests on their own.

The Slippery Slope argument is convincing only if it is supported by predictions of high probability. But then its force comes from empirically supported hypotheses and not from the assumption that certain trends will continue to escalate and intensify or that certain practices will get out of control.

Caution is necessary as long as practitioners involved in novel programs are uncertain of their results. On the other hand they will never be sure of these results if caution prohibits them for ever to take reasonable risks.

Injustice is an important issue but it does not specifically concern embryo research but will have to be dealt with in a broader context. In any case adequate procedures for obtaining informed consent have to be established here as well as in other areas of medicine and research.

5.3
The Need for and Limits of Procedural Solutions to Problems of Substantive Rationality

A characteristic feature of an ethical position is that it seeks to provide justifications for action, and justifications require good reasons. However, one of the main difficulties encountered in ethical debate is that, unlike the situation in the natural sciences, there is considerable disagreement about what constitutes good reasons. In part, such disagreement is a function of espousing different basic values, and the most controversial part of ethics concerns whether or not value judgements (as against their applications) can be rationally or irrationally made. However, the thoughts that follow are, in the first instance, not addressed to presenting a position on this issue (though they will be so applied towards the end). Instead, they are about the limitations and powers of reason in ethical debate (with particular reference to issues that arise in relation to embryo research) apart from (or in addition to) the question of the rationality of basic value judgements themselves. Thus, most pointedly, they are focussed on problems of rationality that arise even when it is held (as a few philosophers, most famously Kant, do) that there are values or ethical principles that all persons must hold on pain of violating the most basic tenet of rationality in general, the prohibition on contradicting oneself in one's assertions.

Suppose, then, that the values/principles to be applied (which we will refer to as "the master principles") are not contested. (Suppose even that this is on the ground that the principles are clearly rationally necessary or true). On these assumptions, permissible action must not violate these principles or be contrary to these values. However, there are at least two reasons why, even in such a case, it must be admitted that answers to ethical questions may be indeterminate, and that, consequently, if a rational answer to an ethical question is defined as one that is logically required by, or logically consistent with, the master principles, then rationality will not be able to settle all ethical questions.

First, there might be instances where two incompatible answers are both consistent with (neither logically required, nor logically prohibited, by) the master principles; but where the master principles nevertheless do not permit the two incompatible answers to be applied indiscriminately. An obvious example concerns whether persons should drive on the right-hand side or the left-hand side of the road. Short of convincing empirical evidence (which, as far as we know is altogether lacking) that the fact that most persons are left-brained and right-handed makes it safer, on average, for them to drive on (say) the right-hand side of the road, we do not see how the matter can be other than morally optional. What is not optional, however (assuming that the master principles attach high value to personal safety and bodily well-being), is that persons going in the same direction must be made to drive on the same side of the road within any specific geographical location. To allow persons to drive on any side of the road they choose is certainly not morally optional.

Second, it might be the case that even if, in principle, specific answers are required by the master principles, the determination of these answers might be attended by enormous complexity or lack of knowledge about non-ethical matters that provide premises in applying the master principles to the issues at hand, in consequence of which the most knowledgeable persons acting in utmost good faith, combined with scrupulous attention to the requirements of logic, will nevertheless be capable of disagreeing. In such cases, where lack of knowledge is the cause, the indeterminacy might be only contingent and capable of being remedied at a later date. However, it might be a function of some of the non-ethical matters required to apply the master principles being essentially metaphysical (meaning that the propositions concerned can neither be known empirically nor does their assertion involve the upholder in any necessary self-contradiction). This is a bit abstract, so an example might help.

Suppose that the master principle specifies that all rational agents (beings capable of acting for purposes that are their freely chosen reasons for acting) have various rights equally. Only an agent ("A") can know directly and with certainty that A is a rational agent. As far as others are concerned, A can know only that they are capable of acting as though they are rational agents. But this does not mean that they are rational agents. They could be automata with no "minds" at all. Conversely, the fact that dogs and cats do not display the abilities expected of rational agents does not mean that they are not agents. They could be, even though this appears to be very unlikely. This means (certainly where morality is regarded as setting categorically binding requirements on action) that apparent agents must be treated as agents and precaution must be exercised in dealing with apparent non-agents. While apparent non-agents cannot be treated as rational agents when they do not exhibit capacities to display the behaviour of such beings, rationality requires us to have concern for their welfare in proportion to the probability that they might be agents, and this probability is determined by how closely their behaviour approximates to that of rational agents (for further details see, e.g., Beyleveld and Brownsword 2001, Chapter 6).

On this basis, we have duties not to harm the embryo and fetus that increase from near zero at conception until the new-born displays the behaviour expected of rational agency. We say "from near zero", because within such a frame it is not impossible (just exceedingly poorly indicated) that the fetus is an agent from the moment of conception, for to establish that it is impossible we would have to prove categorically that reincarnation, metempsychosis and various other metaphysical doctrines are false (which we do not believe can be done).

When, for any reason, our supposed master moral principles inherently leave what they prescribe in application in doubt, then we might say that (within the theory involved) there is no substantively rational solution to the moral problems involved. However, this does not mean that there can be no rational resolution of such problems. The German jurist Hans Kelsen (1967: 195–198) distinguished two ways in which principles can authorise actions/norms. The first way is direct authorisation, or authorisation "according to the static principle", (which occurs when the authorised action/norm can be deduced from a more general principle). For example, "Persons ought not to be burnt alive at the stake" is directly authorised by "Great pain ought not to be caused to people", given that burning them alive causes them great pain. The second way is indirect authorisation, or authorisation "according to the dynamic principle", (which occurs simply when the authorised norm is the result of an authorised procedure). For example, "Whatever norms the British Parliament enacts as laws ought to be applied and obeyed" will authorise whatever norms the British Parliament enacts as laws.

In a moral context, where master principles do not substantively determine answers to moral problems, they might justify procedures for indirect authorisation of answers to moral problems. Thus, they might justify/require the answer to be accepted that is produced randomly (e.g., by the toss of a coin) (which is, arguably, appropriate when the matter is truly optional or in principle indeterminate), or by a body of appropriate persons acting in good faith with proper safeguards against conflicts of interest, having to give reasons, and having their decisions subject to review (which is, arguably, appropriate where the source of the problem is essentially complexity). Such procedural solutions (recourse to procedural rationality) seem to be unproblematic when working within an agreed moral theory. However, we suggest they are not without their uses when dealing with disagreements that stem from the espousal of different moral theories/values etc.

According to what has been discussed above, one of the most serious sources of disagreement about the permissibility of embryo research is undoubtedly different views about the inherent moral status of the human embryo. These, in turn, derive from different views as to the quality or qualities that confer an inherent moral status on beings. By a "moral status", we mean a status that is responsible for persons having duties of non-harm or assistance to the being with this status (which must be distinguished from duties of non-harm/assistance that derive from the relation that a being has to others with moral status). And these different views derive, in turn, from different moral theories. In the example given above, the qualifying quality consists in having the capacities for rational agency (which characterises Kantian and Contractarian, or Contractualist, theories as well as Preference Utilitarianism of the kind that measures utility in terms of the choices made by agents). Other theories, however, maintain that what is critical is simply the capacity to experience pain or pleasure (which is characteristic of, e.g., hedonistic utilitarianism). Then there are theories that maintain that it is necessary and sufficient for inherent moral status to be a member of the human species (characteristic of many religion-based ethics). Whether or not moral theories are capable of being true or rationally required, it has to be recognised that that there is no consensus about what is the "right" moral theory. We suggest, therefore, that if moral disputes about embryo research are to be "managed" in a fully rational manner, this must at some point be procedural, if it is to be possible at all in practice.

The essence of a procedural solution to a moral dispute is that it requires persons to accept decisions as binding that they disagree with morally. If moral requirements are taken, as they often are (perhaps, characteristically), as categorically binding, or at any rate as overriding all conflicting non-moral requirements, then acceptance of a procedural solution is only coherent if the persons involved have moral reasons to accept decisions that they disagree with morally. In short, what is required is that the consequences of not reaching a decision must be regarded as threatening more important values than the values at stake in the primary dispute. Successful procedural solutions at this level, thus, presuppose a set of core values between the involved parties that is shared between their competing moral theories. The extent to which such a consensus-set exists, limits the efficacy of procedural solutions.

This point can be elaborated by looking briefly at two models of what might be called "procedural ethics". The first of these is to be found in the writings of the German social theorist Jürgen Habermas (seminally 1970). In essence, Habermas presents the view that if answers to moral problems are arrived at in an "ideal speech situation" then these answers are to be accepted as binding (though he even more controversially states that they are to be accepted as true) (for a discussion of which see McCarthy 1973, 150). The ideal speech situation is defined by a number of procedural conditions: in particular, the disputants must be able to present their viewpoints in a way that is free of disparities of power that preclude each of them from having an equal say; the debate must be conducted sincerely, orientated towards the truth without ulterior motives; governed by rules of logical consistency; the disputants must be prepared to yield from their starting position; and answers must be revisable when new "facts" come to light, etc.

The second of these is to be found in the writings of the legal philosopher Ronald Dworkin (1978, 248–252), in his elucidation of what he calls "a discriminatory moral position". Such a position is one

1. for which reasons are given; but which excludes:
   (a) prejudices (postures that take into account considerations excluded by "our conventions");
   (b) mere emotional reactions (feelings presented in a way that makes them immune support by reasons);
   (c) rationalisations (which include: appeals to false or irrational beliefs; considerations that cannot be plausibly be connected to their conclusions; and considerations that the proponent would not be prepared to countenance if proffered by others); and
   (d) parroting (repeating what others have said without critical understanding);
2. that is held sincerely;
3. that is consistent with the other beliefs of the proponent;
4. that is not held arbitrarily (i.e., not presented as self-evident when it is generally accepted that reasons for it are required).

We suggest that Habermas's position is overly formalistic (or too purely procedural). The ideal speech situation does not unconditionally provide sufficient reason for persons to accept its conclusions. It only does so if they believe that there are no "right" answers substantively to moral issues, which involves, ultimately, the espousal of relativism of a kind that entails that moral debate is at the substantive level, in the final analysis, not a matter of reason at all, merely of commitment (which renders rational substantive debate pointless as well as disingenuous). On the other hand, while Dworkin's view has a substantive element built in implicitly in the exclusion of prejudice, this merely serves to highlight the fact that the real problem might be that there may be no consensus over "our conventions". This is seriously problematic if "our conventions" refers not only to shared rules of reasoning but also to at least a range of substantive beliefs, if the model is intended to provide neutral reasons for accepting answers that everyone who is rational must accept. Thus, the existence of "conventions" (equivalent to a degree of consensus) should be viewed merely as setting limits to which a rational solution of ethical disputes is possible in practice.

Viewed in the proper manner, we suggest that the models we have attributed to Habermas and Dworkin can be combined to form the formal component of a procedural position that can take us forward. In order to complete the picture, it is necessary, however, to determine whether there are any substantive conditions over which there is sufficient consensus to render such an approach useful and efficacious. An approximation to such a consensus currently exists in the Human Rights Conventions, especially the European Convention on Human Rights (ECHR), to which all EU Member States are party. While there is nothing explicit in the ECHR about embryos, this does not mean that nothing can be inferred about the moral status of embryos from the Convention. However, what is central to such an interpretation is the meaning given to "being human". If being human is "being a member of the species Homo sapiens" then the embryo must be accorded human rights under the Convention. However, there is a problem with this, because many of the rights, which are civil and political rights, are not rights that can be exercised by (thus assertorically attributed to) embryos, and even the right that could, most clearly, in principle, be held to apply to the embryo, the Article 2 right to life, has not been extended by the European Court of Human Rights to the unborn. However, even if, as has been argued in Beyleveld and Brownsword 2001 (80–81), "being human" most plausibly means "being a rational agent" under the Convention, this does not mean that duties to the unborn cannot be inferred from the Convention rights. Such duties can be derived using the precautionary and proportionality reasoning mentioned above.

However, there is no doubt that this itself is controversial, and consensus on this is not to be expected. Indeed, we do not agree on this between ourselves. The role of the Convention rights, however, is no more than to indicate that there is a core of values that rational persons who assent to that core will not be willing to sacrifice because they disagree over matters that they must, if they are sincere and rational, admit that others who are sincere and rational can disagree with them about.

At this point, it might be objected that ethics and politics are different, and that we have now entered the realm of politics. Indeed, we have. In principle, ethics is about what ought to be accepted; politics is about what will be excepted. However, practical ethics cannot ignore features of political reality altogether. Because "ought" implies "can", there is, in general, a difference between what a moral theory requires to be done in ideal circumstances and what it requires when the circumstances place obstacles in the way of its ideal requirements being observed. As one aspect of this, what "ought" to be from an ethical standpoint when all are agreed about the standards that govern moral judgement might not be the same as what ought to be from that ethical standpoint when all are not agreed that this standpoint is the one to employ. In consequence, practical or applied ethics must incorporate a political element in the form of a rational political morality, and we suggest that such a morality (which, in shape, follows the lines of a constitutional democracy) is a procedural ethic with limits set by a core substantive consensus.

Once this is appreciated, the question (from a moral point of view) of whether harmonisation on embryo research is needed or possible in the EU (which question is dealt with in the legal chapter from several perspectives) invites the following response. Harmonisation is not needed so long as there is a political consensus on values (e.g., those of the ECHR) which are considered more important than the values at dispute in the disagreements about the status of the embryo. At the same time, harmonisation is not possible unless the disagreements about the status of the embryo can be uncontroversially derived from this core consensus, which, at least in practice, has proved not to be the case. If the core consensus were to break down, the

EU would break down. So, assuming that it will not break down (at least for the reason of disputes about the status of the embryo), what ethics should be concerning itself with is not so much what the status of the embryo is, but with ensuring that the mechanisms that are in place for decisions about what might be done with embryos have the right features for a procedural ethics to be rationally conducted on the model we have sketched. In other words, we ought to address our structures: are they designed to maximise accountability and transparency, reviewability, and reason giving? Are there adequate safeguards against conflicts of interests and outright corruption? In short, are our procedures designed to ensure that, as far as this is possible, decisions on these matters are conducted in good faith and competently? For, in the final analysis, that is all that we can ask of rationality in practical ethics once we depart from what will be agreed at the substantive level.

6 Recommendations

On the basis of our understanding of the current state of scientific knowledge, presumed medical potential, existing regulations and the sociological and ethical considerations that have been presented in this study, we make the following recommendations on the moral status of human embryos, safety measures that need to be established if embryo research is to be permitted, and possible regulation.

(Note: We generally refer here to embryos less than 14 days post fertilisation, or prior to primitive streak formation.)

> 1. Human embryos must be distinguished from other human biological material: they deserve greater respect and should be handled with precaution.

This claim is based on the following considerations.

1.1 Human *embryos* have an inherent value by virtue of their potential. This potential refers to their theoretical ability to give rise to all cell types that are found in the body and/or to develop into a human being. The greater the potential, the greater the value. For example, an embryo that is a result of fertilisation may be considered of greater value than one that has been activated to develop parthenogenetically. This inherent value is also high because human embryos are difficult to obtain. (N.B. Embryos from other animals also have an intrinsic value in both respects.) If embryos had no potential, they would have no value, and if they were common and readily available they would be less valuable. It is partly for this reason that oocytes are considered more valuable than sperm, although neither has the potential to develop into a newborn without the other. Nevertheless, oocytes are also important because their cytoplasm contains factors required for early development, or for reprogramming after nuclear transfer and because they have realisable potential if they are activated or fertilised.

1.2 The manipulation of human embryos can cause harm in some circumstances. This should be taken into account through rigorous assessment of costs versus benefits. It is therefore essential to monitor the outcome of procedures as well as people's perceptions of them, and to make such records widely available.

   1.2.1 Resulting offspring can be harmed directly by manipulation techniques. When embryos have been created for reproductive purposes, their future interests need to be protected even though these interests are not yet current. With respect to cost versus benefit, many techniques such as gamete donation, IVF and embryo freezing, seem innocuous, while ICSI and PGD carry a slight risk. Reproductive cloning currently carries a very high risk and little or no benefit.

1.2.2 W*omen* providing oocytes might find themselves in a particularly vulnerable situation when undergoing medical treatment in reproductive medicine or are in a generally difficult financial and/or social situation, which puts them at risk for exploitation of their procreative capacities. (N.B. This is similar to the selling of other body parts, much more extreme than hair, less so than kidneys.) Proper counselling and informed consent are essential.

1.2.3 Decisions regarding the treatment of human embryos might cause undesired side effects for other *family* members and/or *society*. For example, the family could be subject to negative discrimination or adverse publicity.

1.2.4 Decisions regarding the legal regulation of the treatment of human embryos might bring about *social* and *political* side effects, including inconsistencies or avoidable tensions, that might endanger the entire system of regulations of biomedical research and practice.

---

2. Embryo research is morally admissible, or even a moral imperative, under the condition that the legitimate interests of persons who suffer from as yet untreatable or insufficiently treatable disabilities or diseases (including infertility) and of researchers working in basic and/or clinical research to meet the needs of such patients, outweigh society's general interest in protecting embryos.

2.1 The interests of these parties have to be weighed against each other in accordance with the principle of precaution. On the basis of rational arguments we consider that early human embryos and mature persons do not have the same moral status and for this reason we consider it legitimate to conclude that human embryos need not be given the full protective rights of adult persons. Although in some European countries the dominant position is that human embryos and mature persons have the same moral status, we argue that mere biological facts, like belonging to the human species, showing a certain genetic make-up etc. can provide a necessary but not a sufficient condition for being owed the same protective rights that we owe persons with capacities like being sentient, being able to have interests or to act autonomously.

2.2 It is important to note that this applies only to embryos that do not yet show any of these morally relevant capacities and that are not going to be implanted for reproductive purposes. In cases in which embryos are created to develop into human persons, their future interests need to be protected at all stages of their development, and this does not permit any experiments that endanger these future interests for reasons other than for the benefit of the embryo itself.

2.3   Basic understanding of human biology is often a pre-requisite for arriving at the best treatments, whether of infertility or disease, but it is also sufficient justification in its own right, as long as the factors discussed above with respect to value, harm and moral value are taken into account. It follows that it would be very difficult to justify such experiments on embryos older than 14 days or after implantation.

2.4   Supernumerary embryos produced in the course of IVF treatment that otherwise will be destroyed without any previous use and that are given freely for research by the gamete donors can be used for research purposes that meet the usual scientific standards. Such a position makes scientific sense and is also supported by a majority of Europeans.

2.5   The creation of embryos specifically for research is considered by some to be morally contentious. There are also some practical problems in that currently the number of oocytes available and donated for such purposes is low (although this might change in the future). The creation of embryos for research purposes should therefore only be permitted if it is justified by the nature of the research, and if it can be argued that the use of supernumerary embryos is inappropriate or insufficient and important scientific and/or medical results are to be anticipated. Any recommendation on this needs to be flexible to allow for future advances in research, for example, if it becomes possible to make oocytes in vitro (from human ES cells) then it may be considered, morally and practically, better to use these to create embryos for research than supernumerary embryos.

2.6   Reproductive cloning should be banned, at least in the interim. This is not just because there are a number of (as we have shown not necessarily rationally convincing) moral arguments against human reproductive cloning which at the present lead to a virtually universal rejection by public opinion. It is, mainly, because current data on reproductive cloning in animals indicate that in addition to a tremendous loss of embryos it is very likely that clones would be born abnormal.

---

3.   It is not advisable to limit science to the use of adult stem cells or of already derived embryonic stem cell lines.

3.1   We do not consider the claim to be proved that adult stem cells or established stem cell lines can serve all research purposes and clinical applications. Decisions on which source of cells is appropriate will have to be made on a case-by-case basis. If equivalent results can be expected, a socially less troubling approach should be chosen. Otherwise, the option that promises the better biomedical results should be used.

3.2   For many reasons it is necessary to derive new human ES cell lines and this should be permitted. For example, in order to avoid the possibility of transferring species-specific endogenous viruses, at latest when research enters the clinical practice, it will be necessary to establish conditions in which human ES cells are isolated, cloned and/or propagated using human fibroblasts as feeder layers or, even better, using a completely defined medium without feeders. To achieve this goal, extensive research on the isolation of human ES cells needs to be done, and this work cannot be done if research is limited to certain arbitrarily chosen and already established cell lines.

3.3   Human ES cells could potentially provide a valuable means to study genetic diseases in vitro and for use in screens to search for molecules that could alleviate the disease and to check for toxicity. More broadly, different people respond to pharmaceuticals and other agents in different ways, and this will often depend on their genetic makeup, and panels of ES cells may be useful in research on this. For both reasons it will be necessary to derive ES cell lines corresponding to specific individuals (or genotypes).

3.4   The reasons set out in the last paragraph might provide the first justification for the use of cloning techniques to derive human ES cells (so-called "therapeutic cloning"). This may be the only way to derive an ES cell line (or any cell line) carrying a particular genetic defect or combination of genes. If it is proven that ES cells are useful in therapies, this may also be the best way to ensure that an appropriate tissue match can be found to prevent rejection of engrafted ES cell-derivatives.

3.5   As part of this recommendation, it therefore follows that use of cloning techniques specifically for the derivation of human ES cell lines should be considered an acceptable aim and that provisions should be made in any regulations to permit this.

---

4.   Strong precautionary measures should be established to ensure that research on human embryos is only conducted in accordance with measures to prevent any of the interested parties, in particular the gamete donors, from being harmed or scientific and economic resources being wasted.

4.1   In order to meet this requirement, research on human embryos should only be conducted:
   (i)   After the informed consent of the gamete donors has been obtained.
   (ii)   In licensed institutions by competent scientists.
   (iii)   After being approved by a scientific/ethics advisory board operating to ensure that research will be conducted in accordance with requirements of good clinical practice, and the law.

4.2   To ensure that procreative liberty of gamete donors is respected and to prevent exploitation in IVF treatment and in obtaining egg cells for research purposes, the informed consent of the donors should be obtained concerning:
   (i)   How long their embryos may be kept in storage.
   (ii)   Whether their embryos may be used for research at all.
   (iii)   The research purposes their embryos may be used for.

4.3   Only licensed institutions or clinics should be allowed to conduct research on human embryos.

4.4   Human ES cells are not embryos, they are cell lines and do not need to be regulated in the same way, although local ethical committee approval should be sought. The reintroduction of ES cells into human embryos would fall under regulation of human embryo research, where it is unlikely to be permitted. The use of ES cell derivatives in therapies would fall under the remit of other regulations, for example those concerning transplantation.

4.5 Independent scientific advisory boards and ethics committees need to be established that act in a similar way to those already operating in most countries that deal with proposals to use human tissues or animals in research.

4.6 Organisations involved in general decision-making processes, and in particular bodies responsible for licensing institutions that conduct research and the research itself need to be established on a European level. This should be done as soon as possible to ensure that decisions on morally contentious issues, especially those that utilize valuable resources as with embryo research, are conducted in good faith, competently and in ways that will maximise accountability and transparency. Furthermore, these organisations should ensure the quality of any review process, provide adequate safeguards against conflicts of interests and minimise any possibility of corruption. These organisations should also facilitate gathering of information with respect to embryo research and its outcomes.

---

5. On a political level, procedural solutions and education of the public need to be promoted.

5.1 Political decision making in morally sensitive areas like embryo research needs to respect common sense as far as possible, because policies that do not respect public beliefs and opinions jeopardise the widespread support they need in order to be effective.

5.2 Since there is evidence that, for a large majority of Europe's population, attitudes and responses to embryo research are based not on knowledge, but on images of fear, stereotypes and beliefs about the moral status and rights of the embryo, the public needs to be better educated first of all about the biological nature of the embryo and, in addition, their moral concerns should be addressed through rational debate.

5.3 Any political decision in a pluralistic society needs to be made in accordance with legitimate procedures to ensure that policies are consistent and can generally be respected by all citizens.

5.4 It must be appreciated, however, that respect for a procedural solution is only attainable if the persons involved have reasons to accept political decisions that they disagree with morally. This requirement is most likely to be met if decisions are based on the best scientific evidence available and deliberations take into account the interests of all parties concerned. Furthermore, for a procedural solution to be successful, the consequences of not reaching a decision must be considered to threaten more important values than the values at stake in the primary dispute. Successful procedural solutions presuppose a set of core values between the involved parties that is shared in spite of otherwise competing moral theories. The extent to which such a consensus exists, limits the efficacy of procedural solutions.

6. Future legislation concerning embryo research needs to be both well informed and flexible. Furthermore, while at present it is only possible to regulate embryo research on a national level, legislation should either take into account the possibility of international harmonisation or allow for the consequences of disparate views and laws.

6.1 From our survey of public attitudes, an evolution towards a common European policy is a clear and dominant aspiration among European citizens. The EU needs to consider this when considering policy and legislation on embryo research.

6.2 The EU currently has no legal competence to regulate on ethics as such. Regulation on embryo research only falls within EU competence if it is for the purposes of the single market or for human health (and the latter can not be applied plausibly to research conducted *in vitro*). Nevertheless, the EU could take some steps towards the future harmonisation of regulations on embryo research by analysing the issues and recommending general guidelines for ethics committees and scientific associations with a European reach.

6.3 The economic and social pressures are in the direction of harmonisation for permission. Therefore, positive action is required if non-harmonisation is to be maintained. Such action might have to include ways to reduce the competitive disadvantage that would accrue to countries that had taken a prohibitive stance.

6.4 Laws governing research on human embryos should in general be Regulative rather than Proscriptive. The former includes the UK system and those being adopted by a number of countries where research is prohibited unless a licence is granted.

6.5 It is important that the regulations are flexible to allow for new developments in technology, in scientific understanding and in public attitudes. Such developments are likely to occur more rapidly than can be accommodated by usual forms of law-making.

7 Glossary

Achodroplasia: dominant genetic condition characterised by extreme short statue but not other problems.

Acrosome: the packet of enzymes in a sperm's head that allows the sperm to dissolve a hole in the coating around the egg, which allows the sperm to penetrate and fertilize the egg.

Adenovirus/
Adeno-associated virus: viral vectors used for gene transfer in vitro and in vivo.

Adipocytes: fat storing cells.

Adult stem cells: stem cells found in the adult organism that replenish tissues in which cells often have limited life spans. They are more differentiated than embryonic stem cells or embryonic germ cells.

Allele: Alternate forms of a gene. Each gene has at least two copies in a diploid cell.

Allogenic: Genetically different organisms of one species.

Amino acids: The subunits (monomers) from which proteins (polymers) are assembled. Each amino acid consists of an amino functional group, and a carboxyl acid group, and differs from other amino acids by the composition of an R group.

Amniocentesis: a procedure, usually carried out between 14 and 18 weeks of pregnancy. in which about 20 millilitres of the amniotic fluid is withdrawn through a needle inserted through the abdomen and the uterine wall into the amniotic say in which the foetus is developing. The fluid and the foetal cells it contains may be tested for genetic disease in the foetus.

Amniotic fluid: the fluid filling the cavity between the embryo and the amnion. Secreted initially by the amnion, it is later supplemented by foetal urine.

Anergy: State of non-responsiveness especially of T-cells, important in maintaining immunological tolerance.

Angiogenesis:	Formation of new blood vessels.
Angiogenetic factors:	factors promoting angiogenesis.
Aneuploidy:	Variation in chromosome number involving one or a small number of chromosomes; commonly involves the gain or loss of a single chromosome.
Anti-apoptotic genes:	genes that are involved in prevention of cellular apoptosis.
Antigene:	molecules or structures which can evoke an immune response.
Apoptosis:	cellular suicide, can be induced by external signals.
ART (assisted reproduction technology):	medical treatments or procedures which involve handling human eggs and sperm for the purpose of inducing a pregnancy. Types of ART include in vitro fertilisation, gamete or zygote transfer, embryo cryopreservation, egg or embryo donation, and surrogate birth.
Autologous:	autologous cells can be transplanted without immunosuppression as they (or in the context of somatic nuclear transfer at least the nuclei) are genetically identical to the recipient. Cells are autologous to a certain individuum if they (or as a minimum the nuclei) are derived from this individuum.
Autosomes:	Chromosomes other than the sex chromosomes X and Y.
Autosomal:	pertaining to any chromosome that occurs in the nucleus, except for the sex chromosomes.
Autosomal dominant disorders:	disorders where the mutation of one allele (inherited from one parent only, or arising anew during egg or sperm formation) can be sufficient for an individuum to be affected.
Autosomal recessive disorders:	disorders where mutation of both allels of a gene is necessary to affect an embryo. The parents can be unaffected since both may have a single copy of the relevant gene.
ß-cells:	insulin secreting cell type of the pancreas.
Basophils:	certain type of white blood cells containing gran-

ules that stain with basic dyes.

Blastocyst: The developmental stage of the fertilised ovum by the time it is ready to implant; formed from the morula and consists of an inner cell mass, an internal cavity, and an outer layer of cells (the trophoblast).

Blastomere: one of cells into which the egg divides after it is fertilised; one of the cells resulting from the division of the fertilised ovum.

B-lymphocytes: Type of lymphocyte responsible for antibody-mediated immunity; mature in the bone marrow and circulate in the circulatory and lymph systems where they transform into antibody-producing plasma cells when exposed to antigens.

CD-marker: "cluster of differentiation", classification of cell surface molecules allowing molecular characterisation identification of different cell types.

CD34+ hematopoetic
stem cells: stem cells of the bone marrow forming all kinds of blood cells and maybe other cell types within the body.

Cell: the basic unit of all living organisms. Complex organisms such as humans are composed of somatic (body) cells and germ line (reproductive) cells.

Chemokines: small cytokines that are involved in the migration and activation of cells. They play a central role in inflammatory responses.

Chemotactic agents: compounds that can attract cells.

Chimera: an individual composed of cells or tissue of diverse genetic constitution.

Chondrocytes: cartilage-forming cells.

Chondrogenic: ability to form chondrocytes and cartilage.

Chorion villus sampling: a procedure usually effected between 8 and 12 weeks of pregnancy, by which a small amount of the chorionic villi, which is a microscopic finger-like fronds of the outer membrane (chorionic) tissue of the primitive placenta which surrounds the embryo, amnion and yolk sac, is biopsied for genetic analysis.

Chromatin: A complex of DNA and protein in eukaryotic cells that is dispersed throughout the nucleus during interphase and condensed into chromosomes during meiosis and mitosis.

Chromosomal deletions: mutations which are based on the erasure of parts of a chromosome.

Chromosomal duplications: mutations which are based on the duplication of parts of a chromosome.

Chromosomal translocation: transfer of parts of chromosomes to improper locations.

Chromosomes: Structures in the nucleus of a eukaryotic cell that consist of DNA molecules that contain the genes.

Cleavage stage: the stage of cell division between fertilisation and the development of cell differentiation to form the blastocyst.

Cloning: the precise production of genetically identical copy of cells, tissues, plants or animals (including humans) by nucleus substitution or by mechanical division of a cleaving zygote.

CNS (Central Nervous System):

Coagulation system: system of cells, peptides and proteins resulting in tightly regulated blood clotting.

Connective tissue: tissue composed of cells embedded in a matrix. Includes loose, dense, and fibrous connective tissues that provide strength (bone, cartilage), storage (bone, adipose), and flexibility (tendons, ligaments).

Cryopreservation: the protective storage, usually in liquid nitrogen at – 196 C, of cells, gametes, embryos and women's eggs which were obtained following superovulation.

Cystic fibrosis: a disorder of the mucus-secreting glands of the lungs, the pancreas, the mouth, and the gastrointestrinal tract.

Cytokines: proteins made by cells and affect the behaviour of other cells (messenger-compounds).

Cytoplasm: The viscous semiliquid inside the plasma membrane of a cell; contains various macromolecules and organelles in solution and suspension.

Cytoplasts: enuleated ES cells.

Cytoskeleton: A three-dimensional network of microtubules and filaments that provides internal support for the

	cells, anchors internal cell structures, and functions in cell movement and division.
Cytoxic T cells:	T cells that destroy other cells e.g. infected by viruses or bacteria; they also attack bacteria, fungi, parasites, and cancer cells and will kill cells of transplanted organs if they are recognized as foreign; also known as killer T cells.
CVS:	Central Nervous System.
DNA:	deoxyribonucleic acid, the major constituent of the chromosomes, and the hereditary material of most living organisms.
Donor-reactive cells:	leukocytes, especially T-cells that specifically attack transplanted allogeneic or xenogeneic cells.
Dopaminergic neurons/ dopaminergic cells:	Dopamin releasing neurons.
Electroporation:	voltage-dependent in vitro approach to deliver DNA into viable cells.
Embryonic Stem (ES) cells:	cells that are derived from the inner cell mass of a blastocyst embryo.
Embryonic Germ (EG) cells:	cells that are derived form precursors of germ cells from a fetus.
Endoderm:	innermost of the three primary layers of the embryo; origin of the digestive tract, liver, pancreas, and lining of the lungs.
Endothelial cell:	cell type forming the inner surface of all blood vessels.
Endocrine glands:	glands which release their products (hormones) directly into the body fluids.
Endothelial progenitor cells (EPC):	bone marrow derived progenitor cells of endothelial cells.
Enucleation:	removement of the nucleus of a cell.
Eosinophils:	certain type of white blood cells containing granula staining with the dye Eosin.
Erythocytes:	red blood cells.
Estrogen:	A female sex hormone that performs many important functions in reproduction.

Exocrine glands: glands that release their secretory products into the body cavities or into the outside world.

FACS-sorting: purification of single cell types using fluorescence activated cell sorting.

FASL (also: FasL):   a member of the TNF family of cytokines and cell surface molecules, binds to FAS, can induce apoptosis.

Fibroblast: major cell type of connective tissues, capable of forming collagen fibres.

Fluorescence activated cell sorter (FACS):   instrument allowing the characterisation and/or purification of different cell types using several parameters including cell size, granularity and fluorescence.

Fluorescent dyes: dyes which release light after excitation with a shorter wavelenght.

Fluorescent in situ hybribisation (FISH):   approach allowing the detection of mRNA-expression directly in tissue-sections.

Follicle: a fluid-filled cavity containing cystic space which develops in the ovary in which the oocyte develops and is nourished.

Follicle stimulating hormone (FSH):   A hormone secreted by the anterior pituitary that promotes gamete formation in both males and females.

Gamete: any germ cell, whether ovum or spermatozoon, or a mature male or female reproductive cell.

Gametogenesis: germ cell based formation of gametes.

Gene: the unit of inheritance; that element of DNA in which the amnio-acid sequence of a protein is encoded. Everyone inherits two copies of each gene.

Genome: the basic set of genes in the chromosomes in any cell, organism or species. Humans have a genome of 23 chromosomes.

Genotype: The genetic (alleleic) makeup of an organism with regard to an observed trait.

Gestation: Period of time between fertilisation and birth of an animal. Commonly called pregnancy.

GMP: Good medical practice.

Granulocyte colony stimulating   protein necessary for growth of white blood cells.
factors (GCSF):

H1 cell line: cell line of human embryonal stem cells established at the University of Wisconsin.

H9 cell line: cell line of human embryonal stem cells established at the University of Wisconsin.

Haploid Cells: cells that contain only one member of each homologous pair of chromosomes. At fertilisation, two haploid gametes fuse to form a single cell with a diploid number of chromosomes.

Haploinsufficiency: condition caused by the absence or mutation of one of the two alleles and when the remaining normal allele is insufficient to maintain normal function.

Hematopoietic   cells able to form colonies of red and white blood
colony-forming cells: cells

Hematopoietic stem cells: early precursor cells giving rise to all cells in the blood and lymphatic organs.

Hepatocytes: liver cells.

Heterologous: refers to a transplant from an unmatched donor.

Heteroplasmy: more than one type of mitochondria within a cell.

Heterozygous: Having two different alleles (one dominant, one recessive) of a gene pair.

HFEA: Human Fertilisation and Embryology Act.

HLA: human leukocyte antigen – the main factor in tissue matching for organ transplantation.

Homologue: A pair of chromosomes in which one member of the pair is obtained from the organism's maternal parent and the other from the paternal parent; found in diploid cells.

Homozygous: Having identical alleles for a given gene.

Human chorionic   peptide hormone secreted by the chorion that pro-
gonadotrophin (HCG):   longs the life of the corpus luteum and prevents the breakdown of the uterine lining.

Huntington's disease (HD):   A progressive and fatal disorder of the nervous system that develops between the ages of 30 and

50 years; caused by an expansion of a trinu-
cleotide repeat and inherited as a dominant trait.

Hybrid: an individual produced by the crossing of two different strains or sometimes species.

Immunogenicity: ability to provide an immune response.

Immunosurgery: use of antisera to eliminate some cells or tissues

Imprinting: a process which differentially marks the genome
of male and female mammalian germ cells result-
ing in their differentiated functioning. It is necessity for both of them to be present to secure normal development.

Inner cell mass: (or basal cell mass) is a clump of cells growing
within and to one side of the blastocyst from
which the embryo develops.

Intra Cytoplasmic Sperm   a micromanipulation technique. A variation of

Injection (ICSI):   IVF treatment where a single sperm is injected
into the inner cellular structure of the egg. This
technique is used for couples in which the male
partner has severely impaired or few sperm.

Inversion: reversal in the order of genes on a chromosome
segment.

In Vitro Fertilisation (IVF):   process by which a woman's eggs are extracted
and fertilised in the laboratory and transferred
after they reach the embryonic stage into the
woman's uterus.

In vitro: Refers to processes taking place in test tubes or
similar containers.

In vivo: Refers to processes taking place in an organism.

Ischaemic muscle: muscle whose blood supply is acutely or chroni-
cally compromised.

Karyotype: chromosome characteristics of an individual cell
or cell line, the microscopic appearance of a set of
chromosomes, including their number, shape and
size.

Keratinocytes: The basic cell type of the epidermis; produced by
basal cells in the inner layer of the epidermis.

Knock-out mice:	mice produced by different methods of genetic manipulation which result in the elimination of a specific gene.
Langerhans' cells:	Epidermal cells that participate in the inflammatory response by engulfing micro-organisms and releasing chemicals that mobilise immune system cells.
Large-calf syndrome:	a culture-induced effect leading to abnormal growth of cattle embryos (also reported in sheep)
Lentiviral vectors:	viral vectors used in gene therapy.
LIF (leukemia inhibitor factor):	a cytokine allowing maintanance of undifferentiated mouse ES-cells without feeder-cells.
Macrophages:	A type of white blood cell derived from monocytes that engulf invading antigenic molecules, viruses, and micro-organisms and then display fragments of the antigen to activate helper T cells; ultimately stimulating the production of antibodies against the antigen.
Major Histocompatibility Complex (MHC):	mouse equivalent of HLA.
Meiosis:	the process by which a diploid cell nucleus divides into four nuclei each with half the number of chromosomes of the parent nucleus.
Microarray technology:	techniques which enable the determination of which genes are expressed in a given tissue and at what level.
Mitochondria:	They provide the energy for the cellular processes including that of meiosis. They have their own hereditary disposition.
Mesenchymal stem cells (MSC):	stem cells isolated from the bone marrow stroma and able to differentiate into many different cell types.
Methylation pattern:	DNA modification which determines whether a given gene is expressed.
Mitochondrial heteroplasmy:	see heteroplasmy.
Monocytes:	White blood cells that clean up dead viruses, bacteria, and fungi and dispose of dead cells and debris at the end of the inflammatory response.

Monosomy:	condition when only one of the two homologous chromosomes is present.
Morbus Parkinson	
(Parkinson's Disease):	degenerative brain disease caused by a decreased level of dopamine in the brain.
Mosaicism:	condition when cells with two different genotypes are present in the same organism.
Multinucleated:	Cells having more than one nucleus per cell.
Multiple sclerosis:	degenerative neurological disease.
Mutation:	Any heritable change in the nucleotide sequence of DNA; can involve substitutions, insertions, or deletions of one or more nucleotides.
Myocardium:	heart muscle.
NK-cells:	natural killer cells, part of the immune system.
Nucleus:	the part of the cell which contains the genetic material DNA; it is separated from the other cell contents by a nuclear membrane.
Oligodendrocytes:	cells found in the nervous tissue (brain).
Oligospermy:	insufficient number of sperms for reproductive purposes.
Oocyte:	diploid cells which will undergo meiosis to form an egg or an immature ovum.
Ooplasm:	cytoplasma from oocytes.
Osteoblast:	Bone-forming cells.
Pathogenicity:	ability to cause disease.
Peptide:	Short chains of amino acids.
Phenotype:	The observed properties or outward appearance of a trait. The physical expression of the alleles possessed by an organism.
Placenta:	An organ produced from interlocking maternal and embryonic tissue in placental mammals; supplies nutrients to the embryo and fetus and removes wastes.
Plasma membrane:	membrane which surrounds the cell.

Pluripotent cells:	cell, present in the early stage of embryo development, that can generate all ell types in a fetus and in the adult and that are capable of self-renewal. Pluripotent cells are not capable of developing into an entire organism.
Polar body biopsy:	removal of the polar body in order to determine the genetics of the oocyte.
Polymerase chain reaction (PCR):	method of DNA amplification.
Polyploid:	having more than the normal number of chromosomes.
Preimplantation Genetic Diagnosis (PGD):	use of genetic testing on a live embryo to determine the presence, absence or change in a particular gene or chromosome prior to implantation of the embryo to the uterus of a woman.
Primordial germ cells (PGC):	stem cells which give rise to sperm and oocytes; PGC can be isolated from embryos and fetuses.
Progentior cells:	in contrast to the divison of a stem cell with self-renewing potential, divison of a progenitor cell always results in two daughter cells both different to the original cell.
Pronucleus state:	state of oocytes after penetration of the sperm, but before resolving of the nuclear membrane.
Recessive gene defect:	The defect is only phenotypically visible if both alleles of the gene are affected. If only on allel of the affected gene is mutated, the wild-type allel guarantees normal phenotype
Retinoic acid:	vitamin A derivate.
Retrovirus/Retroviral vectors:	Viruses that contain a double stranded linear RNA genome. After infection of a host cell, the RNA genome is reverse transcribed into double stranded DNA and inserted into the genome of the host cell.
Reverse transcriptase:	retrovirus-derived enzyme catalysing the translation of RNA into DNA.
Ribonucleic Acid (RNA):	Nucleic acid containing ribose sugar and the base Uracil. There are three types of RNA, each is involved in protein synthesis. mRNA is the precursor of the proteins. The single stranded mole-

cule is transcribed from the coding strand of the DNA.

Ribozymes: special RNA molecules able to act enzymatically.

SCID-mice: strain of immunoincompetent mice lacking functional T-cells.

Serotonergic cells: type of neurons releasing serotonin.

Side population cell: cells that have the ability to exclude certain dyes based on the expression of certain transmembrane transporter molecules.

Spermatid: an immature sperm cell.

Superovulation: the medical stimulation of the ovary with hormones to induce the production of multiple egg-containing follicles in a single menstrual cycle.

Somatic cells: (from soma, the body) cells of the body which in mammals and flowering plants normally are made up fulfilling specific functions within the body. The term is usually not used for stem and precursor cells within the body.

Spermatogenesis: The development of sperm cells from spermatocytes to mature sperm, including meiosis.

Stem cells: self renewing cells that have the ability to divide indefinitely and to give rise to more specialised cells as well as to new stem cells with identical potential.

Spermatogonial transfer: Transfer of germ line stem cells from one testis to another, can be be used to restore fertility to an infertile male.

Spindle: Microtubule construction that aligns and segregates chromosomes during eukaryotic cell division.

Stem cell factor (SCF):   Cytokine involved in regulation of proliferation, renewal and differentiation of bone marrow stem cells.

Striatal tissue: part of the basal stem-ganglia in the brain.

Subcutaneous: directly beneath the skin.

Synthetic peptides   short peptides with a sequence of Arginine,

(RGD-peptides):   Glycin and Aspartate, mediating cellular binding via integrins.

T-cells:	The type of lymphocyte responsible for cell-mediated immunity; also protects against infection by parasites, fungi, and protozoans and can kill cancerous cells; circulate in the blood and become associated with lymph nodes and the spleen.
Telomerase:	Telomerase reverse transcriptase, essential to synthesize the telomeres of the chromosomes, which protect the ends of the chromosomes against degradation.
Teratocarcinoma:	germ cell tumor (usually of the testis).
Tetraploid:	cell contains four instead of two copies of each chromosome.
Thalassemia:	autosomal dominant genetic disease resulting in hemolysis.
Therapeutic cloning:	Isolation of ES-cells from embryos produced by nuclear transfer technologies. Cell products derived from those cells are genetically identical to the donor of the nucleus and are not immunologically rejected after transplantation into this individuum.
Thymus:	Organ of T-cell-development and -selection, located above the heart. The thymus degenerates with age and is lost in adult humans.
Tissue non-specific alkaline phospatase (TNAP):	stem cell marker.
TRAIL:	TNF-related apoptosis inducing ligand.
Transgenesis:	Transfer of genetic elements from one individuum to another.
Thrombocytes:	blood cells involved in coagulation.
Trophoectoderm:	see trophoblast.
Trisomy:	a chromosomal abnormality in which the fertilised human egg shows an extra chromosome compared to the normal complement.
Totipotent cells:	have the capacity to differentiate into the embryo and into extraembryonic membranes and tissues. Totipotent cells contribute to every cell type of the adult organism.
Trophoblast:	outmost layer of the developing blastocyst of a mammal. It differentiates into two layers, the cytotrophoblast and syntrophoblast, the latter coming

	into intimate relationship with uterine endometrium with which it establishes a nutrient relationship.
Turner syndrome:	In humans, a genetically determined condition in which an individual has only one sex chromosome (an X). Affected individuals are always female and are typically short and infertile.
Urothelial cells:	epithelial cells lining the urogenital tract.
Vascularisation:	formation of a new blood vessel system within a tissue.
Xenotransplantation:	transplantation of vialbel cells or tissues form one species to another.
Xenozoonose:	transmission of pathogens (parasites, funghi, bacteria, viruses, prions) from one species to another.
Zygote:	cells resulting from fusion of two gametes in sexual reproduction, a fertilised egg (ovum), a diploid cell resulting from union of sperm and ovum, or a developing organism during the first week after fertilisation.

References

Human Developmental Anatomy Center: http://nmhm.washingtondc.museum/collections/hdac/ Glossary_Reference.htm

Holland, S., Lebacqz, K. Zoloth, L. eds., The Human Embryonic Stem Cell Debate. Science, Ethics, and Public Policy, Massachusetts Institute of Technology 2001

On-Line Biology Book: GLOSSARY

All contents ©1992, 1994, 1995, 1997, 1998, 1999, M.J. Farabee, all rights reserved: http:// www.emc.maricopa.edu/faculty/farabee/BIOBK/BioBookglossV.html

8 Appendices

Appendix 1:
Chronology of Regulatory Events in the UK 1978–2002

1. July 25, 1978. Birth of Louise Brown, the 1st IVF baby.[1]

2. In July 1982, the Warnock Committee was established to consider medical and scientific developments related to human fertilisation and embryology and make recommendations.[2]

3. 1984. Birth of 1st baby utilizing a frozen embryo.[3]

4. 1984. The Warnock Report. This made comprehensive recommendations about the provision/permission and regulation of infertility treatments and embryo research. [4]

5. December 5, 1984 Enoch Powell, conservative M.P., piloted a private Bill, The Unborn Children (Protection) Bill. This aimed to prohibit all embryo experimentation.[5]

6. On February 15, 1985 the Bill, at second reading, commanded a 172 majority.[6]

7. March 1985. The Voluntary Licensing Authority was established by the Medical Research Council (MRC) and the Royal College of Obstetricians and Gynaecologists (RCOG), to self-regulate the provision of infertility treatment services.[7]

8. June 7, 1985. The Unborn Children (Protection) Bill was defeated through parliamentary manoeuvring, after having previously received favourable 238:66 vote.[8]

9. November 1985. PROGRESS, a campaign for research into human reproduction was launched by a coalition of patients, doctors, scientists and parliamentarians in favour of embryo research.[9]

---

[1] Steptoe, P. C., Edwards, R. G. (1978), "Birth after replacement of a human embryo", *Lancet* ii, p. 366.

[2] *Report of the Committee of Inquiry into Human Fertilisation and Embryology* (HMSO London, Cmnd 9314, 1984) republished as *A Question of Life: The Warnock Report on Human Fertilization and Embryology* (Basil Blackwell, Oxford, 1985).

[3] See the Progress Educational Trust website at http://www.progress.org.uk/About/history.html.

[4] See note 2.

[5] See J. Gunning and V. English, "Human In Vitro Fertilization", (Dartmouth 1993), p. 42.

[6] Ibid., p. 42-43.

[7] Ibid., p. 43.

[8] See Repromed (Centre for Reproductive Medicine, University of Bristol) website at http://www.repromed.org.uk/history/20th_uk.htm and the Progress Educational Trust website at http://www.progress.org.uk/About/history.html.

[9] See the Progress Educational Trust website at http://www.progress.org.uk/About/history.html.

10. October 21, 1986. Ken Hargreaves, conservative M. P., re-introduced into Parliament the Unborn Children (Protection) Bill. The purpose was to convince the Government that such a Bill had popular support and force it to sponsor similar legislation in the next session as insufficient parliamentary time was left to have it passed into law. It received a favourable vote of 229 to 129 votes. [10] The Government, however, never introduced such a Bill. [11]

11. In 1986, the Government introduced a consultation paper[12] and in 1987 a White Paper[13] containing its proposals for legislation.

12. In 1988, the Government tabled the Human Fertilisation and Embryology Bill. It contained two mutually exclusive clauses, one permitting and one prohibiting embryo research, inviting parliament to choose one on a "free conscience vote".[14]

13. In July 1988,[15] the Polkinghorne Committee was established to review the Guidance on the Research Use of Fetuses and Fetal Material which had been issued in 1972 by the Peel Committee.[16]

14. In 1989, the Voluntary Licensing Authority's name was changed to the Interim Licensing Authority to reflect the fact that it would give way to a statutory licensing authority.[17]

15. In July 1989, the Polkinghorne Report[18] was published. It adopted the Warnock Report time limit on embryo research and recommended that any research on the fetus[19] after 14 days, whether inside or outside the womb, should only take place under conditions applicable to research on fully-developed human subjects[20]

16. In 1990, the 1st PIGD baby was born.[21]

---

[10] See Repromed (Centre for Reproductive Medicine, University of Bristol) website at http://www.repromed.org.uk/history/20th_uk.htm.

[11] See M. Kirejczyk, "Parliamentary Cultures and Human Embryos: The Dutch and British Debates Compared." *Social Studies of Science* 29/6 (December 1999) 889-912 at 891-892.

[12] Legislation of Human Infertility Services and Embryo Research, December 1986 Cm 46.

[13] Human Fertilisation and Embryology: a Framework for Legislation Cm 259.

[14] See M. Kirejczyk, "Parliamentary Cultures and Human Embryos: The Dutch and British Debates Compared." *Social Studies of Science* 29/6 (December 1999) 889-912 at 892.

[15] See Review of the Guidance on the Research Use of Fetuses and Fetal Material Cm 762 HMSO 1989, p. 2.

[16] See "The Use of Fetuses and Fetal Material for Research" (The Peel Report) HMSO 1972.

[17] See J. Gunning and V. English, "Human In Vitro Fertilization", (Dartmouth 1993), p. 88.

[18] Review of the Guidance on the Research Use of Fetuses and Fetal Material Cm 762 HMSO 1989.

[19] Along with "fetal tissue" defined in the report to include respectively the whole or part of the embryo or fetus from implantation in the womb until the time gestation ends, whether alive or dead, and whether inside or outside the womb. See Review of the Guidance on the Research Use of Fetuses and Fetal Material Cm 762 HMSO 1989, p. 2.

[20] Stated as being that intervention on a living fetus should carry only minimal risk of harm, or if a greater risk than that is involved, the action is, on balance for the benefit of the fetus. See Review of the Guidance on the Research Use of Fetuses and Fetal Material Cm 762 HMSO 1989, p. 6.

[21] See Repromed (Centre for Reproductive Medicine, University of Bristol) website at http://www.repromed.org.uk/history/20th_uk.htm.

17. In 1990, the Human Fertilisation and Embryology Act (HFE Act)[22] was passed.

18. On 1 August 1991, the Human Fertilisation and Embryology Authority (HFEA) took up responsibilities as required by the HFE Act. This licensing Authority took over from the Interim Licensing Authority and began regulating certain assisted reproduction techniques and all human embryo research.

19. In January 1994, the HFEA commenced a public consultation on *"Donated Ovarian Tissue in Embryo Research and Assisted Conception"*.

20. On 12 April 1994, before the consultation was completed, Dame Jill Knight M. P. sponsored an amendment to the HFE Act via section 156 of the Criminal Justice and Public Order Act 1994 c. 33, prohibiting the use of germ cells derived from an embryo or foetus in fertility treatments.

21. July 1994, the *"Donated Ovarian Tissue in Embryo Research and Assisted Conception: Report"* was published. The Report recommended that fetal ovarian tissue should not be used in infertility treatment, but could be used for embryo research subject to the need to obtain informed consent prior to the woman's death or undergoing abortion.

22. 1996. Dolly the sheep was cloned at the Roslin Institute in Edinburgh, using the technique of Cell Nuclear Transfer (CNR).[23]

23. February 1997. The announcement of the birth of Dolly.[24]

24. On the day of the announcement, the House of Commons Science and Technology Committee convened an inquiry into the issue of cloning.[25]

25. 20 March 1997. The Committee publish a report *"The Cloning of Animals from Adult Cells"* and the Government published its response in December 1997.

26. In 1998, the HFEA and the Human Genetics Advisory Commission (HGAC)[26] jointly set up a public consultation on the issues in response to widespread public interest.

27. December 1998. The joint Report *"Cloning Issues in Reproduction, Science and Medicine"* was published with a Government response on 25 June 1999.

28. 1999. The Government mandated the Chief Medical Officer, Sir Liam Donaldson, to convene an expert advisory group to make recommendations on the issue of new regulations.[27]

---

[22] Human Fertilisation and Embryology Act 1990 c.37.

[23] See Roslin Institute website at http://www.roslin.ac.uk/public/cloning.html#technology.

[24] See *Viable Offspring Derived from Foetal and Adult Mammalian Cells* Nature, 385, 881: 1997 and Roslin Institute website at http://www.roslin.ac.uk/public/cloning.html#technology. CNR is described on their website as "transferring the nucleus of a body cell into an egg from which the nucleus has been removed".

[25] See Department of Health: *Cloning Issues in Reproduction, Science and Medicine* at website http://www.doh.gov.uk/cloning.htm.

[26] A non-statutory body established in 1996 to take a broad view of developments in human genetics and foster public confidence in its application. In 1999 the HGAC was merged with the Advisory Committee on Genetic Testing and the Advisory Group on Scientific Advances in Genetics to form the Human Genetics Commission.

[27] See *Government Response to the report by the Human Genetics Advisory Commission and the Human Fertilisation and Embryology Authority on Cloning Issues in Reproduction, Science and Medicine,* Cm 4387 at http://www.doh.gov.uk/cloning.htm.

29. 2000, The "Donaldson Report", *Stem Cell Research: Medical Progress with Responsibility* was published. The Report supported regulated stem cell research and recommended additional purposes for embryo research be legislated.[28]

30. 2000 The Government made a positive response to the Donaldson Report and laid regulations laid before parliament.

31. 31 January 2001. The Human Fertilisation and Embryology (Research Purposes) Regulations 2001 SI 2001 No.188 came into force.

32. 15 November 2001 The High Court rules that CNR Embryos are not covered by the HFE Act.[29]

33. 2001. The Government appeals the High Court Decision.

34. 2001. The Government passes the Human Reproductive Cloning Act 2001 c. 23, prohibiting reproductive cloning.

35. 18 January 2002. The Court of Appeal overturns the high Court judgment, holding that CNR embryos are covered under the HFE ACT and subject to regulation.[30]

36. 13 February 2002. The Report of the House of Lords Select Committee on Stem Cell Research is published. This gives the green light to continued stem cell research on both adult and ES cells, subject to close regulatory oversight.[31]

Appendix 2:
Chronology of Regulatory Events in Spain[32]

1. 1984; Victoria Ana is the first child born in Spain using human assisted reproduction techniques[33]. These techniques have been applied in Spain for the first time in the Hospital Dexeus (Barcelona) and since them more than 10.000 children have been brought to life in Spain using human assisted reproduction techniques.

2. 1985; the Spanish Parliament set up a Special Commission on the Study of In Vitro Fertilisation and of Human Artificial Insemination (Chairman: M. Palacios[34]. This commission approved a Report containing 155 recommendations[35], in order to elaborate the two proposed bills under mentioned.

---

28 Available at http://www.doh.gov.uk/cegc/stemcellreport.htm.
29 *R v. Secretary of State for Health, ex parte Bruno Quintavalle (on behalf of Pro-life Alliance)* [2001], EWHC Admin 918.
30 See *R (Quintavalle) v. Secretary of State for Health* [2002] EWCA Civ 29.
31 Available at http://www.parliament.the-stationery-office.co.uk/pa/ld200102/ldselect/ldstem/ 83/8301.htm.
32 Information provided by Carlos Romeo Casabona and Asier Urruela-Mora.
33 See a short report on this fact at http://db.separ.es/cgi-bin/wdbcgi.exe/separ/perscepticemia. plantilla?pident=209.
34 See Diary of Sessions of the Congress of Deputes, II Legislature, 1985, no. 334, 346, 357, 367, 376, 385, etc. (the name of the Special Commission has varied during its activity).
35 The Report was approved by the Plenary Session of the Congress of Deputes on 10th April 1986.

3. May 1987; the Socialist Parliamentary Group in the Spanish Parliament put forward two proposed bills on "Assisted Reproduction Techniques" and on "Donation and use of human embryos and fetuses or of its cells, tissues and organs" in order to be discussed.

4. 23. March 1988; the Spanish Catholic Church (Conferencia Episcopal Española) published a document on the proposed bills on "Assisted Reproduction Techniques" and on "Donation and use of human embryos and fetuses or of its cells, tissues and organs" (which are to be discussed in the Spanish Parliament in the following months). The document criticises the initiative adopted by the Socialist Parliamentary Group[36].

5. 22. November 1988. The Spanish Parliament passed Law 35/1988, of 22 November, on Assisted Reproduction Techniques[37].

6. 28. November 1988. The Spanish Parliament passed Law 42/1988, of 28 December, on donation and use of human embryos and fetuses or of its cells, tissues and organs[38].

7. 1989; 63 deputies of the Popular Party Group in the Spanish Parliament lodged an appeal to the Constitutional Court (Tribunal Constitucional) (appeal to the Constitutional Court no. 376/89) against Law 35/1988, of 22 November, as they considered that several articles of the aforementioned Law where against the Spanish Constitution of 1978.

8. 1989; 78 deputies of the Popular Party Group in the Spanish Parliament lodged an appeal to the Constitutional Court (appeal to the Constitutional Court no. 596/89) against Law 42/1988, of 28. December, as they considered that several articles of the aforementioned Law where against the Spanish Constitution of 1978.

9. 23. November 1995. The Spanish Parliament passed Law 10/1995, of 23. November, on Criminal Code[39]. For the first time in Spain the Criminal Code includes some offences related with genetic manipulation and the banning of the fertilisation of human eggs with purposes other than human reproduction (Book II, Title V, arts 159-162).

10. Royal Decree 412/1996, of 1. March, establishes the compulsory protocols (medical records) of donors and users of assisted reproduction techniques and which regulates the creation and organisation of the National Register of Donors of Gametes and Pre-embryos with human assisted reproduction purposes[40].

---

[36] See http://www.conferenciaepiscopal.es/doctrina/rtf/reproduccion_asistida.rtf.
[37] B.O.E. (Spanish official gazette) n° 282 of 24th November 1988 and B.O.E. no. 284 of 26th November 1988 (correction of mistakes).
[38] B.O.E. no. 314 of 31. December 1988.
[39] B.O.E. no. 281, of 24. November 1995; B.O.E. no. 54, of 2. March 1996 (correction of mistakes).
[40] B.O.E. no. 72 of 23. March 1996.

11. Royal Decree 413/1996, of 1. March, establishes the technical and functional requirements for the authorisation and official approval of medical centres and services[41] related with human assisted reproduction techniques[42].

12. Order of 25. March 1996, establishes the regulation of the National Register of Donors of Gametes and Pre-embryos[43].

13. Decision of the Spanish Constitutional Court 212/1996, of 19. December 1996[44], resolving appeal no. 596/89, lodged by 78 deputies of the Popular Party in the Spanish Parliament against Law 42/1988, of 22. December 1988. Basically, the decision of the Spanish Constitutional Court established that the treatment afforded to embryo by Law 42/1988 is adjusted to the Spanish Constitution of 1978.

14. Royal Decree 415/1997, of 21. March, creates the National Commission on Human Assisted Reproduction[45] [46].

15. 4. April 1997. Opening for signature by the States of the Convention of the Council of Europe for the Protection of Human Rights and Dignity of the Human Being with regard to the Application of Biology and Medicine: Convention on Human Rights and Biomedicine, in Oviedo (Asturias-Spain).

16. 11. November 1997, first Meeting of the National Commission on Human Assisted Reproduction.

17. 1998; Spain signed, ratified and published the first Protocol to the Convention on Human Rights and Biomedicine, on the prohibition of human reproductive cloning[47].

18. December 1998, first Report of the National Commission on Human Assisted Reproduction[48]. This report focuses on certain aspects related with the use of human assisted reproduction techniques. In principle, the creation of the National Commission on Human Assisted Reproduction was stated in Law 35/1988, of 22. November, on Assisted Reproduction Techniques, but the aforementioned Commission did not started his work until 9 years afterwards. One of the subjects analysed by the Commission was the increasing number of cryoconserved embryos existing in Spain, due to the legal regulation in force in this country. Many of the surplus embryos had been cryo-preserved longer than the term fixed in Spanish law (5 years) for the maintain of the embryos cryo-conserved. The Commission in the First report pointed out the existence of a problem and the need to decide whether this embryos have to be destroyed or not

[41] See Spanish Health and Consumption Ministry website at http://www.msc.es/salud/epidemiologia/ies/repro_asistida/centros.htm.
[42] B.O.E. no. 72 of 23. March 1996.
[43] B.O.E. no. 106 of 2. May 1996.
[44] B.O.E. 22. January 1997.
[45] B.O.E. no. 70 of 22. March 1997.
[46] See Spanish Health and Consumption Ministry web site at http://www.msc.es/salud/epidemiologia/ies/repro_asistida/indice_comision.htm.
[47] Signed in Paris on 12. January 1998, ratified by Ratification Instrument of 24. January 2000 and published in B.O.E. of 1st March 2001.
[48] See a short summary of the Report at the Spanish Health and Consumption Ministry web site at http://www.msc.es/salud/epidemiologia/ies/repro_asistida/resumen_anual.htm. The full report is available at http://www.msc.es/salud/epidemiologia/ies/repro_asistida/r_asist.pdf.

(i.e. used for research and experimentation purposes), but did not adopt an specific position on this problem and did not give any recommendation to the Spanish Government on this subject.

19. Decision of the Spanish Constitutional Court 116/1999, of 17. June 1999[49], resolving appeal no. 376/89, lodged by 63 deputies of the Popular Party in the Spanish Parliament against Law 35/1988, of 22. November 1988. Basically, the decision of the Spanish Constitutional Court established that the treatment afforded to embryo by Law 35/1988 is adjusted to the Spanish Constitution of 1978.

20. 1. January 2000, coming into effect in Spain of Convention of the Council of Europe for the Protection of Human Rights and Dignity of the Human Being with regard to the Application of Biology and Medicine: Convention on Human Rights and Biomedicine[50].

21. 2000, Second Report of the National Commission on Human Assisted Reproduction.[51] In this report the National Commission has analysed the situation of the surplus embryos left over from assisted reproduction techniques. In Spain the number of surplus embryos is increasing yearly, and most of them have already exceeded the legal term of five years fixed by law 35/1988, in order to maintain them cryo-conserved. Due to the fact that there are only two possibilities for these embryos once the legal term of five years has been reached (to destroy them or to use them for research and experimentation), but the Spanish legislation does not admit its use for research purposes, the National Commission has recommended the modification of the Spanish law in order to admit their use for research and experimentation purposes.

22. 23. November 2001; The Socialist Group in the Spanish Parliament introduces a proposed bill in order to modify Law 35/1988, of 22. November on Assisted Reproduction Techniques[52]. The purpose of this proposed bill is to allow research on surplus embryos left over from assisted reproduction techniques which have overcome the legal term of five years. Notwithstanding it was rejected by the Congress of Deputies on 12 February 2002.

23. September 2002; the Communist Group (Izquierda Unida) in the Spanish Parliament introduces a proposed bill in order to modify Law 42/1988, of 28. December, on donation and use of human embryos and fetuses or of its cells, tissues and organs, and Law 10/1995, of 23. November, on Criminal Code. Notwithstanding it was rejected by the Congress of Deputies on 15 October 2002[53].

---

[49] See http://www.aeds.org/jurisprudencia/constituembrion.htm.
[50] See http://www.fiscalia.org/legis/derinter/cbiom.htm#C14 and B.O.E. 20th October 1999 (Instrumento de ratificación).
[51] Available in Spanish at http://www.cnice.mecd.es/tematicas/genetica/2001_12/2001_12_01. html. A summary in Spanish is available at http://www.cnb.uam,es/ transimp/IIinforme CNRHA.pdf.
[52] Boletín Oficial de las Cortes Generales-Congreso de los Diputados, VII. Legislatura, 23 de noviembre de 2001, no. 172-1. See at http://www.congreso .es/public_oficiales/L7/CONG/ BOCG/B/B_172-01.PDF.
[53] See http://www.iucadiz.org/iufederal/article314.html.

24. 9. January 2003, The Socialist Group in the Parliament of Andalusia introduced a proposed bill in order to regulate the research in Andalusia with human surplus non viable pre-embryos coming from in vitro fertilisation[54].

25. Royal Decree 120/2003, of 31. January, establishes the requirements in order to develop controlled experiments with reproductive purposes, of fecundation of ovocites or ovarical tissues previously cryo-conserved, related with human assisted reproduction techniques[55].

26. 7. March 2003, Report on "Stem cell research" of the National Advisory Committee for Ethics of Research in Science and Technology. It has recommended to the Spanish Government to review the current legislation to allow research with embryonic cells derived from surplus embryos under some strict requirements.[56]

27. 26. March 2003; The Spanish Parliament rejected a proposed bill introduced by the Socialist Group in order to modify Law on Assisted Reproduction Techniques, so as to admit the use of surplus pre-embryos coming from reproduction techniques in order to be used in research[57].

Appendix 3:
The legislation on embryo research in the Federal Republic of Germany[58]

1. 1959; The 62nd Deutsche Ärztetag *(German Medical Congress)* regards homologous insemination based on the approval of the person affected as admissible in exceptional cases but unanimously rejects heterologous insemination based on the approval of the person affected.[59]

2. 1970; The 73rd Deutsche Ärztetag now regards homologous insemination based on the approval of the person affected as justifiable yet does not argue in favour of it because of the associated legal problems.[60]

3. Interdisciplinary experts' talk at the Bundesministerium für Bildung und Forschung *(Federal Ministry of Research and Technology)* in September 1983[61].

4. 1983; At the 25th Annual Meeting of the German Female Lawyers' Association female lawyers deal with the questions of genetic engineering and reproductive

---

54   See http://www.psoe-aonline.com/imprimir.html.
55   B.O.E. 15th January 2003.
56   Available in Spanish at http://www.imasd-tecnologia.com/InformeCM.htm. and http://www.imasd-tecnologia.com/imasd/mar03/0303vi2htm.
57   See http://actualidad.wanadoo.es/noticias/89443.html.
58   See *Keller/Günther/Kaiser*, Kommetar zum Embryonenschutzgesetz, Introduction B III and v. Bühlow, Embryonenschutzgesetz, margin nos. 303-380 in: Winter/Fenger/Schreiber, Genmedizin und Recht, München 2001.
59   *Keller/Günther/Kaiser*, Kommetar zum Embryonenschutzgesetz, Introduction B III, 1, margin no. 1.
60   *Keller/Günther/Kaiser*, Kommetar zum Embryonenschutzgesetz, Introduction B III, 1, margin no. 2.
61   v. Bühlow, Embryonenschutzgesetz, margin no. 307 in: Winter/Fenger/Schreiber, Genmedizin und Recht, München 2001.

medicine for the first time; in 1985 a project team passes the 10 theses worked out by then.[62] These call for the production of embryos for research purposes to be prohibited and for embryo protection sanctioned by criminal law, amongst other things.[63]

5. 1984; The Bundesjustizministerium and Bundesforschungsministerium *(Federal Ministers of Justice and of Research and Technology)* appoint an interdisciplinary "Arbeitsgruppe Invitrofertilisation, Genomeanaylse und Gentherapie".[64]

6. 1984; In the same year the Bundestag establishes a commission of inquiry named "Chances and Risks of Genetic Engineering".[65]

7. 1985; The 88th Deutsche Ärztetag approves of extracorporeal fertilisation and embryo transfer in order to treat fertility disorders. Upon recommendation of the Congress some individual medical associations of the Bundesländer *(Federal States)* modify their professional rules accordingly.[66]

8. Statement of the Protestant Church in Germany.[67]

9. 1985; The Ministerrat des Bundeslandes Bayern *(Council of Ministers of the Land of Bavaria)* establishes an interministerial project team supposed to examine if and – in the case of need – how to stipulate extracorporeal fertilisation and embryo transfer by statute. The final report contains a draft for an Act of the Federal State. It makes the use of human embryos punishable if they may hereby be damaged or destroyed.

10. 1985; Motion of the Land Baden-Württemberg to the Bundesrat *(Upper House of Parliament)* to pass a resolution against misuse with regard to extracorporeal fertilisation.[68]

11. 1985; Motion of Bavaria concerning genetic engineering and reproductive technology.[69]

12. December 1985; Guideline of the Bundesärztekammer *(Federal Medical Association)* concerning research on early human embryos[70] admitting research on cryopreserved surplus embryos and – in case those would not emerge anymore – even considered exceptions to the prohibition of production for research purposes.

---

[62] Juristenzeitung (JZ) 1986, p. 777.

[63] Thesis 1, JZ 1986, p. 777.

[64] The former President of the Federal Constitutional Court, Ernst Benda, becomes Chairman, by referring to him the commission was then just called the "Benda Commission", see v. Bühlow, Protection of Embryos Act, margin no. 307 in: Winter/Fenger/Schreiber, Genmedizin und Recht, München 2001. *Keller/Günther/Kaiser*, Kommetar zum Embryonenschutzgesetz, Introduction B III, 1, margin no. 10.

[65] Further references in *Keller/Günther/Kaiser*, Kommetar zum Embryonenschutzgesetz, Introduction B III, 1, margin no. 12.

[66] *Keller/Günther/Kaiser*, Kommetar zum Embryonenschutzgesetz, Introduction B III, 1, margin no. 3.

[67] Evangelische Kirche Deutschland, Text 11, ed. Kirchenamt im Auftrag des Rates der Evangelischen Kirche Deutschland, 1985.

[68] *Keller/Günther/Kaiser*, Kommetar zum Embryonenschutzgesetz, Introduction B III, 1, margin no. 18; BR-Drs. 361/85.

[69] *Keller/Günther/Kaiser*, Kommetar zum Embryonenschutzgesetz, Introduction B III, 1, margin no. 18; BR-Drs. 382/85.

[70] Deutsches Ärzteblatt, 3757-3764.

13. 1986; Because of these motions the Bundesrat passes a resolution on extracorporeal fertilisation.[71] Amongst other things it says consuming scientific experiments on embryos are to be rejected and punishable.

14. 1986; The "Interministerial Project Group on the Reviewing of Bioethical Issues – Reproductive Medicine" of the Ministry of Justice of the Land of Rhineland-Palatinate presents a "Provisional and Initial Draft for an Act of the Land Governing Reproductive Medicine".

15. 1986; The German Association of Judges passes its theses on reproductive medicine and human genetics.[72]

16. 1986; For the first time the 56th Deutsche Juristentag *(German Congress of Jurists)* deals with the subject of reproductive medicine in its civil law section. Two partial expert's reports are produced for this reason. In its resolutions the Congress acknowledges artificial fertilisation as a therapy to be justified medically, approved of ethically and recognised legally. The Congress explicitly decides that embryos may merely be produced for implantation[73] and operations may only be carried out in its own interest.[74]

17. 1986; Guiding principles of the Federal Working Group of Christian-Democratic Jurists (BACDJ)[75].

18. Statement of the Unified Lutheran Church of Germany.[76]

19. 1986; The Bundesjustizministerium *(Federal Minister of Justice)* presents a draft of an act for the Embryonenschutzgesetz – ESchG (Embryo Protection Act) – for discussion which is based on the results of the work of the Benda Commission.[77] The draft prohibits extracorporeal fertilisation of an egg cell if it is not supposed to be transferred to a woman. In addition, misuse is deemed a criminal offence. The draft intended to pass the competence to decide on the use of surplus embryos produced extracorporeally to one of the supreme authorities of the respective Land.[78]

20. 1986; The Bundesjustizministerium *(Federal Minister of Justice)* establishes a project team on reproductive medicine consisting of representatives of the Bundesländer as well as of the Federal Government because of the request contained in the resolution of the Bundesrat. If it concludes the State has to act, it is meant to establish solutions.[79]

---

[71]   BR-Drs. 210/86.
[72]   DRiZ *(specialist journal for German judges)* 1986, 229.
[73]   NJW 1986, 3069, 3070: 56th Deutscher Juristenkongress, Sektion I, Zivilrecht, VII, 3.
[74]   NJW 1986, 3069, 3070: 56th Deutscher Juristenkongress, Sektion I, Zivilrecht, VII, 7.
[75]   *Keller/Günther/Kaiser*, Kommetar zum Embryonenschutzgesetz, Introduction B III, 1, margin no. 9.
[76]   vel kd *(i.e. abbreviation for* Unified Lutheran Church of Germany) 32/1986.
[77]   v. Bühlow, Embryonenschutzgesetz, margin no. 319 in: Winter/Fenger/Schreiber, Genmedizin und Recht, München 2001.
[78]   *Keller/Günther/Kaiser*, Kommetar zum Embryonenschutzgesetz, Introduction B III, 7, margin no. 21 who describe this solution employing a criminal sanction but giving permission as halfhearted.
[79]   *Keller/Günther/Kaiser*, Kommetar zum Embryonenschutzgesetz, Introduction B III, 1 margin no. 22; see also v. Bühlow, Embryonenschutzgesetz, margin nos. 324-329 in: Winter/Fenger/Schreiber, Genmedizin und Recht, München 2001.

21. 1987; The commission of inquiry of the German Bundestag presents its final report.[80] It deals exclusively with genetic engineering and not yet with reproductive technology.[81] The commission recommends a prohibition of genetically manipulating the human germ line sanctioned by criminal law.[82]

22. 1987; Statement of the Catholic Church[83]. It calls for protecting beginning human life and respecting the dignity of reproduction.

23. 1988; Final report of the joint commission (see no. 20).[84] It calls for a prohibition of embryos for research purposes without exception.

24. 1988; Initial draft of the Bundesjustizminister[85] on the prohibition of research on surplus embryos.

25. 1990; Adoption of the Embryonenschutzgesetz.[86]

26. 1995; The German Medical Association passes the guidelines on gene transfer into human body cells[87] regulating clinical treatment in detail.

27. May 1997; The project team "Somatic Genetic Therapy" consisting of representatives of the Bundesländer as well as the Federal Government presents its final report. The team finds a sufficiently dense network of regulations exists and no further regulations are needed.[88]

28. 1997; The German Bundestag asks the Federal Government to be informed on the question if the legislature is required to modify the Protection of Embryos Act because of the techniques employed for cloning animals and the prospective further development.[89]

29. June 1998: Answer by the Federal Government.[90] The report advises to establish criminal liability for cloning if a human embryo has emerged by any other way than fertilisation. Accordingly, totipotential combinations of tissue and cells developing from extracted totipotential cells should be regarded as embryos under the ESchG.

---

[80] BT-Drs. 10/6775.

[81] *Keller/Günther/Kaiser*, Kommetar zum Embryonenschutzgesetz, Introduction B III, 1, margin no. 12.

[82] v. Bühlow, Embryonenschutzgesetz, margin no. 315 in: Winter/Fenger/Schreiber, Genmedizin und Recht, München 2001.

[83] *Announcements by the Apostolic Chair, Number 74.*

[84] A summary of the contents is to be found in *Keller/Günther/Kaiser*, Kommetar zum Embryonenschutzgesetz, Introduction B III, 1, margin nos. 24-25.

[85] *Keller/Günther/Kaiser*, Kommetar zum Embryonenschutzgesetz, Introduction B III, 1, margin no. 29.

[86] Bundesgesetzblatt I, p. 2746.

[87] Deutsches Ärzteblatt 92/1995, A-789-794. The guidelines are also contained in: Winter/Fenger/Schreiber, Genmedizin und Recht, München 2001, p. 743.

[88] A summary is to be found in v. Bühlow, Abschlußbericht der Bund-Länder-Arbeitsgruppe "Somatische Gentherapie" 1997 in: Winter/Fenger/Schreiber, Genmedizin und Recht, München 2001.

[89] BT-Drs. 13/7243.

[90] BT-Drs. 13/11263. A summary is to be found in *v. Bühlow*, Embryonenschutzgesetz, margin no. 356 in Winter/Fenger/Schreiber, Genmedizin und Recht, München 2001.

30. The DFG, Deutsche Forschungsgemeinschaft, passes its statement "Human Embryonic Stem Cells".[91]

31. 2000; Appointment of the commission of inquiry of the German Bundestag called "Ergänzende Kommission Recht und Ethik in der Medizin".[92]

32. 2001; Set up of the Nationaler Ethikrat *(National Ethics Council)*[93]

33. 2001; The majority of the National Ethics Council votes for the import of embryonic stem cells.[94]

34. 2002; The German Bundestag permits the import of embryonic stem cells under strict conditions. 360 MPs vote for and 190 against it. There are 9 abstentions.

35. 2002; The commission of inquiry hands over its final report to the Bundestag.[95]

36. 2002; On 10 July the Federal Government passes the Regulation pursuant to StZG, section 7, in which the Robert Koch Institute is determined as competent authority and the members of the Central Ethics Commission are appointed.[96]

Appendix 4:
Regulations and Associated Laws in the EU, NAS and Selected Other Countries

Australia (Federal)

SOURCES OF REGULATIONS

Acts of the Legislature
- Gene Technology Act 2000 (GTA)[97]

Official Regulations
- None

Official Guidelines
- "Ethical guidelines on assisted reproductive technology" – National Health and Medical Research Council 1996 (NHMRC 1996).[98] (In States which have legislation regulating assisted reproductive technology (ART), currently South Australia, Victoria and Western Australia, their legislative provisions apply. Where both State law and the guidelines apply, State law prevails. [NHMRC 1996

---

91 Press release no. 10 by the DFG from 19 March 1999. A summary is to be found in *v. Bühlow*, Embryonenschutzgesetz, margin no. 377 in Winter/Fenger/Schreiber, Genmedizin und Recht, München 2001.

92 Established by the Resolution of the German Bundestag from 24 March 2000, BT-Drs. 14/3011, see also http://www.bundestag.de/gremien/medi/.

93 http://www.ethikrat.org/ueber_uns/einrichtungserlass.html.

94 http://www.ethikrat.org/publikationen/Stell_stammzellen/inhalt.html.

95 German Bundestag (editor), Commission of Inquiry: Law and Ethics in Modern Medicine, Berlin 2002, available at http://dip.bundestag.de/btd/14/090/1409020.

96 Press release no. 140 available at http://www.bmbf.de/presse01/671.html, p. 2 and appendix.

97 Legislation available online at http://www.austlii.edu.au/au/legis/ and at http://www.lawlex.com.au/.

98 Available online at http://www.health.gov.au/nhmrc/publications/pdf/e28.pdf.

guideline 1.1] The guidelines apply fully in all the other States.)

- The National Statement on Ethical Conduct in Research Involving Humans 1999 (NHMRC 1999a).[99]
- Guidelines for Ethical Review of Research Proposals for Human Somatic Cell Gene Therapy and Related Therapies (1999) (NHMRC 1999b).[100]

International Conventions

- Universal Declaration on the Human Genome and Human Rights. (UDHGHR)[101]

Recommendations

- Report of the Parliament of the Commonwealth of Australia "Human cloning: scientific, ethical and regulatory aspects of human cloning and stem cell research" August 2001 (Australian Parliament 2001).[102]

OVERVIEW OF REGULATIONS

Embryo Research

Definition of "embryo"

"[U]se of the term ["embryo"] … includes embryos created:

- naturally;
- as a result of artificial reproductive technologies (including in vitro fertilisation); and
- by asexual reproduction such as by somatic cell nuclear transfer for the purpose of research or (in the future) possible use in medical treatment." (Australian Parliament 2001, 98).

Research

Permitted with conditions.

Sources of embryos

Supernumerary from in vitro fertilization techniques. (NHMRC 1996, guideline 3.2.5). However procedures which create embryos surplus to the needs of the infertility treatment should be discouraged. (NHMRC 1996, Guideline 6.5). The creation of embryos for research is viewed as unnecessary at this time (Australian Parliament 2001, 121).

Time limit for research

Culturing of an embryo in vitro for more than 14 days is prohibited. (NHMRC Guideline 11.2)

Other conditions governing research

Consent procedures

The gamete providers, and their spouses or partners, must consent to the specific

---

99   Available online at http://www.health.gov.au/nhmrc/publications/pdf/e35.pdf.
100   Available online at http://www.health.gov.au/nhmrc/publications/pdf/e38.pdf.
101   Adopted by consensus, UNESCO General Conference, 11 November, 1997.
102   Available online at http://www.aph.gov.au/house/committee/laca/humancloning/report.pdf.

form of research (NHMRC 1996, Guidelines 3.2.5 and 6.4).

Licenses

Reproductive Medicine Units (RMUs) offering Assisted Reproductive Technology (ART) must obtain accreditation from a recognised accreditation body. Accreditation must include consideration of:

- Compliance with National Health and Medical Research Council (NHMRC) Guidelines.
- Compliance with the Code of Practice of the accreditation or licensing body.
- Certification and maintenance of appropriate professional standards for all personnel involved in relevant clinical and laboratory work.
- Maintenance of quality assurance programs for both laboratory and clinical work. (NMHRC 1996 Guidelines 2.1–2.1.4).

Ethics Committees

Research is governed by Institutional Ethics Committees (IECs), which are composed in accordance with Supplementary Note 1 of the *NHMRC Statement on Human Experimentation*,[103] and which otherwise comply with its provisions (NHMRC 1996, vi). Any research which does not benefit the embryo must be approved (NHMRC 1996, Guidelines 6.3 and 6.4).

Substantive (Must the embryo benefit?)

Embryo experimentation should normally be limited to therapeutic procedures which leave the embryo, or embryos, with an expectation of implantation and development. (NHMRC 1996, Guideline 6.2).

Non-therapeutic research which does not harm the embryo may be approved by an Institutional Ethics Committee (IEC). (Ibid., Guideline 6.3).

Non-therapeutic research which involves the destruction of the embryo, or which may otherwise not leave it in an implantable condition, should only be approved by an IEC in exceptional circumstances. Approval requires:

- a likelihood of significant advance in knowledge or improvement in technologies for treatment as a result of the proposed research;
- that the research involves a restricted number of embryos; and
- the gamete providers, and their spouses or partners, to have consented to the specific form of research (NHMRC 1996, Guideline 6.4).

Storage/Disposal of embryos

- Storage of gametes and embryos should be in accordance with relevant State legislation and the Code of Practice of the accreditation body. Consent procedures dealing with consent for storage, as provided for in guidelines 3.2.6 and 3.2.7, must be followed (NHMRC 1996, guideline 7.1).

---

[103] This has now been replaced by the National Statement on Ethical Conduct in Research Involving Humans. (NHMRC 1999a). See "Official Regulations" under the heading "SOURCES OF REGULATIONS".

- Embryos may be kept for a period not exceeding 10 years, following which, if not used by the couple, they may be donated or allowed to succumb. These arrangements may be varied on compassionate grounds with approval by an IEC (NHMRC 1996, guideline 7.2).
- Embryos may be allowed to succumb by withdrawal of support. Each clinic is to have protocols in place for this purpose. If indicated in their consent, the preferences of a couple who generate an embryo are to be respected in this matter (NHMRC 1996, guideline 7.4).
- Where no consent exists for storage of embryos, or for withdrawal of support (see also guidelines 3.2.6 and 3.2.7), the embryo should remain in storage until the expiry of the maximum period of storage and may then be allowed to succumb (NHMRC 1996, guideline 7.5).

Cloning

Reproductive

Prohibited (GTA s.192B)

Non-reproductive

A three year moratorium on the creation and use of embryos created by Somatic Cell Nuclear Transfer was recommended in 2001. It was also recommended that after this period the issue should be re-examined by the Australian Health Ethics Committee (AHEC)[104] (Australian Parliament 2001, 122).

Gene Therapy

Somatic

Permitted. "Proposals for gene therapy research undergo a process of review and approval which involves a Human Research Ethics Committee, the NHMRC's Gene and Related Therapies Research Advisory Panel (GTRAP) and the Therapeutic Goods Administration (Australia), and may involve an Institutional Biosafety Committee (IBC) and the Genetic Manipulation Advisory Committee (GMAC)." (NHMRC 1999b, 8). Proposals are also subject to the National Statement on Ethical Conduct in Research Involving Humans (NHMRC 1999a), in particular the section on Clinical trials (NHMRC 1999b, 9)

Germ-line

Prohibited. (NHMRC 1999, Guideline 3 (d) (i)).

Pre-Implantation Genetic Diagnosis

Not regulated federally. (NHMRC 1996, p.v).

Stem Cell Therapy

Stem cell research is being pursued and has reaped some encouraging results. There is a mixture of public and private funding. (Australian Parliament 2001, 57–59).

Pre-Natal Diagnosis/Abortion

Not regulated federally. (United Nations 2001).

---

[104] This committee falls under the NHMRC.

SPECIAL REMARKS

The three States that currently have legislation prohibiting cloning, South Australia, Victoria and Western Australia, each define cloning differently from the federal definition and from each other. New South Wales (Human Reproductive Cloning and Trans-Species Fertilization Bill 2001) and Queensland (Cloning of Humans Prohibition Bill 2001) have proposed legislation which would define cloning in the same terms.

Australia: South Australia (SA)

SOURCES OF REGULATIONS

Acts of the Legislature

- Reproductive Technology Act 1988 (RTA).[105]
- Criminal Law Consolidation Act 1935 as amended (CLCA). [106]

Official Regulations

- Reproductive Technology (Code of Ethical Research Practice) Regulations 1995 (RTR). [107]

Official Guidelines

- See Federal table.

International Conventions

- See Federal table.

Recommendations

- See Federal table

OVERVIEW OF REGULATIONS

Embryo Research

Definition of "embryo"

"a human embryo" (RTR Schedule regulation 1).

Research

Permitted with conditions.

Sources of embryos

By implication supernumerary from IVF (RTR Schedule regulation 32 (a) (vi)).

Time limit for research

Research prohibited on an embryo of a developmental age of more than 14 days after fertilization (RTR Schedule regulation 5).

---

[105] Legislation available online at http://www.austlii.edu.au/au/legis/ and at http://www.lawlex. com.au/.
[106] Ibid.
[107] Ibid.

Other conditions governing research

Consent procedures

Consent to research must be obtained from the gamete donors (RTR Schedule s.16

Licenses

A Licence is required from the South Australian Council on Reproductive Technology to conduct embryo research (RTA s.14).

Ethics Committees

The South Australian Council on Reproductive Technology established under section 5 of the RTA is required to maintain a code of ethical practice that governs inter alia research on human embryos and gametes (RTA s.10(1) (a) (ii)).

Substantive (Must the embryo benefit?)

Research that may be detrimental to an embryo is prohibited (RTA s.14 (2) (b)).

Storage/Disposal of embryos

Embryos may be stored for up to 10 years (RTR Schedule regulation 4 (b))

Cloning

Reproductive

Cloning by embryo splitting is prohibited (RTR Schedule regulation 6). It is the view of the South Australian Council on Reproductive Technology that RTR Schedule regulation 9 prohibits cloning by somatic cell nuclear transfer (Australian Parliament 2001, p.135). There is a proposed new definition of cloning which would read "Cloning is defined as the practice of forming an embryo or an entity capable of embryogenesis which is genetically identical to, or substantially identical to, another human being, living or deceased." It is still being considered (Ibid.).

Non-reproductive

Cloning of organs or tissues is not prohibited (Australian Parliament 2001, 135). The proposed new definition in (a) would leave open the possibility of using somatic cells for cloning with methods that do not require human oocytes (Australian Parliament 2001, p.136).

Gene Therapy

Somatic

Prohibited in relation to embryos. (RTR Schedule regulation 8). See Federal table for general position.

Germ-line

Prohibited in relation to embryos. (RTR Schedule regulation 8). See Federal table for general position.

Pre-Implantation Genetic Diagnosis

Permitted. (RTA s.13 (3)(b)(ii) and RTR 11 (1)(b)(i)(B)).

Stem Cell Therapy

See Federal table.

Pre-Natal Diagnosis/Abortion

Abortion permitted if two legally qualified medical practitioners are of the opinion:

- *that continuance of pregnancy would involve greater risk to the life or physical or mental health of the pregnant woman, than abortion; (CLCA s.82A (1)(a)(i)) or*

- *that there is a substantial risk that, the child would suffer from such physical or mental abnormalities as to be seriously handicapped, (CLCA s..82A (1)(a)(ii)) and*

  it is carried out in a prescribed hospital; or

- a legally qualified medical practitioner is of the opinion, abortion is immediately necessary to save the life, or to prevent grave injury to the physical or mental health, of the pregnant woman (CLCA s.82A (1) (b)).

- For a woman to be eligible for an abortion under 1(a), the woman must have resided in South Australia for at least two months prior to the abortion. (CLCA s.82A (2)).

- In assessing the risk of injury to the physical or mental health of a pregnant woman in subsection (1) (a) (i), account may be taken of the pregnant woman's actual or reasonably foreseeable environment. (CLCA s.82A (3)).

Australia: Victoria (VIC)

SOURCES OF REGULATIONS

Acts of the Legislature
- Infertility Treatment Act 1995 (ITA)[108]
- Crimes Act 1958 (CA)[109]

Official Regulations
- Infertility Treatment Regulations 1997 (ITR)[110]

Official Guidelines
- See Federal table.

International Conventions
- See Federal table.

Recommendations
- See Federal table.

---

[108] Legislation available online at http://www.austlii.edu.au/au/legis/ and at http://www.lawlex. com.au/.
[109] Ibid.
[110] Legislation available online at http://www.austlii.edu.au/au/legis/ and at http://www.lawlex. com.au/.

OVERVIEW OF REGULATIONS

Embryo Research

Definition of "embryo"

"any stage of human embryonic development at and from syngamy" (ITA s.3(1))

„zygote" means "the stages of human development from the commencement of penetration of an oocyte by sperm up to but not including syngamy." (ITAs. 3(1)).

Research

Permitted with conditions.

Sources of embryos

In Vitro Fertilisation. Research can only be carried out if: (i) the zygote or embryo was formed for use in a treatment procedure; and (ii) in relation to a zygote, the zygote is not required for use in a treatment procedure; (ITA s.27.(2)(a) & (b)).

Time limit for research

In respect of embryos, none is specified. In respect of zygotes research must not continue until syngamy, at which time the zygote becomes an embryo (ITA s.26).

Other conditions governing research

Consent procedures

Each person who contributed a gamete must give specific consent t0 the particular research procedure (ITA s.27).

Licenses

A licence must be obtained from the Infertility Treatment Authority for all research. (ITA s.22 (1)).

Ethics Committees

Licence may not be granted unless the applicant, in order to approve and monitor the licensed activities at the premises to be licensed, has established an ethics committee, or is able to use the services of an ethics committee established by another licensee (ITA s.97(4)(b)).

Substantive (Must the embryo benefit?)

A person must not carry out research, outside the body of a woman, involving the use of an embryo –

(a) if the embryo is unfit for transfer to a woman; or

(b) in the case of an embryo which is fit for transfer to a woman, if the research would –

   (i)   harm the embryo; or

   (ii) make the embryo unfit for transfer to a woman; or

   (iii) reduce the likelihood of a pregnancy resulting from the transfer of the embryo (ITA s.24).

However in the case of the zygote, if it is supernumerary destructive research may be carried out on it as research involving the formation or use of a zygote is only permitted if the zygote will not be permitted to develop to syngamy. (ITA ss.26 & 27(2)).

Storage/Disposal of embryos

The maximum period of storage is 5 years unless a longer period is approved by the Infertility Treatment Authority. (ITA s.52 (4)). Embryo may be disposed of by allowing it to stand in its container at room temperature for at least 24 hours (ITR s.12 pursuant to ITA s.53 (2)(b)).

Cloning

Reproductive

Prohibited. (ITR s.47).

Non-reproductive

Prohibited. (ITR s.47).

Gene Therapy

Somatic

See Federal table.

Germ-line

See Federal table.

Pre-Implantation Genetic Diagnosis

Permitted. (ITA s.8 (3) (b)). Sex selection is prohibited (ITA s.50).

Stem Cell Therapy

See Federal table.

Pre-Natal Diagnosis/Abortion

Abortion is prohibited. (CA s.65). However case law has created exceptions and permits abortion to preserve the mother's life or physical or mental health. [111]

Australia: Western Australia (WA)

SOURCES OF REGULATIONS

Acts of the Legislature

- Human Reproductive Technology Act 1991 (HRTA)[112]
- Health Act 1911 (HA)[113]
- Criminal Code 1913 (CC)[114]

---

[111] See United Nations (2001); Abortion Policies: A Global Review. Available at http://www.un.org/esa/population/publications/abortion/index.html

[112] Legislation available online at http://www.austlii.edu.au/au/legis/ and at http://www.lawlex.com.au/.

[113] Legislation available online at http://www.austlii.edu.au/au/legis/ and at http://www.lawlex.com.au/.

[114] Ibid.

Official Regulations
- Human Reproductive Technology Act Directions (HRTD)[115]

Official Guidelines
- See Federal table.

International Conventions
- See Federal table.

Recommendations
- See Federal table.

OVERVIEW OF REGULATIONS

Embryo Research

Definition of "embryo"

"a live human embryo in the stage of development which occurs from –

i) the completion of the fertilisation of the egg; or

ii) the initiation of parthenogenesis,

iii) to the time when, excluding any period of storage, 7 completed weeks of the development have occurred" (HRTA s.3).

Note. The Act often regulates procedures involving an "embryo or egg in the process of fertilisation".

Research

Permitted with conditions.

Sources of embryos

In vitro fertilisation. (HRTA s.17(b) and HRTD 6.3)

Time limit for research

An embryo created by in vitro fertilisation may not be maintained or kept outside the body of a woman beyond 14 days excluding any storage period (HRTA s 7 (c))

Other conditions governing research

Consent procedures

Consent must be specific to a particular embryo or egg in the process of fertilisation, and must be related to the use in treatment of that embryo or egg in the process of fertilisation. (HRTA s.22(5)). Consent must be obtained for each procedure, diagnostic test or experiment that is subject to the approval of Council. (HRTD 3.8).

Licenses

Licences are required and may be granted by the Commissioner of health on the advice of the Western Australia Reproductive Technology Council established under section 8 of the HRTA (HRTA s.27).

---

[115] Available online at http://numbat.murdoch.edu.au/RTC/directions.rtf.

Ethics Committees

Research must be approved by the Western Australia Reproductive Technology Council established under section 8 of the HRTA. The Council takes into account deliberations of Institutional Ethics Committees (HRTA s.20 (6)).

Substantive (Must the embryo benefit?)

Research must benefit the embryo or egg in the process of fertilisation and must be with a view to its future implantation in a particular woman (HRTA ss.14 (2) & 17 (b) and HRTD 8.6 & 9.4).

Storage/Disposal of embryos

"The primary purpose stated in any consent to the storage of an egg in the process of fertilisation or any embryo must relate to the probable future implantation of that egg or embryo" (HRTA s.24 (1)(a)). The maximum storage period for embryos is 3 years (HRTA s.24 (4)(b)).

Cloning

Reproductive

Prohibited (HRTA ss.3 & 7 (1) (d) (i))

Non-reproductive

Due to the definition of "embryo" which encompasses embryos created otherwise than by fertilisation, and the strict regulation of embryo research, non reproductive cloning is effectively prohibited even though the definition of cloning would not rule it out (Australian Parliament 2001, p.144).

Gene Therapy

Somatic

Prohibited in relation to the embryo or egg in the process of fertilization (HRTA s.7 (1)(e)). See Federal table for general position.

Germ-line

Prohibited in relation to the embryo or egg in the process of fertilization (HRTA s.7 (1)(e)). See Federal table for general position.

Pre-Implantation Genetic Diagnosis

Permitted (HRTA s.23(a) (ii)).

Stem Cell Therapy

See Federal table.

Pre-Natal Diagnosis/Abortion

Abortion permitted:

- Up to 20 weeks if –
  (a) the woman concerned has given informed consent; or
  (b) the woman concerned will suffer serious personal, family or social consequences if the abortion is not performed; or
  (c) serious danger to the physical or mental health of the woman concerned will result if the abortion is not performed; or

(d) the pregnancy of the woman concerned is causing serious danger to her physical or mental health.

(e) (b), (c) or (d) do not apply unless the woman has given informed consent or in the case of paragraphs (c) or (d) it is impracticable for her to do so. (HA s.334 (3) & (4)).

- After 20 weeks if –

(a) 2 medical practitioners who are specially appointed for the purposes of this section agree that the mother, or the unborn child, has a severe medical condition that, in the clinical judgement of those 2 medical practitioners, justifies the procedure; and

(b) the abortion is performed in an approved facility. (HA s.334 (7)).

Section 334 justifies abortion which would otherwise be unlawful under the CC s 199(1).

AUSTRIA

SOURCES OF LEGISLATION

Acts of the Legislature

- Criminal Code[116] of 23 January 1974[117] in the version of 1 July 2000[118]
- Reproductive Medicine Act (FMedG)[119] of 4 June 1992[120]
- Genetic Engineering Act (GTG) of 1994[121]

International Conventions

- Universal Declaration on the Human Genome and Human Rights (UDHGHR)[122] Recommendations
- Austrian Bioethics Committee on financing of stem cell research within the 6th European research programme framework, 8 May 2002[123]

---

[116] Available in German at http://www.sbg.ac.at.ssk/docs/stgb/stgb_index.htm. An English translation of these sections can be found at http://cyber.law.Harvard.edu/population/abortion/Austria.abo.htm.

[117] Bundesgesetzblatt *(Federal Law Gazette)* (BGBl.) 1974/60, available in English at http://cyber.law.harvard.edu/population/abortion/Austria.abo.htm.

[118] BGBl. I 62/2002.

[119] Available in German at http://www.ris.bka.gv.at/bgbl/.

[120] BGBl. No. 275/1992 STO 175, in force since 1 July 1992.

[121] BGBl. No. 510/1994, with the amendments made by 73rd Federal Act in the version of the 94th Federal Act, BGBl. I Nr. 94/2002 issued on 25 June 2002; Available in German at http://www.bmbwk.gv.at/.

[122] Adopted unanimously and by acclamation at the 29th Session of UNESCO's General Conference, 11 November 1997 (29 C/Resolution 17 – see http://www.unesco.org/ibc/) and endorsed by the United Nations' General Assembly, 85th Plenary Meeting on 9 December 1998 (A/RES/53/152 – see http://www.un.org.documents/).

[123] European Group on Ethics in Science and New Technologies to the European Commission, Brigitte Gratton, National regulations in the European Union regarding research on human embryos, July 2002, p. 7. Available at. http://europa.eu.int. – see also www.bka.gv.at/bka/bioethik.

- Austrian Bioethics Committee on reproductive cloning, protection of and research on embryos, pre-natal diagnosis and further questions of reproductive medicine, 12 February 2003

OVERVIEW OF REGULATIONS

Embryo Research

Definition of "embryo"

There is no definition of "embryo". Fertilised egg cells and cells developed from them are to be regarded as capable of development (Art. 1 section 1 para. 3 FMedG).

Research

Research is not permitted. Cells capable of development may only be examined and treated insofar as this is necessary to effect a pregnancy (FMedG, section 9 para. 1).[124] Research for other purposes than the embryo's own benefit is therefore inadmissible. This follows from the FMedG, section 17 para. 2 as well. Pursuant to this provision, sperms, egg cells and cells capable of development must not be returned to the persons from whom they originate and not be handed over to other persons or institutions.

Sources of embryos

See *Research*. Pursuant to the FMedG, section 10, a fertilisation outside of the body of a woman is only allowed to cover as many egg cells as are necessary according to the state of medical science for a promising and reasonable medically supported reproduction within one cycle. Nevertheless surplus embryos exist.[125]

Time limit for research

See *Research*.

Other conditions governing research

Consent procedures

See *Research*.

Licenses

See *Research*.

Ethics Committees

See *Research*.

Substantive (Must the embryo benefit?)

See *Research*.

Storage/Disposal of embryos

Storage permitted for one year (section 17 para. 1 FMedG).

---

[124] Thus examinations of "cells not capable of development" are not prohibited explicitly, see: Deutscher Bundestag, Schlußbericht der Enquete-Kommission Recht und Ethik in der modernen Medizin, 2002, p. 201.

[125] Taupitz, Rechtliche Regelung der Embryonenforschung im internationalen Vergleich, Berlin, Heidelberg, New York, 2003, p. 155.

Cloning

Reproductive

The law does not explicitly refer to cloning. FMedG, section 1 paras 1 & 2 define methods of medically supported reproduction. If this list is exhaustive, reproductive cloning is impermissible because the respective technique (transfer of the nucleus of a cell) is not mentioned. Others argue, that section 9 para. 1 FMedG prohibits cloning, because cells capable to develop are only allowed to be used for medical assisted procreation. Another opinion sees cloning prohibited by section 9 para. 2 FMedG, because every intervention into the human germ line is prohibited.[126]

Non-reproductive

Therapeutic cloning is impermissible.[127] (Compare FMedG, section 9, since sperm and egg cells may not be used for any purposes other than to effect a pregnancy.)

Gene Therapy

Somatic

Somatic gene therapy may only be used

- for the purpose of therapy or to prevent severe diseases (GTG section 74 no. 1); and

- provided that modification of the genetic information of the germ-line can be precluded according to the state of science and technology. If this risk cannot be precluded absolutely the anticipated advantage of somatic gene therapy for the health of the human being must outweigh this risk and somatic gene therapy may only be applied on human beings who will definitely not have offspring.

- Germ-line cells of a person who has been treated by somatic gene therapy must not be used to produce embryos outside of the body of a woman (GTG, section 74 no. 2).[128]

Germ-line

Interference with the germ-line is prohibited pursuant to the FMedG, section 9 para. 2.[129]

---

[126] Ibid., p. 159.

[127] Zweiter Zwischenbericht der Enquete-Kommission Recht und Ethik in der modernen Medizin – Teilbericht Stammzellforschung, BT-Drs. 14/7546, p. 129, available in German at http://www.bundestag.de/gremien/medi/2zwischen.pdf; Kopetzki argues that the FMedG does not prohibit therapeutic cloning, because "cells capable to develop" are not those, which are created by transfer of the nucleus, see Schulz, Schleichende Harmonisierung der Stammzellforschung in Europa, ZRP 2001, 526, 528 and European Group on Ethics in Science and New Technologies to the European Commission, Brigitte Gratton, National regulations in the European Union regarding research on human embryos, July 2002, p. 6. Available at. http://europa.eu.int.

[128] Pursuant to the GTG, somatic gene therapy is only allowed to be carried out by a doctor at an institution which has been admitted by the Federal Minister of Health, Sports and Consumer Protection after hearing the competent scientific committee of the Commission on Genetic Engineering. The Medicines Act (AMG), section 76, applies to clinical attempts.

[129] The GTG, section 64, repeats this prohibition.

Pre-Implantation Genetic Diagnosis

By the wording of the FMedG, section 9 para.1, this has to be considered as inadmissible because the provision only permits examinations that are necessary to effect a pregnancy.

Stem Cell Therapy

Research on stem cells is legal.[130] There is no information on the current state of research.

Pre-Natal Diagnosis/Abortion

- Pre-natal diagnosis is legal. Pursuant to the GTG, section 65 para. 3, a genetic examination is permitted if it is medically advisable. Before the test is carried out, the pregnant woman has to certify in writing that she has been informed by an expert on human genetics on the nature, consequences and implications of genetic analysis and the risks of the operation, and must have consented to the genetic analysis.

- Termination of pregnancy within the first 3 months is not punishable if it is carried out by a doctor after consulting a doctor (StGB [Criminal Code], section 97, para. 1 no. 1). Termination is not punishable if it is necessary in order to avert a serious threat that cannot be averted or to prevent severe damage to the health of the pregnant woman or if a threat exists for the child to be severely mentally or physical handicapped or the pregnant woman was of unsound mind at the time of the fathering. In all these cases the termination of the pregnancy has to be carried out by a doctor (StGB, section 97 para. 1 no. 2). Without time limit, the pregnancy may also be terminated if it threatens the life of the pregnant woman and medical assistance could not be obtained in time (StGB, section 97 para. 1 no. 3).

SPECIAL REMARKS

The FMedG shall be amended in 2003 by an prohibition of reproductive cloning.[131]

FURTHER READING

Zacherl, Legal Aspects of Gene Therapy – The Austrian Model of Regulation, in: Müller/Simon/Vesting (editor), Interdisciplinary Approaches to Gene Therapy, Legal Ethic and Scientific Aspects, Berlin, Heidelberg, New York, 1997, p 117

Körtner, Ulrich H. J. Forschung an embryonalen Stammzellen – Zur Diskussion und Gesetzeslage in Österreich, http://www.bka.gv.at/bka/bioethik/beitrag_koertner3.pdf.

Bernat, Erwin, (edit.) Fortpflanzungsmedizin, Wertung und Gesetzgebung, Beiträge zum Entwurf eines Fortpflanzungshilfegesetzes, Wien 1991

---

[130] Zweiter Zwischenbericht der Enquete-Kommission Recht und Ethik in der modernen Medizin – Teilbericht Stammzellforschung, BT-Drs. 14/7546, p. 129, available in German at http://www.bundestag.de/gremien/medi/2zwischen.pdf.

[131] European Group on Ethics in Science and New Technologies to the European Commission, Brigitte Gratton, National regulations in the European Union regarding research on human embryos, July 2002, p. 6. Available at. http://europa.eu.int.

Bernat, Erwin, (edit.) Die Reproduktionsmedizin am Prüfstand von Recht und Ethik, Wien, 2000.

Belarus

SOURCES OF LEGISLATION

Acts of Legislation

* Criminal Code from 9 July 1999, Act No. 275-3[132]
* Health Protection Act (GSG) from 18 June 1993 as promulgated on 6 December 2001[133]

Official Regulations

* No information obtained.

Official Guidelines

* No information obtained.

International Conventions

* Universal Declaration on the Human Genome and Human Rights (UDHGHR)[134]

Recommendations

* No information obtained.

OVERVIEW OF REGULATIONS

Embryo Research

Definition of "embryo"

No information obtained.

Research

No information obtained.

Sources of embryos

No information obtained.

Time limit for research

No information obtained.

Other conditions governing research

No information obtained.

---

[132] In force since 1 January 2001, published in the National Register of Laws of the Republic of Belarus, 1999, No. 762/50.

[133] In force since 11 January 2002 (through Act No. 91-3), available in Russian at http://ncpi.gov.by.

[134] Adopted unanimously and by acclamation at the 29th Session of UNESCO's General Conference, 11 November 1997 (29 C/Resolution 17 – see http:www.unesco.org/ibc/), and endorsed by the United Nations' General Assembly, 85th Plenary Meeting on 9 December 1998 (A/RES/53/152 – see http://www.un.org.documents/).

Consent procedures

No information obtained.

Licenses

No information obtained.

Ethics Committees

No information obtained.

Substantive (Must the embryo benefit?)

No information obtained.

Storage/Disposal of embryos

No information obtained.

Cloning

Reproductive

No information obtained.

Non-reproductive

No information obtained.

Gene Therapy

No information obtained.

Somatic

No information obtained.

Germ-line

No information obtained.

Pre-Implantation Genetic Diagnosis

Pursuant to the GSG, Art. 41, medical genetic examinations are permitted in order to prevent hereditary diseases.

Stem Cell Therapy

No information obtained.

Pre-Natal Diagnosis/Abortion

- A doctor with higher medical education in this special area who unlawfully terminates a pregnancy commits an offence and will be punished by fine or barred from certain professional activities or functions (Criminal Code, Art. 156 para. 1).
- A person without higher medical education in this special area who unlawfully terminates a pregnancy commits an offence and will be punished by fine or up to 2 years' imprisonment or barred from activities as in paragraph one (Criminal Code, Art. 156 para. 2).
- The termination of pregnancy is permitted upon request of the pregnant woman up until the 12th week of the pregnancy. It may be carried out until the 22nd week of the pregnancy if there is a medical or social emergency (GSG, Art. 35).

SPECIAL REMARKS

The GSG regulates the conditions for examinations on human beings in Art. 31, artificial fertilisation in Art. 33 and the conditions for genetic examinations in Art. 41.

Belgium

SOURCES OF LEGISLATION

Acts of Legislation

- Code Pénal (Criminal Code) from 8 June 1867, consolidated version from 7 May 2002[135]
- Royal Decree from 7 May 2000[136]
- Decree from 9 December 1993[137]
- Law on Research on Embryos (LRE) in vitro from 3 April 2003[138]

Official Regulations

- No information obtained

Official Guidelines

- No information obtained

International Conventions

- Universal Declaration on the Human Genome and Human Rights (UDHGHR)[139]

Recommendations

- Advice Note No. 2 of the Comité consultatif de Bioethique de Belgique[140] 7July 1997on the Council of Europe's Convention on Human Rights and Biomedicine
- Advice Note No. 3 of the Comité consultatif de Bioethique de Belgique 17. November 1997 on the choice of sex

---

[135] Dossier-No. 1867-06-08/01 in the version from 7 May 2002; available in French at http://www.cass.be/cgi_loi/legislation.pl.

[136] Doss.-No- 2000-05-07/31, in force since 26 May 2000; available in French at http://www.cass.be/cgi_loi/legislation.pl.

[137] Simon; Survey about Regulation on Gene therapy in Some European countries in: Mueller/Simon/Vesting (editor.), Interdisciplinary Approaches to Gene Therapy, Legal Ethic and Scientific Aspects, Berlin, Heidelberg, New York, 1997, 130 & 136.

[138] This law has to be santioned by Royal Approval, see Doc 50-2182/009 Doc. 50-0429/001; Documents législatifs nos. 2-87/1, 2-92/1, 2-114/1, 2-321/1, 2-686/1, 2-695/1, 2-716/1, 2-726/1: Consultations on all bills were held on 10 June 2002 and passed on to the Bioethics Commission; available in French at http://www.Senate.be.

[139] Adopted unanimously and by acclamation at the 29th session of UNESCO's General Conference, 11 November 1997 (29 C/Resolution 17 - see http://www.unesco.org/ibc/), and endorsed by the United Nations' General Assembly, 85th Plenary Meeting at 9 December 1998 (A/RES/53/152 – see http://www.un.org.documents/).

[140] All available in French at http://www.health.fgov.be/bioeht/fr/avis/avis-index.htm.

- Advice Note No. 6 of the Comité consultatif de Bioethique de Belgique 8 June 1998 on the ethic bases for the improvement of supply and the conditions of the Centres for In Vitro Fertilisation[141]
- Advice Note No. 10 of the Comité consultatif de Bioethique de Belgique 14 June 1999 on the reproductive cloning of human beings
- Advice Note No. 18 of the Comité consultatif de Bioethique de Belgique 16. September 2002 on research on embryos in vitro
- Advice Note No. 19 of the Comité consultatif de Bioethique de Belgique 14. October 2002 on the destination of cryo-preserved embryos

OVERVIEW OF REGULATIONS

Embryo Research

Definition of "embryo"

A cell or an organic group of cells able and developing to a human being (Art. 2 para. 1 LRE).

Research

Permitted under conditions (Art. 3 LRE).

Sources of embryos

Creation of embryos for research in vitro is prohibited, except in cases that surplus embryos cannot be used (Art. 4 para. 1 LRE).

Time limit for research

Research on embryos is prohibited after 14 days (Art. 3 para. 5 LRE).

Other conditions governing research

Consent Procedures

Concerned persons must give written consent to the use of gametes or embryos for research (Art. 8 LRE).

Licenses

Research is only permitted in laboratories linked to an university programme of reproductive medicine or human genetics.

Non-university units need a special accreditation (Art. 3 para. 3 LRE).

Ethics committees

Every research has to be permitted by the local ethics committee and afterwards by the Federal Ethics Committee (Art. 7 section 1, Art. 9 LRE).

Substantive (Must the embryo benefit?)

No (Art. 3 para. 1, 2 LRE).

Cloning

Reproductive

---

[141] In this Advice Note the Comité Consultatif de Bioethique de Belgique explicitly only deals with matters of organisation regarding in vitro fertilisation.

Reproductive cloning is prohibited (Art. 6 LRE)

Non-reproductive

The law does not contain neither a definition nor a regulation of non therapeutic cloning (see Art. 2 LRE).[142]

Gene Therapy

Somatic

Permitted by a 1993 decree. [143] It takes the form of experimental treatments which must be approved by the Council for Biological Safety.

Pre-Implantation Genetic Diagnosis

Permitted (Art. 5 para 2 LRE).

Pre-Natal Diagnosis/Abortion

- Pre-natal diagnosis is unregulated.

- A term of imprisonment of up to three months or a fine shall be imposed on any person who terminates a pregnancy with the consent of the pregnant woman. (Code Pénal, Art. 350 para. 1).[144]

- Abortion is permitted if the pregnant woman is in an emergency situation and if the pregnancy is terminated by a doctor under the following conditions:

  (a) the pregnancy has to be terminated before the end of the 12th week (Code Pénal, Art. 350 para. 2 no. 1a). [145]

  (b) it is carried out professionally by a doctor in an institution where the woman can receive the necessary information on the risks of the operation and on her rights regarding assistance for herself, her children and her family, as well as on the possibility of an adoption and which can render assistance in addressing the psychological and social problems of the pregnant woman. (Code Pénal, section 350, para. 2 no 1b). The operation may not take place before six days after the first consultation. (Code Pénal, section 350 para. 2 no. 3).

- The pregnancy may be terminated at any time after the 12th week where it poses a danger to the health of the woman or it is certain that the child would suffer from a particularly severe incurable disease and a second doctor has approved the termination (Code Pénal, section 350 para. 2 no. 4).[146]

---

[142] Therefore by using the *a contrario* argument therapeutic cloning is permitted, see: European Group on Ethics in Science and New Technologies to the European Commission, Brigitte Gratton, National regulations in the European Union regarding research on human embryos, July 2002, p. 57. Available at. http://europa.eu.int.

[143] For details see Simon; Survey about Regulation on Gene therapy in some European countries in: Mueller/Simon/Vesting (editor.), Interdisciplinary Approaches to Gene Therapy, Legal Ethic and Scientific Aspects, Berlin, Heidelberg, New York, 1997, 130 & 136.

[144] Inserted into the Code Pénal by the Act L 1990-04-03/30, Art. 2, 002, in force since 15 April 1990.

[145] There are no precedents regarding this issue yet. As regards the discussion at the Belgian Parliament see Pattinson 2002, Appendix 1.

[146] For an English translation see International Digest of Health Legislation 1990.

SPECIAL REMARKS

- The draft laws point out the lack of legal regulations in their statements of reasons.[147]
- In recent years several bills on embryo protection (in vitro) were introduced.[148]

FURTHER READING

On the termination of pregnancy: Rapport á l'attention du Parlement (1 January to 31 December 1999)[149]

Bulgaria

SOURCES OF REGULATIONS

Acts of the Legislature

- Decree No. 2 of 1 February 1990 (Darzaven Vestnik, 9 February 1990, No. 12, 4–9.) (D2).[150]

Official Regulations

- No information obtained

Official Guidelines

- No information obtained

International Conventions

- Convention on Human Rights and Biomedicine ETS 164, Oviedo, 4.IV.1997 (CHRB) (Signed 31/05/01)[151]
- Universal Declaration on the Human Genome and Human Rights. (UDHGHR)[152]

Recommendations

- No information obtained.

OVERVIEW OF REGULATIONS

Embryo Research

Definition of "embryo"

None.[153]

Research

Not regulated.[154]

---

[147] Ingrid van Kessel et consorts, Document legislatif No. 2-114/1, S. 1; accordingly Philippe Monfils, Document legislatif No.2-87/1, S. 2.
[148] Doss.-No- 2000-05-07/31, in force since 26 May 2000; available in French at *http://www.cass.be/cgi_loi/legislation.pl.*
[149] Available in French at *http://www.lachambre.be/documents/860/1.pdf.*
[150] Available in *International Digest of Health Legislation,* 1990, 41 (4).
[151] See http://conventions.coe.int/Treaty/EN/cadreprincipal.htm.
[152] Adopted by consensus, UNESCO General Conference, 11 November, 1997.
[153] Council of Europe 1998, 135.
[154] Ibid., 144 and 173.

Sources of embryos

Not regulated.[155]

Time limit for research

Not regulated.[156]

Other conditions governing research

Consent procedures

See *Research*.

Licenses

See *Research*.

Ethics Committees

None.[157]

Substantive (Must the embryo benefit?)

No.[158]

Storage/Disposal of embryos

From information available, uncertain if there is regulation.[159]

Cloning

Reproductive

Not regulated.[160] Bulgaria has not signed the Additional Protocol to the Convention on Human Rights and Biomedicine banning reproductive cloning.[161]

Non-reproductive

Not regulated.[162]

Gene Therapy

Somatic

As germ-line therapy is unregulated it is likely somatic is as well. See *next*.

Germ-line

Not regulated.[163]

Pre-Implantation Genetic Diagnosis

Not regulated.[164]

---

[155] Ibid., 182.
[156] Ibid., 180.
[157] Council of Europe 1998, 149.
[158] Ibid., 160.
[159] Ibid., 89.
[160] Ibid., 214.
[161] See http://conventions.coe.int.
[162] Council of Europe 1998, 219.
[163] Ibid., 186.
[164] Ibid., 188.

Stem Cell Therapy

No information obtained.

Pre-Natal Diagnosis/Abortion

Abortion Permitted:

- Up to twelve weeks on request if –
  - (i) the woman is not suffering from any disease (listed in Annex 1) whereby termination could endanger her life or health (D2 s.7); and
  - (ii) it is performed in an approved clinics or hospitals (D2 s.10) by an obstetrician/gynaecologist;
- Up to 20 weeks on request if the woman is suffering from a proven disease (listed in Annex 2) that could later in pregnancy or during childbirth endanger the life or health of the mother. If the disease is not listed in Annex 2 an abortion may nevertheless be permitted on an exceptional basis. (D2 s.12).
- After 20 weeks if the woman's life is in danger, or there is evidence of a serious malformation in or genetic damage to the fetus. (D2 s.12).
- Abortions on medical grounds are to be performed by an obstetrician/gynaecologist and a medical specialist in resuscitation, in the presence of if necessary, a physician specializing in the disease that rendered the abortion necessary. (D2 s.16).

Canada

SOURCES OF REGULATIONS

Acts of the Legislature

  – Proposed Legislation –

- Assisted Human Reproduction Act (PLBill C-56[165])

Official Regulations

- None.

Official Guidelines

- None.

International Conventions

- Universal Declaration on the Human Genome and Human Rights. (UDHGHR)[166]

Recommendations

- Canadian Institutes of Health Research (CIHR) Discussion Paper- Human Stem Cell Research: Opportunities for Health and Ethical Perspectives 2001 [167] (CIHR)

---

[165] Available at http://www.parl.gc.ca/PDF/37/1/parlbus/chambus/house/bills/government/C-56_1.pdf. The first reading of the Bill occurred on May 9, 2002.

[166] Adopted by consensus, UNESCO General Conference, 11 November, 1997.

[167] Available at http://www.cihr-irsc.gc.ca/publications/ethics/stem_cell/stem_cell_e.pdf.

OVERVIEW OF REGULATIONS

"Canada currently has no binding guidelines to regulate reproductive and genetic technologies." [168]

Embryo Research

Definition of "embryo"

"A human organism during the first 56 days of its development following fertilization or creation, excluding any time in which its development has been suspended and includes any cell derived from such an organism that is used for the purpose of creating a human being." (PLBill C-56 s.3).

Research where fertilization occurs should be regarded as research on embryos. (TPS s.9(B)). Note that the PLBill C-56 s.3 includes embryos created other than by fertilization.

Research

Permitted with conditions.

Sources of embryos

Creating an embryo solely for the purpose of research permitted, provided the research is to improve or provide instruction on assisted human reproduction procedures (PLBill C-56 s.5(1)(d)). Note however, the TPS has recommended that the creation of embryos for research be prohibited (TPS art. 9.4).

Time limit for research

14 days of development following fertilization[169] or creation, excluding any time when development is suspended (PLBill C-56 s.5(1)(d)).

Other conditions governing research

Consent procedures

Specific written consent of donor to research required in accordance with regulations (PLBill C-56 s.8(3)). Free and informed consent of donors required (TPS Arts.9.4.a & 9.1)

Licenses

Required to obtain, handle, and conduct research on embryos (PLBill C-56 ss.10(2) & (3) and 40 (1) & (2)). Licenses are to be issued by The Assisted Human Reproduction Agency of Canada (AHRAC) to be established under PLBill C-56 s.21.

Ethics Committees

Part of the responsibility of the AHRAC is to foster the application of ethical principles to research (PLBill C-56 s.22(b)).

Substantive (Must the embryo benefit?)

No. However research must be to improve or provide instruction on assisted human reproduction procedures (PLBill C-56 s.5(1)(d)).

Storage/Disposal of embryos

---

[168] See Shanner 2001, 7.
[169] The TPS art. 9.4(d) recommended the limitation of research to 14 days, but only contemplated embryos created by fertilization.

Embryos must be destroyed at the request of the donor[170] in the manner prescribed by the regulations (PLBill C-56 ss.16(3) & 65(1) (r)). The AHRAC may not dispose of embryos without the consent of the donor or of the responsible person, as defined by the regulations. If the donor does not give consent to disposal, the embryo must be returned to the donor (PLBill C-56 ss.54 (2) & (3) and 65(1) (x)).

Cloning

A voluntary moratorium on the cloning of human embryos has existed since July 1995[171] but several of its provisions have been disregarded.[172]

Reproductive

Prohibited (PLBill C-56 s.5(1)(a), TPS arts. 9.3 & 9.5 and UDHGHR Art.11).

Non-reproductive

Prohibited. (PLBill C-56 ss.3 & 5(1)(c)) / (TPS Art. 9.5).

Gene Therapy

Somatic

Permitted by implication (PLBill C-56 s.5(1)(f)).

Germ-line

Prohibited (PLBill C-56 s.5(1)(f)) / (TPS Art. 8.5).

Pre-Implantation Genetic Diagnosis

Permitted to prevent, diagnose or treat sex-linked diseases or disorders (PLBill C-56 s.5(1)(e)).

Stem Cell Therapy

Stem cell research is ongoing and unregulated (CIHR 2001, 7-10).

Pre-Natal Diagnosis/Abortion

- Abortion unregulated. In 1988 the Supreme Court declared s.251 of the Federal Code under which abortion was regulated by hospital abortion committees unconstitutional, on the grounds that it violated the security and liberty of the pregnant woman guaranteed by Article 7 of the Canadian Charter of Rights and Freedoms.[173] The Supreme Court has not yet decided whether the foetus enjoys protection under the Article 7 of the Charter.[174]

SPECIAL REMARKS

There have been two previous abandoned bills: Bill C-47 "Human Reproductive and Genetic Technologies Act" 1996 and a private members bill, Bill C-247 "An Act to amend the Criminal Code (Genetic Manipulation)" 1997.

---

[170] The regulations are going to define "donor" (PLBill C-56 s. 65(1)(a).

[171] See Shanner 2001, 7 & 44.

[172] See Pattinson 2002, Appendix 4 and Shanner 2001, 7.

[173] R v Morgentaler [1988] 1 S.C.R. 30.

[174] See *Borowski v Canada (Attorney-General)* [1989] 1 SCR 342, where the Court held that the plaintiff lacked standing, and *Tremblay v Daigle* [1989] 2 SCR 530, where the Court held that the Charter was intended to limit government action and did not apply to disputes between private individuals, both highlighted in Pattinson 2002, Appendix1.

FURTHER READING

Shanner, *Laura 2001*. "Embryonic stem cell research: Canadian policy and ethical considerations a report for Health Canada, policy division". *Available at* http://www.msu.edu/~hlnelson/fab/Shanner_Stem_Cell_Policy.pdf

China (Hong Kong Special Administrative Region)

SOURCES OF REGULATIONS

Acts of the Legislature

- Human Reproductive Technology Ordinance (HRTO)
- Offences Against the Person Ordinance (OPO)

Official Regulations

- No information obtained

Official Guidelines

- Council on Human Reproductive Technology Code of Practice (CP)

International Conventions

- Universal Declaration on the Human Genome and Human Rights. (UDHGHR)[175]

Recommendations

- No information obtained

OVERVIEW OF REGULATIONS

Embryo Research

Definition of "embryo"

"Except where otherwise stated –

i)   embryo means a live human embryo where fertilization is complete; and

ii) reference to an embryo include an egg in the process of fertilization,

and for this purpose, fertilization is not complete until the appearance of a 2 cell zygote." (HRTO s.2(7)).

"'embryo research' means any research involving the creation, use or manipulation of an embryo, whether or not the embryo is to be implanted in the body of a woman" and also any procedure specified as such under s.2(a)(i)." (HRTO s.2 (1))

Research

Permitted with conditions.

Sources of embryos

Supernumerary from IVF. Embryos cannot be created for research (HRTO s.15(1)(a)).

Time limit for research

---

[175] Adopted by consensus, UNESCO General Conference, 11 November, 1997.

Up to the appearance of the primitive streak, deemed to be not later than 14 days after gametes have been mixed, excluding any storage period (HTRO s.15(1)(b) & (4).

Other conditions governing research

Consent procedures

No information

Licenses

Required (HRTO s.2(1) & 21).

Ethics Committees

No information

Substantive (Must the embryo benefit?)

No Information

Storage/Disposal of embryos

No information

Cloning

Reproductive

Not permitted if it involves CNR of embryo cell or cloning of an embryo (HRTO s.15(1)(e) & (f)).

Non-reproductive

Not permitted if it involves CNR of embryo cell or cloning of an embryo (HRTO s.15(1)(e) & (f)).

Gene Therapy

Somatic

No information

Germ-line

No information

Pre-Implantation Genetic Diagnosis

Sex selection prohibited except to prevent any of the 73 sex-linked diseases listed in Schedule 2. Two registered medical practitioners must confirm the diagnosis in writing (HRTO s.15 (3)(a) & (b)).

Stem Cell Therapy

Research proceeding on the Chinese mainland.[176] No information on Hong Kong.

Pre-Natal Diagnosis/Abortion

Abortion permitted:

a) up to 24 weeks where two registered medical practitioners concur that –

---

[176] See http://www.newscientist.com/hottopics/cloning/cloning.jsp?id=ns99992012.

i) the continuance of the pregnancy would involve risk of injury to the physical or mental health of the pregnant woman greater than if the pregnancy were terminated; or

ii) there is a substantial risk that if the child were born it would suffer from serious physical or mental handicap; or

iii) the woman is with child before attaining the age of 16; or

iv) the woman is the victim of incest, or intercourse by force or fraud, and she made a report within 3 months of the date of the offence.

b) up to birth where it is absolutely necessary to save the pregnant woman's life. The termination must be performed by registered doctors in government specified hospitals or at the Family Planning Association of Hong Kong (OPO s.47).[177] Illegal abortion is a criminal offence.[178]

SPECIAL REMARKS

1. Access to reproductive technologies limited to married persons or to persons who were married to each other when the treatment started, or to a person carrying out a surrogacy arrangement (HRTO ss.14 (a) ; 15 (5) & (6).

2. Breach of the CP not a criminal offence or civil wrong, but may result in variation, revocation or failure to obtain renewal, of a licence.

Czech Republic

SOURCES OF REGULATIONS

Acts of the Legislature

- Bill of Fundamental Rights and Freedoms 1992 (BFRF)
- Act No 66/1986 Coll., on artificial interruption of pregnancy as amended (AIP)[179]

Official Regulations

- Notice of the Ministry of Health No. 75/1986 Coll. (N75)

Official Guidelines

- None.

International Conventions

- Convention on Human Rights and Biomedicine ETS 164, Oviedo, 4.IV.1997 (CHRB) (Signed 24/06/98, Ratified 22/06/01, Entered into force 01/10/01)[180]
- Additional Protocol to the Convention on Biomedicine concerning the prohibition of cloning ETS 168, Paris, 12.I.1998 (APC) (Signed 24/06/98, Ratified 22/06/01, Entered into force 01/10/01)[181]

---

[177] See the Family Planning Association of Hong Kong Website at http://www.famplan.org. hk/en/sexual/what.asp.
[178] Ibid.
[179] Available at http://cyber.law.harvard.edu/population/abortion/Czech.abo.html.
[180] See http://conventions.coe.int/Treaty/EN/cadreprincipal.htm.
[181] Ibid.

- Universal Declaration on the Human Genome and Human Rights. (UDHGHR)[182]

Recommendations

OVERVIEW OF REGULATIONS[183]

Embryo Research

Definition of "embryo"

None.

Research

May be permitted by default. However this is uncertain as BFRF Art. 6(1) provides, "Everybody has the right to life. Human life is worthy of protection even before birth" and Art. 6(2) states, "Nobody may be deprived of life."

Sources of embryos

Supernumerary from IVF. Embryos may not be created for research (CHRB, Art. 18(1)).

Time limit for research

Not regulated

Other conditions governing research

Consent procedures

No information.

Licenses

No information.

Ethics Committees

No information.

Substantive (Must the embryo benefit?)

No information.

Storage/Disposal of embryos

Not regulated

Cloning

Reproductive

Prohibited. (APC Art 1 & UDHGHR Art.11).

Non-reproductive

No information

Gene Therapy

Somatic

Permitted. (CHRB Art. 13).

---

[182] Adopted by consensus, UNESCO General Conference, 11 November, 1997.
[183] Information on the state of regulation provided by Lukas Prudil.

Germ-line

Prohibited. (CHRB Art. 13).

Pre-Implantation Genetic Diagnosis

Permitted only for health purposes subject to genetic counselling (CHRB Art. 12). Sex selection only permitted to avoid serious hereditary sex-linked disease (CHRB Art. 14)

Stem Cell Therapy

Not regulated. Uncertain if research is proceeding.

Pre-Natal Diagnosis/Abortion

- Abortion Permitted:
  a) up to 12 weeks if the woman makes a written request to this effect, and there are no contraindications on health grounds (AIP s.4).
  b) up to birth with the woman's consent, or at her instigation, if her life or health or the healthy development of the foetus are endangered, or if foetal development manifests genetic anomalies (AIP s.5).

    If woman under 16, abortion in accordance with Section 4 may be performed with the consent of her legal representative or of the person who has been assigned responsibility for bringing her up (AIP s.6(1)).

    If abortion in accordance with Section 4 has been performed on a woman between 16 and 18 her legal representative must be notified (AIP s.6(2)).

    If a woman is refused an abortion she is entitled to have that decision reviewed twice at successively higher levels of the health establishment after which the decision is final (AIP s.8).

    Health grounds against abortion outlined in N75. Abortion prohibited if there was a previous termination within previous 6 months (N75 § 1). Notice also stipulates a list of illnesses, syndromes and states of health which are grounds for termination. Abortions have been decreasing in recent years.[184]

FURTHER READING

Exter den, A. P. and Prudil, L. 2001. "The Czech Republic" In H Nys, ed., International Encyclopaedia of Laws, The Hague: Kluwer Law International.

Prudil, L.: Patients' Rights in the Czech Republic in the Last Decade (in press)

Cyprus

SOURCES OF LEGISLATION

Acts of the Legislature

- Criminal Code
- Medical Registration Act

---

[184] Exter and Prudil 2001, 57.

Official Regulations
- No information obtained

Official Guidelines
- No information obtained

International Conventions
- Convention on Human Rights and Biomedicine ETS Oviedo, 4 April 1997 (CHRB) (signed on 30 September 1998/ ratified on 20 March 2002/ in force since 1 July 2002)[185]
- Additional Protocol to the Convention on Biomedicine Concerning the Prohibition of Cloning ETS 168, Paris, 12 January 1998 (APC) (signed on 30 September 1998/ ratified on 20 March 2002/ in force since 1 July 2002)[186]
- Universal Declaration on the Human Genome and Human Rights (UDHGHR)[187]

Recommendations
- No information obtained

OVERVIEW OF REGULATIONS

Embryo Research

Definition of „embryo"

There is no legislation defining the embryo.[188]

Research

There is no legislation on embryo research.[189]

Sources of embryos

No information obtained.

Time limit for research

No information obtained.

Other conditions governing research

Consent procedures

No information obtained.

---

185 See http://conventions.coe.int.
186 See http://conventions.coe.int.
187 Adopted unanimously and by acclamation at the 29th Session of UNESCO's General Conference, 11 November 1997 (29 C/Resolution 17 – see http:www.unesco.org/ibc/), and endorsed by the United Nations' General Assembly, 85th Plenary Meeting on 9 December 1998 (A/RES/53/152 – see http://www.un.org.documents/).
188 Council of Europe: CLONING – COMPARATIVE STUDY ON THE SITUATION IN 44 STATES, Strasbourg, 4 June 1998, p. 135, available in English at http://legal.coe.int/bioethics/gb/pdf/txtprep2b.pdf.M.
189 Council of Europe: CLONING – COMPARATIVE STUDY ON THE SITUATION IN 44 STATES, Strasbourg, 4 June 1998, 160, 165, & 173, available in English at http://legal.coe.int/bioethics/gb/pdf/txtprep2b.pdf.M.

Licenses

No information obtained.

Ethics Committees

No information obtained.

Substantive (Must the embryo benefit?)

No information obtained.

Storage/Disposal of embryos

No information obtained.

Cloning

Reproductive

Prohibited. (APC Art 1 & UDHGHR Art.11).

Non-reproductive

No information obtained.

Gene Therapy

Somatic

Appears unregulated as germ-line therapy is unregulated as outlined in (b).

Germ-line

Therapeutic interference with the germ-cell line is not governed by law.[190]

Pre-Implantation Genetic Diagnosis

There is no legislation on pre-implantation diagnosis.

Stem Cell Therapy

No information obtained.

Pre-Natal Diagnosis/Abortion

PND:

There is no legislation regulating pre-natal diagnosis.[191]

Abortion:

- Every attempt to terminate a pregnancy – whether or not the woman is actually pregnant – carries a term of imprisonment of 7 years for all persons involved in the attempt including the woman who is alleged to be or is actually pregnant (Criminal Code, sections 167, 168).

- Aiding and abetting termination carries a term of imprisonment of 3 years (Criminal Code, section 169).

---

[190] Council of Europe: CLONING – COMPARATIVE STUDY ON THE SITUATION IN 44 STATES, Strasbourg, 4 June 1998, 169, available in English at http://legal.coe.int/bioethics/gb/pdf/txtprep2b.pdf.M.

[191] Council of Europe: CLONING – COMPARATIVE STUDY ON THE SITUATION IN 44 STATES, Strasbourg, 4 June 1998, 140, available in English at http://legal.coe.int/bioethics/gb/pdf/txtprep2b.pdf.M.

- Abortion is lawful if an approved doctor terminates the pregnancy under the following conditions:
  - the pregnancy has been caused by rape or other circumstances which would endanger the social status of the woman or her family if the pregnancy were not terminated (Criminal Code, section 169 A a); or
  - two approved doctors certify that the continuation of the pregnancy would endanger the life of the pregnant woman; or
  - if the continuation of the pregnancy would cause the woman or any existing children greater physical, mental or psychological harm than if the pregnancy were terminated; or (Criminal Code, section 169 A b)
  - if there is a considerable risk that the child would suffer from such severe physical or psychical abnormalities that it would be seriously handicapped (Criminal Code, section 169 A b).

Czech Republic

SOURCES OF REGULATIONS

Acts of the Legislature
- Bill of Fundamental Rights and Freedoms 1992 (BFRF)
- Act No 66/1986 Coll., on artificial interruption of pregnancy as amended (AIP)[192]

Official Regulations
- Notice of the Ministry of Health No. 75/1986 Coll. (N75)

Official Guidelines
- None

International Conventions
- Convention on Human Rights and Biomedicine ETS 164, Oviedo, 4.IV.1997 (CHRB) (Signed 24/06/98, Ratified 22/06/01, Entered into force 01/10/01)[193]
- Additional Protocol to the Convention on Biomedicine concerning the prohibition of cloning ETS 168, Paris, 12.I.1998 (APC) (Signed 24/06/98, Ratified 22/06/01, Entered into force 01/10/01)[194]
- Universal Declaration on the Human Genome and Human *Rights*. (UDHGHR)[195]

Recommendations
- None

---

[192] Available at http://cyber.law.harvard.edu/population/abortion/Czech.abo.html.
[193] See http://conventions.coe.int/Treaty/EN/cadreprincipal.htm.
[194] Ibid.
[195] Adopted by consensus, UNESCO General Conference, 11 November, 1997.

OVERVIEW OF REGULATIONS[196]

Embryo Research

Definition of "embryo"

None.

Research

May be permitted by default. However this is uncertain as BFRF art. 6(1) provides, "Everybody has the right to life. Human life is worthy of protection even before birth" and art 6(2) states, "Nobody may be deprived of life."

Sources of embryos

Supernumerary from IVF. Embryos may not be created for research (CHRB, art. 18(1)).

Time limit for research

Not regulated.

Other conditions governing research

Consent procedures

No information obtained.

Licenses

No information obtained.

Ethics Committees

No information obtained.

Substantive (Must the embryo benefit?)

No information obtained.

Storage/Disposal of embryos

Not regulated

Cloning

Reproductive

Prohibited. (APC Art. 1 & UDHGHR Art. 11).

Non-reproductive

No information obtained.

Gene Therapy

Somatic

Permitted. (CHRB Art. 13).

Germ-line

Prohibited. (CHRB Art. 13).

---

[196] Information on the state of regulation provided by Lukas Prudil.

Pre-Implantation Genetic Diagnosis

Permitted only for health purposes subject to genetic counselling (CHRB Art. 12). Sex selection only permitted to avoid serious hereditary sex-linked disease (CHRB Art. 14).

Stem Cell Therapy

Not regulated. Uncertain if research proceeding.

Pre-Natal Diagnosis/Abortion

* Abortion Permitted:
    a) up to 12 weeks if the woman makes a written request to this effect, and there are no contraindications on health grounds (AIP s.4).
    b) up to birth with the woman's consent, or at her instigation, if her life or health or the healthy development of the foetus are endangered, or if foetal development manifests genetic anomalies (AIP s.5).

    If woman under 16, abortion in accordance with Section 4 may be performed with the consent of her legal representative or of the person who has been assigned responsibility for bringing her up (AIP s.6(1)).

    If abortion in accordance with Section 4 has been performed on a woman between 16 and 18 her legal representative must be notified (AIP s.6(2)).

    If a woman is refused an abortion she is entitled to have that decision reviewed twice at successively higher levels of the health establishment after which the decision is final (AIP s.8).

    Health grounds against abortion outlined in N75. Abortion prohibited if there was a previous termination within previous 6 months (N75 § 1). Notice also stipulates a list of illnesses, syndromes and states of health which are grounds for termination. Abortions have been decreasing in recent years.[197]

FURTHER READING

* Exter den, A. P. and Prudil, L. 2001. "The Czech Republic" In H Nys, ed., International Encyclopaedia of Laws, The Hague: Kluwer Law International

* *Prudil, L.: Patients' Rights in the Czech Republic in the Last Decade (in press)*

Denmark

SOURCES OF LEGISLATION

Acts of the Legislature

* Act No. 350 from 13 June 1973 on the Termination of Pregnancy[198]

* Act No. 503 from 24 June 1992 on a Scientific Ethics Committee System and the Examination of Biomedical Research Projects[199]

---

[197] Exter and Prudil 2001, 57.
[198] Lovtidende for Kongerit Danmark, Part A, 6 July 1973, No. 32 993-995), available in English at http://cyber.law.Harvard.edu/population/abortio/Denmark.abo.htm.
[199] See http://firewall.unesco.org/opi/eng/bio98/.

- Act No. 460 on Artificial Fertilisation from 10 June 1997[200]
- Act No. 430 from 31 May 2000, amending the Act on the Termination of Pregnancy[201]
- Act No. 499 from 1996 concerning research in conjunction with PND[202]

Official Regulations

- Order No. 728 from 17 September 1997[203] on artificial fertilisation
- Order No.758 from 30 September 1997[204] on the reporting on in vitro fertilisation treatments etc and pre-implantation diagnosis
- Order No. 576 from 22 June 2000[205] amending the Order on the termination of pregnancy

Official Guidelines

- Danish National Board of Health's Guideline No. 109 from 13 June 1994 on the Introduction of New Methods of Treatment in Reproductive Technology

International Conventions

- Convention on Human Rights and Biomedicine ETS 164, Oviedo, 4 April 1997 (CHRB), (Signed on 4 April 1997/ Ratified on 10 August 1999/ in force since 1 December 1999)[206]
- Additional Protocol to the Convention on Biomedicine concerning the prohibition of cloning ETS 168, Paris, 12 January 1998 (APC) (signed 12 January 1998)[207]
- Universal Declaration on the Human Genome and Human Rights (UDHGHR)[208]

---

[200] Available in English at http://www-nt.who.int/idhl/en/Consult/IDHL.cfm. Further information is based on the translation of the Act by Maja Lisa Axen.

[201] Lovtidende, 2000, Part A, June, 2 2000, No. 86, p. 2680. Available in English at http://www-nt.who.int/idhl/en/Consult/IDHL.cfm.

[202] Pattinson 2002 Appendix 1.

[203] Lovtidende, 1997, Part A, 26 September 1997, No. 138, 3926-3928; http://www-nt.who.int/idhl/en/Consult/IDHL.cfm.

[204] Lovtidende, 1997, Part A, 10 October 1997, No. 145, 4074-4075; http://www-nt.who.int/idhl/en/Consult/IDHL.cfm.

[205] Lovtidende 2000, Part A, 30 June 2000, No. 97, 3425; Available in English at http://www-nt.who.int/idhl/en/Consult/IDHL.cfm.

[206] See http://conventions.coe.int, with reservation contained in the instrument of ratification which refers to Art 10 para. 2 concerning the right of registered persons to information and declaration which affects Art 20 para. 2, sub-para. ii, concerning the removal of regenerative tissue and territorial application ("Denmark declares that until further notice the Convention shall not apply to the Faroe Islands and Greenland.") (see http://conventions.coe.int/Treaty/EN/cadreprincipal.htm).

[207] See http://conventions.coe.int.

[208] Adopted unanimously and by acclamation at the 29th Session of UNESCO's General Conference, 11 November 1997 (29 C/Resolution 17 – see http:www.unesco.org/ibc/), and endorsed by the United Nations' General Assembly, 85th Plenary Meeting on 9 December 1998 (A/RES/53/152 – see http://www.un.org.documents/).

Recommendations

- Statements of the Danish Ethics Committee[209] on

  - Research on Human Gametes, Fertilised Ova, Embryos and Fetuses

  - Fetal Diagnosis

  - Assisted Reproduction

  - Genetic Engineering and Cloning

OVERVIEW OF REGULATIONS

Embryo Research

Definition of „ embryo "

A definition of the embryo has not been found. The terms of the law are fertilised eggs, pre-embryos[210] and gametes.

Research

Permitted under conditions, Act No. 460, section 25.

Sources of embryos

Eggs must not be fertilised and collected for other purposes than those stated in the Act No. 460, section 25 para. 1. According to the introduction to Act No. 460, section 25, the experiments mentioned are only allowed to be carried out on human eggs and gametes which are meant to be fertilised. Consequently, research is only legal on embryos having been created through IVF.[211]

Time limit for research

Fertilised eggs may be stored outside of the uterus for a maximum of fourteen days excluding any storage time (Act No. 460, section 26).

Other conditions governing research

Consent procedures

No information obtained.

Licenses

No information obtained.

Ethics Committees

Research projects have to be assessed and approved by a scientific ethics commission (Act No. 460, section 27 para. 2).[212]

---

[209] Available in English at http://etiskraad.dk/english/index.html.

[210] The Danish Ministry of Interior and Health does not want to use this word, meanwhile the European Group on Ethics in Science and New Technologies to the European Commission, Brigitte Gratton, National regulations in the European Union regarding research on human embryos, July 2002, p. 13. Available at. http://europa.eu.int. says it is generally used.

[211] This also results from Denmark having ratified the Convention on Human Rights and Biomedicine with reservation.

[212] There are seven regional ethics committees and the Central Ethics Committee, see European Group on Ethics in Science and New Technologies to the European Commission, Brigitte Gratton, National regulations in the European Union regarding research on human embryos, July 2002, p. 13, footnote 4. Available at. http://europa.eu.int.

Substantive (Must the embryo benefit?)

The purposes of experiments with fertilised eggs and gametes are limited to:

- improving in vitro fertilisation or similar techniques to effect pregnancies (Act No. 460, section 25 para. 1 no. 1)
- improve pre-implantation genetic examinations of fertilised eggs for severe hereditary disease or chromosomal anomalies. (Act No. 460, section 25 para. 1 no. 2). "Benefit" is not a condition.[213]

Storage/Disposal of embryos

Cryopreservation permitted for two years. After this period or in the case that one of the couple dies, the couple divorces or end the cohabitation fertilised ova shall be destroyed. (sections 2, 4, 15 para. 1 and 2).

Cloning

Reproductive

Prohibited. Experiments aiming at creating genetically identical human beings are unlawful.[214] (Act No. 460, section 28 no. 1).[215] It is also prohibited to to implant identical fertilised or unfertilised ova (Act. No. 460, section 4)

Non-reproductive

Not regulated but regarded as prohibited.[216]

Gene Therapy

Somatic

Permitted by implication and regarded as ordinary medical treatment.[217] See *next*.

Germ-line

The Act No. 460, section 27 prohibits implanting genetically modified fertilised eggs in a woman. In vitro fertilisation is only permitted with egg and sperm cells which have not been genetically modified, Act No. 460, section 2.

Pre-Implantation Genetic Diagnosis

Permitted to prevent:

- a severe hereditary disease (Act No. 460, section 7 para. 1).

---

[213] Information provided by Bent Rasmussen.

[214] The Danish Ethics Council had explicitly denounced the cloning of human beings in 1997 and supported a world wide ban, Working Paper on Cloning, 20 May 1997, J No. ER 1997-3.2-22 or Statement on Cloning from 29 May 1997 respectively, http://www.etiskraad.dk/english/cloning.htm.

[215] The Act No. 503 on the Scientific Ethics Committee System and the Examination of Biomedical Research Projects (Lovtidende, 1992, Part A, 26 June 1992, No. 84, 2017-2020), section 15 para. 1, already prohibited these experiments ("Experiments whose purpose is to enable genetically identical individuals to be produced"), available at http://www.all.org/abac/clontx08htm.

[216] European Group on Ethics in Science and New Technologies to the European Commission, Brigitte Gratton, National regulations in the European Union regarding research on human embryos, July 2002, p. 15. Available at. http://europa.eu.int.

[217] Information provided by Bent Rasmussen.

- chromosomal abnormalities (Act No. 460, section 7 para. 2).
- sex linked severe hereditary diseases (Act No. 460, section 8).

New methods of diagnosis require the approval of the Minister of Health, see Act No. 460, section 21 para. 1.[218]

Stem Cell Therapy

There is no explicit regulation in general on stem cells.[219] Research with fertilised eggs is permitted under the conditions set out Act No. 480 on MAP, section 25.[220]

Pre-Natal Diagnosis/Abortion

- PND is permitted for women older than 35 and women with having a child with risk of hereditary diseases.[221]
- The termination of pregnancy is permitted:
  - until the 12th week after consultation with a doctor or a special institution
  - after the 12th week without special permission if on medical grounds it is necessary because of a risk to the women's life or of a serious deterioration in her physical or mental state exists
  - after the 12th week if;
  - the pregnancy, prospective birth and care of the child threaten to damage the health of the mother because of a physical or mental disease or other aspects of her personal situation and – if the child is already viable – these reasons are so strong as to justify an abortion (social indication, Act No. 350, section 3 para. 1, supplemented by Act No. 430 from 31 May 2000),
  - the woman was raped or the child to be expected was fathered through an act of incest (Act No. 350, section 3 para. 2),
  - there is a danger that the child will suffer severe physical or mental damage due to a hereditary disease or a disease in the course of the pregnancy (Act No. 350, section 3 para. 3),
  - the woman will not be able to take care of the child adequately because of physical or mental handicaps (Act No. 350, section 3 para. 4) or because she is too young (Act No. 350, section 3 para. 5),
  - social circumstances suggest it is reasonable (Law No. 350, catalogue, section 3 para 6).

SPECIAL REMARKS

On 21 February 2001 the Danish Ethics Council recommended that the issue of therapeutic cloning should be debated.

---

[218] There is a legal duty to give advice, the couples have to consent in writing and the number of eggs to be fertilised is not limited, see Deutscher Bundestag, Schlußbericht der Enquete-Kommission Recht und Ethik in der modernen Medizin, 2002, p. 199.

[219] European Group on Ethics in Science and New Technologies to the European Commission, Brigitte Gratton, National regulations in the European Union regarding research on human embryos, July 2002, p. 14. Available at. http://europa.eu.int.

[220] See also European Science Foundation Policy Briefing: Human Stem Cell Research, June 2001, p. 6, available at http://www.esf.org/publication/89/ESPB14.pdf. The Danish government has announced to regulate the use of stem cells in 2003.

[221] Pattinson, 2002, Appendix 1, footnote 172.

Estonia

SOURCES OF LEGISLATION

Acts of the Legislature

- Penal Code (PC), passed on 6 June 2001, entered into force on 1 September 2002[222]
- Criminal Code (CC) from 7 May 1992[223], entered into force on 1 June 1992
- Artificial Insemination and Embryo Protection Act, passed on 11 June 1997, entered into force on 17 July 1997[224]
- Patents Act (PA) from 16 March 1994[225] the version from 21 February 2001[226]
- Human Genes Research Act from 13 December 2000, entered into force on 8 January 2001[227]

Official Regulations

- No information obtained.

Official Guidelines

- No information obtained.

International Conventions

- Convention on Human Rights and Biomedicine ETS 164, Oviedo, 4 April 1997 (CHRB), signed on 4 April 1997/ ratified on 8 February 2002/ entered into force on 1 June 2002[228]
- Additional Protocol to the Convention on Biomedicine concerning the Prohibition of Cloning ETS 168, Paris, 12 January 1998 (APC), signed on 12 January 1998/ ratified on 8 February 2002/ entered into force on 1 June 2002[229]
- Universal Declaration on the Human Genome and Human Rights (UDHGHR)[230]

Recommendations

- No information obtained.

---

[222] Riigi Teataja (RT-State Gazette) I 2001, 61, 364; available in English at http://www.legaltext.ee/text/en/X30068.htm.

[223] Riigi Teataja (RT-State Gazette) I 1992, 20, 288 in the version from 1 January 2002, RT I 2001, 87, 526; available in English at http://www.legaltext.ee/text/en/X300022K4.htm.

[224] Riigi Teataja (RT-State Gazette) I 1997, 51, 824 (not available in English).

[225] Riigi Teataja (RT-State Gazette) I 1994, 25, 406; 1998, 74, 1227, available in English at http://www.legaltext.ee/text/en/X40034K1.htm.

[226] Riigi Teataja (RT-State Gazette) I 2001, 27, 152.

[227] Riigi Teataja (RT-State Gazette) I 2001, 104, 685. See also III. Special Remarks.

[228] See http://conventions.coe.int/.

[229] Ibid.

[230] Adopted unanimously and by acclamation at the 29th Session of UNESCO's General Conference, 11 November 1997 (29 C/Resolution 17 – see http://www.unesco.org/ibc/), and endorsed by the United Nations' General Assembly, 85th Plenary Meeting on 9th December 1998 (A/RES/53/152 – see http://www.un.org.documents/).

OVERVIEW OF REGULATIONS

Embryo Research

Definition of "embryo"

There is no statutory definition of the embryo.[231]

Research

Permitted with conditions.[232]

Sources of embryos

Embryos are only permitted to be created in vitro where the intention is to transfer them to a woman. Infringements are punishable by fine. The storage of embryos for a period exceeding the statutory limit or illegal transactions also carry a fine, PC, section 131.

Time limit for research

14 days.[233]

Cloning

Reproductive

Prohibited and carries a fine or up to three years' imprisonment pursuant to the P C, section 130. The Patents Act prohibits the protection of techniques leading to the cloning of human beings, PA, section 7 para. 2 no. 1.

Non-reproductive

Prohibited by CC section 120.[234]

Gene Therapy

Somatic

No information obtained.

Germ-line

The Patents Act prohibits the grant of patents for technologies which allow interference with the germ line genetic identity of human beings, PA, section 7 para. 2 no. 2.

---

[231] Council of Europe: CLONING – Comparative Study on the situation in 44 countries, Strasbourg, 4 June 1998, p. 135, available in English at http://legal.coe.int/bioethics/gb/pdf/txtprep2b.pdf.

[232] Council of Europe: CLONING – Comparative Study on the situation in 44 countries, Strasbourg, 4 June 1998, p. 144, available in English at http://legal.coe.int/bioethics/gb/pdf/txtprep2b.pdf.

[233] Council of Europe: CLONING – Comparative Study on the situation in 44 countries, Strasbourg, 4 June 1998, p. 180, available in English at http://legal.coe.int/bioethics/gb/pdf/txtprep2b.pdf.

[234] It prohibits ".. substitution of the nucleus of a fertilised ovum by a somatic cell of another embryo, foetus or living or deceased person in order to create a human embryo with genetic information identical compared with that of such embryo, foetus or living or deceased person ...".

Pre-Implantation Genetic Diagnosis

Unregulated by statute.[235] Medical institutions need particular approval to carry out the procedure.[236]

Stem Cell Therapy

Unregulated by statute.

Pre-Natal Diagnosis/Abortion

Pre-Natal Diagnosis is unregulated. [237]

* **Abortion is unlawful if performed:**
    - against the will of the pregnant woman (Penalty of imprisonment of 3–12 years imprisonment), Penal Code, section 125,
    - by an unauthorized person upon the request of the pregnant woman (Penalty of a fine or imprisonment of up to 3 years), Penal Code, section 126 para. 1,[238]
    - by an unauthorized person upon request of the pregnant woman after the 21st week of the pregnancy (Penalty of imprisonment of up to 5 years), Penal Code, section 126 para. 2,[239]
    - by an authorized person after the legal deadline (Penalty of fine or imprisonment of up to 1 year), Penal Code, section 127.[240]

The woman on whom the abortion is performed may be fined for consenting to it, if it is carried out by an unauthorized person or after the legal deadline, Penal Code, section 128.

SPECIAL REMARKS

The Human Genes Research Act only regulates the establishment and maintenance of a gene bank and does not refer to research on embryos.

Finland

SOURCES OF REGULATIONS

Acts of the Legislature

* Medical Research Act No. 488 of 1999[241] (MRA)

---

[235] Council of Europe: CLONING – Comparative Study on the situation in 44 countries, Strasbourg, 4 June 1998, p. 140, available in English at http://legal.coe.int/bioethics/gb/pdf/txtprep2b.pdf.
[236] Council of Europe: CLONING – Comparative Study on the situation in 44 countries, Strasbourg, 4 June 1998, p. 142, available in English at http://legal.coe.int/bioethics/gb/pdf/txtprep2b.pdf.
[237] Council of Europe: CLONING – Comparative Study on the situation in 44 countries, Strasbourg, 4 June 1998, p. 140, available in English at http://legal.coe.int/bioethics/gb/pdf/txtprep2b.pdf.
[238] The CC laid down a fine or imprisonment of up to 2 years for this, CC, section 120 para. 2.
[239] The CC laid down imprisonment of up to 4 years for this, CC, section 120, para. 3.
[240] Pursuant to the CC a gynaecologist who after the legal deadline performs an abortion at the request of the pregnant woman may be fined or barred from certain forms of practice.
[241] For an English translation see Bulletin of Medical Ethics February 2000.

- Law No. 239 of 24 March 1970, as amended (ABO)[242]

Official Regulations
- None

Official Guidelines
- None

International Conventions
- Convention on Human Rights and Biomedicine ETS 164, Oviedo, 4.IV.1997 (CHRB) (Signed 04/04/97)[243]
- Additional Protocol to the Convention on Biomedicine concerning the prohibition of cloning ETS 168, Paris, 12.I.1998 (APC) (Signed 12/01/98)[244]
- Universal Declaration on the Human Genome and Human Rights. (UDHGHR)[245]

Recommendations
- None

OVERVIEW OF REGULATIONS

Embryo Research

Definition of "embryo"

"A living group of cells resulting from fertilisation not implanted in a woman's body." (MRA s.2.2)

Research

Permitted with conditions.

Sources of embryos

Supernumerary from IVF. Creation for research prohibited (MRA s.13).

Time limit for research

14 days from formation excluding any time spent frozen (MRA s.11).

Other conditions governing research

Consent procedures

Consent in writing of gamete donors required (MRA s.12). Donors must be advised prior to consenting, of the purpose and nature of the research, any risks and harm possible, and of their rights to withdraw their consent at any time prior to the completion of the research (MRA ss.6 & 12).

Licenses

Required from the National Authority for Medico-legal Affairs (terveydenhuollon oikeusturvakesus) (MRA s.11).

---

[242] Amended by Law No. 18 of 1971, No. 564 of 1978, No. 572 of 1985, and No. 1085 of 1995.
[243] See http://conventions.coe.int/Treaty/EN/cadreprincipal.htm.
[244] Ibid.
[245] Adopted by consensus, UNESCO General Conference, 11 November,1997.

Ethics Committees

Approval required (MRA ss.3 & 17). Ethics Committees must examine whether the research plan takes account of the MRA, data protection legislation, international obligations covering the status of research subjects and the guidelines that govern medical research (MRA s.17).

Substantive (Must the embryo benefit?)

No. However Ethics Committees must evaluate research projects and give a reasoned view on whether the research is ethically acceptable (MRA s.17).

Storage/Disposal of embryos

Embryos may be stored for up to 15 years after which they must be destroyed (MRA s.13).

Cloning

Reproductive

Prohibited (MRA s.26)

Non-reproductive

No information obtained.

Gene Therapy

Somatic

Appears is permitted by default in relation to embryos in light of MRA s.15. See *next*.

Germ-line

Research prohibited in relation to embryos, unless it aims to find a cure or to prevent a severe hereditary disease (MRA s.15).

The working group (set up by the Ministry of Justice) reporting on the use of gametes and embryos in assisted fertilisation, had proposed that no gametes or embryos be used in assisted fertilisation where the genetic heritage has been modified.[246]

Pre-Implantation Genetic Diagnosis

Proposed legislation exists.[247]

A working group report, given to the Ministry of Justice in 1997, recommended that assisted reproduction should not be allowed for the purpose of choosing a child's sex or characteristics, except to avoid serious hereditary sex-related disease.[248]

Stem Cell Therapy

No information obtained.

---

[246] See Pattinson 2002, Appendix 5.
[247] Ibid, Appendix 2.
[248] Ibid.

Pre-Natal Diagnosis/Abortion

Abortion must be performed as early as possible (ABO s.5).

Permitted:

- up to 16 weeks (s.5);
  - to prevent notable strain on the pregnant woman in relation to her living conditions and those of her family and other circumstances (s.1(2));
  - if the pregnancy was caused by rape or other specified crime. A substantiated report must have been made (ss.1(3) & 3);
  - if the woman was under 17 or was 40 or had already given birth to 4 children;
  - if there is a reason to assume that the child would be "mentally retarded," or would have or develop a "severe disease or physical defect" (ss.1(5) & 6(3)); or
  - if the parents through disease, mental disturbance or other comparable cause are severely limited in their ability to care for the child (s.1(6)).
- between 16 and 20 weeks; if the woman has a disease or physical deformity;
  - was under 17 at the time of conception or for another specific reason, subject to the permission of the National Board of Health (s.5); and
  - up to birth, where the woman's life or health is endangered (s.1(1) & 5).
- Before an abortion is performed, the woman must be advised of the significance and effects of the procedure (ABO s.4). Where possible the father of child must be given opportunity to state his opinion on the matter (ABO s.7). When an opinion of the national Board of Health is sought, it should be informed by persons possessing legal, psychiatric, obstetric, genetic and social expertise (ABO s.10).

France

SOURCES OF LEGISLATION

Acts of the Legislature

- Code Pénal (Criminal Code), Art. 511-1 to 511-5[249]
- Code Civil (Civil Code), Art. 16[250]
- Code de la Santé Publique[251]

---

[249] Inserted into the Code Pénal by the Act No. 94-653 from 29 July 1994 art. 9 Journal Officiel from 30. July 1994, amended by Ordonnance No. 2000-916 from 16 September 2000 art. 3 Journal Officiel from 22 September 2000 – in force since 1 January 2002, available in French at: http://www.legifrance.gouv.fr/html/frame_codes1.htm.

[250] Inserted into the Code Civil by the Act No. 94-653 from 29 July 1994 art. 9 Journal Officiel from 30. July 1994, http://www.legifrance.gouv.fr/html/frame_codes1.htm.

[251] Available in French at: http://www.legifrance.gouv.fr/html//frame_codes1.htm.

- Draft for a Bioethics Act[252]

Official Regulations
- No information obtained.

Official Guidelines
- Code de Déontologie Medicale from 6 September 1995[253]

International Conventions
- Convention on Human Rights and Biomedicine ETS 164, Oviedo, 4 April 1997 (CHRB), (Signed on 4 April 1997)[254]
- Additional Protocol to the Convention on Biomedicine concerning the Prohibition of Cloning ETS 168, Paris, 12 January 1998 (APC) (signed on 12 January 1998)[255]
- Universal Declaration on the Human Genome and Human Rights (UDHGHR)[256]

Recommendations
- CCNE[257], Advice Note No. 22–13 December 1990 on Gene Therapy,
- CCNE, Advice Note No. 27–2 December 1991 the human genome should not be used for commercial purposes. Report. Thoughts relating to ethic problems of human genome research,
- CCNE, Advice Note No. 52–11 March 1997 on the creation of human embryonic organ and tissue collections and their use for scientific purposes,
- CCNE, Advice Note No. 53–11 March 1997 on the establishment of collections of human embryonic cells and their use for therapeutic or scientific purposes,
- CCNE, Advice Note No. 60–25 June 1998 Review of the Bioethics Act,
- CCNE, Advice Note No. 67–18 January 2001 Statement of opinion on the preliminary Draft reviewing the Bioethics Act,
- CCNE, Advice Note No. 72–4 July 2002, Reflections on an extension of preimplantation diagnosis

---

[252] Adopted by the National Assembly in the First Reading as texte adopte No. 763 on 22 January 2002, and examined by the Senate in its First Reading and transmitted for the Second Reading to the National Assembly on 30. January 2003. The full text version of the First Reading is available in French at: http://www.assemblee-nationale.fr/legislatures/11/pdf/ta/ta0763.pdf. The present version is available at htp://www.assemblee-nationale.fr/12/projets/pl0593.asp.

[253] Available in French at: http://www.legifrance.gouv.fr/html/frame_codes1.htm.

[254] See http://conventions.coe.int/.

[255] Ibid.

[256] Adopted unanimously and by acclamation at the 29th Session of UNESCO's General Conference, 11 November 1997 (29 C/Resolution 17 – see http:www.unesco.org/ibc/), and endorsed by the United Nations' General Assembly, 85th Plenary Meeting on 9 December 1998 (A/RES/53/152 – see http://www.un.org.documents/).

[257] Comité Consultatif National d'Ethique pour les sciences de la vie et de la santé – National Ethics Committee for Life Sciences and Health - All statements of opinion are available in English and French at http://www.ccne.ethique.org/.

OVERVIEW OF REGULATIONS

Embryo Research

Definition of "embryo"

There is no definition of the term embryo. The Code Civil, Art. 16-1, protects the "human being" from the beginning of life.

Research

All experiments with embryos are prohibited (Code de la Santé Publique, Art. 2141-8).[258] However, parents may consent in writing to the examination of their embryo. Examination must be for medical purposes and must not harm the embryo.[259]

Sources of embryos

It is explicitly prohibited to create embryos in vitro for research purposes (Code de la Santé Publique, Art. L 2141-8 and R 2141-14).

However, embryos having been created in vitro they may be examined after the gamete donors have consented in writing. This examination is only allowed for medical purposes and must not harm the embryo.[260]

Time limit for research

The law does not lay down a deadline. In the report to the European Parliament a time limit of 7 day is mentioned.[261]

Other conditions governing research

Consent procedures

The parents (the man and the woman who form the couple) have to consent in writing, L-2141-8 Code de la Santé Publique.

Licenses

For every research a licence of ministry of health is needed. Code de la Santé Publique, R2141-15)

---

[258] The new Draft of the Bioethics Act is supposed to change this.

[259] According to the Draft of the Bioethics Act the creation of embryos for research purposes remains prohibited, Art. 19 of the Draft revising the Code de la Santé Publique, Art. L 2151.

[260] Pursuant to Art. 19 of the Draft of the Bioethics Act revising the Code de la Santé Publique surplus ("qui ne font plus l'object d'un projet parental" – which are not subject to planned parenthood anymore) embryos having been created in vitro for the purpose of medically supported fathering may be used for research purposes. Furthermore, the Code de la Santé Publique, new Art. L 2151-3-1 permits the import of embryonic tissue and cells. The Minister for Research is only allowed to grant the necessary approval if the tissue and the cells have been extracted in accordance with the Guiding Principles of the Code Civil, Art. 16 to 16-8.

[261] European Parliament, Report on the Ethic, Legal, Economic, and Social Effects of Human Genetics, A 5-0391/2001, p. 157; accordingly Pattinson, Appendix 3; Deutscher Bundestag, Zweiter Zwischenbericht der Enquete-Kommission Recht und Ethik in der modernen Medizin – Teilbericht Stammzellforschung, BT-Drs. 14/7546, p. 126, available in German at http://www.bundestag.de/gremien/medi/2zwischen.pdf.

Ethics Committees

The Provisions set out by the National Commission on Bioethics[262] govern the examinations that are permitted (pursuant to the Code de la Santé Publique, Art. L 2113-1), Code de la Santé Publique, Art. 2141-8-6.[263]

Substantive (Must the embryo benefit?)

The examinations must be for medical purposes and must not damage the embryo, Code de la Santé Publique, Art. L 2141-8-5).

Storage/Disposal

Embryos created for procreation can be stored for a period up to five years (Code de la Santé Publique, Art. 2141-3).

In Art. 19 the Draft of the Bioethics Act governing the Code de la Santé Publique, Art. 2151-3, explicitly determines that embryos on which research has been carried out must not be implanted for the purpose of pregnancy.[264]

Cloning

Reproductive

The law does not mention cloning explicitly,[265] but prohibited by Art. L 2141-8 Code de la Santé Publique). The National Ethics Commission of France (CCNE) in 1997 declared cloning to be prohibited but recommended an explicit legislative prohibition in the Code de la Santé Publique.[266]

Non-reproductive

See reproductive.

---

[262] This commission is meant to be renamed to "l'Agence de la procreation, de l'embryology et de la génétique humaines" (Procreation, Embryology and Human Genetic Agency), see Art. 17 of the Draft of the Bioethics Act.

[263] Art. 19 of the Draft of the Bioethics Act summarises the conditions in the Code de la Santé Publique, Art. L2151-3: The couple has to consent in writing after a period for reflection of three months after having been instructed on the possibilities of conservation or the implantation of the embryo with a different couple. The consent is revocable at any time without a statement of reasons. In addition, the Procreation, Embryology and Human Genetic Agency has to consent to the research. Upon its decision it has to bear in mind scientific plausibility, the conditions of research with reference to ethic issues and the interest in public health. The Agency has to present the documents on the research to the Ministers for Health and Research who may prohibit or suspend the project if they regard it as not scientifically plausible or irreconcilable with fundamental ethical principles.

[264] Projet de Lois (no. 593), Art. L2141-4: In general five years after the creation of the embryo or when the couple separates or one of both dies. The couple can demand a longer period.

[265] European Group on Ethics in Science and New Technologies to the European Commission, Brigitte Gratton, National regulations in the European Union regarding research on human embryos, July 2002, p. 25. Available at. http://europa.eu.int.

[266] Le Monde, 30 April 1997, 32. Information available at: http://www.All.org/abac/clontx08.htm. The new Draft of the Bioethics Act does so by proposing a new Art. 16-4 for the Code Civil which is meant to contain another paragraph: Pursuant to it every operation shall be prohibited aiming at effecting the birth of a child or the development of an embryo not having directly emerged from the gametes of a man and a woman (Art. 15 of the Draft of the Bioethics Act, texte adopté no. 763). The Senat (Projet de loi, no. 593) changed that into: Pursuant .. effecting the birth of a child with the same genetic identity of another person, alive or dead.

Gene Therapy

Somatic gene

No information obtained.

Germ-line

Pursuant to the Code Civil, Art. 16-4, interference with the "human species" is prohibited. Eugenic techniques for selection – even if intended to prevent or treat genetic diseases– are prohibited.

Pre-Implantation Genetic Diagnosis

Pre-implantation diagnosis is permitted pursuant to the Code de la Santé Publique, Art. 2131-4. A doctor practising at a Centre for Pre-Natal Diagnosis must certify that a child with a genetic disease for which there is no treatment at the time of diagnosis and which is pre-determined by his/ her descent would be born. The abnormality potentially causing such a disease must have been previously detected in one of the parents. The parents have to consent to the diagnosis in writing and it may only be carried out to detect, prevent or treat the disease.[267]

Stem Cell Therapy

No information.

Pre-Natal Diagnosis/Abortion

- Pre-natal diagnosis is only permitted in approved institutions (Art. L 2131-1) and may only be carried out for the purpose of diagnosing a particularly severe ailment of the embryo. It has to be preceded by a genetic consultation (Art. L 2131-1).

- The termination of pregnancy is permitted before the end of the 12th week, Code de la Santé Publique, Art. L 2212-1[268], if it is carried out by a doctor at a specific institution. The termination of pregnancy has to be preceded by a consultation at least on week before (Art. L 2212-3)[269] The one week requirement for personal consideration may be reduced if the 12-week-deadline could not be met if it is observed (Art. L 2212-4 and L 2212-5).

Germany

SOURCES OF LEGISLATION

Acts of the Legislature

- Criminal Code (StGB) as promulgated on 13 November 1998[270]
- Embryo Protection Act (ESchG) from 13 December 1990[271]

---

[267] Currently three Centres have the required licence, see: Deutscher Bundestag, Schlußbericht der Enquete-Kommission Recht und Ethik in der modernen Medizin, 2002, p. 199.

[268] Act No. 2001-588 from 4 July 2001 art. 1 art 2 Journal Officiel from 7 July 2001.

[269] Latest amendment by Act No. 2001-588 from 4 July 2001 art. 1 art. 4 Journal Officiel from 7 July 2001.

[270] BGBl. *(Federal Law Gazette)* I p. 3322.

[271] BGBl. I p. 2746.

- Act for the Safeguarding of Embryo Protection in Conjunction with the Import and Use of Human Embryonic Stem Cells – Stem Cell Act (StZG) from 28 June 2002[272]
- Act on the Circulation of Medicines (Medicines Act-AMG) from 24 August 1976 as promulgated on 11 December 1998[273]

Official Regulations

- (Sample) Professional Rules for German Doctors[274]

Official Guidelines

- Guidelines for Pre-Natal Diagnoses of Diseases and Dispositions for Diseases[275]
- Guidelines for Conducting Assisted Reproduction[276]
- Guidelines for the Transfer of Genes into Human Body Cells[277]

International Conventions

- Universal Declaration on the Human Genome and Human Rights (UDHGHR)[278]

Recommendations

- None

OVERVIEW OF REGULATIONS

Embryo Research

Definition of "embryo"

The fertilised human egg cell capable of development is already deemed to be an embryo, pursuant to the ESchG, section 8 para. 1. Furthermore, this applies to any totipotent cell extracted from an embryo which may divide and develop into an individual human being once the necessary further conditions are provided, ESchG, section 8 para. 1, StZG, section 3 no. 4.

Research

It is prohibited to create embryos without the intention to effect a pregnancy pursuant to the ESchG, section 1 para. 1 nos. 2, 6, para. 2. It is also prohibited to use an

---

[272] BGBl. I p. 2277, in force since 1 July 2001.
[273] BGBl. I p. 3586.
[274] Bundesärztekammer *(Federal Medical Association),* Deutsches Ärzteblatt *(leading medical journal)* (DÄBl.) 85, 1988, A-3601. Available in German at http://www.bundesaerztekammer.de.
[275] Bundesärztekammer, Deutsches Ärzteblatt (DÄBl.) 95, 1998, A-3236, available in German at http://www.aerzteblatt.de or aerztekammer.de, Stichwort *(cue):* Richtlinien.
[276] Bundesärztekammer, Deutsches Ärzteblatt (DÄBl.) 95, 1998, A-3166, available in German at http://www.aerzteblatt.de or aerztekammer.de, Stichwort: Richtlinien.
[277] Bundesärztekammer, Deutsches Ärzteblatt (DÄBl.) 95, 1998, A-789, available in German at http://www.aerzteblatt.de or aerztekammer.de, Stichwort: Richtlinien.
[278] Adopted unanimously and by acclamation at the 29th Session of UNESCO's General Conference, 11 November 1997 (29 C/Resolution 17 – see http://unesco.org/ibc/). and endorsed by the United Nations General Assembly, 85th Plenary Meeting on 9th December 1998 (A/RES/53/152 – see http://www.un.org.documents/).

embryo for any other purpose than its own preservation, pursuant to the ESchG, section 2. Thus, embryo research for the purpose of serving other objectives is prohibited. The StZG does not change this position because pursuant to sections 3 no. 1, 4 no. 2, it only admits the import of stem cells for research purposes that have been extracted before a fixed deadline.

Sources of embryos

It is only permitted to import embryonic stem cells which have been produced in accordance with the current legal situation in the country of origin and originate from surplus embryos not needed for an in vitro fertilisation, pursuant to the StZG, section 4.

Time limit for research

It requires no time limit since embryos are only permitted to be produced with the intention of achieving a pregnancy.

Other conditions governing research

Consent procedures

See *Sources of embryos*.

Licenses

See *Sources of embryos*.

Ethics Committees

The import of and the research on embryonic stem cells have to be approved by the competent authority pursuant to the StZG, section 6. This authority is subordinate to the Federal Ministry of Health pursuant to the StZG, section 7, and has to consult an interdisciplinary Central Ethics Commission which is appointed by the Federal Government pursuant to the StZG, sections 6 para. 3, 8 para. 1.

Substantial (Must the embryo benefit?)

The embryo may only be subjected to procedures for its own benefit. [279]

Storage/ Disposal of Embryos

There are no explicit statutory rules. Anything not serving the preservation of the embryo is prohibited.

Cloning

Reproductive

The ESchG prohibits artificially creating a human embryo with genetic information identical to that of a different embryo, foetus, human being or a deceased person pursuant to the ESchG, section 6 para. 1. Consequently, reproductive as well as therapeutic cloning is prohibited.[280]

---

[279] As regards the dividing line between research and treatment, see Pattinson 2002, Appendix 3 comment no. 1.

[280] Concerning the discussion if a human embryo with identical or only almost the same genetic information emerges when transplanting a nucleus, see Country Report Germany.

Beyond this, it is prohibited to transfer such an embryo to a woman.[281]

Non-reproductive

Prohibited. See *Reproductive.*

Gene Therapy

Somatic

It is not prohibited by the ESchG. It is governed by Guidelines issued by the Bundesärztekammer which are compulsory for doctors.[282] Furthermore the AMG, sections 40, 41, have to be applied in the clinical examination.[283]

Germ-line

Interference with the human germ-line is prohibited pursuant to the ESchG, section 5 para. 1, as is the use of modified human germ-cells pursuant to the ESchG, section 5 para. 2.

Pre-Implantation Genetic Diagnosis

The destruction of a totipotent cell, i.e. an embryo within the meaning of the ESchG, which is associated with the diagnosis is prohibited pursuant to the ESchG. The Professional Rules (MBO) also prohibit the diagnosis before the transfer of the embryo unless to exclude severe gender-specific diseases within the meaning of the ESchG, section 3, (MBO C IV. Nr. 15).[284]

Stem Cell Therapy

The StZG makes research on imported stem cells possible.

Pre-Natal Diagnosis/Abortion

* Prenatal diagnosis is not specifically regulated by statute. The Guidelines of the Bundesärztekammer apply[285]

* The termination of pregnancy is an offence and subject to imprisonment or a fine. Actions before the nidation of the fertilised egg are not regarded as termination of pregnancy within the meaning of the law pursuant to the StGB, section 218 para

---

[281] Keller/Günther/Kaiser, Embryonenschutzgesetz, Kommentar, section 6 margin no. 11, criticise the inherent obligation to kill which is even sanctioned by criminal law.

[282] Ibid., section 5 margin no. 6; Bundesärztekammer, Deutsches Ärzteblatt (DÄBl.) 92, 1995, A-789, available in German at http://www.aerzteblatt.de or aerztekammer.de: Stichwort Richtlinien.

[283] Vesting, Jan, Somatische Gentherapie – Regelung und Regelungsbedarf in Deutschland, Zeitschrift für Rechtspraxis *(Journal for Legal Practice)* (ZRP) 1997, 21-22.

[284] The Deutsche Ärztetag *(German Medical Congress)* has just recently supported a prohibition of PID, compare http://www.bundesaerztekammer.de/ - BÄK-INTERN.

[285] Bundesärztekammer, Richtlinien zur pränatalen Diagnostik von Krankheiten und Krankheitsdispositionen, Deutsches Ärzteblatt (DÄBl.) 95, 1998, A-3236, available in German at http:// www.aerzteblatt.de.

Abortion is however permitted up to 12 weeks if the pregnant woman consulted a competent person three days before the termination, and it is performed by a doctor (StGB, section 218 a para. 1).

After 12 weeks abortion is permitted if it is indicated upon consideration of the current and future personal situation of the woman and (or?) according to medical expertise ii is required to avert a threat to her life or of severe damage to her physical or mental health and the danger may not reasonably be averted in any other way.

Up until the 22nd week the pregnancy may be lawfully terminated by a doctor if the pregnant woman has consulted a competent person, StGB, section 218 a, para. 4.

After 22 weeks, it is possible abortion may be permitted if the pregnant woman was exposed to particular pressure at the time of the termination of the pregnancy.

FURTHER READING

- Beljin/Engsterhold/Fenger/Schmitz, Rechtliche Aspekte der Gentechnik – Ein Überblick in Raem/Braun/Fenger/Michaelis/Nikol/Winter, GenMedizin – Eine Bestandaufnahme, Berlin, Heidelberg, New York 2001, p 525.
- v. Bülow, Embryonenschutzgesetz, margin nos. 303-380 in: Winter/Fenger/Schreiber, Genmedizin und Recht, Munich 2001
- Deutscher Bundestag, Schlußbericht der Enquete-Kommission Recht und Ethik in der modernen Medizin, Berlin 2002, available at http://dip.bundestag.de/btd/14/090/1409020.
- Keller/Günther/Kaiser, Embryonenschutzgesetz, Kommentar, Stuttgart, Berlin, Cologne, 1992
- Lilie, Biorecht und Politik am Beispiel der internationalen Stammzelldiskussion, in: Amelung/Beulke/Lilie/Rosenau/Rüping/Wolfslast (edit.), Strafrecht-Biorecht-Rechtsphilosophie, Festschrift für Hans-Ludwig Schreiber, Munich, 2003, p. 729, 730.
- Lilie/Albrecht, Strafbarkeit im Umgang mit Stammzellinien aus Embryonen und damit imZusammenhang stehender Tätigkeiten nach deutschem Recht, Neue Juristische Wochenschrift (New Weekly Law Journal) 2001, 2774
- Merkel, Forschungsobjekt Embryo – Verfassungsrechtliche und ethische Grundlagen der Forschung an menschlichen embryonalen Stammzellen, München 2003
- Mildenberger, Der Streit um die um die Embryonen – Warum ungewollte Schwangerschaften, Embryoselektion und Embryonenforschung grundsätzlich unterschiedlich behandelt werden müssen, Medizinrecht 2002, 293

Greece

SOURCES OF LEGISLATION

Acts of the Legislature

- Act 1609/1986 on Voluntary Termination of Pregnancy[286]
- Criminal Code, sections 304, 305[287]
- Act 1036/1980 on Family Planning

Official Regulations

- Guidelines of the Health Authority[288]
- Decision No. 2 by the 23rd Plenary Meeting of the Central Council of Health in 1985[289]

Official Guidelines

- Guidelines of the Central Council of Health on Medical Assisted Procreation, 1998.[290]

International Conventions

- Convention on Human Rights and Biomedicine ETS 164, Oviedo, 4 April 1997 (CHRB), (signed on 4 April 1997/ ratified on 6 October 1998/ entered into force on 1 December 1999)[291]
- Additional Protocol to Convention on Biomedicine Concerning the Prohibition of Cloning ETS 168, Paris, 12 January 1988 (APC) (signed on 12 January 1998/ratified on 22 December 1998 entered into force on 1 March 2001)[292]
- Universal Declaration on the Human Genome and Human Rights (UDHGHR)[293]

Recommendations

- Recommendation of the National Bioethics Commission 21 December 2001 on the use of stem cells in biomedicine and clinical medicine.[294]
- Recommendation of the National Bioethics Commission 28 February 2003 on reproductive cloning[295]

---

[286] Translated in: International Digest of Health Legislation, 1986, 37 (4): 793.
[287] Available in English at http://cyber.law.Harvard.edu/population.
[288] European Parliament, Paper No. 14/2001, EP No. 303.112, May 2001, 2.9.
[289] Pattinson 2002, Appendix. 1.
[290] European Group on Ethics in Science and New Technologies to the European Commission, Brigitte Gratton, National regulations in the European Union regarding research on human embryos, July 2002, p. 35. Available at. http://europa.eu.int.
[291] See http://conventions.coe.int.
[292] See http://conventions.coe.int.
[293] Adopted unanimously and by acclamation at the 29th Session of UNESCO's General Conference, 11 November 1997 (29 C/Resolution 17 – see http://www.unesco.org/ibc/) and endorsed by the United Nations' General Assembly, 85th Plenary Meeting on 9 December 1998 (A/RES/53/152 – see http://www.un.org.documents/).
[294] Available in English at http://www.bioethics.gr.
[295] Available in English at http://www.bioethics.gr.

OVERVIEW OF REGULATIONS

Embryo Research

Definition of "embryo"

There is no statutory definition of the term embryo.[296]

Research

Embryo research is not regulated explicitly but it is legal.[297]

Sources of embryos

Surplus embryos from in vitro fertilisation.[298]

Time limit for research

Within 14 days after conception.[299]

Consent procedures

Gamete donors must consent.[300]

Licenses

See *Research*.

Ethics Committees

The competent ethics committee must approve the research.[301]

Substantive (Must the embryo benefit?)

No information obtained.

Storage/Disposal of embryos

The embryo can be stored for a period of one year and has to be destroyed after that time.[302]

Cloning

Reproductive

The declaration by the Health Authority explicitly prohibits the cloning of human beings[303] and according to the Additional Protocol (APC).

Non-reproductive

No information obtained.

---

[296] Council of Europe, CLONING – COMPARATIVE STUDY IN 44 STATES, Strasbourg, 4 June 1998, 135, available in English at http://legal.coe.int/bioethics/gb/pdf/txtprep2b.pdf.

[297] European Parliament, Paper No. 14/2001, EP No. 303.112, May 2001, 2.9.

[298] Ibid.

[299] Ibid. Pattinson, Appendix 3; European Group on Ethics in Science and New Technologies to the European Commission, Brigitte Gratton, National regulations in the European Union regarding research on human embryos, p. 35 (excluding the time of storage). Available at http://europa.eu.int.

[300] European Parliament, Paper No. 14/2001, EP No. 303.112, May 2001, 2.9.

[301] Ibid.

[302] European Group on Ethics in Science and New Technologies to the European Commission, Brigitte Gratton, National regulations in the European Union regarding research on human embryos, July 2002, p. 35. Available at. http://europa.eu.int.

[303] European Parliament, Paper No. 14/2001, EP No. 303.112, May 2001, 2.9. According to a different view cloning is merely prohibited implicitly, see http:www.glphr.org/genetic/europe/htm.

Gene Therapy

No information obtained.

Somatic

No information obtained.

Germ-line

No information obtained.

Pre-Implantation Genetic Diagnosis

Is not regulated by law.[304]

Stem Cell Therapy

Research on stem cells is not regulated by law.[305]

Pre-Natal Diagnosis/Abortion

* PND may only be performed in a state hospital or a university or military clinic[306]
* Termination of pregnancy is subject to imprisonment of at least 6 months, Criminal Code, section 304 para. 2a. The maximum sentence for the woman is one year, para. 3

If the woman wants the termination of the pregnancy and it is carried out by a special doctor in an approved institution it is lawful if one of the following conditions has been met:

– it is carried out within the first 12 weeks of the pregnancy;
– there is an indication that the foetus suffers from severe abnormalities which would cause severe congenital defects and the pregnancy has not lasted longer than 24 weeks;
– there is an unavoidable risk to the life of the woman and of severe permanent harm to her physical or mental health and this has been confirmed by a medical certificate;
– the pregnancy resulted from rape or sexual intercourse with a minor or defenceless and has not progressed beyond 19 weeks.

---

[304] Deutscher Bundestag, Schlußbericht der Enquete-Kommission Recht und Ethik in der modernen Medizin, 2002, p. 199. PID Centres require an official licence. An examination is subject to conditions which are the instruction and consent of the couple in question.

[305] European Science Foundation Policy Briefing, Human Stem Cell Research, June 2001, available at http://www.esf.org/publication/89/ESB14.pdf.

[306] Pattinson 2002, Appendix 1 (footnote 172).

Hungary

SOURCES OF REGULATIONS

Acts of the Legislature

- Law No. 154 of 15 December 1997 on Public Health (Magyar Kozlony, 23 December 1997, No. 119) (LPH)[307]
- Law No. 79 of 17 December 1992 on the Protection of the Life of the Fetus. (Magyar Kozlony, 23 December 1992, No. 132, 4705-4708)(PLF)[308]

Official Regulations

- No information obtained

Official Guidelines

- No information obtained

International Conventions

- Convention on Human Rights and Biomedicine ETS 164, Oviedo, 4.IV.1997 (CHRB) (Signed 07/05/99, Ratified 09/01/02, Entered into force 01/05/02)[309]
- Additional Protocol to the Convention on Biomedicine concerning the prohibition of cloning ETS 168, Paris, 12.I.1998 (APC) (Signed 07/05/99, Ratified 09/01/02, Entered into force 01/05/02)[310]
- Universal Declaration on the Human Genome and Human Rights. (UDHGHR)[311]

Recommendations

No information obtained.

OVERVIEW OF REGULATIONS

Embryo Research (LPH ss.180-182)[312]

Definition of "embryo"

"every living human embryo following the completion of fertilisation" (LPH s.3).[313]

Research

Permitted with conditions.

Sources of embryos

Surplus embryos from MAP. Embryos cannot be created for research.[314]

Time limit for research

Fourteen days.[315]

---

[307] Brief summary of the law available at http://www-nt.who.int/idhl/en/ConsultIDHL.cfm.
[308] See *International Digest of Health Legislation*, 1993, 44 (2) 249.
[309] See http://conventions.coe.int/Treaty/EN/cadreprincipal.htm.
[310] Ibid.
[311] Adopted by consensus, UNESCO General Conference, 11 November, 1997.
[312] See summary at note 1.
[313] Ibid. and Sandor 1998, 2.
[314] Sandor 1998, 8.
[315] Ibid.

Other conditions governing research

Consent procedures

No information obtained.

Licenses

No information obtained.

Ethics Committees

The Human Reproduction Commission, (LPH s.186), authorizes medical research on embryos on the basis of the submitted research proposal, and the opinion of a competent scientific body set up in a separate Act.[316]

Substantive (Must the embryo benefit?)

No information. However, research can only be done on embryos for the purposes the LPH sets with regard to medical research.[317]

Storage/Disposal of embryos

No information obtained.

Cloning

Reproductive

Prohibited. (APC art 1 & UDHGHR art.11).

Non-reproductive

No information obtained.

Gene Therapy

Somatic

Permitted, to prevent or treat a foreseeable disorder in the unborn child (LPH ss.162 & 182(2)).

Germ-line

Prohibited. (LPH s.162).

Pre-Implantation Genetic Diagnosis

Permissible to diagnose or prevent sex-linked hereditary disease or to prevent or treat other genetic characteristics which would cause a foreseeable disorder in the unborn child (LPH ss.162 & 182).

Stem Cell Therapy

No information obtained.

Pre-Natal Diagnosis/Abortion

Abortion Permitted

General principle, termination only permitted where risk exists (PLF s.5).

---

[316] Sandor 1998, 10.
[317] Sandor 1998, 8.

Permitted:

- Up to 12 weeks where,
  - pregnancy is serious risk to the woman; or
  - fetus shows risks of malformation or serious lesions; or
  - pregnancy results from a crime; or
  - pregnant woman is in serious crisis (PLF s.6(1)).
- Up to 18 weeks where conditions in subsection 1 fulfilled and pregnant woman, totally or partially incapacitated; or
  - failed to recognize pregnancy in time due to health reasons, medical error or administrative negligence (PLF s.6(2)).
- Up to 20 weeks (or 24 weeks if there is a delay in diagnosis), if probability of genetic or teratological lesion exceeds 50% (PLF s.6 (3)).
- Irrespective of duration where,
  - life of pregnant woman endangered; or
  - fetus has malformation making postnatal life impossible (PLF s.6(4)).
- Despite the existence of the general principle in section 5, abortion not based on health grounds may be permitted at the request of the pregnant woman (PLF s.7).
- PLF sections 9-12 outline procedural requirements prior to abortion being carried out depending on the circumstances such as counselling, a waiting period, a signed request, and certification by the relevant authorities of health problems or the fact of the pregnancy being the result of a crime.

FURTHER READING

Sandor Judit. 1998. Reproductive Rights in Hungary, Central European University Political Sciences Department.

INDIA

SOURCES OF REGULATIONS

Acts of the Legislature

- The Medical Termination of Pregnancy Act, 1971 (Act No. 34 of 1971) (Enacted 10th August 1971) (MTP) [318]
- The Prenatal Diagnostic Techniques (Regulation and Prevention of Misuse) Act, 1994 (PDT)[319]

Official Regulations

- None

---

[318] Available at http://www.indialawinfo.com/bareacts/mtp.html. There is a Medical Termination of Pregnancy (Amendment) Bill 2002 available at http://indiacode.nic.in/incodis/whatsnew/Medical.htm. It is "aimed at eliminating abortion by untrained persons and in unhygienic conditions, thus reducing maternal morbidity and mortality.".

[319] Available at http://www.indialawinfo.com/bareacts/prenatal.html.

Official Guidelines
- Indian Council of Medical Research New Delhi (ICMR): Ethical Guidelines for Biomedical Research on Human Subjects 2000 (EG 2000)[320]

International Conventions
- Universal Declaration on the Human Genome and Human Rights. (UDHGHR)[321]

Recommendations
- No information obtained.

OVERVIEW OF REGULATIONS

Embryo Research

Definition of "embryo"

"Embryonic state": Between 15 days and 8 weeks post-conception of a pregnancy. Up to 14 days after conception æ pre-embryo (EG 2000, 56).

Research

Permitted (EG 2000, 67).

Sources of embryos

No information.

Time limit for research

14 days after fertilization excluding the period during which the [pre-] embryo was frozen. (EG 2000, 67).

Other conditions governing research

Consent procedures

No information with respect to embryos. [See (EG 2000, 17–18) in relation to human subjects].

Licenses

No information obtained.

Ethics Committees

No information with respect to embryos. [See (EG 2000, 11–16) in relation to human subjects].

Substantive (Must the embryo benefit?)

No information obtained.

Storage/Disposal of embryos

Maximum storage period of 10 years and a 5 yearly review of embryo deposits as practised in other countries e.g. U. K (EG 2000, 67).

---

[320] Available at http://www.icmr.nic.in/ethical.pdf.
[321] Adopted by consensus, UNESCO General Conference, 11 November, 1997.

Cloning

Reproductive

Prohibited. (EG 2000, 48)

Non-reproductive

Permitted by implication (EG 2000, 48).

Gene Therapy

Somatic

Permitted. However restricted to alleviation of life threatening or seriously disabling genetic disease in individual patients and should not change normal human traits. (EG 2000, 44). The preparation of any 'gene construct' to be administered is to be regulated by the National Bioethics Committee under the Department of Biotechnology. The evaluation of the efficacy and safety of the administered 'gene construct' is to be regulated by the local institutional Ethics Committee (IEC) and the Central Ethical Committee (CEC) of the ICMR. (EG 2000, 45).

Germ-line

Prohibited given the present state of knowledge (EG 2000, 45).

Pre-Implantation Genetic Diagnosis

Permitted. (EG 2000, 47–48). New draft legislation seeks to ban its use for sex-selection.[322]

Stem Cell Therapy

Use of foetal tissue for any such therapies permitted subject to:

- ethics committee approval at local and national level;
- scientific committee approval at local level;
- informed consent of mother to donation of tissue, separate from the informed consent to the abortion (if subsequent use of stored tissue differs from the original objective a fresh sanction is required from the scientific and ethical committees); and

subject further to certain prohibitions (for example) against:

- deliberate conception and abortion to obtain material for research; and
- the involvement of commercial benefit in obtaining embryos, fetuses or their tissue; and
- the dictation by the mother regarding who should receive foetal tissue taken for transplant; and
- patenting cells obtained from fetuses for commercial exploitation. (EG 2000, 56–59).

Pre-Natal Diagnosis/Abortion

- Pre-natal Diagnosis permitted only to determine if:
  - foetus has a chromosomal abnormality or congenital anomaly, a genetic

---

[322] Mudur 2002, 385.

metabolic or sex-linked genetic disease or haemoglobinopathies or other disease specified by the Central supervisory Board; and

- the pregnant woman is over 35, or has undergone two or more spontaneous abortions, or has been exposed to potentially teratogenic agents, or has a family history of mental retardation or physical deformities such as spasticity or any other genetic disease, or for such other condition as specified by the Central Supervisory Board. (PDT Chap III s.4(2) & (3)).

Pre-natal diagnosis to determine the sex of a foetus is prohibited. (PDT Chap III s.6). Use of prenatal diagnosis should be limited to situations where there is a clear benefit. (EG 2000, 48).

- Abortion permitted:

- up to 12 weeks where in the opinion of one registered medical practitioner the continuance of the pregnancy would involve risk to the life of the pregnant woman or of grave injury to her physical or mental health; or there is a substantial risk that the child would be seriously physically or mentally handicapped. (MTP s.3(2)(a)(i) & (ii)).

- between 12 and 20 weeks where in the opinion of two registered medical practitioner the continuance of the pregnancy would involve risk to the life of the pregnant woman or of grave injury to her physical or mental health; or there is a substantial risk that the child would be seriously physically or mentally handicapped. (MTP s.3(2)(b)(i) & (ii)).

Pregnancies resulting from rape and in the case of married women, the failure of contraceptive devices are considered capable of causing grave injury to the mental health of the pregnant woman (MTP, Explanations 1 & 2).

Written consent to termination is required in the cases of pregnant women under 18 or who are mentally retarded. (MTP 3(4)).

Republic of Ireland

SOURCES OF REGULATIONS

Acts of the Legislature

- Eighth Amendment to the Constitution Art. 40.3.3. (EAC)
- Offences Against the Person Act 1861 (OAPA)

  – Proposed Legislation –

- Twenty-fifth amendment to the Constitution (Protection of Human Life in Pregnancy) Bill 2001 (PLPHLP)[323]

Official Regulations

- None

---

[323] Available from the Irish Department of Health Website at http://www.doh.ie/publications/propprot.html.

Official Guidelines
- Medical Council's Ethical Guidelines 1998 (MCEG). [Only apply to registered medical practitioners]

International Conventions
- Universal Declaration on the Human Genome and Human Rights. (UDHGHR)[324]

Recommendations

No information obtained.

OVERVIEW OF REGULATIONS

Embryo Research

The Government set up a Commission on Assisted Reproduction in 2000 to report on, *inter alia,* embryo research.

Definition of "embryo"

None

Research

Prohibited by implication under (EAC). It states, "The State acknowledges the right to life of the unborn and, with due regard to the equal right to life of the mother, guarantees in its laws to respect, and, as far as practicable, by its laws to defend and vindicate that right."

Also prohibited by the MCEG. "Any fertilised ovum must be used for normal implantation and must not be deliberately destroyed" (MCEG para 26.4). It would be professional misconduct to create embryos for experimental purposes (MCEG para. 26.2).[325]

Sources of embryos

See *Research.*

Time limit for research

See *Research.*

Other conditions governing research

Consent procedures

See *Research.*

Licenses

See *Research.*

Ethics Committees

See *Research.*

Substantive (Must the embryo benefit?)

See *Research.*

---

[324] Adopted by consensus, UNESCO General Conference, 11 November, 1997.
[325] Pattinson 2002, Appendix 3.

Storage/Disposal of embryos

See *Research*.

Cloning

Reproductive cloning

The legislative position is unsure. Use of the Dolly technique might be affected by the Control of Clinical Trials Act 1987 and the Control of Clinical Trials and Drugs Act 1990, if they are not "too vague in their nature to include the procedure."[326]

If reproductive cloning is seen as creating a "new form of life" and "experimental" it would be prohibited by the MCEG. "The creation of new forms of life for experimental purposes or the deliberate and intentional destruction of human life already formed is professional misconduct." (MCEG para.26.1).

Non-reproductive

No information obtained.

Gene Therapy

Somatic

Permitted by default. Manipulation of sperm or eggs is limited to the "improvement of health." (MCEG para. 26.2).

Germ-line

Unregulated. Manipulation of sperm or eggs is limited to the "improvement of health." (MCEG para. 26.2)

Pre-Implantation Genetic Diagnosis

Prohibited if pre-implantation embryo seen as "unborn". "The State acknowledges the right to life of the unborn and, with due regard to the equal right to life of the mother, guarantees in its laws to respect, and, as far as practicable, by its laws to defend and vindicate that right." (EAC).

– Proposed Legislation –

Implicitly permitted. Abortion, defined as "the intentional destruction by any means of an unborn human life after implantation in the womb of a woman". (PLPHLP s.1).

Stem Cell Therapy

No information obtained.

Pre-Natal Diagnosis/Abortion

* Abortion prohibited (OAPA ss.58 & 59;EAC), unless "there is a real and substantial threat to the life, as distinct from the health, of the mother." (Attorney General v. X [1992] 1 IR 1, 53-54, per Finlay CJ.) Threat to life of mother includes the risk of suicide. (Attorney General v. X [1992] 1 IR 1). Based on the strict proscriptions on Abortion, Pre-natal Diagnosis, likely to be prohibited.

---

[326] Sheikh 1997, 95.

- **Proposed Legislation –**

  Would prevent the "risk of suicide" from being viewed as a "real and substantial threat to the life of the mother" and hence remove it as a ground for abortion (PLPHLP s.1(2)). Would however permit women to travel abroad to procure abortions on grounds which are legal in other countries, but which could not be legally obtained in Ireland (PLPHLP s.4).

  PLPHLP to be decided on by a referendum.[327]

SPECIAL REMARKS

The Assisted Human Reproduction Commission was set up by the Government in February 2000, with the following terms of reference:

"to prepare a report on the possible approaches to the regulation of all aspects of assisted human reproduction and the social, ethical and legal factors to be taken into account in determining public policy in this area."[328]

The Commission may therefore include in its report recommendations regarding embryo research and cloning.

Israel

SOURCES OF REGULATIONS

Acts of the Legislature

- Prohibition of Genetic Intervention (Human Cloning and Genetic Manipulation of Reproductive Cells) Law, No. 5759-1999 (PGIL)[329]
- Proposed law for the regulation of the donation of eggs for purposes of IVF
- Criminal Law Amendment (Interruption of Pregnancy) 1977 (CLA)
- National Health Insurance Law 1995 (NHIL)
- Official Regulations
- Public Health (Extra-Corporeal Fertilization) Regulations, 1987 (PHR)[330]

Official Regulations

- No information obtained.

Official Guidelines

- No information obtained.

International Conventions

- Universal Declaration on the Human Genome and Human Rights. (UDHGHR)[331]

---

[327] See Explanatory memorandum on the Irish Department of Health Website at note 1.
[328] Available on the Irish Department of Health Website relevant page http://www.doh.ie/pres-room/pr20000225a.html.
[329] Copy of legislation provided by Vardit Ravitsky.
[330] See International Digest of Health Legislation, 1987 38 (4) 779.
[331] Adopted by consensus, UNESCO General Conference, 11 November 1997.

Recommendations

- Report of the Bioethics Advisory Committee of The Israel Academy of Sciences and Humanities on The Use of Embryonic Stem Cells for Therapeutic Research, August 2001[332] (RBAC)

OVERVIEW OF REGULATIONS

Embryo Research

Definition of "embryo"

No information obtained.

Research

Unregulated. Embryo research is being conducted[333].

Sources of embryos

The PHR (which prescribes conditions for the retrieval, fertilisation, freezing and implantation of fertilised eggs) bans the removal of ovum save for the purposes of fertilisation and subsequent implantation into a woman's womb (PHR s.3).

No mention is made as to the fate of supernumerary embryos or frozen embryos at the end of the freezing period.

The proposed law for the regulation of the donation of eggs for IVF also fails to address the possibility of research (RBAC section 7 note 46).

The RBAC supports the donation of supernumerary embryos for research as well as the creation of embryos by nuclear transfer for research. However it is not in favour of creating embryos for research by fertilization of an ovum with sperm as it a) might be viewed by the public as a misuse of IVF technology; b) may offend the gamete donors; and c) may lead to abuse and commercialisation of gamete donations (RBAC section 6 notes 30–37).

Time limit for research

The very early stages of embryonic development. The present limit is 14 days (Period actually stated is 2 weeks) (RBAC section 8 note 10).

Other conditions governing research

Consent procedures

Free and informed consent to donation required. Donation must be regulated to ensure it is made with respect for the human dignity, autonomy and liberty of the donors. The possibility of donation should be mentioned at the beginning of the IVF process, but donors should be advised of the alternative of freezing the embryos. Regulation should require the clear separation of the IVF treatment and research teams in accordance with present IVF regulations and organ transplantation in genera, and protect the rights of parents who find embryo research unacceptable (RBAC section 8 notes 6,7 & 9).

---

[332] Available at: http://www.academy.ac.il/bioethics.html.

[333] Differentiation of human embryo stem cells into cardiac muscle cells has been achieved in Israel. See RBAC section 3 note 8).

Licenses

Apparently not required at present, as research is unregulated.

Ethics Committees

The Supreme Helsinki Committee for Genetics appointed under the Public Health (Medical Experiments on Humans) Regulations 5741-1980 operates under the PGIL (RBAC ss.2 & 4). It examines on a case-by-case basis, approving or rejecting applications for genetic research projects, including research on pre-implantation embryos. The RBAC including its recommendations were aimed at providing Guidelines fro the work of this committee (RBAC section 7 note 46).

Substantive (Must the embryo benefit?)

No. However:

- the research must be subject to strict supervision. The research and possible applications must be justifiable in terms of the benefit it offers to humanity (RBAC section on "Ethical Restraints and Conduct of Human Stem Cell Research" (ehscr) note 9).
- research involving derivation of embryonic stem cells should be meticulously scrutinized to avoid non-scientific or unethical aims (RBAC ehscr note 10).

The medical applications of stem cell derived replacement tissues for transplantation must be restricted to therapeutic, not cosmetic or eugenic aims (RBAC ehscr note 12).

In all aspects of embryo-related research particular importance should be given to respect of human dignity and the moral safeguards set out as international principles in the UDHGHR (RBAC ehscr note18).

Storage/Disposal of embryos

At present the PHR allow the discarding of frozen embryos after 5 years, unless the parents instruct otherwise (RBAC section 8 note 9a).

Cloning

Reproductive

Not permitted (PGIL s.3 (1) & UDHGHR art.11). There is a five year moratorium (PGIL s.8).

Non-reproductive

Permitted by implication. Human cloning defined as "the creation of a complete human being chromosomally and genetically absolutely identical to another person or fetus, living or dead". (PGIL s.2).

Gene Therapy

Somatic

Permitted by implication. PGIL s.3(2) only prohibits germ line gene therapy.

Germ-line gene

Prohibited (PGIL s.3(2)).

Pre-Implantation Genetic Diagnosis

Practiced (RBAC section 8 note 9b).

Stem Cell Therapy

There are several lines of research proceeding. Sources of stem cells being pursued are IVF embryos, adult stem cells and embryos created by nuclear transfer (RBAC section 6 notes 31, 35-37 & 39). In respect of embryos created by nuclear transfer an alternative technology being considered is the reprogramming of somatic cell nuclei by transfer into enucleated cells from pre-existing human ES cell lines to circumvent the need for enucleated oocytes (RBAC section 8, part entitled "ES cells obtained using cloning technologies" note 5). A source that was being contemplated is embryonic germ cells from cadaveric fetal tissues, though it was recognised that ethical safeguards would need to be in place to prevent elective abortions to provide a supply (RBAC section 6 notes 30–39).

A wide range of possible applications are envisaged including treatment of diseases of the nervous system, heart, bone and cartilage and immune system. Treatments are also forecast for cancer, diabetes and for transplantation to overcome graft-versus-host disease (RBAC section 3 notes 7–12).

Pre-Natal Diagnosis/Abortion

- NHIL requires health care organisations to supply a "standardised basket" of medical services within reasonable time and distance from the insured persons home. These include "preventive medicine and health education (i.e. early diagnosis of embryo abnormalities...counselling for pregnant women...)"[334]
- CLA permits abortions:
  - To save the woman's life
  - If pregnancy may cause her physical or mental harm
  - If woman under the age of marriage or over 40
  - If pregnancy is the result of a sexual offence, incest or extra-marital sexual intercourse
  - If the child is likely to have a physical or mental malformation

Penalty of five years imprisonment or fine for a person performing an illegal abortion.

No time limit is specified by law.[335]

SPECIAL REMARKS

The definition of "Human cloning" in the PGIL needs to be amended as not even identical twins are chromosomally and genetically absolutely identical due to the inevitable occurrence of slight gene mutations.

The RBAC also recommended that:

A National Committee such as the Helsinki Committee for Genetics should be established to oversee research and eventually approve specific research proposals (RBAC ehscr note 15).

Public discussions should be encouraged (RBAC ehscr note 16).

---

[334] See Israeli Ministry foreign Affairs website. "National Health Insurance" Available at: http://www.israel-mfa.gov.il/mfa/go.asp?MFAH00km0.

[335] Abortion Policies: A Global Review United Nations (2001). Israeli country profile available at http://www.un.org/esa/population/publications/abortion/profiles.htm.

Italy

SOURCES OF LEGISLATION

Acts of the Legislature

*   There is no particular legislation on the research on embryos[336]
*   The draft of a Bioethics Act No. 1514/ XIV Legislatura is currently under consultation in the Senate[337]
*   Act No. 194/78 from 22 May 1978 on the Social Protection of Motherhood and the Voluntary Termination of Pregnancy[338]

Official Regulations

*   Order by the Minister of Health from 5 March 1997[339]
*   Order by the Minister of Health of 25. July 2001 on Prohibition of import and export of human gametes and embryos[340]
*   Order of the Minister of 18. December 2001 banning the commercialisation of human gametes and embryos (prolongation of the order of 25 July 2001)[341]
*   Order of the Minister of Health banning any kind of research related to cloning of humans (third prolongation of the order of 5 March 1997)[342]

International Conventions

*   Convention on Human Rights and Biomedicine ETS 164, Oviedo, 4 April 1997 (CHRB), (signed on 4 April 1997)[343]
*   Additional Protocol to the Convention on Biomedicine Concerning the Prohibition of Cloning ETS 168, Paris, 12 January 1998 (APC) (signed on 12 January 1998)[344]

---

[336] European Parliament, Paper No. 14/2001, EP No. 303.112, May 2001, Embryos, Scientific Research and European Legislation, 2.10; available at http://europarl.eu.int/stoa/publi/default_en.htm; European Group on Ethics in Science and New Technologies to the European Commission, Brigitte Gratton, National regulations in the European Union regarding research on human embryos, July 2002, p. 39. Available at. http://europa.eu.int.

[337] Disegno di Legge: Norme in materia di procreazione medicalmente assistita, Approved by the Chamber of Deputies on 18. June 2002 and transmissed to the Senate on 19. June 2002. Session No. 144 of the Senate Commission on 25. June 2003. A full text version of the draft in Italian is available at http://www.parlamento.it/att/ddl/home.htm.

[338] Legge sull'interruzione volontatia della gravidanza, Gazzetta Uficiale della Repubblica Italiana, Part. I, 2 May 1978, No. 140, 3642-3646; available in English at http:/cyber.law.Harvard.edu/population/abortion/Italy.abo.htm.

[339] See at http://firewall.unesco.org/opi/eng/bio98/.

[340] European Group on Ethics in Science and New Technologies to the European Commission, Brigitte Gratton, National regulations in the European Union regarding research on human embryos, July 2002, p. 39. Available at. http://europa.eu.int.

[341] European Group on Ethics in Science and New Technologies to the European Commission, Brigitte Gratton, National regulations in the European Union regarding research on human embryos, July 2002, p. 39. Available at. http://europa.eu.int.

[342] Gazzetta Ufficiale (official gazette) no. 30 of 5. February 2002. Available also at: http://www.cecos.it/html/normativa/21dicembre.php.

[343] See http://conventions.coe.int.

[344] See http://conventions.coe.int.

- Universal Declaration on the Human Genome and Human Rights (UDHGHR)[345]

Recommendations

- Report on the Identity and Status of the Human Embryo, 22 June 1996[346]
- Report on cloning, 17. October 1997
- Report on the Therapeutic Use of Stem Cells, 27 October 2000[347]
- Statement of Opinion by the National Bioethics Committee on Reproductive Cloning from 17 January 2003
- Report of the National Bioethics Committee on Research by using human embryos and stem cells[348]

OVERVIEW OF REGULATIONS

Embryo Research

Embryo research is not regulated by law.[349] The Chamber of Deputies has passed Act No. 1514 which has been under consideration of the Senate since 19 June 2002.[350] At present only the Orders of the Minister of Health apply to embryo research.[351]

Definition of "embryo"

None.

Research

According to the draft which is currently being discussed experiments on embryos are prohibited, art. 13 no. 1 draft act no. 1514. Research is permitted for diagnostic and therapeutic reasons, art. 13 no. 2 draft act no. 1514.

Source of embryos

Do not exist: see Research and Storage/Disposal.

Time limit for research

See Research.

---

[345] Adopted unanimously and by acclamation at the 29th Session of UNESCO's General Conference, 11 November 1997 (29 C/Resolution 17 – see http:www.unesco.org/ibc/), and endorsed by the United Nations' General Assembly, 85th Plenary Meeting on 9 December 1998 (A/RES/53/152 – see http://www.un.org.documents/).

[346] Available in English at http://www.governo.it/bioetica/eng/opinion.html.

[347] European Group on Ethics in Science and New Technologies to the European Commission, Brigitte Gratton, National regulations in the European Union regarding research on human embryos, July 2002, p. 40. Available at. http://europa.eu.int.

[348] Available in Italian at http://www.palazzochigi.it/bioetica.

[349] European Parliament, Paper No. 14/2001, EP No. 303.112, May 2001, Embryos, Scientific Research and European Legislation, 2.10; available at http://europarl.eu.int/stoa/publi/default_en.htm.

[350] Last Session was No. 144 of the Senate Commission on 25. June 2003. A full text version of the draft in Italian is available at http://www.parlamento.it/att/ddl/home.htm.

[351] See also Taupitz, Die rechtliche Regelung der Embryonenforschung im internationalen Vergleich, Heidelberg 2003, pp. 105, 106.

Substantive (Must the embryo benefit?)

Yes. See *Research*.

Storage/Disposal of embryos

Cryoconservation and Disposal of embryos are prohibited, art. 14 no. 1 draft act no. 1514.

Cloning

Reproductive

There is no statutory prohibition of cloning. However, the government on 21 March 1997 declared its opposition to cloning[352] On 5 March 1997 the Minister of Health prohibited by Decree all experiments which could potentially lead to the cloning of human beings.[353] The National Bioethics Committee opposes cloning. The Draft of the Bioethics Act also prohibits cloning, art. 13 no. 3 lit. c draft act no. 1514.[354]

Non-reproductive

Prohibited, art. 13 no. 3 lit. a draft act no. 1514.

Gene Therapy

Somatic

It is not regulated by law. The National Bioethics Committee has argued in favour of somatic gene therapy in a published paper.[355] Somatic gene therapy is applied for attempted treatment on a vague legal basis. Approval by the Instituto Superiore di Sanita has to be sought as well as by the ethics commission.

Germ-Line

The government has expressed its opposition to human germ-line modification.[356] just as well as the National Bioethics Committee has also opposed such intervention.[357] Art. 13 no. 3 lit. b draft act no. 1514 prohibits the manipulation of the genotype of embryo or gametes except for diagnostic ant therapeutic reasons.

---

[352] European Parliament, Paper No. 14/2001, EP No. 303.112, May 2001, Embryos, Scientific Research and European Legislation, 2.10; available at http://europarl.eu.int/stoa/publi/default_en.htm.

[353] European Group on Ethics in Science and New Technologies to the European Commission, Brigitte Gratton, National regulations in the European Union regarding research on human embryos, July 2002, p. 39. Available at. http://europa.eu.intrefers to decision of the Court of Verona, that stipulated in 1999 that a mere order cannot prohibit scientific research and concludes, that research on human cloning could be allowed by default.

[354] Pattinson,2002, Appendix 4.

[355] Pattinson 2002, Appendix 5. See also Simon: Survey on Regulation of Gene Therapy in some European Countries, in: Müller/Simon/Vesting (editor), Interdisciplinary Approaches to Gene Therapy, Legal Ethic and Scientific Aspects, Berlin, Heidelberg, New York, 1997, p. 129.

[356] European Parliament, Paper No. 14/2001, EP No. 303.112, May 2001, Embryos, Scientific Research and European Legislation, 2.10; available at http://europarl.eu.int/stoa/publi/default_en.htm.

[357] Pattinson 2002, Appendix 5.

Pre-Implantation Genetic Diagnosis

Presently unregulated by law.[358] The Ethical Code of Practice restricts tests to specific diseases.[359] However, it is disputed whether this code is legally binding.[360]

Stem Cell Therapy

No information obtained.

Pre-Natal Diagnosis/Abortion

- Pre-Natal Diagnosis is not regulated by law. It is widely practised.[361]
- **Abortion Permitted.**

(a) Within the first 90 days a pregnancy may be terminated if

- the continuation of the pregnancy, the birth or motherhood would constitute a considerable threat to the mental or physical health of the woman due to her state of health, her economic, social or family situation, or the the circumstances of the conception, or
- there is a significant likelihood that the child would be born with damages or deformities, Act No. 194, section 4.

The doctor has to inform the pregnant woman about her rights and possible means of assistance; the operation is only permitted to be carried out 7 days after the consultation, Act No. 194, section 5.

(b) After the first 90 days the pregnancy may be terminated if

- the pregnancy or the birth threaten the life of the woman,
- it constitutes a serious threat to the physical or mental health of the woman and the foetus has abnormalities and deformities which have been diagnosed (Act No. 194, section 6). However this only applies if the foetus is not yet viable (Act No. 194, section 7).

---

[358] Council of Europe: CLONING – COMPARATIVE STUDY ON THE SITUATION IN 44 STATES, Strasbourg, 4 June 1998, pp 140, 141; available in English at http://legal.coe.int/bioethics/gb/pdf/txtprep2b.pdf. By now no case of PID has become known, Deutscher Bundestag, Schlußbericht der Enquete-Kommission Recht und Ethik in der modernen Medizin, 2002, p. 200.

[359] Deutscher Bundestag, Schlußbericht der Enquete-Kommission Recht und Ethik in der modernen Medizin, 2002, p. 200 adds that the Draft for the Act on the Regulation of Assisted Reproduction does not relate to PID but merely prohibits any form of eugenic selection.

[360] Pattinson 2002, Appendix, 2, who reports that a court decided contrary to a provision of the Code in its judgment in 2000. Accordingly Simon: Survey on Regulation of Gene Therapy in some European Countries in: Müller/Simon/Vesting (editor), Interdisciplinary Approaches to Gene Therapy, Legal Ethic and Scientific Aspects, Berlin, Heidelberg, New York, 1997, 129.

[361] Council of Europe: CLONING – COMPARATIVE STUDY ON THE SITUATION IN 44 STATES, Strasbourg, 4 June 1998, p. 140; available in English at http://legal.coe.int/bioethics/gb/pdf/txtprep2b.pdf.

Japan

SOURCES OF REGULATIONS

Acts of the Legislature

- The Law Concerning Regulation Relating to Human Cloning Techniques and other Similar Techniques (Adopted 30th November 2000. In force June 2001) (RRHCT)[362]
- Criminal Code 1908, Chapter 29, Articles 212-216 (CC)
- Maternal Protection Law 1996 (MPL)

Official Regulations

- None

Official Guidelines

- Guidelines concerning the derivation and the use of Human Embryo Stem Cells. In force from September 2001. (GDUHESC)[363]
- The Guidelines for Handling of a Specified Embryo. In force from December 2001 (GHSE)[364]
- "Guidelines for Genetic Counseling and Prenatal Diagnosis" Japan Society of Human Genetics, December 1994 (GGCPD) [365]
- "Guidelines on Pre-implantation Genetic diagnosis" Japan Society of Obstetrics and Gynecology (JSOG) October 1998

International Conventions

- Universal Declaration on the Human Genome and Human Rights. (UDHGHR)[366]

Recommendations

- No information obtained.

OVERVIEW OF REGULATIONS

Embryo Research

Definition of "embryo"

A cell (except for a germ cell) or cells which has/have potential to grow into an individual through the process of development in utero of a human or an animal and has/have not yet begun formation of a placenta. Article 2 RRHCT

---

362 Translation available from the Ministry of Education, Culture, Sports, Science and Technology (MEXT) website at http://www.mext.go.jp/a_menu/shinkou/seimei/eclone.pdf. Guidance on legislation and guidelines provided by Tomoko Hamada.
363 Translation available at http://www.mext.go.jp/a_menu/shinkou/seimei/2001/es/020101.pdf.
364 Translation available at http://www.mext.go.jp/a_menu/shinkou/seimei/2001/hai3/31_shi shin_e.pdf.
365 Review of guidelines available at: http://www.biol.tsukuba.ac.jp/~macer/EJ64/EJ64I.html.
366 Adopted by consensus, UNESCO General Conference, 11 November, 1997.

Research

Permitted with conditions.

Sources of embryos

Research confined to supernumerary embryos from IVF. Article 6(1) GDUHESC. An animal-human chimeric embryo is permitted to be created for the purpose of "research concerning production of human cells-derived organs transplantable to a human being" (Article 2(1) GHSE) only when scientific knowledge which could not be acquired from the animal embryos/cells alone is acquired from the production (Article 1 GHSE).

Time limit for research

Fourteen days (excepting the period when frozen) Article 6(4) GDUHESC & Article 7 GHSE).

Other conditions governing research

Consent procedures

Informed consent of donors required Article 6(2) (Details Articles 22-24) GDUHESC & Article 3 GHSE.

Licenses

No information obtained.

Ethics Committee

Research must be approved both by the particular Institutional Review Board and the National Committee.

Substantive (Must the embryo benefit?)

No information obtained.

Storage/Disposal of embryos

Embryos must be stored frozen Article 6(3) GDUHESC.

Cloning

Reproductive

Prohibited.

Article 3 RRHCT prohibits the transfer of a human somatic clone into the uterus of a human or animal.

Non-reproductive

RRHCT does not expressly prohibit the creation of a human somatic clone embryo. It is included on the list of nine specified embryos. However the GHSE which were issued under Article 4 of the Act only allow for the creation of human-animal chimeric embryos at the present time. (On advice) the Council for Science and Technology Policy wish to postpone decisions with regard to the other eight whilst they deliberate "the method of handling of a human fertilized embryo". The result being that although human somatic clone embryos could potentially be created under the Act, the Guidelines serve to temporarily restrict their creation until further research and debate on ethical issues.[367]

---

[367] Information provided by Ms. Tomoko Hamada ("Hatopoppo").

Gene Therapy

Somatic

No information obtained.

Germ-line

Production of germ cells from human ES cells is prohibited. Article 27(4) GDUH-ESC.

Pre-Implantation Genetic Diagnosis

Guidelines (JSOG) sanction in vitro fertilisation for purposes other than the treatment of infertility[368].

Guidelines are not specific as to which diseases are to be tested for.

Each case will be assessed on its merits by a specialist committee of doctors and geneticists.

Various factors will be assessed before the society gives approval for a genetic test, including:

– the type of targeted disease,
– the genetic histories of the couple, and
– the facilities at institutions carrying out the test[369]

Sex selection prior to conception is only permissible to prevent the conception of fetuses with severe sex-linked recessive genetic disorders (a joint decision made by the JSOG and Ethical Committee of the Japan Medical Association (JMA) in September 1996).[370]

Stem Cell Therapy

Embryonic stem cell research is limited to:

– basic research on life science aiming to investigate the mechanism of human birth, as well as the function and differentiation and regeneration of the human body
– medical research directed to develop new methods of diagnosis and treatment[371]

Until specific criteria other than guidelines are in place, it is not permitted to transplant the embryonic stem cells or their differentiated cells and tissues into human bodies or utilise them in medicine and it's associated fields. Article 2 GDUHESC.

Pre-Natal Diagnosis/Abortion

• Prenatal diagnosis may be carried out during the first half of pregnancy when there is the possibility the foetus has a serious genetic disease and highly accu-

---

[368] Seouka 1999,.
[369] Saegusa 1998, 110 and Shirai 2001.
[370] Macer 2000.
[371] Website Parkinson's Disease Patients in search of Cure and Comfort on the page Medical Information: Regulations on Regenerative Medicine in Japan located at http://kterrace.cside6.com/pd/pd_contents_english/medinfo_english/regulations_regenerativemed.html. This page is said to be edited and translated by Ms Tomoko Hamada ("Hatopoppo") from publications issued by the Ministry of Education, Culture, Sports, Science and Technology and is stated as updated to July 6, 2001.

rate diagnostic information can be obtained by a particular method.

Amniocentesis and Chorioinic Villous Sampling are considered in the following cases:

- either parent is a carrier of a chromosomal abnormality
- if there is an existing child with a chromosomal abnormality
- advanced maternal age
- the woman is heterozygous for a serious x-linked disease
- the man and the woman are heterozygous for a serious autosomal recessive disease
- either is heterozygous for a serious autosomal dominant disease
- risk of serious foetal abnormality

With the exception of (4), the sex of the foetus should not be revealed (GGCPD).

- The CC prohibits the performance of abortions. However the MPL (amending the Eugenic Protection Law) allows legal abortions on the grounds of:
  - saving the woman's life
  - preserving her physical health
  - the pregnancy being the result of rape or incest
  - economic or social reasons

No mention is made of the preservation of mental health or foetal impairment but abortion on those grounds may be covered under social reasons.

The time limit for the performance of an abortion is that of the viability of the foetus (a notice from the Ministry of Health, Labour and Welfare sets the current limit at 22 weeks).[372]

Luxembourg

SOURCES OF REGULATIONS

Acts of the Legislature

- Law of 1978
- – Proposed Legislation –
- Law No. 4567(29 April 1999) (PL4567)

Official Regulations

- None

Official Guidelines

- None

---

[372] United Nations (2001).

International Conventions
- Convention on Human Rights and Biomedicine ETS 164, Oviedo, 4.IV.1997 (CHRB) (Signed 04/04/97)[373]
- Additional Protocol to the Convention on Biomedicine concerning the prohibition of cloning ETS 168, Paris, 12.I.1998 (APC) (Signed 12/01/98)[374]
- Universal Declaration on the Human Genome and Human Rights. (UDHGHR)[375]

Recommendations
- None.

OVERVIEW OF REGULATIONS

Embryo Research

Definition of "embryo"

No information.

Research

Permitted with conditions (PL4567).

Sources of embryos

Supernumerary embryos from IVF.[376] Creation of embryos for commercial, industrial, study or experimentation purposes prohibited (PL4567 Art. 8).[377]

Time limit for research

Fourteen days.[378]

Other conditions governing research

Consent procedures

Written consent of gamete donors required.[379]

Licenses

Only non-profit organizations can collect, treat and conserve gametes (PL4567 Art. 15). [380]

It is a criminal offence to obtain embryos by payment (PL4567 Art. 28).[381]

Ethics Committees

The assent of the National Committee of Reproductive Medicine and Biology required.[382]

---

[373] See http://conventions.coe.int/Treaty/EN/cadreprincipal.htm.
[374] Ibid.
[375] Adopted by consensus, UNESCO General Conference, 11 November, 1997.
[376] See European Parliament 2001b, 4.
[377] See European Parliament 2001a, 17.
[378] See European Parliament 2001b, 8.
[379] See European Parliament 2001a, 17.
[380] Ibid.
[381] Ibid.
[382] Ibid.

Substantive (Must the embryo benefit?)

Yes. Research must have a medical aim and may not endanger the embryo.[383]

Storage/Disposal of embryos

No information.

Cloning

Reproductive

Prohibited (UDHGHR Art.11).

Non-reproductive

No information.

Gene Therapy

Somatic

No information.

Germ-line

No information.

Pre-Implantation Genetic Diagnosis

No information.

Stem Cell Therapy

No information.

Pre-Natal Diagnosis/Abortion

• Abortion Permitted:

(a) before 12 weeks

    i) if the continuation of the pregnancy or the conditions of life that would enable birth would threaten the woman's physical and mental health

    ii) if there is a substantial risk that the child, if it were born, would be very sick or be physically or mentally seriously handicapped

    iii) if the pregnancy can be considered the result of rape

(b) after 12 weeks, if two medical doctors ascertain in a written statement that birth of the child

(c) presents a serious risk to the health of the pregnant woman or the child to be born.[384]

Malta

SOURCES OF REGULATIONS

Acts of the Legislature

• Criminal Code (CAP 9) 1854 (CC) [385]

---

[383] See European Parliament 2001b, 4.

[384] See MacKellar 1997, 19.

[385] Available at http://justice.magnet.mt/dir2-laws/toppage.asp.

Official Regulations

• None

Official Guidelines

• None

International Conventions

• Universal Declaration on the Human Genome and Human Rights. (UDHGHR)[386]

Recommendations

• Malta Bioethics Consultative Committee Report 1999 (MBCC)[387]

OVERVIEW OF REGULATIONS

Embryo Research

Definition of "embryo"

None.

Research

Recommendation for legislative prohibition (MBCC rec. 8(2)). Under article. 1 of the general considerations section of the MBCC it is stated that "Since human life exists from the moment of conception, it deserves the respect that is due to a human being at all stages of development."

Sources of embryos

Recommendation that creation of embryos for experimentation be made illegal (MBCC rec. 8(1)).

Time limit for research

See *Research*.

Other conditions governing research

Consent procedures

See *Research*.

Licenses

See *Research*.

Ethics Committees

The MBCC reports to the Minister for Health.

Substantive (Must the embryo benefit?)

See *Research*.

Storage/Disposal of embryos

Recommended that embryos should not be stored for future use (MBCC Art. 12).

---

[386] Adopted by consensus, UNESCO General Conference, 11 November, 1997.
[387] MBCC 1999. Available at http://www.thesynapse.net/. Registration will be necessary to access document.

Cloning

Reproductive

Prohibited. (UDHGHR art.11).

The Minister of Health has publicly stated that the Government intends to sign the European Convention on Biomedicine and the additional Protocol on Cloning. However, a decision has not yet been finalised. [388]

Non-reproductive

Not regulated.

Gene Therapy

Somatic

Permitted provided it is intended solely for the benefit of the embryo, there is no other way to correct specific abnormalities and the mother's consent has been given. It must be ensured that there is no undue risk to the embryo or mother and the expected benefits justify the risks associated with the procedure (MBCC Art 21).

Germ-line

Recommendation for legislative prohibition (MBCC Art. 22 & rec. 8(4)).

Pre-Implantation Genetic Diagnosis

Permitted provided it is intended solely for the benefit of the embryo, there is no other way to correct specific abnormalities and the mother's consent has been given. It must be ensured that there is no undue risk to the embryo or mother and the expected benefits justify the risks associated with the procedure (MBCC Art 21). Sex selection permitted only to prevent sex-linked disease (MBCC Art. 2 & explanatory memorandum re art.2).

Stem Cell Therapy

Not regulated.

Pre-Natal Diagnosis/Abortion

- The Criminal Code prohibits abortion under all circumstances.

   A woman who procures her own or consents to an abortion, liable on conviction to imprisonment for 18 months to 3 years (CC s.241(2)).

   A physician, surgeon, obstetrician or apothecary who knowingly prescribes or administers the means by which an abortion is procured liable on conviction to imprisonment for 18 months to 4 years and to perpetual interdiction from the exercise of his profession (CC s.243).

   Any one who through imprudence, negligence or non-observance of regulations causes a woman to miscarry a child, liable on conviction to maximum 6 months imprisonment and fine of 1000 liri (CC s.243a).

---

[388] Information provided by Professor Maurice Cauchi. Dr Pierre Mallia further advises that the delay is due to the fact that Ireland did not sign and their reasons for not doing so are being further investigated.

SPECIAL REMARKS

The MBCC's report has been submitted to the Government.

In the section "General considerations" other relevant recommendations include:

1. Research on embryonic tissue obtained after a natural miscarriage is acceptable provided that such research has been approved by the appropriate ethics research committee and the consent of the couple has been obtained in writing (MBCC Art. 26).

2. The use of embryonic tissues for therapeutic purposes is acceptable provided it is the result of a natural miscarriage, and all the criteria laid down in the document "Ethical Guidelines Relating to Transplantation will be observed (MBCC Art. 27).

Apart from those already indicated a call for a ban on the following procedures was included in the recommendations for Legislation:

– cross-species embryo transfer involving human beings.
– donation or preservation of human embryos.
– breeding hybrid embryos (chimeras) involving human cells.
– commercialising and profit making from donation of gametes.
– surrogacy.
– induction of post-menopausal pregnancy.
– donation of stored sperm from dead donor (including husband).

Netherlands

SOURCES OF REGULATIONS

Acts of the Legislature

• Pregnancy Termination Act of 1 May 1981(LTP)[389]
• Penal code (PC)

  – Proposed Legislation –

• Bill containing rules relating to the use of gametes and embryos (Embryos Bill) 2000 (PLEB)[390]

Official Regulations

• No information obtained

Official Guidelines

• No information obtained

---

[389] See Health Council of the Netherlands 1997, 50. English translation available at http://cyber.law.harvard.edu/population/abortion/Nether.abo.htm.

[390] Available at http://www.minvws.nl/documents/Health/eng_embryowettekst.pdf.

International Conventions

- Convention on Human Rights and Biomedicine ETS 164, Oviedo, 4.IV.1997 (CHRB) (Signed 04/04/97)[391]
- Additional Protocol to the Convention on Biomedicine concerning the prohibition of cloning ETS 168, Paris, 12.I.1998 (APC) (Signed 04/05/98)[392]
- Universal Declaration on the Human Genome and Human Rights. (UDHGHR)[393]

Recommendations

- Health Council Recommendations 1997 (HCR97)
- Health Council Recommendations 1998 (HCR98)

OVERVIEW OF REGULATIONS

Embryo Research

– Proposed Legislation –

Definition of "embryo"

"A cell or a complex of cells with the capacity to develop into a human being" (PLEB Art. 1c.).

Research

Presently permitted by default.[394]

Sources of embryos

Surplus embryos from IVF (PLEB Art. 8 & HCR98, 61).

Would expressly permit embryos from IVF to be donated for research (PLEB Art. 8).

PLEB would initially prohibit the creation of embryos specifically for research (EB Art. 24(a)). Provision however made for Royal Decree to be passed repealing article 24 by five years after article 24(a) takes effect (PLEB Art. 32). This Royal decree would bring article 11 into force, which, though it would generally prohibit the creation of embryos specifically for research, would exceptionally permit their specific creation for research, if the scientific research is reasonably likely to lead to the identification of new insights in the field of infertility, in the field of artificial reproduction techniques, in the field of congenital diseases or in the field of transplant medicine, and the research could only be performed by making use of embryos created specifically for research (PLEB Arts.32 & 11). Health Council recommendations prefer the use of surplus embryos on moral grounds but approver the creation of embryos for research where research cannot be conducted using surplus embryos or there are insufficient surplus embryos (HER98, 59).

---

[391] See http://conventions.coe.int/Treaty/EN/cadreprincipal.htm. The Netherlands made a special reservation as follows: *"In relation to Article 1 of the Protocol, the Government of the Kingdom of the Netherlands declares that it interprets the term "human beings" as referring exclusively to a human individual, i.e., a human being who has been born.".*

[392] Ibid.

[393] Adopted by consensus, UNESCO General Conference, 11 November, 1997.

[394] See Gunning and English 1993, 172.

Time limit for research

An embryo should not be allowed to develop outside the human body for longer than 14 days (PLEB Art. 24(c)). The Health Council 1998 report noted that while the international standard is 14 days, some persons call for a later limit of 6 weeks. It refrained from pronouncing on this especially as current knowledge does not allow embryos to be cultured beyond 7 days and the research that the report was concerned with (Preimplantation Genetic Diagnosis & Assited Reproduction) could be conducted in a few days (HCR 98, 61).

Other conditions governing research

Consent procedures

"Adults who are capable of making a reasonable assessment of their interests in this regard" may provide embryos for research (EB Art. 8). Consent is required and must be informed and in writing (PLEB Arts. 8(2)& (3) & 6). If difference of opinion between parties donating embryo, embryo will not be used and consent is revocable until onset of use (PLEB Art. 8(2)).

Licenses

No information.

Ethics Committees

Approval required from the Central Committee which apart from ensuring the scientific necessity and quality of the research must also ensure that the research "satisfies requirements which might reasonably be made of it in other respects" (PLEB Art. 10).

Substantive (Must the embryo benefit?)

No. However the Central Committee shall only deliver a positive opinion on a research protocol concerning scientific research with embryos with which a pregnancy is not generated if:

a) it is reasonably likely that the research will lead to the identification of new insights in the field of medical science;

b) it is reasonably likely that the identification referred to under a. cannot take place via any forms of scientific research methods other than research with the embryos in question or via a less invasive form of research;

c) the research satisfies the requirements of correct scientific research methodology;

d) the research is carried out by, or under the direction of, persons who are expert in the relevant area of scientific research;

e) the research also satisfies requirements which might reasonably be made of it in other respects. (PLEB Art. 10).

Other substantive conditions governing research

a) research must involve a major health issue;.

b) set up and execution of the research must be scientifically sound;

c) envisaged results must not be obtainable any other way;

d) Research must not use any more embryos than strictly scientifically necessary (HCR98, 58).

e) embryos which have been the subject of research may not be transferred to a woman (HCR, 61–62).

Storage/Disposal of embryos

Embryos may be stored for as long as agreed during the consent procedure. Persons keeping embryos and persons providing them may jointly extend period of storage. (PLEB Arts. 8 & 6(2)). Embryos must be destroyed after expiration of agreed period or upon revocation of consent. (PLEB Arts. 8 & 7). Unclear what happens if one of the parties who donated the embryo dies unless that party had consented in writing to its use after his/her death (See (PLEB Arts. 8 & 7).

Cloning

Reproductive

Prohibited (PLEB art. 24(d))

Non-reproductive

The reservation to the APC would permit non-reproductive cloning. In 1998 the government proposed to pass legislation prohibiting the cloning of human beings, but permitting the use of cloning techniques in embryo research (before 14 days after conception).[395]

Gene Therapy

Somatic

Permitted by implication. See (PLEB Art. 24(e)).

Germ-line

Prohibited (PLEB Art. 24(e)). The Health Council has recommended a moratorium on human germ-line gene therapy.[396]

Pre-Implantation Genetic Diagnosis

Presently permitted by default.[397]

Sex selection prohibited, unless to prevent serious sex linked hereditary disease (PLEB Art. 26(2)). Health Council also approves of PGD to prevent particular genetic abnormalities in the child or where the child will have carrier status, (even where the condition is recessive), to prevent future suffering (HCR98, 29–30).

Stem Cell Therapy

No information.

PRE-NATAL DIAGNOSIS/ABORTION

- Abortion permitted:
  - only if woman is in distress and there is no other choice. Woman must be counselled as to options apart from termination, and physician must be sat-

---

[395] Pattinson 2002, Appendix 4.
[396] Ibid., Appendix 5.
[397] Schenker 1997, 178.

isfied woman has made request of her own free will, conscious of her responsibilities to her unborn child and of the consequences to herself and those nearest to her. (LTP s.5).

A period of 6 days must elapse between the woman's expression of intent to her physician (LTP s.5), unless the woman's life or health is in imminent danger (LTP s.16(2)).

– Once fetus is viable PC applies. PC states that "The term 'to take the life of another person or of a child at the time of birth or shortly after birth' includes the killing of a fetus which may reasonably be assumed to be capable of remaining alive outside the mother's body". (PC 82(a)). Fetus now considered viable at 24 weeks. Even though PC applies after 24 weeks and not the LTP, the defence of "force majeure" or necessity would avail as a defence to homicide or murder if the life of the mother were in danger. It is uncertain whether this defence would apply if viable fetus aborted due to severe abnormalities.[398]

New Zealand

SOURCES OF REGULATIONS

Acts of the Legislature

- Crimes Act 1961(as amended) (CA)[399]
- Contraception, Sterilisation, and Abortion Act 1977 (CSA)[400]
- The Medicines (Restricted Biotechnical Procedures) Amendment Act 2002 (MRBPAA)

    – Proposed Legislation –

- Human Assisted Reproductive Technology Bill 1996 (PLHART) (Modelled on British, Canadian and Australian Legislation
- Assisted Human Reproduction Bill 1998)(PLAHR)[401]

Official Regulations

- No information obtained

Official Guidelines

- No information obtained

International Conventions

- Universal Declaration on the Human Genome and Human Rights. (UDHGHR)[402]

---

[398] See Braake 2000, 388 & 391 for discussion.
[399] Available at http://rangi.knowledge-basket.co.nz/gpacts/actlists.html.
[400] Ibid.
[401] The New Zealand Government has indicated that both the PLHART and the PLAHR need updating due to technological and scientific advances since they were drafted. Information provided by Jenny Hawes of the New Zealand Ministry of Health.
[402] Adopted by consensus, UNESCO General Conference, 11 November, 1997.

Recommendations
- No information obtained.

OVERVIEW OF REGULATIONS

Embryo Research

Definition of "embryo"

In this Act, unless the context otherwise requires, „embryo" means a live human embryo where fertilisation is complete and, for this purpose, fertilisation is not complete until the appearance of a two cell zygote. (PLHART s.4(1)).

References in this Act to gametes, eggs or sperm, except where otherwise stated, are to live human gametes, eggs or sperm; but references hereinafter to gametes or eggs do not include eggs in the process of fertilisation. (PLHART s.4(4)).

Research

Permitted with conditions.

Sources of embryos

Embryos may be created for research (PLHART s.4(1)).

Time limit for research

Until the appearance of the primitive streak which is deemed to be not later than 14 days from the mixing of the gametes excluding any storage time (PLHART ss.19(1)(a) & (2)).

Other conditions governing research

Consent procedures

One guiding principle of PLHART includes the right of informed consent to experimentation (PLHART s.3(b)). It is uncertain whether this extends to experimentation on embryos. (See PLHART s.22(b)).

Licenses

Required (PLHART ss.8; 18(1)(c); 20 & Second Schedule part 3 & Third Schedule s.3).

Ethics Committees

Human Assisted Reproductive Technology Authority would be established under PLHART s. 11 and charged with maintaining a code of practice (PLHART s.14(c)).

The PLAHR s 6(1) proposes to establish the National Ethics Committee on Assisted Human Reproduction. However the PLAHR does not contemplate embryo research.

Substantive (Must the embryo benefit?)

No (PLHART Second Schedule part 3). However research is only permitted where it is necessary or desirable for

i) promoting advances in the treatment of infertility and the techniques of contraception (PLHART Second Schedule part 3 (2)(a) & (d));

ii) increasing knowledge about the causes of congenital disease and of miscarriage (PLHART Second Schedule part 3 (2)(b & (c));

iii) detecting gene/chromosome abnormalities in embryos before implantation (PLHART Second Schedule part 3 (2)(e)); and

iv) any other purpose specified in regulations which may only be to increasing knowledge about the creation and development of embryos or about disease and to enable the application of such knowledge (PLHART Second Schedule part 3 (3)).

Storage/Disposal of embryos

Embryos may be stored for a maximum of 5 years (PLHART Second Schedule part 2 s.(3).

Cloning

Reproductive

Prohibited unless authorisation was granted or recommended by the Minister of Health, who would need to be satisfied that:

(a) the conduct of the procedure or class of procedure does not pose an unacceptable risk to the health or safety of the public

(b) any risks posed by the conduct of the procedure or class of procedure will be appropriately managed

(c) any ethical issues have been adequately addressed

(d) any cultural issues have been adequately addressed

(e) any spiritual issues have been adequately addressed (MRBPAA)[403]

Prohibited. (PLHART s.19(1)(f)).

Prohibited where 1 or more genetically identical human cells or organisms are derived from a single parent, by means of a biological mechanism of parthogenic reproduction (PLAHR ss.2(1) & 4(1)).

Non-reproductive

Prohibited (PLHART s.19(1)(f)).

Prohibited where 1 or more genetically identical human cells or organisms are derived from a single parent, by means of a biological mechanism of parthogenic reproduction (PLAHR ss.2(1) & 4(1)).

Gene Therapy

Somatic

Permitted by implication (PLAHR s.19(1)(g)).

Germ-line

Prohibited (PLAHR s.19(1)(g)).

Pre-Implantation Genetic Diagnosis

National Ethics Committee on Assisted Human Reproduction (NECAHR) allows genetic testing research on early embryos (IBAC 2001, 15).

---

[403] Information provided by Jenny Hawes of the New Zealand Ministry of Health.

Stem Cell Therapy

No stem cell research on human embryos was being carried out up to 2001.[404]

Pre-Natal Diagnosis/Abortion

- Abortion permitted:
  - a) up to 20 weeks where –
    - i) the continuance of the pregnancy would endanger the life, or the physical or mental health, of the woman; or
    - ii) there is a substantial risk that the child, if born would be so physically or mentally abnormal as to be seriously handicapped; or
    - iii) the pregnancy is the result of sexual intercourse between close relatives or rape; or
    - iv) the woman is mentally handicapped
  - b) up to birth, where the miscarriage is necessary to save the life of the woman or to prevent serious permanent injury to her physical or mental health. (CA s.187A)

    Must be authorised by 2 certifying consultants (CSA s.33).

Norway

SOURCES OF REGULATIONS

Acts of the Legislature

- The Act Relating to the Application of Biotechnology in Medicine. Law No. 56 of 5 August 1994, as amended in February 1998 (L56)
- The Act Concerning Termination of pregnancy and Regulations for the implementation of the Act 1978

Official Regulations

- No information

Official Guidelines

- No information

International Conventions

- Convention on Human Rights and Biomedicine ETS 164, Oviedo, 4.IV.1997 (CHRB) (Signed 04/04/97)[405]
- Additional Protocol to the Convention on Biomedicine concerning the prohibition of cloning ETS 168, Paris, 12.I.1998 (APC) (Signed 12/01/98)[406]
- Universal Declaration on the Human Genome and Human Rights. (UDHGHR)[407]

---

[404] IBAC 2001, 11.
[405] See http://conventions.coe.int/Treaty/EN/cadreprincipal.htm.
[406] Ibid.
[407] Adopted by consensus, UNESCO General Conference, 11 November, 1997.

Recommendations

• No information

OVERVIEW OF REGULATIONS

Embryo Research

Definition of "embryo"

Not defined. Reference made throughout to "fertilised eggs" (L56).

Research

Not permitted (L56 chap. 3.1).

Sources of embryos

See *Research*.

Time limit for research

See *Research*.

Other conditions governing research

Consent procedures

See *Research*.

Licenses

See *Research*.

Ethics Committees

See *Research*.

Substantive (Must the embryo benefit?)

See *Research*.

Storage/Disposal of embryos

Fertilized eggs may not be stored for more than 3 years (L56 chap. 2.12).

Cloning

Reproductive

Prohibited. (L56 chap.3a.1).[408]

Non-reproductive

No information obtained.

Gene Therapy

Somatic

Permitted for treatment or prevention of serious diseases (L56 chap. 7.1). Treatment must be approved by the Biotechnology Board established under L56 chap.8.4. (L56 chap.7.2). Written consent must be obtained from the person to be treated or the parent or guardian of a child under 16 who is to be treated. (L56 chap. 7.3).

Germ-line

Prohibited (L56 chap. 7.1).

---

[408] Introduced by amendment adopted February 1998.

Pre-Implantation Genetic Diagnosis

Permitted only in special cases involving serious hereditary disease with no possibility of treatment (L56 chap. 4.2). Sex selection is prohibited except in special cases involving serious hereditary sex-linked diseases (L56 s. 4.3). Before authorization, woman or couple must receive genetic counselling and information. (L56 chap. 4.4).

Stem Cell Therapy

No information obtained.

Pre-Natal Diagnosis/Abortion

- Prenatal Diagnosis permitted provided:
  a) it is approved by the Ministry after an opinion has been received from the Biotechnology Board (L56 chap. 5.2);
  b) the woman or couple is given detailed genetic counselling and information on the possible consequences for all parties including the child (L56 chap. 5.3);

  Information about the sex of the fetus obtained from prenatal diagnosis performed before the 12th week of pregnancy to be disclosed to woman only if she is carrier of a serious sex-linked disease. (L56 chap. 5.4).

- Abortion Permitted
  a) before the end of the 12th week, on application of the woman to a doctor, provided there are no serious medical reasons proscribing it.
  b) after the 12th and before the end of the 18th week with the permission of a medical board if:
     i) the pregnancy, childbirth, or care of child may result in unreasonable strain on the physical or mental health of the woman, account being taken of whether she is predisposed to illness;
     ii) the pregnancy, childbirth or care of child may place the woman in a difficult life situation;
     iii) there is a major risk that the child will suffer from a serious disease as a result of its genotype, or disease or harmful influence during pregnancy;
     iv) the pregnancy is the result of rape;
     v) the woman is suffering from severe mental illness or retardation.
  c) after the 18th week, only if there are particularly important grounds for doing so. If there is reason to believe the fetus is viable termination will not be authorised.
  d) at anytime prior to birth where pregnancy constitutes an impending risk to woman's life or health.

SPECIAL REMARKS

Maximum penalty under L56 is only 3 months imprisonment (L56 chap. 8.5).

Poland

SOURCES OF LEGISLATION

Acts of Legislation

- Act from 6 June 1997[409] promulgating the Penal Code
- Act from 7 January 1993 on Family Planning, the Protection of Emerging Life and the Conditions for the Legal Termination of Pregnancy[410]
- Penal Code from 6 June 1997[411]

Official Regulations

No information obtained.

Official Guidelines

No information obtained.

International Conventions

- Convention on Human Rights and Biomedicine ETS Oviedo, 4 April 1997 (CHRB) (signed on 7 May 1999)[412]
- Additional Protocol to the Convention on Biomedicine Concerning the Prohibition of Cloning ETS 168, Paris, 12 January 1998 (APC) (signed on 7 May 1999)[413]
- Universal Declaration on the Human Genome and Human Rights (UDHGHR)[414]

Recommendations

No information obtained.

OVERVIEW OF REGULATIONS

Embryo Research

Definition of "embryo"

There is no legislation defining the embryo.[415]

---

[409] Dziennik Ustaw Rzeczypospolitej Polskiej,2 August 1997, No. 88, 2677-2716, a summary is available at *http://www-nt.who.int/idhl/*.

[410] Dziennik Ustaw Rzeczypospolitej Polskiej, No. 17, poz. 78.

[411] Dziennik Ustaw Rzeczypospolitej Polskiej, No. 88, poz.553.

[412] See http://conventions.coe.int.

[413] See http://conventions.coe.int.

[414] Adopted unanimously and by acclamation at the 29th Session of UNESCO's General Conference, 11 November 1997 (29 C/Resolution 17 – see http:www.unesco.org/ibc/), and endorsed by the United Nations' General Assembly, 85th Plenary Meeting on 9 December 1998 (A/RES/53/152 – see http://www.un.org.documents/).

[415] Council of Europe: CLONING – COMPARATIVE STUDY ON THE SITUATION IN 44 STATES, Strasbourg, 4 June 1998, 135, available in English at http://legal.coe.int/ bioethics/gb/pdf/txtprep2b.pdf.M. M. Nesterowicz, Prawo Medyczne *(Medical Law)*, 5. edition, Torun, 2001, p.74, who establishes that Poland lacks statutory regulations in the field of bioethics.

Research

It is permitted in view of specific diseases (hereditary diseases and serious risks).[416]
The Act on the Medical Profession, Art. 26 para. 3, prohibits experimental examinations referring to emerging life.[417]

Sources of embryos

Embryos must not be created solely for research purposes.[418]

Time limit for research

No regulations.

Other conditions governing research

No regulations.

Consent procedures

No information obtained.

Licenses

No information obtained.

Ethics Committees

No information obtained.

Substantive (Must the embryo benefit?)

Storage/Disposal of embryos

No information obtained.

Storage/ Disposal of embryos

No regulations.

Cloning

Reproductive cloning

No regulations.

Non-reproductive

No regulations.

Gene Therapy

Somatic

No regulations.

Germ-line

Therapeutic interference with the germ cell line is prohibited in general.[419]

---

[416] Council of Europe: CLONING – COMPARATIVE STUDY ON THE SITUATION IN 44 STATES, Strasbourg, 4 June 1998, p. 135, available in English at http://legal.coe.int/bioethics/gb/pdf/txtprep2b.pdf.M.

[417] M. Nesterowicz, Prawo Medyczne *(Medical Law)*, 174.

[418] Council of Europe: CLONING – COMPARATIVE STUDY ON THE SITUATION IN 44 STATES, Strasbourg, 4 June 1998, 135, available in English at http://legal.coe.int/bioethics/gb/pdf/txtprep2b.pdf.M.

[419] Council of Europe: CLONING – COMPARATIVE STUDY ON THE SITUATION IN 44 STATES, Strasbourg, 4 June 1998, 135, available in English athttp://legal.coe.int/bioethics/gb/pdf/txtprep2b.pdf.M.

Pre-Implantation Genetic Diagnosis

There are no regulations on in vitro fertilisation in Poland.[420] Consequently, no rules on pre-implantation genetic diagnosis exist. At present only the Code of Medical Ethics, applies. In Art. 37 it states the general duties of the doctor with respect to emerging life.[421]

Stem Cell Therapy

No regulations.

Pre-Natal Diagnosis/Abortion

Pre-natal diagnosis

Is permitted pursuant to the Penal Code, section 23 b if the parents suffer from a hereditary disease, if a hereditary disease can be treated during the pregnancy and if there is a fear that the embryo is seriously deformed.[422]

Abortion

Anyone who, contrary to the regulations, with the pregnant woman's consent performs an abortion, or helps or encourages her to have an abortion, commits an offence and is liable to imprisonment for a maximum 3 years. If the foetus was able to exist outside of the woman's body, the maximum term of imprisonment is 8 years (Penal Code, Art. 152).[423] The Constitutional Court has deemed abortion for social reasons. The corresponding regulation permitting such abortions has therefore been null and void since 23 December 1997.[424]

Portugal

SOURCES OF REGULATIONS

Acts of the Legislature

- Constitution of the Portuguese Republic (CON)[425]
- Penal Code Chapter II (PC)

Official Regulations

- Order 5411/97(O97)

Official Guidelines

- No information obtained

---

[420] M. Nesterowicz, Prawo Medyczne, 184 & 201.

[421] Art. 37 Kodeksu Etyki Lekarskiej, see Nesterowicz, Prawo Medyczne, 185.

[422] Council of Europe: CLONING – COMPARATIVE STUDY ON THE SITUATION IN 44 STATES, Strasbourg, 4 June 1998, p. 135, available in English at http://legal.coe.int/bioethics/gb/pdf/txtprep2b.pdf.M. Pursuant to the Act from 7 January 1993, Art. 2, the government and the territorial self-governing bodies are obliged to provide pre-natal care of emerging life (para. 1) and to grant free access to pre-natal examinations (para. 2a).

[423] Inserted into the Penal Code by Act No. 88 from 2 August 1997.

[424] Dziennik Ustaw Rzeczypospolitej Polskiej, No. 157, poz. 1040.

[425] Available at http://www.parlamento.pt/leis/ constituicao_ingles/crp_uk.htm.

International Conventions

- Convention on Human Rights and Biomedicine ETS 164, Oviedo, 4.IV.1997 (CHRB) (Signed 04/04/97, Ratified 13/08/01, Entered into force 01/12/01)[426]
- Additional Protocol to the Convention on Biomedicine concerning the prohibition of cloning ETS 168, Paris, 12.I.1998 (APC) (Signed 12/01/98, Ratified 13/08/01, Entered into force 01/12/01)[427]
- Universal Declaration on the Human Genome and Human Rights. (UDHGHR)[428]

Recommendations

- Report-Opinion on Experimentation on the Human Embryo 1995 (CNEV95)
- Opinion 21/CNEV/97 from National Council on Ethics for the Life Sciences (CNEV 97)

OVERVIEW OF REGULATIONS

Embryo Research

Definition of "embryo"

None.

Research

Apparently permitted by default as a 1998 Law on Assisted Procreation which would have banned embryo research, was vetoed by the President.[429] However appears no research is being performed.[430] Research could also be prohibited by Con Art. 26 which states "The law shall guarantee the personal dignity and genetic identity of the human being, particularly in the creation, development and use of technology and in scientific experimentation", and also, CON Art. 67.2(e), which places a duty on the State to protect the family by "regulating assisted procreation, in such terms as safeguard human dignity."

Sources of embryos

Production of embryos for research prohibited (CHRB art. 18(1) & CNEV95).

Time limit for research

No information obtained.

Other conditions governing research

Consent procedures

No information obtained.

Licenses

No information obtained.

---

[426] See http://conventions.coe.int/Treaty/EN/cadreprincipal.htm.
[427] Ibid.
[428] Adopted by consensus, UNESCO General Conference, 11 November, 1997.
[429] See Europarl. 2001, 6.
[430] See EGESNT 2000, 11.

Ethics Committees

No information obtained.

Substantive (Must the embryo benefit?)

No information obtained.

Storage/Disposal of embryos

No information obtained.

Cloning

Reproductive

Prohibited. (APC Art 1, UDHGHR Art.11 & CNEV97).

Non-reproductive

No information obtained.

Gene Therapy

Somatic

Permitted (CHRB Art. 13).

Germ-line

Prohibited (CHRB Art. 13). Such therapy may also be inconsistent with Con Art. 26 which states "The law shall guarantee the personal dignity and genetic identity of the human being, particularly in the creation, development and use of technology and in scientific experimentation", and also, CON Art. 67.2(e), which places a duty on the State to protect the family by "regulating assisted procreation, in such terms as safeguard human dignity."

Pre-Implantation Genetic Diagnosis

Permitted only for health purposes subject to genetic counselling (CHRB art. 12). Sex selection only permitted to avoid serious hereditary sex-linked disease (CHRB Art. 14).

Stem Cell Therapy

No information obtained.

Pre-Natal Diagnosis/Abortion

- *Pre-natal Diagnosis*

  Embryos and fetuses have some constitutional protection under CON Art. 24 (1), which states "Human life is inviolable".[431] Further CON Art. 26(3) states, "The law shall guarantee the personal dignity and genetic identity of the human being, particularly in the creation, development and the use of technology and in scientific experimentation."

  CON Art. 13 prohibits any form of discrimination against anyone by reason of their sex. Therefore though sex selection is not expressly prohibited, since the general abortion provisions also apply, sex selection except to avoid a serious X-linked disease would be unlawful.

---

[431] See Pereira 2001, 495.

In O97, prenatal diagnosis is defined as "a set of procedures that are carried out to determine whether an embryo or foetus has or does not have a congenital abnormality" (O97Art. 1). For the test to be performed there must be a "large probability" that a serious genetic disease will be detected, and their must be genetic counselling before the test (O97Art. 3). "Serious genetic disease" is not defined.[432]

- *Abortion Permitted:*
  a) up to12 weeks, if it is deemed necessary to remove the danger of death or serious and irreversible damage to the women's body, physical or mental health (PC Art.1(b)).
  b) up to 16 weeks, if the pregnancy results from a crime against sexual freedom and self-determination (PC art. 1(d)).
  c) up to 24 weeks, if there are strong medical reasons indicating that the unborn child has a serious incurable illness or congenital malformation (PC Art. 1(c)).
  d) up to birth if it is the only means of removing] risk of death or serious and irreversible damage to the body or physical or mental health of the pregnant woman (PC art. 1(a)), or if the fetus is not viable (PC art. 1(c)).
  e) up to birth, (Art. 142, no. 1 of the Penal Code).

Pregnant woman or her best available representative should whenever possible sign a consent to the abortion three days prior to the procedure (PC art. 3(a) & (b)).

Romania

SOURCES OF REGULATIONS

Acts of the Legislature
- Law No. 140, November 1996
- Law No. 7, 22/02/2001

Official Regulations
- No information obtained

Official Guidelines
- No information obtained

International Conventions
- Convention on Human Rights and Biomedicine ETS 164, Oviedo, 4.IV.1997 (CHRB) (Signed 04/04/97, Ratified 24/04/01, Entered into force 01/08/01)[433]
- Additional Protocol to the Convention on Biomedicine concerning the prohibition of cloning ETS 168, Paris, 12.I.1998 (APC) (Signed 12/01/98, Ratified 24/01/00, Entered into force 01/08/01)[434]

---

[432] Ibid, 496.
[433] See http://conventions.coe.int/Treaty/EN/cadreprincipal.htm.
[434] Ibid.

- Universal Declaration on the Human Genome and Human Rights. (UDHGHR)[435]

Recommendations
- No information obtained

OVERVIEW OF REGULATIONS

Embryo Research

Definition of "embryo"
None.[436]

Research
Not regulated[437] information.

Sources of embryos
Not regulated.[438]

Time limit for research
Not regulated.[439]

Other conditions governing research

Consent procedures
See *Research.*

Licenses
See *Research.*

Ethics Committees
No.[440]

Substantive (Must the embryo benefit?)
No.[441]

Storage/Disposal of embryos
Not regulated.[442]

Cloning

Reproductive
Prohibited (Law 7/2001).[443]

---

[435] Adopted by consensus, UNESCO General Conference, 11 November, 1997.
[436] Council of Europe 1998, 135.
[437] Ibid., 144 and 173.
[438] Ibid., 182.
[439] Ibid., 180.
[440] Ibid., 149.
[441] Ibid., 160.
[442] Ibid., 89.
[443] This law is concerned with implementing the Convention on Human Rights and Biomedicine. There are no specific laws regarding gene therapy, stem cells or PGD. Information provided by Ioana Ispas.

Non-reproductive

It is not know whether law 7/2001 addresses non-reproductive cloning.

Gene Therapy[444]

Somatic gene

No regulation.

Germ-line

No regulation.

Pre-Implantation Genetic Diagnosis[445]

No regulation.

Stem Cell Therapy[446]

No regulation.

Pre-Natal Diagnosis/Abortion

- Abortions are available upon request during the first fourteen weeks of pregnancy. (Law 140)[447]
- Abortion carried out after the fourteen weeks of pregnancy must be for therapeutic purposes. If performed for any other reason the doctor performing the abortion is liable to imprisonment for a maximum of three years, if the woman had consented. If the woman had not consented, the maximum term of imprisonment is seven years and the doctor is also liable to suspension from practice.

SPECIAL REMARKS

Romania's' abortion rate is the highest in Europe (78 abortions per 1000 women aged 15–44 in 1996.

Russia

SOURCES OF LEGISLATION

Acts of Legislation

- Criminal Code from 13 June 1996 as promulgated on 25 July 2002[448]
- Act on the Official Regulation Regarding the Handling of Techniques of Genetic Engineering from 1996[449]

---

[444] Information provided by Ioana Ispas.
[445] Information provided by Ioana Ispas.
[446] Ibid.
[447] United Nations 2001.
[448] Federal Act No. 63, published in the Legislative Collection of the Russian Federation, 1996, No. 25, p. 2954, available in Russian at http://www.duma.gov.ru.
[449] Zweiter Zwischenbericht der Enquete-Kommission Recht und Ethik in der modernen Medizin – Teilbericht Stammzellforschung, BT-Drs. 14/7546, p. 129 & 130, available in German at http://www.bundestag.de/gremien/medi/2zwischen.pdf. which received the relevant information courtesy of Alexander Butrimenko from the Department of Science of the Embassy of the Russian Federation in Germany.

- Addition to Regulatory Act on Genetic Therapy from 1996 (2000)[450]
- Transplantation Act[451]
- Government Decree on the Registration of all Genetically Modified Organisms[452]
- Provisional Prohibition of Cloning of Human Beings from 20 May 2002 (Prohibition of Cloning Act)[453]
- Draft of a Health Protection Act[454]

Official Regulations

- No information obtained

Official Guidelines

- No information obtained

International Conventions

- Universal Declaration on the Human Genome and Human Rights (UDHGHR)[455]

Recommendations

- No information obtained

OVERVIEW OF REGULATIONS

Embryo Research

Definition of "embryo"

The human foetus in its development up to the 8th week is regarded as an embryo pursuant to the Prohibition of Cloning Act, Art. 2.

Research

Research on embryos is not regulated by law.[456] Sale and purchase are prohibited according to the rules of the Transplantation Act.[457]

Sources of embryos

No information obtained.

Time limit for research

No information obtained.

---

450 See footnote no. 433.
451 See footnote no. 433.
452 See footnote no. 433.
453 Federal Act No. 54, published in the Legislative Collection of the Russian Federation, 2002, No. 21, 1917, available in Russian at http://www.duma.gov.ru.
454 Information provided by Maria Tkachova.
455 Adopted unanimously and by acclamation at the 29th Session of UNESCO's General Conference, 11 November 1997 (29 C/Resolution 17 – see http:www.unesco.org/ibc/), and endorsed by the United Nations' General Assembly, 85th Plenary Meeting on 9 December 1998 (A/RES/53/152 – see http://www.un.org.documents/).
456 Council of Europe: CLONING - COMPARATIVE STUDY ON THE SITUATION IN 44 STATES, Strasbourg, 4. June 1998, S. 165, available in English at *http://legal.coe.int/bioethics/gb/pdf/txtprep2b.pdf.* See also footnote 2.
457 See footnote 2.

Other conditions governing research

Consent procedures

No information obtained.

Licenses

No information obtained.

Ethics Committees

No information obtained.

Substantive (Must the embryo benefit?)

No information obtained.

Storage/Disposal of embryos

No information obtained.

Cloning

In Art. 3 the Prohibition of Cloning Act prohibits the cloning of human organisms as well as the import and export of embryos having been created by cloning for a period of 5 years.[458]

Reproductive

No information obtained.

Non-reproductive

No information obtained.

Gene Therapy

Somatic

No information obtained.

Germ-line

Therapeutic interference with the germ cell line is not regulated by law.[459]

Pre-Implantation Genetic Diagnosis

It is not regulated by law.[460] PID is carried out in 14 clinics. In two of them the sex of the embryo or possible genetic diseases can be determined.[461]

Stem Cell Therapy

No information obtained.

---

[458] The Russian Institute of Molecular Genetics had already called for an Act prohibiting the cloning of human beings on 12 January 1998, see Reproductive Cloning: UNESCO Director General's Message, available at http://firewall.unesco.org/opi/eng/bio98/.

[459] Council of Europe: CLONING – COMPARATIVE STUDY ON THE SITUATION IN 44 STATES, Strasbourg, 4. June 1998, S. 169, available in English at *http://legal.coe.int/ bioethics/gb/pdf/txtprep2b.pdf.*

[460] Council of Europe: CLONING - COMPARATIVE STUDY ON THE SITUATION IN 44 STATES, Strasbourg, 4. June 1998, 137, available in English at *http://legal.coe.int/bioethics/ gb/pdf/txtprep2b.pdf.*

[461] See footnote 433.

Pre-Natal Diagnosis/Abortion
- Pre-natal diagnosis is not regulated by law.[462]
- Abortion

A person without higher medical education in this subject area who terminates the pregnancy of a woman will be punished by fine or a social work order pursuant to the Criminal Code, section 123.

Recidivists are liable to a maximum imprisonment of 2 years.

SPECIAL REMARKS

In 1996 the Russian Duma (Parliament) adopted the first Act on Genetic Engineering.[463] However, it does not apply to human beings. Such an Act is still to be passed.[464]

Research on embryos is almost exclusively state sponsored. The existing laws are quite permissive.

SLOVENIA

SOURCES OF REGULATIONS

Acts of the Legislature
- Law of 7 October 1977 (L77)
  - *Proposed Legislation –*
- Gene Technology Bill (PLGTB)
- Bill on Treatment of Infertility and Fertilisation with Biomedical Assistance (1997) (PLBTIF)

Official Regulations
- No information obtained

Official Guidelines
- No information obtained

International Conventions
- Convention on Human Rights and Biomedicine ETS 164, Oviedo, 4.IV.1997 (CHRB) (Signed 04/04/97, Ratified 05/11/98, Entered into force 01/12/99)[465]
- Additional Protocol to the Convention on Biomedicine concerning the prohibition of cloning ETS 168, Paris, 12.I.1998 (APC) (Signed 12/01/98, Ratified 05/11/98, Entered into force 01/03/01)[466]

---

[462] Council of Europe: CLONING – COMPARATIVE STUDY ON THE SITUATION IN 44 STATES, Strasbourg, 4. June 1998, 137, available in English at *http://legal.coe.int/ bioethics/gb/pdf/txtprep2b.pdf.*

[463] Federal Act of the Russian Federation on the Official Regulation Regarding the Handling of Techniques of Genetic Engineering, June 1996.

[464] EFB Task Group on Public Perception of Biotechnology – Biotechnology Legislation in Central and Eastern Europe, EFB Briefing, Paper 9, June 1999, 5 & 9, available at http://www.kluyver.stm.tudelft.nl/efb/tgppb/eng9.htm.

[465] See http://conventions.coe.int/Treaty/EN/cadreprincipal.htm.

[466] Ibid.

- Universal Declaration on the Human Genome and Human Rights. (UDHGHR)[467]

Recommendations

- No information obtained

OVERVIEW OF REGULATIONS

Embryo Research

Definition of "embryo"

The preimplantation embryo is defined as „early embryo" (PLBTIF).[468]

Research

Permitted with conditions.

Sources of embryos

Surplus embryos.[469]

Time limit for research

Up to 14 days, or the stage of the primitive crest.[470]

Other conditions governing research

Consent procedures

No information obtained.

Licenses

No information obtained.

Ethics Committees

National Medical Ethics Committee has to authorize research. Additionally the State Committee for fertilisation with biomedical assistance monitors all activities that concern MAP.[471] After approval of some research projects their follow up is left to Institutional Review Boards which are not represented on the National Medical Ethics Committee.[472]

Substantive (Must the embryo benefit?)

No, provided the research is on surplus embryos and within the 14 day limitation period.[473]

Storage/Disposal of embryos

Maximum period where embryos created for MAP is 6 years. The National Medical Ethics Committee may however approve the use of such embryos for research after this period.[474]

---

[467] Adopted by consensus, UNESCO General Conference, 11 November, 1997.
[468] Counsel of Europe 1998, 136.
[469] Ibid., 164.
[470] Ibid., 164 & 180.
[471] Counsel of Europe 1998, 150.
[472] Ibid.,157.
[473] Ibid., 172.
[474] Ibid., 90.

Cloning

Reproductive

Prohibited (PLBTIF). [475]

Non-reproductive

Prohibited (PLBTIF). [476]

Gene Therapy

Somatic

Permitted provided there is no risk of introducing a genetic modification to patient's descendants (PLGTB).[477]

Germ-line

Prohibited, until new developments in medical technology make it safe, without risking the introduction of unwanted changes in genetically transmissible traits (PLGTB).[478]

Pre-Implantation Genetic Diagnosis

May be practiced with special authorization.[479] No information on what this entails.

Stem Cell Therapy

No legislation[480] No information as to whether they are being pursued.

Pre-Natal Diagnosis/Abortion

- Abortion permitted
  a) up to 10 weeks, on request. If woman a minor approval of her parents or guardian required unless she has been fully recognized as fully competent to earn her own living;
  b) after 10 weeks, on the authorization of a commission if it concludes that the procedure entails a risk to the woman's life, health, or future motherhood that is less than the risk to the woman or the child associated with the continuation of the pregnancy or childbirth.[481]

Spain

SOURCES OF REGULATION

Acts of the Legislature

- Ley 35/1988 sobre Técnicas de Reproducción Asistida (Assisted Reproduction Techniques Act (ARTA) of 22. November 1988[482]

---

[475] Ibid, 217.
[476] Ibid.
[477] Counsel of Europe 1998, 170.
[478] Ibid.
[479] Ibid., 142.
[480] ESFPB 2001, 7.
[481] United Nations 2001.
[482] Boletín Oficial del Estado (BOE, State Gazette) No. 282 of 24. November1988, correction of errors in BOE No. 284 of 26 November 1988. Available at http://www.congreso.es.

- Ley 42/1988 de donación y utilización de embriones y fetos humanos o de sus celulas, tejidos u organos (Donation and Utilization of Human Embryos and Foetus Act – DUA) of 28. December 1988[483]
- Penal Code, 1995
- Ley 9/1985 de la despenalización parcial del aborto (Partial Decriminalisation Act) of 5 July 1985[484]
- Real Decreto 2409/1986 sobre centros sanitarios acreditados (Royal Decree on medical centres) of 21. November 1986[485]

International Conventions

- Universal Declaration on the Human Genome an Human Rights (UDHGHR)[486]
- Convention on Human Rights and Biomedicine ETS 164, Oviedo, 4 April 1997 (CHRB), (Signed 4. April1997/Ratified 1. September 1999/Entered into force 1 January 2000)[487]

Recommendations

- First Report of the National Commission for Assisted Human Reproduction on IVF 1999[488]
- Second Report of the National Commission for Assisted Human Reproduction on Research with surplus Embryos, 2000[489]
- Report on Stem Cell Research of the National Advisory Committee for Ethics of Research in Science and Technology, 2003[490]

OVERVIEW OF REGULATIONS

Embryo Research

Definition of "embryo"

There is no legal definition of embryo.[491]

Research

Research or experimentation is permitted under certain conditions (art. 15.3 ARTA).

---

[483] Boletín Oficial del Estado (BOE, State Gazette) No. 314, of 31. December 1988.
[484] Boletín Oficial del Estado (BOE, State Gazette) No. 166, of 12. July 1985.
[485] Boletín Oficial del Estado (BOE, State Gazette) No. 281, of 24. November 1986.
[486] Adopted unanimously and by acclamation at the 29th session of UNESCO's General Conference, 11 November 1997 (29 C/Resolution 17 - see *http://www.unesco.org/ibc/*), and endorsed by the United Nations General Assembly, 85th Plenary Meeting at 9 December 1998 (A/RES/53/152 – see *http://www.un.org.documents/*).
[487] See *http:conventions.coe.int /Treaty/EN/cadreprincipal.htm.*
[488] See a short summary of the Report at the Spanish Health and Consumption Ministry web site at http://www.msc.es/salud/epidemiologia/ies/repro_asistida/resumen_anual.htm. The full report is available at http://www.msc.es/salud/epidemiologia/ies/repro_asistida/r_asist.pdf.
[489] Available in Spanish at http://www.cnice.mecd.es/tematicas/genetica/2001_12/2001_12_01.html. A summary in Spanish is available at http://www.cnb.uam,es/ transimp/IIinformeCN-RHA.pdf.
[490] Available in Spanish at http://www.actualderechosanitario.com/informecelulastroncales.pdf. and http://www.imasd-tecnologia.com/InformeCM.htm.
[491] See Romeo Casabona (ed.), Código de Leyes sobre Genetica, § 1.1 II, p. 23.

Sources of embryos

The possibility to obtain supernumerary embryos from in-vitro fertilisation is not excluded, because cryopreservation of embryos is permitted.[492] Creation of embryos for research is prohibited, art. 20 B lit. a ARTA.

The fertilisation of human ovules for other purposes than procreation constitutes a crime and can be punished with one year up to five years in prison, art. 161 Penal Code.[493]

Time limit for research

During the first 14 days after the fertilisation of ovum, deducting the time of cryop-reservation, art. 15.1 b ARTA.

Other conditions governing research

Consent procedures

A written informed consent of the genetic parents is needed in cases of applied research (diagnostic or therapeutic purposes) or fundamental research on non-viable embryos, art. 15.1 lit. a ARTA, while there is no alternative for example on animals, art. 16 2 ARTA.[494]

Substantive (Must the embryo benefit?)

Research or experimentation on biologically non-viable embryos is permitted under certain conditions. Aborted preimplantory embryos are considered dead or non-viable and can be used above all for research, art. 17.4 ARTA.

Research on viable in vitro embryos is allowed, when it can be applied to diagnosis with therapeutic or preventive purposes, that means: for the benefit of the embryo. Furthermore modification of genetic non-pathological heritage of embryo is ensured , art. 15.2 ARTA.[495]

Storage/Disposal of embryos

Storage of cryo-preserved embryos in authorised banks is permitted up to five years, art. 11.3 ARTA; embryos not coming from donors remain at the disposal of the bank after two years of cryo-preservation, art. 11.4 ARTA.

---

[492] Conclusion drawn by C. Romeo Casabona, Embryonic Stem Cell Research and Therapy at European Level: Is a common legal framework needed? Revista de Derecho y Genoma Humano, No. 15, 2001, p. 121, 127.

[493] The National Commission on Assisted Human Reproduction presented its report which proposes permitting research using supernumerary embryos from assisted reproduction procedures. See National Commission on Assisted Reproduction (Spain), Second Report: Research using surplus embryos, May 2000, p. 39.

[494] See also the comparison in: Zweiter Zwischenbericht der Enquete-Kommission Recht und Ethik in der modernen Medizin des Deutschen Bundestages, BT-DrS. 14/7546,S. 129, 131 available at *http://www.bundestag.de/gremien/medi/2zwischen.pdf*.

[495] See C. Romeo Casabona, Embryonic Stem Cell Research and Therapy at European Level: Is a common legal framework needed?, Revista de Derecho y Genoma Humano, No. 15, 2001, p. 121, 127.The non-viable embryo has no legal protection under the Spanish Constitution.

Cloning

Reproductive cloning.

Is prohibited and constitutes a crime, which can be punished with prison from one year up to five years, art. 161.2 Penal Code, art. 20 B lit. k.

Non-reproductive cloning

Also not permitted.[496]

Gene Therapy

Germ line

Changing the genotype for other reasons than to eliminate or reduce diseases constitutes a crime and can be punished with prison from two year up to six years, art. 159.1 Penal Code. See also art. 15.2 lit. b ARTA, which only permits investigations if they do not modify the non-pathological genotype.

Pre-Implantation Genetic Diagnosis

Permitted, art. 12.1 and 13.1 and 3 ARTA. The only reason to intervene in an embryo is to evaluate its viability and to detect hereditary diseases to cure them or not to suggest to implant them.

Investigation on embryos is permitted for diagnosis and therapeutic or preventive reasons, art. 15.2 lit. a ARTA.

Pre-Natal Diagnosis/Abortion

Pre-natal diagnosis is permitted, art. 12.2 and 13.2 ARTA and art. 5.1.DUA.

Intervention is permitted for diagnostic and therapeutic reasons.

Abortion is prohibited, art. 145 Penal Code except in cases provided by law. The actor of the crime can be punished with prison from 1 year up to three years, the woman with prison from 6 month up to 1 year or a fine. In case of rape abortion is permitted in accredited medical centres up to 12 week, of grave risk to mental or physical health of the mother up to 22 week, also when there is a defected child to be born. Approval of two medicines is needed, Law 9/1985 and Royal Decree 2409/1986.

Further Reading

Carlos Lema Añón, Los problemas pendientes de la regulación jurídica española sobre reproducción assistida: la sentencia de Tribunal Constitucional y el I. informe de la Comisión Nacional de Reproducción Humana Asistida (Parte I y II) in: Revista de Derecho y Genoma Humano (Law and the Human Genome Review), No. 12, 2000, p.47, 103

Carlos María Romeo Casabona, Los llamados delitos relativos a la manipulación genetica, Escuela Judicial – Consejo General del Poder Judicial/Ilustre Colegio Oficial de Médicos de Madrid, Genética y derecho, Separata, 36, 2002, p. 331

---

[496] European Group on Ethics in Science and New Technologies to the European Commission, Brigitte Gratton, National regulations in the European Union regarding research on human embryos, July 2002, p. 57. Available at. http://europa.eu.int.

Carlos María Romeo Casabona, El derecho y la bioethica ante los limites de la vida humana, Madrid 1994

Carlos María Romeo Casabona, Del gen al derecho, Bogotá 1996

C. Romeo Casabona, Embryonic Stem Cell Research and Therapy at European Level: Is a common legal framework needed?, Revista de Derecho y Genoma Humano, No. 15, 2001, p. 121

European Group on Ethics in Science and New Technologies to the European Commission, Brigitte Gratton, National regulations in the European Union regarding research on human embryos, July 2002, p. 55. Available at. http://europa.eu.int

Sweden

SOURCES OF REGULATIONS

Acts of the Legislature

- Law No.115 of 14 March 1991(L115)
- Abortion Act 1974 as amended by Law No.66 of 18 May 1995 (AA)

Official Regulations

- Rulings by National Board of Health and Welfare (NBHW), under L115, s.5

Official Guidelines

- National Board on Health and Welfare guidelines in area of "Gene Ethics and Gene Technology in Health Care" (1999).[497] (GEGTHC)

International Conventions

Convention on Human Rights and Biomedicine ETS 164, Oviedo, 4.IV.1997 (CHRB) (Signed 04/04/97)[498]

Additional Protocol to the Convention on Biomedicine concerning the prohibition of cloning ETS 168, Paris, 12.I.1998 (APC) (Signed 12/01/98)[499]

Universal Declaration on the Human Genome and Human Rights (UDHGHR).[500]

Recommendations

None.

OVERVIEW OF REGULATIONS

Embryo Research

Definition of "embryo"

Law refers to "fertilised oocytes of human origin" (L115 s.1).

Research

Permitted with conditions.

---

[497] See Pattinson 2002, appendix 4.
[498] See http://conventions.coe.int/Treaty/EN/cadreprincipal.htm.
[499] Ibid.
[500] Adopted by consensus, UNESCO General Conference, 11 November, 1997.

Sources of embryos

Surplus embryos from IVF. In 2002 a Parliamentary Commission was established to consider inter alia whether somatic cell nuclear transfer which involves creating embryos for research should be allowed.[501]

Time limit for research

Fourteen days from the time of fertilization excluding any storage period (ss.(2) & (3) L115).

Other conditions governing research

Consent procedures

Consent of oocyte and sperm donors required before undertaking any "measures involving fertilized oocytes of human origin" (s.(1) L115).

Licenses

No information obtained.

Ethics Committees

In practice embryo research often approved by an Ethics Committee.[502]

Substantive (Must the embryo benefit?)

No.

Storage/Disposal of embryos

Fertilised ovum may be stored for 5 years or longer as determined by National Board of Health and Welfare under s. (5) L115 (s. (3) L115 as amended in 1998).[503]

Cloning

Reproductive

Prohibited by implication, if it involves "fertilisation". (ss.(2) & (4) L115)[504]. The National Council on Medical Ethics and the National Board of Gene Technology point out that cloning by Cell Nuclear Replacement (CNR) may not be prohibited.[505]

Non-reproductive

Permitted, provided its purpose is not to achieve hereditary genetic effects (s.(2) L115).

Gene Therapy

Somatic

Permitted by implication. (s.(2) L115).

Germ-line

Prohibited. (s.(2) L115).

---

[501] Information provided by Elizabeth Rynning.
[502] See Pattinson 2002, Appendix 4.
[503] Ibid.
[504] See UNESCO 1998,11; and HGAC and HFEA 1998, Annex E, interpreting this section.
[505] See Pattinson 2002, Appendix 4.

Pre-Implantation Genetic Diagnosis

Permitted only for diagnosis of serious, progressive, incurable and untreatable hereditary disease, resulting in premature death. (Parliamentary declaration, though not legislated).[506] Ethical review or Ethics consultation is recommended (GEGTHC).[507]

Stem Cell Therapy

A parliamentary commission was established in 2002 to consider inter alia stem cell research.[508]

Pre-Natal Diagnosis/Abortion

* Abortion Permitted:
    a) before the expiry of 18 weeks, provided there is no serious danger to woman's life or health due to any illness she may have (AA s. 1). If physician refuses to perform abortion issue must be referred to the NBHW, which may refuse permission if there is reason to suppose embryo is viable (AA s. 4).
    b) up to birth, if owing to woman's illness or bodily defect, pregnancy entails a serious danger to her life or health (AA s.6). If after 18 weeks permission from he NBHW must be obtained ,which may only be granted if there are special reasons for the abortion (AA s.3). Decisions of the NBHW are not subject to appeal (AA s.7). If after 18 weeks, the woman must be offered a supportive interview, before and after any abortion (AA ss.2 & 8).

Switzerland

SOURCES OF LEGISLATION

Acts of the Legislature

* Federal Constitution of the Swiss Confederation (SchwBV)[509] Art 119: Reproductive Medicine and Genetic Engineering as Applied to Human Beings[510]
* Swiss Criminal Code from 21 December 1937 as promulgated on 26 March 2002[511]
* Reproductive Medicine Act (FMedG)[512]
* Regulation on Reproductive Medicine (FMedV)[513]
* Federal Act on Medicines and Medical Products (Medicines Act, HMG) from 15 December 2000[514]

---

[506] Ibid.
[507] Ibid.
[508] Information provided by Elizabeth Rynning.
[509] Adopted by Referendum on 18 April 1999, AS 1999, 2556.
[510] Available at http://cloning.ch./cloning/staatlich/ch.html.
[511] AS 54,.757, in force since 1 January 1942, available in German at http://www.admin/ch/ch/d//sr/c311_0.html.
[512] Federal Act on reproduction with medical assistance from 18 December 1998 (as of 28 December 2000) AS 2000, 3055. Available at http://www.admin.ch/ch/d/sr/8/814.90.de.pdf.
[513] From 4 December 2000, as of 28 December 2000, AS 2000, 3068.
[514] AS 2001, 2790, in force since 1 January 2002, available in German at http://www.admin/ch.

- Regulation on Clinical Experiments with Medicines (VKlin) from 17 October 2001[515]
- Federal Resolution on the Monitoring of Transplants from 22 March 1996[516]

Proposed legislation

- Federal Act on the Research on Surplus Embryos and Embryonic Stem Cells (EFG)[517]
- Federal Act on Human Research (Humanforschungsgesetz-HFG) will be discussed in parliament in 2003[518]

Official Guidelines

- Medical-Ethical Guidelines for Somatic Gene Therapy on Human Beings by the Swiss Academy of Medical Sciences (SAMW), 1997[519]

International Conventions

- Convention on Human Rights and Biomedicine ETS Oviedo, 4 April 1997 (CHRB) (signed on 7 May 1999)[520]
- Additional Protocol to the Convention on Biomedicine Concerning the Prohibition of Cloning ETS 168, Paris, 12 January 1998 (APC) (signed on 7 May 1999)[521]

OVERVIEW OF REGULATIONS

Embryo Research

Definition of "embryo"

Embryo: the fruit from conjugation to the completion of organ development (FMedG, Art 2 lit. i).[522]

Research

Research on embryos is not explicitly prohibited.[523] The Federation has not regulated embryo research.[524] Legislation of some cantons has prohibited research

---

[515] AS 2001, 3511, in force since 1 January 2002, available in German at http://www.admin/ch.
[516] AS 1996, 2296, version pursuant to the Medicines Act, Appendix No. II 7, from 15 December 2000, in force since 1 January 2002, available in German at http://www.admin/ch.
[517] A German version is available at http://www.bag.admin.ch/embryonen/bundesgesetz/d/efg _entw.pdf. The Bundesrat *(Federal Council, i.e. executive body)* presented this draft on 20. November 2002. This Act on Embryo Research shall be implemented into the following Act on Human Research. Regulation on embryo research was regarded as that necessary, that it was regulated in a special law before, see Eidgenössisches Departemant des Innern, Erläuternder Bericht zum Entwurf des Embryonenforschungsgesetzes, S. 12. Available at http://www.bag.admin.ch/embryonen/bundesgesetz/d/erl_efg.pdf.
[518] Information provided by Dolores Krapf, Bundesamt für Gesundheit.
[519] Schweizerische Ärztezeitung *(Medical Journal)* 1997, 935-938.
[520] See http://conventions.coe.int.
[521] See http://conventions.coe.int.
[522] The Draft of the EFG, Art. 2 a, uses the same wording.
[523] But seemed to be so, for instance: Zweiter Zwischenbericht der Enquete-Kommission Recht und Ethik in der modernen Medizin – Teilbericht Stammzellforschung, BT-Drs. 14/7546, p. 129, 130, available in German at http://www.bundestag.de/gremien/medi/2zwischen.pdf.
[524] But this does not mean that it is prohibited.

while other did not regulate it at all. The draft of the EFG permits research on embryos and embryonic stem cells.[525]

Sources of embryos

There are none. Embryos must not be created (FMedG, Art. 29 para. 1) for research purposes and may not be preserved (FMedG, Art. 17 para. 3). Embryos preserved when the Act entered into force had to be registered with the competent authority (FMedG, Art. 42 para. 1). Since the Act entered into force[526] they may be stored for a maximum period of 3 years (FMedG, Art. 42 para. 2).

Pursuant to the Draft of the EFG surplus embryos[527] and embryonic stem cells may be used for research (Art 1 para. 1). The prohibition against creating embryos for research is retained (Draft of the EFG, Art. 3 para. 1 lit. a). The import or export of embryos is also prohibited (Draft of the EFG, Art. 3 para. 2 lit. a).

Time limit for research

Currently there is none.

The Draft of the EFG, Art. 3 para. 2 lit. b, prohibits developing a surplus embryo for more than 14 days.

Other conditions governing research

Consent procedures

The couple in question must give their written consent and prior to consenting must have been instructed in writing and orally in a comprehensible form before pursuant to the Draft of the EFG, Art. 10 para. 1. A couple may only be asked to donate embryos which are surplus (Draft of the EFG, Art. 10 para. 2).

The couple or the woman or the man can revoke their consent at any time prior to the commencement of the research activities or the extraction of stem cells (Draft of the EFG, Art. 10 para. 3).[528]

Substantive (Must the embryo benefit?)

No as the embryos are surplus. Furthermore a catalogue containing several requirements, e.g. ethical sustainability, has to be complied with pursuant to the Draft of the EFG, Art. 6 para. 1 lit. a – e.[529]

Storage/ Disposal of Embryos

The preservation of embryos is prohibited (FMedG, Art. 17 para. 3). Embryos in existence when the Act came into force may be stored for a maximum of 3 years (FMedG, Art. 42 para. 2).

Pursuant to the Draft of the EFG, Art. 3 para. 2 lit. c, surplus embryos which have been used for research purposes may not be transfered to a woman. Art. 7 lit. a

---

[525] See "Sources of embryo".

[526] 1 January 2001.

[527] A surplus embryo is an embryo having been created in conjunction with in vitro fertilisation which cannot be used to effect a pregnancy and does not stand a chance to survive for this reason (Draft of the EFG, Art. 2 lit. b).

[528] If the consent is revoked the embryo has to be annihilated immediately pursuant to the Draft of the EFG, Art. 10 para. 4.

[529] The Federal Authority seeks the advice of independent experts in order to evaluate the project scientifically and ethically (Draft of the EFG, Art. 6 para. 2).

requires the licence holder to immediately destroy the embryo after completion of the research activities.

Cloning

Reproductive cloning.

All forms of cloning are illegal (SchwBV, Art. 119 para. 2 lit. a), i.e. reproductive as well as therapeutic cloning. Pursuant to the FMedG, Art. 36, it is a criminal offence punishable by imprisonment to create a clone.[530]

Non-reproductive cloning

See *Reproductive cloning.*

Gene Therapy

Somatic

The HMG, Art. 2 para. 1 lit. c covers somatic gene therapy as listed treatment[531] which is still at the experimental stage. The VKlin, Art. 2 para. 1, applies to it.

Germ-line

Any interference with the genetic information of human germ cells is illegal (SchwBV, Art. 119 para. 2 lit. a). Pursuant to the FMedG, Art. 35 para. 1, 2, it is a criminal offence punishable by imprisonment to interfere with and modify the genetic information of a germ cell or an embryonic cell.[532]

Pre-Implantation Genetic Diagnosis

Cells must not be extracted from the embryo in vitro in order to examine them (FMedG, Art. 5 para. 3).[533]

Stem Cell Therapy

Are not explicitly prohibited. However, since embryonic stem cells must not be created as outlined above they would have to be imported, and the trade in products derived from embryos is prohibited by the SchwBV, Art. 119 para. 2 lit. e.[534]

With regard to adult stem cells, the Federal Resolution on the Monitoring of Transplants applies in general and pursuant to Art. 3 a the HMG, Art. 37, 39 para. 1, 2, 4, 5 and Art. 40, have to be considered by analogy in addition.[535]

---

[530] The Draft of the EFG, Art. 3 para. 1 lit. c repeats this prohibition and adds that stem cells which have been extracted from such a living being must not be used neither.

[531] http://www.bag.admin.ch/heilmitt/gesetz/d/botsch02.pdf. Statement of Reasons for the Act, p. 36. Thus for the first time, Switzerland regulates somatic gene therapy by federal law.

[532] The Draft of the EFG, Art. 3 para. 1 lit. b repeats this prohibition and adds that stem cells which have been extracted from such a living being must not be used either.

[533] On 22 March 2002 the National Council dealt with this subject in its 12th Session. The legislative initiative to admit pre-implantation diagnostics in case of a severe threat won 56 votes, 102 MPs voted against it. Simultaneously 74 MPs voted for the initiative to be referred as a motion to the competent commission, 83 MPs voted against this. See: Amtliches Bulletin-Nationalrat *(Official Bulletin of the National Council)*, 00.455 and 01.3647, available at http://www.parlament.ch/ab/data/d/.

[534] Yet research is permitted pursuant to the provisions of the Draft of the EFG, see II. 1 c – f.

[535] These provisions stipulate the handling of blood and blood products. The Draft refers to them because stem cells count as transplants by their nature as they are not consumed but are essentially handled like blood regarding extraction and transfer and therefore are similarly at risk. See Statement of Reasons for the Act, p. 128f., http://www.bag.admin.ch/heilmitt/gesetz/d/botsch02.pdf.

Pre-Natal Diagnosis/Abortion

- PND is not regulated by law but applied practically.[536]
- Different rules can apply to the termination of pregnancy in the individual Cantons because of the federal system in Switzerland. Restrictions are limited by federal law.[537]

The Criminal Code has federal jurisdiction and provides:

- the termination of pregnancy by the pregnant woman herself is a criminal offence and punishable by imprisonment (SchwStGB, Art. 118 para. 2).
- a person other than the pregnant woman who terminates the pregnancy upon her request or assists her in so doing is liable to imprisonment of up to five years or (SchwStGB, Art. 119 para. 1).
- it is not abortion within the meaning of the law if the pregnancy is terminated by a certified doctor after the pregnant woman has consented in writing. The doctor must have obtained a certificate from another certified medical specialist. The termination must be necessary because the pregnancy poses a threat to the life of the pregnant woman or carries a risk of permanent severe damage to her health which cannot be averted otherwise than by termination of the pregnancy (SchwStGB, Art. 120 para. 1).
- If the pregnancy is terminated because of an emergency situation other than those outlined in Art 120 para. 1., the judge may reduce the sentence of anyone convicted of an offence at his or her own discretion pursuant to the SchwStGB, Art. 120 para. 3.

SPECIAL REMARKS

In order to apply reproduction methods or receives germ cells or impregnated egg cells for preservation, or brokers sperm cells requires a permission by the competent authority (FMedV, Art. 1, 8).

A referendum on the termination of pregnancy took place on 2 June 2002. 72 per cent of the electorate voted for the termination of pregnancy to remain not punishable within the first 12 weeks.[538] The legislature resolved to establish a National Ethics Commission in the FMedG, Art. 28. This commission shall in particular work out supplementing guidelines referring to the FMedG, point out gaps in the legislation, advise the Bundesversammlung (Federal Assembly), the Bundesrat and the Cantons upon request and inform the public on important discoveries and promote the discussion on ethical questions in society (FMedG, Art. 28 para. 3). The Bundesrat has in Regulations laid down the responsibilities and methods of working of the National Ethics Commission.[539]

---

[536] Compare Honsell, Guide on Medical Law, Zurich 1994, 413.
[537] Rahman/Katzvive/Henshaw, A Global Review of Laws on Induced Abortion, 1985-1997 in: Family Planning Perspectives, Volume 24, No. 2, June 1998, available at http://www.agi-usa.org/pubs/journals/2405698.html.
[538] Available at: http://www.parlament.ch/poly/Framesets/D/Frame-D.htm.
[539] Regulation on the National Ethics Commission in the Area of Human Medicine (VNEK) from 4 December 2000, as of 28 December 2000, AS 2000, 3079, in force since 1 January 2001.

The Commission for Science, Education and Culture of the Nationalrat (National Council) rejected a legislative initiative called "Prohibition of Consuming Research on Embryos"[540] by 13 to 12 votes in July 2002. The initiative had demanded the prohibition of the extraction of stem cells from embryos until the Federal Act on the Research on Human Beings came into force.[541]

The Swiss National Fund for the Promotion of Scientific Research supports research on embryonic stem cells in spite of the dissenting vote of the National Ethics Commission.[542]

FURTHER READING

Kettner, Matthias (edit.) Biotechnologie und Menschenwürde, 2003

Turkey

SOURCES OF LEGISLATION

Acts of the Legislature

- Population Planning Law (PPL)[543], Law No. 2827 from 24 May 1983[544]

Official Regulations

- No information obtained.

Official Guidelines

- No information obtained.

International Conventions

- Convention on Human Rights and Biomedicine ETS Oviedo, 4 April 1997 (CHRB) (signed on 4 April 1997)[545]

- Additional Protocol to the Convention on Biomedicine Concerning the Prohibition of Cloning ETS 168, Paris, 12 January 1998 (APC) (signed on 12 January 1998)[546]

- Universal Declaration on the Human Genome and Human Rights (UDHGHR)[547]

---

[540] Moratorium (01.441).

[541] Available at http://www.parlament.ch/D/Pressemitteilungen.

[542] Zweiter Zwischenbericht der Enquete-Kommission Recht und Ethik in der modernen Medizin – Teilbericht Stammzellforschung, BT-Drs. 14/7546, p. 129, 130, available in German at http://www.bundestag.de/gremien/medi/2zwischen.pdf.

[543] Available in English at http://cyber.law.harvard.edu/population/abortion/Turkey.abo.htm.

[544] T.C. Resmi Gazete, 27 May 1983, No. 18059, 3–6.

[545] See http://conventions.coe.int, with the declaration that the government has confirmed the signing of the Convention, done at referendum on 4 April 1997 and the reservation of the right not to apply the provisions of Art. 20, para. of the Convention which authorise the removal of regenerative tissue from a person who does not have the capacity to consent under certain conditions (See http://conventions.coe.int/Treaty/EN/cadreprincipal.htm).

[546] See http://conventions.coe.int.

[547] Adopted unanimously and by acclamation at the 29th Session of UNESCO's General Conference, 11 November 1997 (29 C/Resolution 17 – see http:www.unesco.org/ibc/), and endorsed by the United Nations' General Assembly, 85th Plenary Meeting on 9 December 1998 (A/RES/53/152 – see http://www.un.org.documents/).

Recommendations

• No information obtained.

OVERVIEW OF REGULATIONS

Embryo Research

Definition of "embryo"

There is no statutory definition. The term "embryo" is used in medical practice until the 12th week.[548]

Research

Research is permitted on animals only.[549]

Sources of embryos

See *Research*.

Time limit for research

Not regulated.[550]

Other conditions governing research

Consent procedures

See *Research*.

Licenses

See *Research*.

Ethics Committees

See *Research*.

Substantive (Must the embryo benefit?)

See *Research*.

Storage/Disposal of embryos

See *Research*.

Cloning

Reproductive cloning.

Cloning is prohibited by law.[551]

---

[548] Council of Europe: CLONING – COMPARATIVE STUDY ON THE SITUATION IN 44 STATES, Strasbourg, 4 June 1998, p. 136, available in English at http://legal.coe.int/bioethics/gb/pdf/txtprep2b.pdf.M.

[549] Council of Europe: CLONING – COMPARATIVE STUDY ON THE SITUATION IN 44 STATES, Strasbourg, 4 June 1998, 160-161, available in English at http://legal.coe.int/bioethics/gb/pdf/txtprep2b.pdf.M.

[550] Council of Europe: CLONING – COMPARATIVE STUDY ON THE SITUATION IN 44 STATES, Strasbourg, 4 June 1998, 180, available in English at http://legal.coe.int/bioethics/gb/pdf/txtprep2b.pdf.M.

[551] Council of Europe: CLONING – COMPARATIVE STUDY ON THE SITUATION IN 44 STATES, Strasbourg, 4 June 1998, 214, available in English at http://legal.coe.int/bioethics/gb/pdf/txtprep2b.pdf.M However, the expert's report does not mention any legal sources for this statement.

Non-reproductive cloning

Cloning is prohibited by law.[552]

Gene Therapy

Intervention is not regulated by law and only occurs in experiments on animals.[553]

Stem Cell Therapy

See *Research*.

Pre-Implantation Genetic Diagnosis

Pre-implantation genetic diagnosis is not practised.[554]

Pre-Natal Diagnosis/Abortion

- Rules of Professional Conduct apply here.[555]

  - Without medical indication the pregnancy may be terminated up to the end of the 10th week of the pregnancy pursuant to the PPL, section 5 para. 1.

  - After 10 weeks it may only be terminated if it poses a threat to the life of the mother or the child or it suffers from a serious heritable disease. This has to be established by two specialist doctors (PPL, section 5 para. 2).

United Kingdom

SOURCES OF REGULATIONS

Acts of the Legislature

- Human Fertilisation and Embryology Act 1990 (c.37) (HFEA)[556]
- Human Reproductive Cloning Act 2001 (c.23) (HRCA)[557]
- Abortion Act 1967(c.87), as amended by the HFEA (AA)

Official Regulations

- Human Fertilisation and Embryology (Research Purposes) Regulations 2001(HFERPR)[558] [Made under Sch.2, para.3(2)]

---

[552] Council of Europe: CLONING – COMPARATIVE STUDY ON THE SITUATION IN 44 STATES, Strasbourg, 4 June 1998, 214, available in English at http://legal.coe.int/bioethics/gb/pdf/txtprep2b.pdf.M However, the expert's report does not mention any legal sources for this statement.

[553] Council of Europe: CLONING – COMPARATIVE STUDY ON THE SITUATION IN 44 STATES, Strasbourg, 4 June 1998, p. 180, available in English at http://legal.coe.int/bioethics/gb/pdf/txtprep2b.pdf.M.

[554] Council of Europe: CLONING – COMPARATIVE STUDY ON THE SITUATION IN 44 STATES, Strasbourg, 4 June 1998, 169, available in English at http://legal.coe.int/bioethics/gb/pdf/txtprep2b.pdf.M.

[555] Council of Europe: CLONING – COMPARATIVE STUDY ON THE SITUATION IN 44 STATES, Strasbourg, 4 June 1998, 137 & 139, available in English at http://legal.coe.int/bioethics/gb/pdf/txtprep2b.pdf.M.

[556] Available at http://www.legislation.hmso.gov.uk/.

[557] Ibid.

[558] Statutory Instrument 2001 No. 188. Available at http://www.legislation.hmso.gov.uk/.

- Human Fertilisation and Embryology (Statutory Storage Period for Embryos) Regulations (HFESSR) [559]

Official Guidelines

- Human Fertilisation and Embryology Authority (HFE Authority) Code of Practice (HFEACP) (Fifth Edition April 2001)[560] [Required by s.25 HFEA]
- Gene Therapy Advisory Committee (Guidance on Making Proposals to Conduct Gene Therapy Research on Human Subjects) (GTACGP)[561]

International Conventions

- Universal Declaration on the Human Genome and Human Rights. (UDHGHR)[562]

Recommendations

- House of Lords Stem Cell Committee Report (February 2002) (HLSCC)[563]

OVERVIEW OF REGULATIONS

Embryo Research

Definition of "embryo"

"... a live human embryo where fertilisation is complete, ... and includes an egg in the process of fertilisation, and, for this purpose, fertilisation is not complete until the appearance of a two cell zygote" (HFEA s.1(1)(a)&(b)).

Research

Permitted with conditions

Sources of embryos

Surplus embryos from IVF treatment may be used for research under licence (HFEACP para. 6.8b). Embryos may also be created for research under licence (HFEA Sch. 2, para. 3(1)).

Embryos should not be created for research unless there is a demonstrable and exceptional need which cannot be met by the use of surplus embryos (HLSCC conclusion. ix).

Time limit for research

Up to the appearance of the primitive streak, which is deemed not to be later than fourteen days, after the gametes are mixed excluding any storage time (HFEA s.3(3)(a) & (4)).

Embryos however created, have the same ethical status with respect to their use for research purposes up to the 14 day limit. (HLSCC Conclusion, xi).

Other conditions governing research

---

[559] Statutory Instrument 1996 No. 375. Available at http://www.legislation.hmso.gov.uk/.
[560] Available at http://www.hfea.gov.uk/frame.htm.
[561] Available at http://www.doh.gov.uk/pub/docs/doh/humans.pdf.
[562] Adopted by consensus, UNESCO General Conference, 11 November, 1997.
[563] Available at http://www.parliament.the-stationery-office.co.uk/pa/ld200102/ldselect/ldstem/ 83/8301.htm.

Consent procedures

Consent of donors is necessary and must

a) be informed (HFEA Sch. 3 para. 3(1)(b));

b) be in writing and effective (i.e. not withdrawn) (HFEA Sch.3 para. 1);

c) specifically be agreement to use of the embryo/gamete in research, which may outline conditions for such research or for storage (HFEA Sch.3 paras. 2(1)(c), & 2(2), (3) & (4));

d) be preceded by counseling which specifically advises each donor that consent may be varied or withdrawn before the gamete/embryo is used (HFEA Sch.3 para. 3(1)(a) & (2) & para.4).

Licenses

Required (HFEA s. 11(1)(c) & Sch. 2, para. 3).

Ethics Committees

Research projects must be submitted to a Research Ethics Committee for approval whether they are publicly or privately run, prior to application for a licence (HFEACP parts 11.6 & 11.7).

Substantive (Must the embryo benefit?)

No. However research is only permitted where it is necessary or desirable for

i) promoting advances in the treatment of infertility and the techniques of contraception (HFEA Sch. 2, para. 3(2)(a) & (d));

ii) increasing knowledge about the causes of congenital disease and of miscarriage (HFEA Sch. 2, para. 3(2)(b) & (c));

iii) detecting gene/chromosome abnormalities in embryos before implantation (HFEA Sch. 2, para. 3(2)(e)).

Additionally research may be permitted for the purposes of:

i) increasing knowledge about development of embryos and about serious disease. (HFERPR s. 2(2)(a) & (b));

ii) the application of such knowledge to the treatment of serious disease. (HFERPR s.2(2)(c)).

Storage/Disposal of embryos

Embryos to be used for research may be stored for a maximum period of 5 years, (HFEA s.14(4) & Sch. 2, para. 2(3)), after which they must be allowed to perish. (HFEA s.14(1)(c)). Centres doing embryo research should decide at the outset the duration of the culture period, the method of termination and the procedure which will ensure that development does not continue beyond 14 days, or if earlier, the appearance of the primitive streak. (HFEACP para. 9.14). Embryos to be used for treatment, may be stored for a maximum period of 10 years (HFESSR).

Cloning

Reproductive cloning

No. (HRCA s.1(1)) and (HFEA s. 3(3)(d). See R (Quintavalle) v Secretary of State for Health [2002] EWCA, 18 January 2002).

Non-reproductive cloning

Yes, subject to licence being granted by the HFE Authority. (See R (Quintavalle) v Secretary of State for Health [2002] EWCA, 18 January 2002 and HLSCC Chap 5 para. 5.11).

Gene Therapy

Somatic

Permitted (GTACGP Note 11).

Germ-line

Not permitted. (HFEA Sch.2, para. 1(4)). (See also GTACGP Note 12).

Pre-Implantation Genetic Diagnosis

Permitted. (HFEA Sch. 2, para. 3(2)(e)).

Stem Cell Therapy

Embryo and adult stem cell research including research involving CNR and oocyte nucleus transfer, permitted. (HLSCC conclusions ii, iii, iv, xii & xiii.).

Pre-Natal Diagnosis/Abortion

- Abortion permitted
  a) up to 24 weeks where the continuance of the pregnancy would involve risk of injury to the physical or mental health of the pregnant woman or any existing children of her family greater than if the pregnancy were terminated; (AA s.1(1)(a)) and
  b) at any time up to birth where i) termination is necessary to prevent grave permanent physical or mental injury to the woman; (AA s.1(1)(b)) or ii) continuance of the pregnancy would involve greater risk to the life of the pregnant woman, than termination of the pregnancy; (AA s.1(1)(c)) or iii) there is a substantial risk that if the child were born it would suffer from serious physical or mental handicap (AA s.1(1)(d)).

Ukraine

SOURCES OF REGULATIONS

Acts of the Legislature

- Health Protection Act from 19 November 1992, as promulgated on 22 February 2000[564]
- Kriminalnyi Kodeks *(Criminal Code)* from 1 April 2001, as promulgated on 7 March 2002[565]

---

[564] In force since 15 December 1992, published on 15 December 1992 in Golos Ukraini *(Voice of Ukraine)* No. 2801, available in Russian at http://www.rada.gov.ua.

[565] Published on 19 June 2001 in Golos Ukraini No. 107, available in Russian at http://www.rada.gov.ua.

Official Regulations
- Nakas (Decree) No. 24 by the Ministry for Health Protection from 4. February 1997 on the conditions and methods for carrying out artificial insemination and implantation of embryos[566]

Official Guidelines
- No information obtained

International Conventions
- Convention on Human Rights and Biomedicine ETS Oviedo, 4 April 1997 (CHRB) (signed on 22nd March 2002)[567]
- Universal Declaration on the Human Genome and Human Rights (UDHGHR)[568]

Recommendations
- No information obtained

OVERVIEW OF REGULATIONS

Embryo Research

No information obtained

Storage/Disposal of embryos

Cloning

The experts' report of the Council of Europe says in this respect: "A question concerning human cloning never appears among specialists of that field."[569]

Gene Therapy

Somatic

No information obtained.

Germ-line

Any medical operation interfering with the genetic apparatus of a human being is prohibited pursuant to section 29, para. 2 of the Health Protection Act.

Pre-Implantation Genetic Diagnosis

No information obtained.

---

[566] See Ministry of Justice No. 58/1862 from 3 March 1997, available in Russian at http://rada.gov.ua.

[567] See http://conventions.coe.int.

[568] Adopted unanimously and by acclamation at the 29th Session of UNESCO's General Conference, 11 November 1997 (29 C/Resolution 17 – see http:www.unesco.org/ibc/), and endorsed by the United Nations' General Assembly, 85th Plenary Meeting on 9 December 1998 (A/RES/53/152 – see http://www.un.org.documents/).

[569] Council of Europe: CLONING – COMPARATIVE STUDY ON THE SITUATION IN 44 STATES, Strasbourg, 4 June 1998, pp 216, 217, available in English at http://legal.coe.int/bioethics/gb/pdf/txtprep2b.pdf.M However, the Council of Europe pointed out that the information received from the Ukraine had not been verified officially.

Stem Cell Therapy

No information obtained.

Pre-Natal Diagnosis/Abortion

- PND: No information obtained.

  Abortion: The termination of pregnancy is legal until the 12th week of the pregnancy.

  In emergency situations, i.e. in case of a medical indication (continuation of the pregnancy would pose a threat to the woman) or a social indication the termination of pregnancy is legal until the 28th week pursuant to Art. 50 of the Health Protection Act.

USA

SOURCES OF REGULATIONS

Acts of the Legislature

- Public Law No. 105-277 (1998) (P. L. (1998))

Official Regulations

- None

Official Guidelines

- National Institutes of Health Guidelines for Research Using Human Pluripotent Stem Cells (Effective August 25, 2000, 65 FR 51976) (Corrected November 21, 2000, 65 FR 69951) (NIH 2000)[570]
- Withdrawal Notice – National Institutes of Health Guidelines for Research Using Human Pluripotent Stem Cells [Federal Register: November 14, 2001 (Volume 66, Number 220)] (Withdrew those sections of the NIH Guidelines for Research Using Human Pluripotent Stem Cells)
- White house Fact Sheet on Stem Cell Research 9 August 2001[571] (WHFS)

International Conventions

- Universal Declaration on the Human Genome and Human Rights. (UDHGHR)[572]

Recommendations

- Ethical Issues in Human Stem Cell Research Volume I Report and Recommendations of the National Bioethics Advisory Commission (NBAC 1999)

---

[570] Available online at http://www.nih.gov/news/stemcell/stemcellguidelines.htm.
[571] Available at http://www.whitehouse.gov/news/releases/2001/08/print/20010809-1.html.
[572] Adopted by consensus, UNESCO General Conference, 11 November, 1997.

OVERVIEW OF REGULATIONS

Embryo Research

Definition of "embryo"

Section 511 (b) of the Omnibus Consolidated and Emergency Supplemental Appropriations Act for Fiscal Year 1999[573], contains a definition of "embryo". It reads "[A]ny organism ... derived by fertilization, parthenogenesis, cloning, or any other means from one or more human gametes or human diploid cells." This Act included provisions for federal funding of research on stem cells derived from human embryos. No indication has been found that any new definition has been adopted.

Research

Federal: "Federal funds will only be used for research on existing stem cell lines that were derived: (1) with the informed consent of the donors; (2) from excess embryos created solely for reproductive purposes; and (3) without any financial inducements to the donors." (WHFS). There is no formal regulation of privately funded research.

State: 24 states have no specific laws on embryo research.

Of the remaining 26 only ten of the state laws apply specifically to in vitro embryos. Nine of these ban all research on in vitro embryos.[574]

Sources of embryos

Pursuant to federal policy announced by President Bush, the federal government will only fund research on pre-existing stem cell lines. There is therefore no need on a federal level for any new embryos.

Embryos have been created for the purposes of research where no NIH funds have been used.[575]

Time limit for research

In 1994 the NIH Human Embryo Research Panel supported a 14 day limit for federally funded research.[576] As no more embryos will be permitted to be used for federally funded research at the federal level, a time limit is no longer relevant. No information was obtained on whether there is limit for privately funded research.

Other conditions governing research

Consent procedures

Owing to the federal position (See *Research*), the procedures listed below are no longer relevant. However, the existing cell lines created in America that are currently the focus of federally funded research, would have had to be created from embryos obtained in keeping with these procedures. There is/was no corresponding federal requirement for privately funded research.

No inducements, monetary or otherwise, should be offered for the donation of human embryos for research purposes (NIH 2000 sec II, A. 2(a)).

---

[573] Pub. L. No. 105–277, 112 Stat. 2681 (1998).
[574] Andrews 2000, A-4.
[575] Josefson 2001.
[576] NBAC 1999, Chapter 5.

There should be a clear separation between the decision to create embryos for fertility treatment and the decision to donate human embryos in excess of clinical need for research purposes to derive pluripotent stem cells (NIH 2000 sec II, A.2(b)).

Informed consent should be obtained from those who have decided to donate excess embryos from fertility treatment (NIH 2000 sec II, A.2(e)).

Licenses

As in (i) supra owing to the federal position (See *Research*), the procedures listed below are no longer relevant. However, the existing cell lines created in America that are currently the focus of federally funded research, would have had to be created from embryos obtained in keeping with these procedures. There is/was no corresponding federal requirement for privately funded research.

No licenses are required but anyone applying for NIH funding to use stem cells derived from human embryos must submit the following information before funding will be granted:

An assurance that the stem cells were derived in accordance with Section II.A.1. of the guidelines where the require –

- A sample informed consent document
- An abstract of the scientific protocol used to derive the stem cells.
- Documentation of Institutional Review Board (IRB) approval
- An assurance that stem cells were obtained through a donation or a payment that did not exceed reasonable costs to cover transportation, storage etc.
- Title of the research proposal
- An assurance that the proposed research is not one ineligible for funding under the guidelines
- Consent to disclose the submission as necessary to carry out public review.

Ethics Committees

As in (i) supra owing to the federal position (See *Research*), the procedures listed below are no longer relevant. However, the existing cell lines created in America that are currently the focus of federally funded research, would have had to be created from embryos obtained in keeping with these procedures. There is/was no corresponding federal requirement for privately funded research.

Derivation protocols should have been approved by an Institutional (IRB) established in accordance with 45 CFR 46.107 and 46.108 or FDA regulations at 21 CFR 56.107 and 56.108. (NIH 2000. Section II.A.2(f))

Substantive (Must the embryo benefit?)

No longer relevant. (See *Research*). There is no such requirement for privately funded research.

Storage/Disposal of embryos

No longer relevant. (See *Research*)

Cloning

Reproductive cloning

No Federal legislation exists that directly addresses reproductive cloning.[577]

The National Bioethics Advisory Commission (NBAC) recommended a legislative prohibition on reproductive cloning that would be subsequently reviewed.[578]

The Food and Drug Administration (FDA) has assumed jurisdiction to regulate cloning as its remit covers authority to regulate medical products, including biological products, drugs, and devices. Due to safety concerns, the FDA would not permit human cloning at this time.[579] Research in which human pluripotent stem cells are used in combination with somatic cell nuclear transfer for the purposes of reproductive cloning of a human is ineligible for NIH funding (NIH 2000, Section III (G).

Proposed Federal legislation: There are two Human Cloning Prohibition Bills before the US parliament. One (S. 1899 [2001]) was passed by the US House of Representatives on July 31, 2001. It has yet to be considered by the Senate. It prohibits cloning for any purpose. The other (S. 2439 [2002]) only prohibits reproductive cloning and would therefore permit non-reproductive cloning. Both Bills carry stiff penalties for breaches of up to 10 years imprisonment and fines of at least $1 million dollars. The 2001 Bill also prohibits the importation of products derived from cloned embryos, while the 2002 Bill also provides for forfeiture of property derived from or used to commit or attempt to commit an offence under the Bill.[580]

State legislation:

Over twenty States have either banned or considered banning the cloning of human beings. In some instances however the ban is for a limited time, (e.g. the Californian legislation[581] enacted in 1998 has a 5 year ban), or prohibits the use of State funds in cloning research. [582] These legislative positions could be interpreted as to implying that the bans could be lifted and government funding provided if it is eventually shown that cloning is safe.

Non-reproductive cloning

President Bush's decision in August 2001 (See *Sources of embryos*) means that non-reproductive cloning is only permissible using private funds.

Gene Therapy

Somatic

Permitted by default. No federal legislation exists that explicitly addresses gene therapy. The Food and Drug Administration (FDA) has claimed jurisdiction to regulate gene therapy since 1993[583]. The FDA's Center for Biologics Evaluation and Research (CBER) regulates human gene therapies, which fall under the legal defi-

---

[577] NBAC 1997, 88.
[578] Ibid. 109, Recommendation II.
[579] FDA 2001.
[580] Comparison of bills available at http://www.aau.edu/research/Compare.pdf.
[581] Californian Health and Safety Code, s. 24185; see Harvard Law Review 1998, 2353.
[582] NBAC 1997, 104.
[583] FDA 1993.

nition of a "biologic". For any unapproved biological product that is to be tested on humans, an investigational new drug (IND) application must be filed with the FDA.[584]

The National Institutes of Health (NIH) and its advisory committee, the Recombinant DNA Advisory Committee (RAC), are also involved in the regulation of gene therapy clinical research and work alongside the FDA[585]. However, the NIH only has jurisdiction over those institutes using federal funds.

Germ-line

Not permitted at present.

The Recombinant DNA Advisory Committee of the National Institutes of Health [RAC] at present will not approve germ line alterations. An outright ban was urged by the Council for Responsible Genetics (CRG) in 1985, but the RAC declined. Though the RAC is not prepared to approve such experiments now, it invites researchers to submit protocols that might offer an acceptable risk/benefit balance. If a sound clinically proposal is ever made, the RAC may approve it.[586]

Pre-Implantation Genetic Diagnosis

No federal legislation. Permitted subject to any State legislation.

Stem Cell Therapy

President Bush announced in August 2001 that federal funding will be available for stem cell research on existing cell cultures, but would not be available for research on stem cells that had not been derived from embryos already existing at the time of the announcement. (WHFS). This means that research may be carried out on 64 approved cell lines from public laboratories and private companies based in the United States, Sweden, Australia, Israel and India.[587]

Pre-Natal Diagnosis/Abortion

- In Roe V. Wade[588] the Supreme Court held that women had a constitutional right to an abortion, preventing a State from denying her an abortion during the first trimester. It also held that during the second trimester the State could intervene to protect the mother's health and that at the point of viability the State obtained an interest in the fetus entitling it to restrict abortions provided there are exceptions for endangerment of the woman's life or health.

- In Webster [589] a Missouri statute that banned the use of public funds, employees and buildings for abortions and required tests to determine whether a fetus believed to be 20 weeks old was viable, was upheld.

- Abortion Law is currently governed by the precedent in Casey[590] which while upholding a woman's constitutional right to have an abortion, modified the Roe

---

[584] FDA 1993.
[585] FDA 2000.
[586] NBAC 1997, 97.
[587] Fletcher 2001, 893.
[588] 410 US 113 (1973).
[589] Webster v. Reproductive Health Services 492 US 490 (1989).
[590] Planned Parenthood of Southeastern Pennsylvania v Casey 120 L. ed 2nd 674 (1992).

ruling. It declared that States have interests in protecting the life of the fetus and the health of the woman from the beginning of pregnancy and may therefore regulate abortion throughout pregnancy provided they do not "unduly burden" a woman's right to choose.

- States that accept federal Medicaid funds (health insurance funds) are required to pay for abortions if the pregnancy is life threatening or the result of rape or incest where the woman is a Medicaid recipient.[591]
- Currently there is great debate about the practice of late term abortions. Thirty States have sought to ban late term abortion procedures. In 18 States the legislation has been struck down, while in the other 12 the bans are partially or fully in place.[592]

---

[591] United Nations (2001).
[592] Ibid.

References

Andrews, Lori B. 2000. "State Regulation of Embryo Stem Cell Research." In National Bioethics Advisory Commission (NBAC), *Ethical Issues in Human Stem Cell Research. Volume II. Commissioned Papers*. Rockville, Maryland: NBAC, 2000. Available online at http://bioethics.georgetown.edu/nbac/stemcell2.pdf

Australian Parliament. 2001. *Human Cloning: scientific, ethical and regulatory aspects of human cloning and stem cell research*. Canberra: The Parliament of the Commonwealth of Australia

Braake, Trees A. M. te. 2000. "Later Termination of Pregnancy Because of Severe Foetal Abnormalities: Legal Acceptability, Notification and Review in the Netherlands." *European Journal of Health Law* 7: 387–403

CIHR (Canadian Institutes of Health Research). 2001. *Human Stem Cell Research: Opportunities for Health and Ethical Perspectives–A Discussion Paper*

CNEV95 (National Council of Ethics for the Life Sciences) (Conselho Nacional de Ética Para as Cinêcia da Vida). 1995. *Report-Opinion on Experimentation on the Human Embryo* (15/CNECV/95). 4th October. Lisbon

CNEV97. National Council of Ethics for the Life Sciences (Conselho Nacional de Ética Para as Cinêcia da Vida). *Opinion on the Ethical Implications of Cloning* (21/CNECV/97). 1st April. Lisbon

Council of Europe. 1998. *Medically Assisted Procreation and the Protection of the Human Embryo: Comparative Study of the Situation in 39 States/ Cloning: Comparative study of the situation in 44 States*. Council of Europe: Strasbourg Available at http://www.legal.coe.int/bioethics/gb/pdf/txtprep2b.pdf

EGESNT (European Group on Ethics in Science and New Technologies). 2000. *Ethical Aspects of Human Stem Cell Research and Use*. Opinion No. 15. 14 November. Available at http://www.europa.eu.int/european_group_ethics /index_en.htm)

ESFPB (European Science Foundation Policy Briefing). 2001. Human stem cell research: scientific uncertainties and ethical dilemmas. Available at http://www.esf.org/ publication/89/ESPB14.pdf

Exter den, A. P. and Prudil, L. 2001. "The Czech Republic" In H Nys, ed., *International Encyclopaedia of Laws,* The Hague: Kluwer Law International

European Parliament 2001a. *Fact Sheet on the Temporary Committee on Human Genetics and Other New Technologies of Modern Medicine*. Available at http://www.europarl.eu.int/comparl/tempcom/genetics/fact_sheet_en.pdf

–. 2001b. European Parliament Directorate General for Research Scientific and Technical Options Assessment. *Embryos, Scientific Research and European Legislation*. Briefing Note No. 14/2001. Available at http://www.europarl.eu. int/stoa/publi/pdf/briefings/14_en.pdf

FDA (Food and Drug Administration). 1993. Application of Current Statutory Authorities to Human Somatic Cell Therapy Products and Gene Therapy Products; Notice. *Federal Register.* Vol. 58, No. 197. Available online at http:// www.fda.gov/cber/genadmin/fr101493.pdf

–. „Statement of Jay P. Siegel, Director of Office of Therapeutics Research and Review, Center for Biologics Evaluation and Research, Food and Drug Administration, Department of Health and Human Services, Before the Subcommittee on Public Health, Committee on Health, Education, Labor and pensions, United States Senate. February 2. Available online at http://www.fda.gov/ ola/2000/genetherapy.html

–. 2001. "Statement of Kathryn C. Zoon. Director, Center for Biologics Evaluation and Research, Food and Drug Administration, Department of Health and Human Services Before the Subcommittee on Oversight and Investigations, Committee on Energy and Commerce, United States House of Representatives. March 28. Available online at http://www.fda.gov/ ola/2001/humancloning.html

Fletcher, Liz. 2001. "US Stem Cell Policy Comes Under Fire." *Nature Biotechnology* 19: 893–894

Gunning, Jennifer, and English, Veronica. 1993. *Human In Vitro Fertilization: A Case Study in the Regulation of Medical Innovation*. Aldershot: Dartmouth Publishing

HCR97 (Health Council of the Netherlands Recommendations). 1997.Committee on In Vitro Fertilization. *In Vitro Fertilization (IVF)*. Rijswijk

HCR98 (Health Council of the Netherlands Recommendations). 1998. Committee on In Vitro Fertilization. *IVF-related research.* Rijswijk

HGAC and HFEA (Human Genetics Advisory Commission and Human Fertilisation and Embryology Authority). 1998. *Cloning Issues in Reproduction, Science, and Medicine.* December. London: Her Majesty's Stationary Office

IBAC (Independent Biotechnology Advisory Council). 2001 *Cloning and Stem Cell Research.* Wellington New Zealand. Available at http://www.ibac.org.nz/ stem_cell/Stemcell1.pdf

Josefson, Deborah. 2001. "Embryos created for stem cell research" *British Medical Journal* 323: 127

Lacadena, Juan Ramón. 1996. "Genetic Manipulation Offences in Spain's New Penal Code: A Genetic Commentary." *Law and the Human Genome Review* 5 (July–Dec): 195–203

Macer, Darryl 2000. "International Aspects: National Profiles, Japan" In T. J. Murray and M. J. Mehlman, eds., *Encyclopaedia of Ethical, Legal and Policy Issues in Biotechnology.* John Wiley and Sons. Available at http://www.biol. tsukuba.ac.jp/~macer/Papers/biojap.htm

MacKellar, Callum. 1997. *Reproductive Medicine and Embryological Research: A European Handbook of Bioethical Legislation 1997–8.* Edinburgh: European Bioethical Research

Mudur, Ganapati. 2002. "India plans new legislation to prevent sex selection." *British Medical Journal* 324:385

NBAC (National Bioethics Advisory Commission). 1997. *Cloning Human Beings: Report and Recommendations of the National Bioethics Commission.* Rockville, Maryland: NBAC. Available online at http://bioethics.georgetown. edu/nbac/pubs/cloning1/cloning.pdf

–. 1999. *Ethical Issues in Human Stem Cell Research. Volume I: Report and Recommendations of the National Bioethics Advisory Commission.* Rockville, Maryland: NBAC. Available online at http://bioethics.georgetown.edu/ nbac/stemcell.pdf

Pattinson, Shaun. 2002. *Influencing Traits Before Birth.* Forthcomimg

Pereira, André Dias. 2001. *Country Report Portugal.* Centro de Directo Biomédico, Faculdade de Direito, Universidade de Coimbra

Saegusa, Asako. 1998. „Japan okays test-tube baby gene tests." *Nature* 394:110

Sandor, Judit. 1998. *Reproductive Rights in Hungary.* Central European University Political Sciences Department

Schenker, Joseph G. 1997. "Assisted Reproduction in Europe: Legal and Ethical Aspects." *Human Reproduction Update* 3/2:173–184

Shanner, Laura 2001. *Embryonic stem cell research: Canadian policy and ethical considerations a report for Health Canada, policy division.* Available at http://www.msu.edu/~hlnelson/fab/ Shanner_Stem_Cell_Policy.pdf

Sheikh, Asim. 1997. "'Time to Clone Around'-Human Cloning: A Brave New World for Law, Medicine and Ethics" *Medical Law Journal of Ireland:* 89-96

Shirai, Yasuko. 2001. "Ethical debate over Preimplantation Genetic diagnosis in Japan" *Eubios Journal of Asian and International Bioethics* 11:132–136. Available at http://www.biol.tsukuba.ac.jp/ ~macer/EJ115/ej115d.htm

Sueoka, Kou 1999. "Reproductive Medicine in Japan: Progress and Paradox". *Japan Echo* (February) 26/1: 40-44. Available at http://www.japanecho.co. jp/docs/html/260113.html

UNESCO (United Nations Educational, Scientific, and Cultural Organisation). 1998. "Reproductive Human Cloning: Ethical Questions." (Available at http://firewall.unesco.org/opi/eng/ bio98/#legis)

United Nations 2001 *Abortion Policies: A Global Review.* United Nations: New York. Available at http://www.un.org/esa/population/publications/abortion/ index.htm

9 References

1 Theoretical and practical possibilities in human embryo experimentation

Abbott A, Cyranoski D (2001) China plans 'hybrid' embryonic stem cells. Nature 413: 339

Adjaye J, Daniels R, Monk M (1998) The construction of cDNA libraries from human single preimplantation embryos and their use in the study of gene expression during development. J. Assist. Reprod. Genet. 15: 344–348

Alison MR, Poulsom R, Jeffery R, Dhillon AP, Quaglia A, Jacob J, Novelli M, Prentice G, Williamson J, Wright NA (2000) Hepatocytes from non-hepatic adult stem cells. Nature 406: 257

Amit M, Carpenter MK, Inokuma MS, Chiu C-P, Harris CP, Waknitz MA, Itskovitz-Eldor J, Thomson JA (2000) Clonally derived human embryonic stem cell lines maintain pluripotency and proliferative potential for prolonged periods of culture. Dev. Biol. 227: 271–278

Anversa P, Nadal-Ginard B (2002) Myocyte renewal and ventricular remodelling. Nature 415: 240–243

Aslam I, Fishel S, Green S, Campbell A, Garratt L, McDermott H, Dowell K, Thornton S (1998) Can we justify spermatid microinjection for severe male factor infertility? Hum. Reprod. Update 4: 213–222

Barritt J, Willadsen S, Brenner C, Cohen J (2001a) Cytoplasmic transfer in assisted reproduction. Hum. Reprod. Update 7: 428–435

Barritt JA, Brenner CA, Malter HE, Cohen J (2001b) Mitochondria in human offspring derived from ooplasmic transplantation. Hum. Reprod. 16: 513–516

Beddington R (1985) The development of 12th to 14th day foetuses following reimplantation of pre-and early-primitive-streak-stage mouse embryos. J. Embryol. exp. Morph. 88: 281–91

BenEzra D(2003) In-vitro fertilization and retinoblastoma. Lancet 361:273–274

Bjornson CRR, Rietze RL, Reynolds BA, Magli MC, Vescovi AL (1999) Turning brain into blood: a hematopoietic fate adopted by adult neural stem cells in vivo. Science 283: 534–537

Blackshaw S, Cepko CL (2002) Stem cells that know their place. Nat. Neurosci. 5: 1251–1252

Blau H, Brazelton T, Keshet G, Rossi F (2002) Something in the eye of the beholder. Science 298: 361–362

Blau HM, Brazelton TR, Weimann JM (2001) The evolving concept of a stem cell: entity or function? Cell 105: 829–841

Bonduelle M, Liebaers I, Deketelaere V, Derde MP, Camus M, Devroey P, Van Steirteghem A (2002) Neonatal data on a cohort of 2889 infants born after ICSI (1991-1999) and of 2995 infants born after IVF (1983-1999). Hum. Reprod. 17: 671–694

Bourc'his D, Le Bourhis D, Patin D, Niveleau A, Comizzoli P, Renard J-P, Viegas-Pequignot E (2001) Delayed and incomplete reprogramming of chromosome methylation patterns in bovine cloned embryos. Curr. Biol. 11: 1542–1546

Braude P (2001) Preimplantation genetic diagnosis and embryo research – human developmental biology in clinical practice. Int. J. Dev. Biol. 45: 607–611

Braude P, Bolton V, Moore S (1988) Human gene expression first occurs between the four- and eight-cell stages of preimplantation development. Nature 332: 459–461

Braude PR, De Wert GM, Evers-Kiebooms G, Pettigrew RA, Geraedts JP (1998) Non-disclosure preimplantation genetic diagnosis for Huntington's disease: practical and ethical dilemmas. Prenat. Diagn. 18: 1422–1426

Braun KM, Degen JL, Sandgren EP (2000) Hepatocyte transplantation in a model of toxin-induced liver disease: variable therapeutic effect during replacement of damaged parenchyma by donor cells. Nat. Med. 6: 320–326

Brazelton TR, Rossi FMV, Keshet GI, Blau HM (2000) From marrow to brain: expression of neuronal phenotypes in adult mice. Science 290: 1775–1779

Bredkjaer HE, Grudzinskas JG (2001) Cryobiology in human assisted reproductive technology. Would Hippocrates approve? Early Pregnancy 5: 211–213

Brinster RL (2002) Germline stem cell transplantation and transgenesis. Science 296: 2174–2176

Brüstle O, Jones KN, Learish RD, Karram K, Choudhary K, Wiestler OD, Duncan ID, McKay RD (1999) Embryonic stem cell-derived glial precursors: a source of myelinating transplants. Science 285: 754–756

Bullen PJ, Robson SC, Strachan T (1998) Human post-implantation embryo collection: medical and surgical techniques. Early Hum. Dev. 51: 213–221

Capmany G, Taylor A, Braude PR, Bolton VN (1996) The timing of pronuclear formation, DNA synthesis and cleavage in the human 1-cell embryo. Mol. Hum. Reprod. 2: 299–306

Castro RF, Jackson KA, Goodell MA, Robertson CS, Liu H, Shine HD (2002a) Failure of bone marrow cells to transdifferentiate into neural cells in vivo. Science 297: 1299

Castro RF, Jackson KA, Goodell MA, Robertson CS, Liu H, Shine HD (2002b) Response to "Something in the eye of the beholder". Science 298: 362

Cavazzana-Calvo M, Hacein-Bey S, de Saint Basile G, Gross F, Yvon E, Nusbaum P, Selz F, Hue C, Certain S, Casanova J-L, Bousso P, Le Deist F, Fischer A (2000) Gene therapy of human severe combined immunodeficiency (SCID)-X1 disease. Science 288: 669–672

Chan AW, Dominko T, Luetjens CM, Neuber E, Martinovich C, Hewitson L, Simerly CR, Schatten GP (2000) Clonal propagation of primate offspring by embryo splitting. Science 287: 317–319

Chung YG, Mann MRW, Bartolomei MS, Latham KE (2002) Nuclear-cytoplasmic "tug of war" during cloning: effects of somatic cell nuclei on culture medium preferences of preimplantation cloned mouse embryos. Biol. Reprod. 66: 1178–1184

Cibelli JB, Grant KA, Chapman KB, Cunniff K, Worst T, Green HL, Walker SJ, Gutin PH, Vilner L, Tabar V, Dominko T, Kane J, Wettstein PJ, Lanza RP, Studer L, Vrana KE, West MD (2002) Parthenogenetic stem cells in nonhuman primates. Science 295: 819

Cibelli JB, Kiessling AA, Cunniff K, Richards C, Lanza RP, West MD (2001) Somatic cell nuclear transfer in humans: pronuclear and early embryonic development. e-biomed 2: 25–31

Clarke DL, Johansson CB, Wilbertz J, Veress B, Nilsson E, Karlström H, Lendahl U, Frisén J (2000) Generalized potential of adult neural stem cells. Science 288: 1660–1663

Cohen J, Scott R, Alikani M, Schimmel T, Munne S, Levron J, Wu L, Brenner C, Warner C, Willadsen S (1998) Ooplasmic transfer in mature human oocytes. Mol. Hum. Reprod. 4: 269–280

Cohen J, Simons RF, Fehilly CB, Fishel SB, Edwards RG, Hewitt J, Rowlant GF, Steptoe PC, Webster JM (1985) Birth after replacement of hatching blastocyst cryopreserved at expanded blastocyst stage. Lancet 1: 647

Couzin J (2002) Quirks of fetal environment felt decades later. Science 296: 2167–2169

Cox GF, Burger J, Lip V, Mau UA, Sperling K, Wu BL, Horsthemke B (2002) Intracytoplasmic sperm injection may increase the risk of imprinting defects. Am. J. Hum. Genet. 71: 162–164

D'Amour KA, Gage FH (2002) Are somatic stem cells pluripotent or lineage-restricted? Nat. Med. 8: 213–214

Daniels R, Hall V, Trounson AO (2000) Analysis of gene transcription in bovine nuclear transfer embryos reconstructed with granulosa cell nuclei. Biol. Reprod. 63: 1034–1040

Davies TJ, Gardner RL (2002) The plane of first cleavage is not related to the distribution of sperm components in the mouse. Hum. Reprod. 17: –2379

Davis OK, Rosenwaks Z (2001) Superovulation strategies for assisted reproductive technologies. Semin. Reprod. Med. 19: 207–212

DeBraun MR, Niemitz EL, Feinberg AP (2003) Association of in vitro fertilization with Beckwith-Wiedemann syndrome and epigenetic alteratios of *LIT1* and *H19*. Am. J. Hum. Genet. 72: 156–160

DeWitt N, Knight J (2002) Biologists question adult stem-cell versatility. Nature 416: 354

Donovan PJ, Gearhart J (2001) The end of the beginning for pluripotent stem cells. Nature 414: 92–97

Downing BG, Mohr LR, Trounson AO, Freemann LE, Wood C (1985) Birth after transfer of cryopreserved embryos. Med. J. Aust. 142: 409–411

Drukker M, Katz G, Urbach A, Schuldiner M, Markel G, Itskovitz-Eldor J, Reubinoff B, Mandelboim O, Benvenisty N (2002) Characterization of the expression of MHC proteins in human embryonic stem cells. Proc Natl Acad Sci U S A 99: 9864–9

Eiges R, Schuldiner M, Drukker M, Yanuka O, Itskovitz-Eldor J, Benvenisty N (2001) Establishment of human embryonic stem cell-transfected clones carrying a marker for undifferentiated cells. Curr. Biol. 11: 514–518

Ertzeid G, Storeng R (2001) The impact of ovarian stimulation on implantation and fetal development in mice. Hum. Reprod. 16: 221–225

ESHRE Preimplantation Genetic Diagnosis Consortium (2002) Data collection III (May 2001). Hum. Reprod. 17: 233–246

Eto K, Murphy R, Kerrigan SW, Bertoni A, Stuhlmann H, Nakano T, Leavitt AD, Shattil SJ (2002) Megakaryocytes derived from embryonic stem cells implicate CalDAG-GEFI in integrin signaling. Proc. Natl. Acad. Sci. USA 99: 12819–12824

Evans MJ, Kaufman MH (1981) Establishment in culture of pluripotential cells from mouse embryos. Nature 292: 154–156

Fairburn HR, Young LE, Hendrich BD (2002) Epigenetic reprogramming: how now, cloned cow? Curr. Biol. 12: R68–R70

Finch CE, Kirkwood TBL (2000) Chance, development, and aging. Oxford University Press, New York

Fougerousse F, Bullen P, Herasse M, Lindsay S, Richard I, Wilson D, Suel L, Durand M, Robson S, Abitbol M, Beckmann JS, Strachan T (2000) Human-mouse differences in the embryonic expression patterns of developmental control genes and disease genes. Hum. Mol. Genet. 9: 165–173

Fukuyama F (2002) Our posthuman future. Farrar, Straus & Girou, New York

Gage FH (2000) Mammalian neural stem cells. Science 287: 1433–1438

Gardner DK, Schoolcraft WB (1998) Human embryo viability: what determines developmental potential, and can it be assessed? J. Assist. Reprod. Genet. 15: 455–458

Geraedts J, Handyside A, Harper J, Liebaers I, Sermon K, Staessen C, Thornhill A, Vanderfaeillie A, Viville S (1999) ESHRE Preimplantation Genetic Diagnosis (PGD) Consortium: preliminary assessment of data from January 1997 to September 1998. ESHRE PGD Consortium Steering Committee. Hum. Reprod. 14: 3138–3148

Geraedts J, Handyside A, Harper J, Liebaers I, Sermon K, Staessen C, Thornhill A, Viville S, Wilton L (2000) ESHRE preimplantation genetic diagnosis (PGD) consortium: data collection II (May 2000). Hum Reprod 15: 2673–83

Geraedts JP, Harper J, Braude P, Sermon K, Veiga A, Gianaroli L, Agan N, Munne S, Gitlin S, Blenow E, de Boer K, Hussey N, Traeger-Synodinos J, Lee SH, Viville S, Krey L, Ray P, Emiliani S, Liu YH, Vermeulen S, Kanavakis E (2001) Preimplantation genetic diagnosis (PGD), a collaborative activity of clinical genetic departments and IVF centres. Prenat. Diagn. 21: 1086–1092

Griffin DK, Handyside AH, Harper JC, Wilton LJ, Atkinson G, Soussis I, Wells D, Kontogianni E, Tarin J, Geber S, et al. (1994) Clinical experience with preimplantation diagnosis of sex by dual fluorescent in situ hybridization. J. Assist. Reprod. Genet. 11: 132–143

Griffin DK, Handyside AH, Penketh RJ, Winston RM, Delhanty JD (1991) Fluorescent in-situ hybridization to interphase nuclei of human preimplantation embryos with X and Y chromosome specific probes. Hum. Reprod. 6: 101–105

Gross M (2002) Green light for selected baby. Curr. Biol. 12: R193

Hack M, Fanaroff AA (1999) Outcomes of children of extremely low birthweight and gestational age in the 1990's. Early Hum. Dev. 53: 193–218

Håkelien A-M, Landsverk HB, Robl JM, Skålhegg BS, Collas P (2002) Reprogramming fibroblasts to express T-cell functions using cell extracts. Nat. Biotechnol. 20: 460–466

Handyside AH, Kontogianni EH, Hardy K, Winston RM (1990) Pregnancies from biopsied human preimplantation embryos sexed by Y-specific DNA amplification. Nature 344: 768–770

Hansen M, Kurinczuk JJ, Bower C, Webb S (2002) The risk of major birth defects after intracytoplasmic sperm injection and in vitro fertilization N. Engl. J. Med. 346:725–730

Hardy K, Martin KL, Leese HJ, Winston RM, Handyside AH (1990) Human preimplantation development in vitro is not adversely affected by biopsy at the 8-cell stage. Hum. Reprod. 5: 708–714

Harper JC, Delhanty JDA, Handyside AH (2001) Preimplantation genetic diagnosis. John Wiley & Co., Chichester (UK)

Hassold T, Hunt P (2001) To err (meiotically) is human: the genesis of human aneuploidy. Nat. Rev. Genet. 2: 280–291

Heindryckx B, Rybouchkin A, van der Elst J, Dhont M (2001) Effect of culture media on *in vitro* development of cloned mouse embryos. Cloning 3: 41–50

Hesterlee SE (2001) Recognizing risks and potential promise of germline engineering. Nature 414: 15

Hewitson L, Simerly CR, Schatten G (2002) Fate of sperm components during assisted reproduction: implications for infertility. Hum. Fertil. (Camb.) 5: 110–116

Honaramooz A, Snedaker A, Boiani M, Schöler H, Dobrinski I, Schlatt S (2002) Sperm from neonatal mammalian testes grafted in mice. Nature 418: 778–781

Houghton FD, Hawkhead JA, Humpherson PG, Hogg JE, Balen AH, Rutherford AJ, Leese HJ (2002) Non-invasive amino acid turnover predicts human embryo developmental capacity. Hum. Reprod. 17: 999–1005

Hovatta O (2000) Cryopreservation and culture of human primordial and primary ovarian follicles. Mol. Cell. Endocrinol. 169: 95–97

Hübner K, Fuhrmann G, Christenson LK, Kehler J, Reinbold R, De La Fuente R, Wood J, Strauss III JF, Boiani M, Schöler HR (2003) Derivation of oocytes from mouse embryonic stem cells. Science 300: 1251–1256

Humpherys D, Eggan K, Akutsu H, Friedman A, Hochedlinger K, Yanagimachi R, Lander ES, Golub TR, Jaenisch R (2002) Abnormal gene expression in cloned mice derived from embryonic stem cell and cumulus cell nuclei. Proc. Natl. Acad. Sci. USA 99: 12889–12894

Humpherys D, Eggan K, Akutsu H, Hochedlinger K, Rideout WM, III, Biniszkiewicz D, Yanagimachi R, Jaenisch R (2001) Epigenetic instability in ES cells and cloned mice. Science 293: 95–97

Hunt PA, Hassold TJ (2002) Sex matters in meiosis. Science 296: 2181–2183

Inoue K, Kohda T, Lee J, Ogonuki N, Mochida K, Noguchi Y, Tanemura K, Kaneko-Ishino T, Ishino F, Ogura A (2002) Faithful expression of imprinted genes in cloned mice. Science 295: 297

Jackson KA, Mi T, Goodell MA (1999) Hematopoietic potential of stem cells isolated from murine skeletal muscle. Proc. Natl. Acad. Sci. USA 96: 14482–14486

Jiang Y, Jahagirdar BN, Reinhardt RL, Schwartz RE, Keene CD, Ortiz-Gonzalez XR, Reyes M, Lenvik T, Lund T, Blackstad M, Du J, Aldrich S, Lisberg A, Low WC, Largaespada DA, Verfaillie CM (2002) Pluripotency of mesenchymal stem cells derived from adult marrow. Nature 418: 41–49

Jurisicova A, Varmuza S, Casper RF (1996) Programmed cell death and human embryo fragmentation. Mol. Hum. Reprod. 2: 93–98

Kahraman S, Polat G, Samli M, Sozen E, Ozgun OD, Dirican K, Ozbicer T (1998) Multiple pregnancies obtained by testicular spermatid injection in combination with intracytoplasmic sperm injection. Hum. Reprod. 13: 104–110

Kang Y-K, Koo D-B, Park J-S, Choi Y-H, Chung A-S, Lee K-K, Han Y-M (2001) Aberrant methylation of donor genome in cloned bovine embryos. Nat. Genet. 28: 173–177

Kao RL (2001) Autologous satellite cells for myocardial regeneration. e-biomed 2: 1–8

Kaufman DS, Hanson ET, Lewis RL, Auerbach R, Thomson JA (2001) Hematopoietic colony-forming cells derived from human embryonic stem cells. Proc. Natl. Acad. Sci. USA 98: 10716–10721

Kaufman MH, Robertson EJ, Handyside AH, Evans MJ (1983) Establishment of pluripotential cell lines from haploid mouse embryos. J. Embryol. exp. Morph. 73: 249–261

Kawase E, Yamazaki Y, Yagi T, Yanagimachi R, Pedersen RA (2000) Mouse embryonic stem (ES) cell lines established from neuronal cell- derived cloned blastocysts. Genesis 28: 156–163

Khosla S, Dean W, Reik W, Feil R (2001) Culture of preimplantation embryos and its long-term effects on gene expression and phenotype. Hum. Reprod. Update 7: 419–427

Kim J-H, Auerbach JM, Rodríguez-Gómez JA, Velasco I, Gavin D, Lumelsky N, Lee S-H, Nguyen J, Sánchez-Pernaute R, Bankiewicz K, McKay R (2002) Dopamine neurons derived from embryonic stem cells function in an animal model of Parkinson's disease. Nature 418: 50–56

Kim SS, Battaglia DE, Soules MR (2001) The future of human ovarian cryopreservation and transplantation: fertility and beyond. Fertil. Steril. 75: 1049–1056

Klinger FG, De Felici M (2002) *In vitro* development of growing oocytes from fetal mouse oocytes: stage- specific regulation by stem cell factor and granulosa cells. Dev. Biol. 244: 85–95

Knight J (2001) Biology's last taboo. Nature 413: 12–15

Knight J (2002) An out of body experience. Nature 419: 106–107

Komori S, Kato H, Kobayashi S, Koyama K, Isojima S (2002) Transmission of Y chromosomal microdeletions from father to son through intracytoplasmic sperm injection. J. Hum. Genet. 47: 465–468

Kondo T, Raff M (2000) Oligodendrocyte precursor cells reprogrammed to become multipotential CNS stem cells. Science 289: 1754–1757

Krause DS, Theise ND, Collector MI, Henegariu O, Hwang S, Gardner R, Neutzel S, Sharkis SJ (2001) Multi-organ, multi-lineage engraftment by a single bone marrow-derived stem cell. Cell 105: 369–377

Kusakabe H, Szczygiel MA, Whittingham DG, Yanagimachi R (2001) Maintenance of genetic integrity in frozen and freeze-dried mouse spermatozoa. Proc. Natl. Acad. Sci. USA 98: 13501–13506

Lagasse E, Connors H, Al-Dhalimy M, Reitsma M, Dohse M, Osborne L, Wang X, Finegold M, Weissman IL, Grompe M (2000) Purified hematopoietic stem cells can differentiate into hepatocytes *in vivo*. Nat. Med. 6: 1229–1234

Lanza RP, Cibelli JB, Faber D, Sweeney RW, Henderson B, Nevala W, West MD, Wettstein PJ (2001) Cloned cattle can be healthy and normal. Science 294: 1893–4

Lee KF, Chow JF, Xu JS, Chan ST, Ip SM, Yeung WS (2001) A comparative study of gene expression in murine embryos developed in vivo, cultured in vitro, and cocultured with human oviductal cells using messenger ribonucleic acid differential display. Biol. Reprod. 64: 910–917

Leese HJ (2002) Quiet please, do not disturb: a hypothesis of embryo metabolism and viability. Bioessays 24: 845–849

Leese HJ, Conaghan J, Martin KL, Hardy K (1993) Early human embryo metabolism. Bioessays 15: 259–264

Leese HJ, Donnay I, Thompson JG (1998) Human assisted conception: a cautionary tale. Lessons from domestic animals. Hum. Reprod. 13 Suppl 4: 184–202

Lemischka I (2002) Rethinking somatic stem cell plasticity. Nat. Biotechnol. 20: 425

Levenberg S, Golub JS, Amit M, Itskovitz-Eldor J, Langer R (2002) Endothelial cells derived from human embryonic stem cells. Proc. Natl. Acad. Sci. USA 99: 4391–4396

Lewin M, Carlesso N, Tung C-H, Tang X-W, Cory D, Scadden DT, Weissleder R (2000) Tat peptide-derivatized magnetic nanoparticles allow in vivo tracking and recovery of progenitor cells. Nat. Biotechnol. 18: 410–414

Lewis CM, Pinel T, Whittaker JC, Handyside AH (2001) Controlling misdiagnosis errors in preimplantation genetic diagnosis: a comprehensive model encompassing extrinsic and intrinsic sources of error. Hum. Reprod. 16: 43–50

Li Z, Düllmann J, Schiedlmeier B, Schmidt M, von Kalle C, Meyer J, Forster M, Stocking C, Wahlers A, Frank O, Ostertag W, Kühlcke K, Eckert H-G, Fehse B, Baum C (2002) Murine leukemia induced by retroviral gene marking. Science 296: 497

Liang L, Bickenbach JR (2002) Somatic epidermal stem cells can produce multiple cell lineages during development. Stem Cells 20: 21–31

Lovell-Badge R (2001) The future for stem cell research. Nature 414: 88–91

Lumelsky N, Blondel O, Laeng P, Velasco I, Ravin R, McKay R (2001) Differentiation of embryonic stem cells to insulin-secreting structures similar to pancreatic islets. Science 292: 1389–1394

Mann JR, Gadi I, L. HM, Abbondanzo SJ, Stewart CL (1990) Androgenetic mouse embryonic stem cells are pluripotent and cause skeletal defects in chimeras: Implications for genetic imprinting. Cell 62: 251–260

Marshall E (1998) Claim of human-cow embryo greeted with skepticism. Science 282: 1390–1391

Martin GR (1981) Isolation of a pluripotent cell line from early mouse embryos cultured in medium conditioned by teratocarcinoma stem cells. Proc. Natl. Acad. Sci. USA 78: 7634–7638

May C, Rivella S, Callegari J, Heller G, Gaensler KML, Luzzatto L, Sadelain M (2000) Therapeutic haemoglobin synthesis in b-thalassaemic mice expressing lentivirus-encoded human b-globin. Nature 406: 82–86

McEvoy TG, Sinclair KD, Young LE, Wilmut I, Robinson JJ (2000) Large offspring syndrome and other consequences of ruminant embryo culture in vitro: relevance to blastocyst culture in human ART. Hum. Fertil. (Camb.) 3: 238–246

McGann CJ, Odelberg SJ, Keating MT (2001) Mammalian myotube dedifferentiation induced by newt regeneration extract. Proc. Natl. Acad. Sci. USA 98: 13699–13704

McGrath J, Solter D (1983) Nuclear transplantation in the mouse embryo by microsurgery and cell fusion. Science 220: 1300–1302

McKinney-Freeman SL, Jackson KA, Camargo FD, Ferrari G, Mavilio F, Goodell MA (2002) Muscle-derived hematopoietic stem cells are hematopoietic in origin. Proc. Natl. Acad. Sci. USA 99: 1341–1346

Mezey É, Chandross KJ, Harta G, Maki RA, McKercher SR (2000) Turning blood into brain: cells bearing neuronal antigens generated in vivo from bone marrow. Science 290: 1779–1782

Moll AC, Imhof SM, Cruysberg JRM, Schouten-van-Meeteren AYN, Boers M, van Leeuwen FE (2003) Lancet 361:309–310

Monk M (1988) Pre-implantation diagnosis. Bioessays 8: 184–189

Morshead CM, Benveniste P, Iscove NN, van der Kooy D (2002) Hematopoietic competence is a rare property of neural stem cells that may depend on genetic and epigenetic alterations. Nat. Med. 8: 268–273

Mulhall JP, Reijo R, Alagappan R, Brown L, Page D, Carson R, Oates RD (1997) Azoospermic men with deletion of the DAZ gene cluster are capable of completing spermatogenesis: fertilization, normal embryonic development and pregnancy occur when retrieved testicular spermatozoa are used for intracytoplasmic sperm injection. Hum. Reprod. 12: 503–508

Munne S, Sandalinas M, Escudero T, Fung J, Gianaroli L, Cohen J (2000) Outcome of preimplantation genetic diagnosis of translocations. Fertil. Steril. 73: 1209–1218

Munne S, Weier HU, Grifo J, Cohen J (1994) Chromosome mosaicism in human embryos. Biol. Reprod. 51: 373–379

Munsie MJ, Michalska AE, O'Brien CM, Trounson AO, Pera MF, Mountford PS (2000) Isolation of pluripotent embryonic stem cells from reprogrammed adult somatic cell nuclei. Curr. Biol. 10: 989–992

Naviaux RK, Singh KK (2001) Need for public debate about fertility treatments. Nature 413: 347

Oates RD, Silber S, Brown LG, Page DC (2002) Clinical characterization of 42 oligospermic or azoospermic men with microdeletion of the AZFc region of the Y chromosome, and of 18 children conceived via ICSI. Hum. Reprod. 17: 2813–2824

Ogawa T, Dobrinski I, Avarbock MR, Brinster RL (2000) Transplantation of male germ line stem cells restores fertility in infertile mice. Nat. Med. 6: 29–34

Ogonuki N, Inoue K, Yamamoto Y, Noguchi Y, Tanemura K, Suzuki O, Nakayama H, Doi K, Ohtomo Y, Satoh M, Nishida A, Ogura A (2002) Early death of mice cloned from somatic cells. Nat. Genet. 30: 253–254

Ogura A, Yanagimachi R (1995) Spermatids as male gametes. Reprod. Fertil. Dev. 7: 155-158; discussion 158–159

Ohgane J, Wakayama T, Kogo Y, Senda S, Hattori N, Tanaka S, Yanagimachi R, Shiota K (2001) DNA methylation variation in cloned mice. Genesis 30: 45–50

Ono Y, Shimozawa N, Ito M, Kono T (2001) Cloned mice from fetal fibroblast cells arrested at metaphase by a serial nuclear transfer. Biol. Reprod. 64: 44–50

Orlic D, Kajstura J, Chimenti S, Jakoniuk I, Anderson SM, Li B, Pickel J, McKay R, Nadal-Ginard B, Bodine DM, Leri A, Anversa P (2001) Bone marrow cells regenerate infarcted myocardium. Nature 410: 701–705

Ørstavik KH, Eiklid K, van der Hagen CB, Spetalen S, Kierulf K, Skjeldal O, Butting K (2003) Another case of imprinting defect in a girl with Angelman syndrome who was conceived by intracytoplasmic sperm injection. Am. J. Hum. Genet. 72:218–219

Palermo G, Joris H, Devroey P, Van Steirteghem AC (1992) Pregnancies after intracytoplasmic injection of single spermatozoon into an oocyte. Lancet 340: 17–18

Palermo GD, Takeuchi T, Rosenwaks Z (2002) Technical approaches to correction of oocyte aneuploidy. Hum. Reprod. 17: 2165–2173

Paria BC, Reese J, Das SK, Dey SK (2002) Deciphering the cross-talk of implantation: advances and challenges. Science 296: 2185–2188

Perlingeiro RCR, Kyba M, Daley GQ (2001) Clonal analysis of differentiating embryonic stem cells reveals a hematopoietic progenitor with primitive erythroid and adult lymphoid- myeloid potential. Development 128: 4597–4604

Perry ACF, Wakayama T (2002) Untimely ends and new beginnings in mouse cloning. Nat. Genet. 30: 243–244

Pickering SJ, Taylor A, Johnson MH, Braude PR (1995) An analysis of multinucleated blastomere formation in human embryos. Hum. Reprod. 10: 1912–1922

Piotrowska K, Zernicka-Goetz M (2001) Role for sperm in spatial patterning of the early mouse embryo. Nature 409: 517–521

Piotrowska K, Zernicka-Goetz M (2002) Early patterning of the mouse embryo – contributions of sperm and egg. Development 129: 5803–5813

Pittenger MF, Mackay AM, Beck SC, Jaiswal RK, Douglas R, Mosca JD, Moorman MA, Simonetti DW, Craig S, Marshak DR (1999) Multilineage potential of adult human mesenchymal stem cells. Science 284: 143–147

Porcu E, Fabbri R, Damiano G, Giunchi S, Fratto R, Ciotti PM, Venturoli S, Flamigni C (2000) Clinical experience and applications of oocyte cryopreservation. Mol. Cell. Endocrinol. 169: 33–37

Prockop DJ (2002) Adult stem cells gradually come of age. Nat. Biotechnol. 20: 791–792

Ramiya VK, Maraist M, Arfors KE, Schatz DA, Peck AB, Cornelius JG (2000) Reversal of insulin-dependent diabetes using islets generated *in vitro* from pancreatic stem cells. Nat. Med. 6: 278–282

Rathjen J, Rathjen PD (2001) Mouse ES cells: experimental exploitation of pluripotent differentiation potential. Curr. Opin. Genet. Dev. 11: 587–594

Ray PF, Munnich A, Nisand I, Frydman R, Vekemans M, Viville S (2002) The place of 'social sexing' in medicine and science. Hum. Reprod. 17: 248–249

Rechitsky S, Strom C, Verlinsky O, Amet T, Ivakhnenko V, Kukharenko V, Kuliev A, Verlinsky Y (1999) Accuracy of preimplantation diagnosis of single-gene disorders by polar body analysis of oocytes. J. Assist. Reprod. Genet. 16: 192–198

Reik W, Davies K, Dean W, Kelsey G, Constancia M (2001) Imprinted genes and the coordination of fetal and postnatal growth in mammals. Novartis Found. Symp. 237: 19-31; discussion 31–42

Reik W, Walter J (2001) Genomic imprinting: parental influence on the genome. Nat. Rev. Genet. 2: 21–32

Renard J-P, Chastant S, Chesné P, Richard C, Marchal J, Cordonnier N, Chavatte P, Vignon X (1999) Lymphoid hypoplasia and somatic cloning. Lancet 353: 1489–1491

Reubinoff BE, Itsykson P, Turetsky T, Pera MF, Reinhartz E, Itzik A, Ben-Hur T (2001) Neural progenitors from human embryonic stem cells. Nat. Biotechnol. 19: 1134–1140

Reubinoff BE, Pera MF, Fong C-Y, Trounson A, Bongso A (2000) Embryonic stem cell lines from human blastocysts: somatic differentiation in vitro [published erratum appears in Nat Biotechnol 2000 May;18(5):559]. Nat. Biotechnol. 18: 399–404

Richards M, Fong C-Y, Chan W-K, Wong P-C, Bongso A (2002) Human feeders support prolonged undifferentiated growth of human inner cell masses and embryonic stem cells. Nat. Biotechnol. 20: 933–936

Rideout WM, III, Eggan K, Jaenisch R (2001) Nuclear cloning and epigenetic reprogramming of the genome. Science 293: 1093–1098

Rideout WM, III, Hochedlinger K, Kyba M, Daley GQ, Jaenisch R (2002) Correction of a genetic defect by nuclear transplantation and combined cell and gene therapy. Cell 109: 17–27

Rossant J (2002) A monoclonal mouse? Nature 415: 967–969

Rossi F, Cattaneo E (2002) Neural stem cell therapy for neurological diseases: dreams and reality. Nat. Rev. Neurosci. 3: 401–409

Roy NS, Wang S, Jiang L, Kang J, Benraiss A, Harrison-Restelli C, Fraser RAR, Couldwell WT, Kawaguchi A, Okano H, Nedergaard M, Goldman SA (2000) *In vitro* neurogenesis by progenitor cells isolated from the adult human hippocampus. Nat. Med. 6: 271–277

Schuldiner M, Yanuka O, Itskovitz-Eldor J, Melton DA, Benvenisty N (2000) Effects of eight growth factors on the differentiation of cells derived from human embryonic stem cells. Proc. Natl. Acad. Sci. USA 97: 11307–11312

Schultz RM, Williams CJ (2002) The science of ART. Science 296: 2188–2190

Sekiya I, Vuoristo JT, Larson BL, Prockop DJ (2002) *In vitro* cartilage formation by human adult stem cells from bone marrow stroma defines the sequence of cellular and molecular events during chondrogenesis. Proc. Natl. Acad. Sci. USA 99: 4397–4402

Shamblott MJ, Axelman J, Littlefield JW, Blumenthal PD, Huggins GR, Cui Y, Cheng L, Gearhart JD (2001) Human embryonic germ cell derivatives express a broad range of developmentally distinct markers and proliferate extensively *in vitro*. Proc. Natl. Acad. Sci. USA 98: 113–118

Shamblott MJ, Axelman J, Wang S, Bugg EM, Littlefield JW, Donovan PJ, Blumenthal PD, Huggins GR, Gearhart JD (1998) Derivation of pluripotent stem cells from cultured human primordial germ cells [published erratum appears in Proc Natl Acad Sci U S A 1999 Feb 2;96(3):1162]. Proc. Natl. Acad. Sci. USA 95: 13726–13731

Sinclair KD, Young LE, Wilmut I, McEvoy TG (2000) In-utero overgrowth in ruminants following embryo culture: lessons from mice and a warning to men. Hum. Reprod. 15 Suppl 5: 6–86

Smith A (1998) Cell therapy: In search of pluripotency. Curr. Biol. 8: R802–804

Solter D (1998a) Dolly *is* a clone-and no longer alone. Nature 394: 315–316

Solter D (1998b) Imprinting. Int. J. Dev. Biol. 42: 951–954

Solter D (1999) Cloning and embryonic stem cells: a new era in human biology and medicine. Croat. Med. J. 40: 309–318

Solter D (2000) Mammalian cloning: advances and limitations. Nat. Rev. Genet. 1: 199–207

Solter D (2001) Nonequivalence of parental genomes and the discovery of imprinting. Great Experiments: Development. http://www.ergito.com

Solter D, De Vries WN, Evsikov AV, Peaston AE, Chen FH, Knowles BB (2002) Fertilization and activation of the embryonic genome. In: Rossant J, Tam PPL (eds) Mouse development. Patterning, morphogenesis, and organogenesis. Academic Press, San Diego, pp 5–19

Solter D, Gearhart J (1999) Putting stem cells to work. Science 283: 1468–1470

Steptoe PC, Edwards RG (1978) Birth after the reimplantation of a human embryo. Lancet 2: 366

Stock G (2002) Redesigning humans: our inevitable genetic future. Houghton Mifflin, Boston

Surani MA (2001) Reprogramming of genome function through epigenetic inheritance. Nature 414: 122–128

Tada M, Takahama Y, Abe K, Nakatsuji N, Tada T (2001) Nuclear reprogramming of somatic cells by in vitro hybridization with ES cells. Curr. Biol. 11: 1553–1558

Tamashiro KLK, Wakayama T, Akutsu H, Yamazaki Y, Lachey JL, Wortman MD, Seeley RJ, D'Alessio DA, Woods SC, Yanagimachi R, Sakai RR (2002) Cloned mice have an obese phenotype not transmitted to their offspring. Nat. Med. 8: 262–267

Terada N, Hamazaki T, Oka M, Hoki M, Mastalerz DM, Nakano Y, Meyer EM, Morel L, Petersen BE, Scott EW (2002) Bone marrow cells adopt the phenotype of other cells by spontaneous cell fusion. Nature 416: 542–545

Tesarik J, Kopecny V, Plachot M, Mandelbaum J (1988) Early morphological signs of embryonic genome expression in human preimplantation development as revealed by quantitative electron microscopy. Dev. Biol. 128: 15–20

Thomson JA, Itskovitz-Eldor J, Shapiro SS, Waknitz MA, Swiergiel JJ, Marshall VS, Jones JM (1998) Embryonic stem cell lines derived from human blastocysts. Science 282: 1145–1147

Tropepe V, Coles BLK, Chiasson BJ, Horsford DJ, Elia AJ, McInnes RR, van der Kooy D (2000) Retinal stem cells in the adult mammalian eye. Science 287: 2032–2036

Trounson A, Mohr L (1983) Human pregnancy following cryopreservation, thawing and transfer of an eight-cell embryo. Nature 305: 707–709

Van Blerkom J, Antczak M, Schrader R (1997) The developmental potential of the human oocyte is related to the dissolved oxygen content of follicular fluid: association with vascular endothelial growth factor levels and perifollicular blood flow characteristics. Hum. Reprod. 12: 1047–1055

Van Blerkom J, Davis P (1995) Evolution of the sperm aster after microinjection of isolated human sperm centrosomes into meiotically mature human oocytes. Hum. Reprod. 10: 2179–2182

Van de Velde H, De Vos A, Sermon K, Staessen C, De Rycke M, Van Assche E, Lissens W, Vandervorst M, Van Ranst H, Liebaers I, Van Steirteghem A (2000) Embryo implantation after biopsy of one or two cells from cleavage-stage embryos with a view to preimplantation genetic diagnosis. Prenat. Diagn. 20: 1030–1037

Van Steirteghem A, Bonduelle M, Devroey P, Liebaers I (2002) Follow-up of children born after ICSI. Hum Reprod Update 8: 111–6

Van Steirteghem AC, Nagy Z, Joris H, Liu J, Staessen C, Smitz J, Wisanto A, Devroey P (1993) High fertilization and implantation rates after intracytoplasmic sperm injection. Hum. Reprod. 8: 1061–1066

Van Winkle LJ (2001) Amino acid transport regulation and early embryo development. Biol. Reprod. 64: 1–12

Vanderzwalmen P, Nijs M, Stecher A, Zech H, Bertin G, Lejeune B, Vandamme B, Chatziparasidou A, Prapas Y, Schoysman R (1998) Is there a future for spermatid injections? Hum. Reprod. 13 Suppl 4: 71–84

Veiga A, Sandalinas M, Benkhalifa M, Boada M, Carrera M, Santalo J, Barri PN, Menezo Y (1997) Laser blastocyst biopsy for preimplantation diagnosis in the human. Zygote 5: 351–354

Verlinsky Y, Cieslak J, Freidine M, Ivakhnenko V, Wolf G, Kovalinskaya L, White M, Lifchez A, Kaplan B, Moise J, Valle J, Ginsberg N, Strom C, Kuliev A (1996) Polar body diagnosis of common aneuploidies by FISH. J. Assist. Reprod. Genet. 13: 157–162

Wagers AJ, Sherwood RI, Christensen JL, Weissman IL (2002a) Little evidence for developmental plasticity of adult hematopoietic stem cells. Science 297: 2256–2259

Wagers AJ, Sherwood RI, Christensen JL, Weissman IL (2002b) Response to "Something in the eye of the beholder". Science 298: 362–363

Wakayama T, Tabar V, Rodriguez I, Perry ACF, Studer L, Mombaerts P (2001) Differentiation of embryonic stem cell lines generated from adult somatic cells by nuclear transfer. Science 292: 740–743

Watt FM, Hogan BLM (2000) Out of Eden: stem cells and their niches. Science 287: 1427–1430

Weissman IL (2000) Translating stem and progenitor cell biology to the clinic: barriers and opportunities. Science 287: 1442–1446

Western PS, Surani MA (2002) Nuclear reprogramming – alchemy or analysis? Nat. Biotechnol. 20: 445–446

Whittingham DG (1974) The viability of frozen-thawed mouse blastocysts. J. Reprod. Fertil. 37: 159–162

Whittingham DG, Leibo SP, Mazur P (1972) Survival of mouse embryos frozen to -196 degrees and -269 degrees C. Science 178: 411–414

Williams N (2002) Dolly clouds cloning hopes. Curr. Biol. 12: R79–R80

Wilmut I (2002) Are there any normal cloned mammals? Nat. Med. 8: 215–216

Wilton L (2002) Preimplantation genetic diagnosis for aneuploidy screening in early human embryos: a review. Prenat. Diagn. 22: 512–518

Wininger JD, Kort HI (2002) Cryopreservation of immature and mature human oocytes. Semin. Reprod. Med. 20: 45–49

Wu P, Tarasenko YI, Gu Y, Huang L-YM, Coggeshall RE, Yu Y (2002) Region-specific generation of cholinergic neurons from fetal human neural stem cells grafted in adult rat. Nat. Neurosci. 5: 1271–1278

Wurmser AE, Gage FH (2002) Cell fusion causes confusion. Nature 416: 485–487

Xu C, Inokuma MS, Denham J, Golds K, Kundu P, Gold JD, Carpenter MK (2001a) Feeder-free growth of undifferentiated human embryonic stem cells. Nat. Biotechnol. 19: 971–974

Xu JS, Cheung TM, Chan ST, Ho PC, Yeung WS (2001b) Temporal effect of human oviductal cell and its derived embryotrophic factors on mouse embryo development. Biol. Reprod. 65: 1481–1488

Yamashita J, Itoh H, Hirashima M, Ogawa M, Nishikawa S, Yurugi T, Naito M, Nakao K, Nishikawa S-I (2000) Flk1-positive cells derived from embryonic stem cells serve as vascular progenitors. Nature 408: 92–96

Yanagimachi R (2001) Gamete manipulation for development: new methods for conception. Reprod. Fertil. Dev. 13: 3–14

Yin H, Baart E, Betzendahl I, Eichenlaub-Ritter U (1998) Diazepam induces meiotic delay, aneuploidy and predivision of homologues and chromatids in mammalian oocytes. Mutagenesis 13: 567–580

Ying Q-L, Nichols J, Evans EP, Smith AG (2002) Changing potency by spontaneous fusion. Nature 416: 545–548

Young LE, Fernandes K, McEvoy TG, Butterwith SC, Gutierrez CG, Carolan C, Broadbent PJ, Robinson JJ, Wilmut I, Sinclair KD (2001) Epigenetic change in *IGF2R* is associated with fetal overgrowth after sheep embryo culture. Nat. Genet. 27: 153–154

Yuan L, Liu JG, Hoja MR, Wilbertz J, Nordqvist K, Hoog C (2002) Female germ cell aneuploidy and embryo death in mice lacking the meiosis-specific protein SCP3. Science 296: 1115–1118

Zhang S-C, Wernig M, Duncan ID, Brüstle O, Thomson JA (2001) *In vitro* differentiation of transplantable neural precursors from human embryonic stem cells. Nat. Biotechnol. 19: 1129–1133

2   Adult and Embryonic Stem Cells: Clinical Perspectives

Ahmad I, Tang L, Pham H (2000) Identification of neural progenitors in the adult mammalian eye. Biochem Biophys Res Commun 270: 517–21

Amit M, Carpenter MK, Inokuma MS, Chiu CP, Harris CP, Waknitz MA, Itskovitz-Eldor J, Thomson JA (2000) Clonally derived human embryonic stem cell lines maintain pluripotency and proliferative potential for prolonged periods of culture [In Process Citation]. Dev Biol 227: 271–8

Antonchuk J, Sauvageau G, Humphries RK (2002) HOXB4-Induced Expansion of Adult Hematopoietic Stem Cells Ex Vivo. Cell 109: 39–45

Appelbaum FR, Buckner CD (1986) Overview of the clinical relevance of autologous bone marrow transplantation. Clin Haematol 15: 1–18

Asahara T, Masuda H, Takahashi T, Kalka C, Pastore C, Silver M, Kearne M, Magner M, Isner JM (1999a) Bone marrow origin of endothelial progenitor cells responsible for postnatal vasculogenesis in physiological and pathological neovascularization. Circ Res 85: 221–8

Asahara T, Murohara T, Sullivan A, Silver M, van der Zee R, Li T, Witzenbichler B, Schatteman G, Isner JM (1997) Isolation of putative progenitor endothelial cells for angiogenesis. Science 275: 964–7

Asahara T, Takahashi T, Masuda H, Kalka C, Chen D, Iwaguro H, Inai Y, Silver M, Isner JM (1999b) VEGF contributes to postnatal neovascularization by mobilizing bone marrow-derived endothelial progenitor cells. Embo J 18: 3964–72

Asakura A, Rudnicki MA (2002) Side population cells from diverse adult tissues are capable of in vitro hematopoietic differentiation. Exp Hematol 30: 1339–45

Ashton BA, Allen TD, Howlett CR, Eaglesom CC, Hattori A, Owen M (1980) Formation of bone and cartilage by marrow stromal cells in diffusion chambers in vivo. Clin Orthop 36: 294–307

Atala A (2000) Tissue engineering of artificial organs. J Endourol 14: 49–57

Atala A (2002) Experimental and clinical experience with tissue engineering techniques for urethral reconstruction. Urol Clin North Am 29: 485–92, ix

Badorff C, Brandes RP, Popp R, Rupp S, Urbich C, Aicher A, Fleming I, Busse R, Zeiher AM, Dimmeler S (2003) Transdifferentiation of blood-derived human adult endothelial progenitor cells into functionally active cardiomyocytes. Circulation 107: 1024–32

Bagutti C, Wobus AM, Fassler R, Watt FM (1996) Differentiation of embryonal stem cells into keratinocytes: comparison of wild-type and beta 1 integrin-deficient cells. Dev Biol 179: 184–96

Barnard CN (1967) The operation. A human cardiac transplant: an interim report of a successful operation performed at Groote Schuur Hospital, Cape Town. S Afr Med J 41: 1271–4

Baum BJ, Wang S, Cukierman E, Delporte C, Kagami H, Marmary Y, Fox PC, Mooney DJ, Yamada KM (1999) Re-engineering the functions of a terminally differentiated epithelial cell in vivo. Ann N Y Acad Sci 875: 294–300

Bayes-Genis A, Salido M, Sole Ristol F, Puig M, Brossa V, Camprecios M, Corominas JM, Marinoso ML, Baro T, Vela MC, Serrano S, Padro JM, Bayes de Luna A, Cinca J (2002) Host cell-derived cardiomyocytes in sex-mismatch cardiac allografts. Cardiovasc Res 56: 404–10

Beatty PG, Boucher KM, Mori M, Milford EL (2000) Probability of finding HLA-mismatched related or unrelated marrow or cord blood donors. Hum Immunol 61: 834–40

Beatty PG, Kollman C, Howe CW (1995) Unrelated-donor marrow transplants: the experience of the National Marrow Donor Program. Clin Transpl : 271–7

Beltrami AP, Urbanek K, Kajstura J, Yan SM, Finato N, Bussani R, Nadal-Ginard B, Silvestri F, Leri A, Beltrami CA, Anversa P (2001) Evidence that human cardiac myocytes divide after myocardial infarction. N Engl J Med 344: 1750–7

Betthauser J, Forsberg E, Augenstein M, Childs L, Eilertsen K, Enos J, Forsythe T, Golueke P, Jurgella G, Koppang R, Lesmeister T, Mallon K, Mell G, Misica P, Pace M, Pfister-Genskow M, Strelchenko N, Voelker G, Watt S, Thompson S, Bishop M (2000) Production of cloned pigs from in vitro systems. Nat Biotechnol 18: 1055–9

Billingham RE, Brent L, Medawar PB (1953) Acquired immunological tolerance for foreign cells. Nature 172: 603–606

Billon N, Jolicoeur C, Tokumoto Y, Vennstrom B, Raff M (2002) Normal timing of oligodendrocyte development depends on thyroid hormone receptor alpha 1 (TRalpha1). Embo J 21: 6452–60

Bjornson CR, Rietze RL, Reynolds BA, Magli MC, Vescovi AL (1999) Turning brain into blood: a hematopoietic fate adopted by adult neural stem cells in vivo. Science 283: 534–7

Blusch JH, Patience C, Martin U (2002) Pig endogenous retroviruses and xenotransplantation. Xenotransplantation 9: 242–251

Blyszcuk P, Czyz J, Zuschratter W, St-Onge L, Wobus AM Constitutive expression of Pdx-1 and Pax4 in ES cells promotes pancreaticß cell differentiation Keystone Symposium, Stem Cells: Origins, Fates and Functions, Keystone Resort, Colorado 2002, pp no. 447

Blyszczuk P, Czyz J, Kania G, Wagner M, Roll U, St-Onge L, Wobus AM (2003) Expression of Pax4 in embryonic stem cells promotes differentiation of nestin-positive progenitor and insulin-producing cells. Proc Natl Acad Sci U S A 100: 998–1003

Bodnar AG, Ouellette M, Frolkis M, Holt SE, Chiu CP, Morin GB, Harley CB, Shay JW, Lichtsteiner S, Wright WE (1998) Extension of life-span by introduction of telomerase into normal human cells [see comments]. Science 279: 349–52

Boiani M, Eckardt S, Scholer HR, McLaughlin KJ (2002) Oct4 distribution and level in mouse clones: consequences for pluripotency. Genes Dev 16: 1209–19

Bonner-Weir S (2000) Perspective: Postnatal pancreatic beta cell growth. Endocrinology 141: 1926–9

Booth C, O'Shea JA, Potten CS (1999) Maintenance of functional stem cells in isolated and cultured adult intestinal epithelium. Exp Cell Res 249: 359–66

Bouwens L (1998) Transdifferentiation versus stem cell hypothesis for the regeneration of islet beta-cells in the pancreas. Microsc Res Tech 43: 332–6

Boyan BD, Lohmann CH, Romero J, Schwartz Z (1999) Bone and cartilage tissue engineering. Clin Plast Surg 26: 629-45, ix

Bradley A, Evans M, Kaufman MH, Robertson E (1984) Formation of germ-line chimaeras from embryo-derived teratocarcinoma cell lines. Nature 309: 255–6

Bradley JA, Bolton EM, Pedersen RA (2002) Stem cell medicine encounters the immune system. Nat Rev Immunol 2: 859–71

Braun T, Arnold HH (1994) ES-cells carrying two inactivated myf-5 alleles form skeletal muscle cells: activation of an alternative myf-5-independent differentiation pathway. Dev Biol 164: 24–36

Brill S, Holst P, Sigal S, Zvibel I, Fiorino A, Ochs A, Somasundaran U, Reid LM (1993) Hepatic progenitor populations in embryonic, neonatal, and adult liver. Proc Soc Exp Biol Med 204: 261–9

Brown JP, Wei W, Sedivy JM (1997) Bypass of senescence after disruption of p21CIP1/WAF1 gene in normal diploid human fibroblasts. Science 277: 831–4

Bruder SP, Kraus KH, Goldberg VM, Kadiyala S (1998a) The effect of implants loaded with autologous mesenchymal stem cells on the healing of canine segmental bone defects. J Bone Joint Surg Am 80: 985–96

Bruder SP, Kurth AA, Shea M, Hayes WC, Jaiswal N, Kadiyala S (1998b) Bone regeneration by implantation of purified, culture-expanded human mesenchymal stem cells. J Orthop Res 16: 155–62

Brustle O, McKay RD (1996) Neuronal progenitors as tools for cell replacement in the nervous system. Curr Opin Neurobiol 6: 688–95.

Burkert U, von Ruden T, Wagner EF (1991) Early fetal hematopoietic development from in vitro differentiated embryonic stem cells. New Biol 3: 698–708

Cairo MS, Wagner JE (1997) Placental and/or umbilical cord blood: an alternative source of hematopoietic stem cells for transplantation. Blood 90: 4665–78

Campbell KH, McWhir J, Ritchie WA, Wilmut I (1996) Sheep cloned by nuclear transfer from a cultured cell line. Nature 380: 64–6

Carpenter MK, Rosler E, Xu C, Priest C, Thies RS Human embryonic stem cells: characterisation and differentiation Keystone Symposium: From Stem Cells to Therapy, Steamboat Springs, Colorado 2003

Cascalho M, Platt JL (2001) Xenotransplantation and other means of organ replacement. Nature Rev Immunol 1: 154–60

Castro RF, Jackson KA, Goodell MA, Robertson CS, Liu H, Shine HD (2002) Failure of bone marrow cells to transdifferentiate into neural cells in vivo. Science 297: 1299

Cattaneo E, McKay R (1990) Proliferation and differentiation of neuronal stem cells regulated by nerve growth factor. Nature 347: 762–5

Cattaneo E, McKay R (1991) Identifying and manipulating neuronal stem cells. Trends Neurosci 14: 338–40

Cervantes RB, Stringer JR, Shao C, Tischfield JA, Stambrook PJ (2002) Embryonic stem cells and somatic cells differ in mutation frequency and type. Proc Natl Acad Sci U S A 99: 3586–90

Chamberlain LJ, Yannas IV, Hsu HP, Spector M (2000a) Connective tissue response to tubular implants for peripheral nerve regeneration: the role of myofibroblasts. J Comp Neurol 417: 415–30

Chamberlain LJ, Yannas IV, Hsu HP, Strichartz GR, Spector M (2000b) Near-terminus axonal structure and function following rat sciatic nerve regeneration through a collagen-GAG matrix in a ten-millimeter gap. J Neurosci Res 60: 666–77

Chen U (1992) Differentiation of mouse embryonic stem cells to lympho-hematopoietic lineages in vitro. Dev Immunol 2: 29–50

Cherny RA, Stokes TM, Merei J, Lom L, Brandon MR, Williams RL (1994) Strategies for the isolation and characterization of bovine embryonic stem cells. Reprod Fertil Dev 6: 569–75

Chesne P, Adenot PG, Viglietta C, Baratte M, Boulanger L, Renard JP (2002) Cloned rabbits produced by nuclear transfer from adult somatic cells. Nat Biotechnol 20: 366–9

Chien KR (1995) Cardiac muscle diseases in genetically engineered mice: evolution of molecular physiology. Am J Physiol 269: H755–66

Cibelli JB, Grant KA, Chapman KB, Cunniff K, Worst T, Green HL, Walker SJ, Gutin PH, Vilner L, Tabar V, Dominko T, Kane J, Wettstein PJ, Lanza RP, Studer L, Vrana KE, West MD (2002) Parthenogenetic stem cells in nonhuman primates. Science 295: 819

Clarke DL, Johansson CB, Wilbertz J, Veress B, Nilsson E, Karlstrom H, Lendahl U, Frisen J (2000) Generalized potential of adult neural stem cells. Science 288: 1660–3

Colter DC, Sekiya I, Prockop DJ (2001) Identification of a subpopulation of rapidly self-renewing and multipotential adult stem cells in colonies of human marrow stromal cells. Proc Natl Acad Sci U S A 98: 7841–5

Condorelli G, Borello U, De Angelis L, Latronico M, Sirabella D, Coletta M, Galli R, Balconi G, Follenzi A, Frati G, Cusella De Angelis MG, Gioglio L, Amuchastegui S, Adorini L, Naldini L, Vescovi A, Dejana E, Cossu G (2001) Cardiomyocytes induce endothelial cells to trans-differentiate into cardiac muscle: implications for myocardium regeneration. Proc Natl Acad Sci U S A 98: 10733–8

Cotsarelis G, Cheng SZ, Dong G, Sun TT, Lavker RM (1989) Existence of slow-cycling limbal epithelial basal cells that can be preferentially stimulated to proliferate: implications on epithelial stem cells. Cell 57: 201–9

Dani C, Smith AG, Dessolin S, Leroy P, Staccini L, Villageois P, Darimont C, Ailhaud G (1997) Differentiation of embryonic stem cells into adipocytes in vitro. J Cell Sci 110: 1279–85

Deacon T, Dinsmore J, Costantini LC, Ratliff J, Isacson O (1998) Blastula-stage stem cells can differentiate into dopaminergic and serotonergic neurons after transplantation. Exp Neurol 149: 28–41

Deveraux QL, Schendel SL, Reed JC (2001) Antiapoptotic proteins. The bcl-2 and inhibitor of apoptosis protein families. Cardiol Clin 19: 57–74

Doetschman TC, Eistetter H, Katz M, Schmidt W, Kemler R (1985) The in vitro development of blastocyst-derived embryonic stem cell lines: formation of visceral yolk sac, blood islands and myocardium. J Embryol Exp Morphol 87: 27–45

Donovan PJ, Stott D, Cairns LA, Heasman J, Wylie CC (1986) Migratory and postmigratory mouse primordial germ cells behave differently in culture. Cell 44: 831–8

Drab M, Haller H, Bychkov R, Erdmann B, Lindschau C, Haase H, Morano I, Luft FC, Wobus AM (1997) From totipotent embryonic stem cells to spontaneously contracting smooth muscle cells: a retinoic acid and db-cAMP in vitro differentiation model. Faseb J 11: 905–15

Draper S, Gokhale PJ, Andrews PW Adaptation and karyotypic evolution of human embryonic stem (ES) cells in culture Keystone Symposium: From Stem Cells to Therapy, Steamboat Springs, Colorado 2003

Eguchi G, Kodama R (1993) Transdifferentiation. Curr Opin Cell Biol 5: 1023–8

Eiges R, Schuldiner M, Drukker M, Yanuka O, Itskovitz-Eldor J, Benvenisty N (2001) Establishment of human embryonic stem cell-transfected clones carrying a marker for undifferentiated cells. Curr Biol 11: 514–8

Eschenhagen T, Fink C, Remmers U, Scholz H, Wattchow J, Weil J, Zimmermann W, Dohmen HH, Schafer H, Bishopric N, Wakatsuki T, Elson EL (1997) Three-dimensional reconstitution of embryonic cardiomyocytes in a collagen matrix: a new heart muscle model system. Faseb J 11: 683–94

Evans MJ, Kaufman MH (1981) Establishment in culture of pluripotential cells from mouse embryos. Nature 292: 154–6

Fandrich F, Lin X, Chai GX, Schulze M, Ganten D, Bader M, Holle J, Huang DS, Parwaresch R, Zavazava N, Binas B (2002) Preimplantation-stage stem cells induce long-term allogeneic graft acceptance without supplementary host conditioning. Nat Med 8: 171–8

Farkas G, Karacsonyi S (1985) Clinical transplantation of fetal human pancreatic islets. Biomed Biochim Acta 44: 155–9

Ferrari G, Cusella-De Angelis G, Coletta M, Paolucci E, Stornaiuolo A, Cossu G, Mavilio F (1998) Muscle regeneration by bone marrow-derived myogenic progenitors. Science 279: 1528–30

Forbes SJ, Poulsom R, Wright NA (2002) Hepatic and renal differentiation from blood-borne stem cells. Gene Ther 9: 625–30

Freed CR, Breeze RE, Rosenberg NL, Schneck SA, Kriek E, Qi JX, Lone T, Zhang YB, Snyder JA, Wells TH, et al. (1992) Survival of implanted fetal dopamine cells and neurologic improvement 12 to 46 months after transplantation for Parkinson's disease. N Engl J Med 327: 1549–55

Freed CR, Greene PE, Breeze RE, Tsai WY, DuMouchel W, Kao R, Dillon S, Winfield H, Culver S, Trojanowski JQ, Eidelberg D, Fahn S (2001) Transplantation of embryonic dopamine neurons for severe Parkinson's disease. N Engl J Med 344: 710–9

Friedenstein AJ, Chailakhjan RK, Lalykina KS (1970) The development of fibroblast colonies in monolayer cultures of guinea-pig bone marrow and spleen cells. Cell Tissue Kinet 3: 393–403

Friedenstein AJ, Chailakhyan RK, Gerasimov UV (1987) Bone marrow osteogenic stem cells: in vitro cultivation and transplantation in diffusion chambers. Cell Tissue Kinet 20: 263–72

Friedenstein AJ, Chailakhyan RK, Latsinik NV, Panasyuk AF, Keiliss-Borok IV (1974a) Stromal cells responsible for transferring the microenvironment of the hemopoietic tissues. Cloning in vitro and retransplantation in vivo. Transplantation 17: 331–40

Friedenstein AJ, Deriglasova UF, Kulagina NN, Panasuk AF, Rudakowa SF, Luria EA, Ruadkow IA (1974b) Precursors for fibroblasts in different populations of hematopoietic cells as detected by the in vitro colony assay method. Exp Hematol 2: 83–92

Friedenstein AJ, Gorskaja JF, Kulagina NN (1976) Fibroblast precursors in normal and irradiated mouse hematopoietic organs. Exp Hematol 4: 267–74

Gage FH (2000) Mammalian neural stem cells. Science 287: 1433–8

Gardner MB (1978) Type C viruses of wild mice: characterization and natural history of amphotropic, ecotropic, and xenotropic MuLv. Curr Top Microbiol Immunol 79: 215–59

Geber S, Winston RM, Handyside AH (1995) Proliferation of blastomeres from biopsied cleavage stage human embryos in vitro: an alternative to blastocyst biopsy for preimplantation diagnosis. Hum Reprod 10: 1492–6

Gielen V, Faure M, Mauduit G, Thivolet J (1987) Progressive replacement of human cultured epithelial allografts by recipient cells as evidenced by HLA class I antigens expression. Dermatologica 175: 166–70

Gottlieb DI, Huettner JE (1999) An in vitro pathway from embryonic stem cells to neurons and glia. Cells Tissues Organs 165: 165–72

Green DR, Ferguson TA (2001) The role of Fas ligand in immune privilege. Nat Rev Mol Cell Biol 2: 917–24

Griffith M, Osborne R, Munger R, Xiong X, Doillon CJ, Laycock NL, Hakim M, Song Y, Watsky MA (1999) Functional human corneal equivalents constructed from cell lines. Science 286: 2169–72

Grigoriadis AE, Heersche JN, Aubin JE (1988) Differentiation of muscle, fat, cartilage, and bone from progenitor cells present in a bone-derived clonal cell population: effect of dexamethasone. J Cell Biol 106: 2139–51

Gropp M, Itsykson P, Singer O, Ben-Hur T, Reinhartz E, Galun E, Reubinoff BE (2003) Stable genetic modification of human embryonic stem cells by lentiviral vectors. Mol Ther 7: 281–7

Habibullah CM, Syed IH, Qamar A, Taher-Uz Z (1994) Human fetal hepatocyte transplantation in patients with fulminant hepatic failure. Transplantation 58: 951–2

Hacein-Bey-Abina S, von Kalle C, Schmidt M, Le Deist F, Wulffraat N, McIntyre E, Radford I, Villeval JL, Fraser CC, Cavazzana-Calvo M, Fischer A (2003) A serious adverse event after successful gene therapy for X-linked severe combined immunodeficiency. N Engl J Med 348: 255–6

Hakuno D, Fukuda K, Makino S, Konishi F, Tomita Y, Manabe T, Suzuki Y, Umezawa A, Ogawa S (2002) Bone marrow-derived regenerated cardiomyocytes (CMG Cells) express functional adrenergic and muscarinic receptors. Circulation 105: 380–6

Haynesworth SE, Goshima J, Goldberg VM, Caplan AI (1992) Characterization of cells with osteogenic potential from human marrow. Bone 13: 81–8

Hirschi K, Goodell M (2001) Common origins of blood and blood vessels in adults? Differentiation 68: 186–92

Humpherys D, Eggan K, Akutsu H, Hochedlinger K, Rideout WM, 3rd, Biniszkiewicz D, Yanagi-machi R, Jaenisch R (2001) Epigenetic instability in ES cells and cloned mice. Science 293: 95–7

Itskovitz-Eldor J, Schuldiner M, Karsenti D, Eden A, Yanuka O, Amit M, Soreq H, Benvenisty N (2000) Differentiation of human embryonic stem cells into embryoid bodies compromising the three embryonic germ layers. Mol Med 6: 88–95

Jackson KA, Majka SM, Wang H, Pocius J, Hartley CJ, Majesky MW, Entman ML, Michael LH, Hirschi KK, Goodell MA (2001) Regeneration of ischemic cardiac muscle and vascular endothelium by adult stem cells. J Clin Invest 107: 1395–402

Jiang Y, Jahagirdar BN, Reinhardt RL, Schwartz RE, Keene CD, Ortiz-Gonzalez XR, Reyes M, Lenvik T, Lund T, Blackstad M, Du J, Aldrich S, Lisberg A, Low WC, Largaespada DA, Ver-faillie CM (2002) Pluripotency of mesenchymal stem cells derived from adult marrow. Nature 418: 41–9

Kaihara S, Kim SS, Kim BS, Mooney D, Tanaka K, Vacanti JP (2000) Long-term follow-up of tis-sue-engineered intestine after anastomosis to native small bowel. Transplantation 69: 1927–32

Karnezis AN, Dorokhov M, Grompe M, Zhu L (2001) Loss of p27(Kip1) enhances the transplan-tation efficiency of hepatocytes transferred into diseased livers. J Clin Invest 108: 383–90

Karsner HT, Saphir O, Todd TW (1925) Am J Pathol 1: 351–371

Kato Y, Rideout WM, Hilton K, Barton SC, Tsunoda Y, Surani MA (1999) Developmental poten-tial of mouse primordial germ cells. Development 126: 1823–32

Kaushal S, Amiel GE, Guleserian KJ, Shapira OM, Perry T, Sutherland FW, Rabkin E, Moran AM, Schoen FJ, Atala A, Soker S, Bischoff J, Mayer JE, Jr. (2001) Functional small-diameter neovessels created using endothelial progenitor cells expanded ex vivo. Nat Med 7: 1035–40

Kawaguchi A, Miyata T, Sawamoto K, Takashita N, Murayama A, Akamatsu W, Ogawa M, Okabe M, Tano Y, Goldman SA, Okano H (2001) Nestin-EGFP transgenic mice: visualization of the self-renewal and multipotency of CNS stem cells. Mol Cell Neurosci 17: 259–73

Keefer CL, Baldassarre H, Keyston R, Wang B, Bhatia B, Bilodeau AS, Zhou JF, Leduc M, Downey BR, Lazaris A, Karatzas CN (2001) Generation of dwarf goat (Capra hircus) clones following nuclear transfer with transfected and nontransfected fetal fibroblasts and in vitro-matured oocytes. Biol Reprod 64: 849–56

Kehat I, Kenyagin-Karsenti D, Snir M, Segev H, Amit M, Gepstein A, Livne E, Binah O, Itskovitz-Eldor J, Gepstein L (2001) Human embryonic stem cells can differentiate into myocytes with structural and functional properties of cardiomyocytes. J Clin Invest 108: 407–14

Keller G, Kennedy M, Papayannopoulou T, Wiles MV (1993) Hematopoietic commitment during embryonic stem cell differentiation in culture. Mol Cell Biol 13: 473–86

Kilpatrick TJ, Bartlett PF (1993) Cloning and growth of multipotential neural precursors: require-ments for proliferation and differentiation. Neuron 10: 255–65

Kim SS, Kaihara S, Benvenuto M, Choi RS, Kim BS, Mooney DJ, Taylor GA, Vacanti JP (1999) Regenerative signals for tissue-engineered small intestine. Transplant Proc 31: 657–60

Kimikawa M, Sachs DH, Colvin RB, Bartholomew A, Kawai T, Cosimi AB (1997) Modifications of the conditioning regimen for achieving mixed chimerism and donor-specific tolerance in cynomolgus monkeys. Transplantation 64: 709–16

Kirchhof N, Schmittwolf C, Kirsch RD, Harder F, Hacker C, Zenke M, Mueller AM(1993) Enhancing the developmental potentials of somatic stem cells by epigenetic modification Keystone symposium: From Stem Cells to Therapy, Steamboat Springs, Colorado

Klug MG, Soonpaa MH, Koh GY, Field LJ (1996) Genetically selected cardiomyocytes from dif-ferentiating embronic stem cells form stable intracardiac grafts. J Clin Invest 98: 216–24

Kobayashi N, Fujiwara T, Westerman KA, Inoue Y, Sakaguchi M, Noguchi H, Miyazaki M, Cai J, Tanaka N, Fox IJ, Leboulch P (2000) Prevention of acute liver failure in rats with reversibly immortalized human hepatocytes [see comments]. Science 287: 1258–62

Kocher AA, Schuster MD, Szabolcs MJ, Takuma S, Burkhoff D, Wang J, Homma S, Edwards NM, Itescu S (2001) Neovascularization of ischemic myocardium by human bone-marrow-derived angioblasts prevents cardiomyocyte apoptosis, reduces remodeling and improves cardiac func-tion. Nat Med 7: 430–6

Kofidis T, Akhyari P, Boublik J, Theodorou P, Martin U, Ruhparwar A, Fischer S, Eschenhagen T, Kubis HP, Kraft T, Leyh R, Haverich A (2002) In vitro engineering of heart muscle: Artificial myocardial tissue. J Thorac Cardiovasc Surg 124: 63–69

Krause DS, Theise ND, Collector MI, Henegariu O, Hwang S, Gardner R, Neutzel S, Sharkis SJ (2001) Multi-organ, multi-lineage engraftment by a single bone marrow-derived stem cell. Cell 105: 369–77

LaBarge MA, Blau HM (2002) Biological progression from adult bone marrow to mononucleate muscle stem cell to multinucleate muscle fiber in response to injury. Cell 111: 589–601

Laflamme MA, Myerson D, Saffitz JE, Murry CE (2002) Evidence for cardiomyocyte repopulation by extracardiac progenitors in transplanted human hearts. Circ Res 90: 634–40

Lanza RP, Cibelli JB, Faber D, Sweeney RW, Henderson B, Nevala W, West MD, Wettstein PJ (2001) Cloned cattle can be healthy and normal. Science 294: 1893–4

Lavker RM, Sun TT (1982) Heterogeneity in epidermal basal keratinocytes: morphological and functional correlations. Science 215: 1239–41

Lee KY, Huang H, Ju B, Yang Z, Lin S (2002) Cloned zebrafish by nuclear transfer from long-term-cultured cells. Nat Biotechnol 20: 795–9

Lee SH, Lumelsky N, Studer L, Auerbach JM, McKay RD (2000) Efficient generation of midbrain and hindbrain neurons from mouse embryonic stem cells. Nat Biotechnol 18: 675–9

Lehrman S (1999) Virus treatment questioned after gene therapy death. Nature 401: 517–8

Levy JA (1978) Xenotropic type C viruses. Curr Top Microbiol Immunol 79: 111–213

Li F, Lu S, Vida L, Thomson JA, Honig GR (2001) Bone morphogenetic protein 4 induces efficient hematopoietic differentiation of rhesus monkey embryonic stem cells in vitro. Blood 98: 335–42

Li Z, Dullmann J, Schiedlmeier B, Schmidt M, von Kalle C, Meyer J, Forster M, Stocking C, Wahlers A, Frank O, Ostertag W, Kuhlcke K, Eckert HG, Fehse B, Baum C (2002) Murine leukemia induced by retroviral gene marking. Science 296: 497

Liang L, Bickenbach JR (2002) Somatic epidermal stem cells can produce multiple cell lineages during development. Stem Cells 20: 21–31

Lin Y, Weisdorf DJ, Solovey A, Hebbel RP (2000) Origins of circulating endothelial cells and endothelial outgrowth from blood. J Clin Invest 105: 71–7

Little CW, Cox C, Wyatt J, del Cerro C, del Cerro M (1998) Correlates of photoreceptor rescue by transplantation of human fetal RPE in the RCS rat. Exp Neurol 149: 151–60

Logan JS (2000) Prospects for xenotransplantation. Curr Opin Immunol 12: 563–8

Lois C, Alvarez-Buylla A (1993) Proliferating subventricular zone cells in the adult mammalian forebrain can differentiate into neurons and glia. Proc Natl Acad Sci U S A 90: 2074–7

Lu SJ, Li F, Vida L, Honig GR (2002) Comparative gene expression in hematopoietic progenitor cells derived from embryonic stem cells. Exp Hematol 30: 58–66

Lumelsky N, Blondel O, Laeng P, Velasco I, Ravin R, McKay R (2001) Differentiation of embryonic stem cells to insulin-secreting structures similar to pancreatic islets. Science 292: 1389–94

Ma Y, Ramezani A, Lewis R, Hawley RG, Thomson JA (2003) High-level sustained transgene expression in human embryonic stem cells using lentiviral vectors. Stem Cells 21: 111–7

Machens HG, Berger AC, Mailaender P (2000) Bioartificial skin. Cells Tissues Organs 167: 88–94

Majumdar MK, Thiede MA, Mosca JD, Moorman M, Gerson SL (1998) Phenotypic and functional comparison of cultures of marrow-derived mesenchymal stem cells (MSCs) and stromal cells. J Cell Physiol 176: 57–66

Makino S, Fukuda K, Miyoshi S, Konishi F, Kodama H, Pan J, Sano M, Takahashi T, Hori S, Abe H, Hata J, Umezawa A, Ogawa S (1999) Cardiomyocytes can be generated from marrow stromal cells in vitro. J Clin Invest 103: 697–705

Malouf NN, Coleman WB, Grisham JW, Lininger RA, Madden VJ, Sproul M, Anderson PA (2001) Adult-derived stem cells from the liver become myocytes in the heart in vivo. Am J Pathol 158: 1929–35

Marshall E (1999) Gene therapy death prompts review of adenovirus vector. Science 286: 2244–5

Martin GR, Evans MJ (1974) The morphology and growth of a pluripotent teratocarcinoma cell line and its derivatives in tissue culture. Cell 2: 163–72

Matsui Y, Toksoz D, Nishikawa S, Williams D, Zsebo K, Hogan BL (1991) Effect of Steel factor and leukaemia inhibitory factor on murine primordial germ cells in culture. Nature 353: 750–2

Matsui Y, Zsebo K, Hogan BL (1992) Derivation of pluripotential embryonic stem cells from murine primordial germ cells in culture. Cell 70: 841–7

McKay R (1997) Stem cells in the central nervous system. Science 276: 66–71

Meletis K, Frisen J (2001) Have the bloody cells gone to our heads? J Cell Biol 155: 699–702

Meyerson M, Counter CM, Eaton EN, Ellisen LW, Steiner P, Caddle SD, Ziaugra L, Beijersbergen RL, Davidoff MJ, Liu Q, Bacchetti S, Haber DA, Weinberg RA (1997) hEST2, the putative human telomerase catalytic subunit gene, is up-regulated in tumor cells and during immortalization. Cell 90: 785–95

Michejda M (1987) Treatment of Parkinson's disease in adult patients by transplantation of human fetal brain tissue obtained from elective abortions. Fetal Ther 2: 129–34

Mitalipov SM, Yeoman RR, Nusser KD, Wolf DP (2002) Rhesus monkey embryos produced by nuclear transfer from embryonic blastomeres or somatic cells. Biol Reprod 66: 1367–73

Mizuno H, Emi N, Abe A, Takahashi I, Kojima T, Saito H, Sumi Y, Hata KI, Ueda M (1999) Successful culture and sustainability in vivo of gene-modified human oral mucosal epithelium. Hum Gene Ther 10: 825–30

Morse HC, 3rd, Hartley JW (1982) Expression of xenotropic murine leukemia viruses. Curr Top Microbiol Immunol 98: 17–26

Muller M, Fleischmann BK, Selbert S, Ji GJ, Endl E, Middeler G, Muller OJ, Schlenke P, Frese S, Wobus AM, Hescheler J, Katus HA, Franz WM (2000) Selection of ventricular-like cardiomyocytes from ES cells in vitro. Faseb J 14: 2540–8

Muller P, Pfeiffer P, Koglin J, Schafers HJ, Seeland U, Janzen I, Urbschat S, Bohm M (2002) Cardiomyocytes of noncardiac origin in myocardial biopsies of human transplanted hearts. Circulation 106: 31–5

Mummery C, Ward-van Oostwaard D, Doevendans P, Spijker R, van den Brink S, Hassink R, van der Heyden M, Opthof T, Pera M, Brutel de la Riviere A, Passier R, Tertoolen L (2003) Cardiomyocytes differentiation of human embryonic stem cells induced by coculture with visceral endoderm-like cells Keystone Symposium: From Stem Cells to Therapy, Steamboat Springs

Nakano T, Kodama H, Honjo T (1994) Generation of lymphohematopoietic cells from embryonic stem cells in culture. Science 265: 1098–101

Nelson PT, Kondziolka D, Wechsler L, Goldstein S, Gebel J, DeCesare S, Elder EM, Zhang PJ, Jacobs A, McGrogan M, Lee VM, Trojanowski JQ (2002) Clonal human (hNT) neuron grafts for stroke therapy: neuropathology in a patient 27 months after implantation. Am J Pathol 160: 1201–6

Niemann C, Watt FM (2002) Designer skin: lineage commitment in postnatal epidermis. Trends Cell Biol 12: 185–92

Niklason LE (2000) Engineering of bone grafts. Nat Biotechnol 18: 929–30

Niklason LE, Gao J, Abbott WM, Hirschi KK, Houser S, Marini R, Langer R (1999) Functional arteries grown in vitro. Science 284: 489–93

Notarianni E, Galli C, Laurie S, Moor RM, Evans MJ (1991) Derivation of pluripotent, embryonic cell lines from the pig and sheep. J Reprod Fertil Suppl 43: 255–60

O'Connell J (2002) Fas ligand and the fate of antitumour cytotoxic T lymphocytes. Immunology 105: 263–6

O'Connor NE, Mulliken JB (1981) Grafting of burns with cultured epithelium prepared from autologous epidermal cells. Lancet 1: 75–8

Oh T, Miura T, Bradfute SB, Chi X, Entman ML, Loyd MH, Schwartz RJ, Schneider (2003) MD Cardiac homing, difefrentiation, and fusion by progenitor cells from adult heart Keystone Symposium: From Stem Cells To Therapy, Steamboat Springs, Colorado

Onishi A, Iwamoto M, Akita T, Mikawa S, Takeda K, Awata T, Hanada H, Perry AC (2000) Pig cloning by microinjection of fetal fibroblast nuclei. Science 289: 1188–90

Orlic D, Kajstura J, Chimenti S, Jakoniuk I, Anderson SM, Li B, Pickel J, McKay R, Nadal-Ginard B, Bodine DM, Leri A, Anversa P (2001a) Bone marrow cells regenerate infarcted myocardium. Nature 410: 701–5

Orlic D, Kajstura J, Chimenti S, Limana F, Jakoniuk I, Quaini F, Nadal-Ginard B, Bodine DM, Leri A, Anversa P (2001b) Mobilized bone marrow cells repair the infarcted heart, improving function and survival. Proc Natl Acad Sci U S A 14: 14

Oshima H, Rochat A, Kedzia C, Kobayashi K, Barrandon Y (2001) Morphogenesis and renewal of hair follicles from adult multipotent stem cells. Cell 104: 233–45

Papas KK, Long RC, Sambanis A, Constantinidis I (1999) Development of a bioartificial pancreas: I. long-term propagation and basal and induced secretion from entrapped betaTC3 cell cultures. Biotechnol Bioeng 66: 219–30

Park HJ, Yoo JJ, Kershen RT, Moreland R, Atala A (1999) Reconstitution of human corporal smooth muscle and endothelial cells in vivo. J Urol 162: 1106–9

Perka C, Spitzer RS, Lindenhayn K, Sittinger M, Schultz O (2000) Matrix-mixed culture: new methodology for chondrocyte culture and preparation of cartilage transplants. J Biomed Mater Res 49: 305–11

Perron M, Harris WA (2000) Retinal stem cells in vertebrates. Bioessays 22: 685–8

Petersen BE, Bowen WC, Patrene KD, Mars WM, Sullivan AK, Murase N, Boggs SS, Greenberger JS, Goff JP (1999) Bone marrow as a potential source of hepatic oval cells. Science 284: 1168–70

Pfeifer A, Ikawa M, Dayn Y, Verma IM (2002) Transgenesis by lentiviral vectors: lack of gene silencing in mammalian embryonic stem cells and preimplantation embryos. Proc Natl Acad Sci U S A 99: 2140–5

Pittenger MF, Mackay AM, Beck SC, Jaiswal RK, Douglas R, Mosca JD, Moorman MA, Simonetti DW, Craig S, Marshak DR (1999) Multilineage potential of adult human mesenchymal stem cells. Science 284: 143–7

Platt JL (2000a) Immunobiology of xenotransplantation. Transpl Int 13 Suppl 1: S7–10

Platt JL (2000b) Physiologic barriers to xenotransplantation. Transplant Proc 32: 1547–8

Polejaeva IA, Chen SH, Vaught TD, Page RL, Mullins J, Ball S, Dai Y, Boone J, Walker S, Ayares DL, Colman A, Campbell KH (2000) Cloned pigs produced by nuclear transfer from adult somatic cells. Nature 407: 86–90

Pralong D, Mrozik K, Occhiodoro F, Verma P (2002) Nuclear transfer to mouse ES-like cells: pre- and post-fusion enucleation Keystone Symposium, Stem Cells: Origins, Fates and Functions, Keystone Resort, pp 333

Price LH, Spencer DD, Marek KL, Robbins RJ, Leranth C, Farhi A, Naftolin F, Roth RH, Bunney BS, Hoffer PB, et al. (1995) Psychiatric status after human fetal mesencephalic tissue transplantation in Parkinson's disease. Biol Psychiatry 38: 498–505

Prockop DJ (1997) Marrow stromal cells as stem cells for nonhematopoietic tissues. Science 276: 71–4

Pundt LL, Kondoh T, Conrad JA, Low WC (1996) Transplantation of human fetal striatum into a rodent model of Huntington's disease ameliorates locomotor deficits. Neurosci Res 24: 415–20

Rafii S (2000) Circulating endothelial precursors: mystery, reality, and promise. J Clin Invest 105: 17–9

Ramiya VK, Maraist M, Arfors KE, Schatz DA, Peck AB, Cornelius JG (2000) Reversal of insulin-dependent diabetes using islets generated in vitro from pancreatic stem cells. Nat Med 6: 278–82

Rao MS (1999) Multipotent and restricted precursors in the central nervous system. Anat Rec 257: 137–48

Rao MS, Yeldandi AV, Reddy JK (1990) Stem cell potential of ductular and periductular cells in the adult rat pancreas. Cell Differ Dev 29: 155–63

Reik RA, Noto TA, Fernandez HF (1997) Safety of large-volume leukapheresis for collection of peripheral blood progenitor cells. J Clin Apheresis 12: 10–3

Resnick JL, Bixler LS, Cheng L, Donovan PJ (1992) Long-term proliferation of mouse primordial germ cells in culture. Nature 359: 550–1

Reubinoff BE, Itsykson P, Turetsky T, Pera MF, Reinhartz E, Itzik A, Ben-Hur T (2001) Neural progenitors from human embryonic stem cells. Nat Biotechnol 19: 1134–40

Reyes M, Dudek A, Jahagirdar B, Koodie L, Marker PH, Verfaillie CM (2002) Origin of endothelial progenitors in human postnatal bone marrow. J Clin Invest 109: 337–46

Reyes M, Lund T, Lenvik T, Aguiar D, Koodie L, Verfaillie CM (2001) Purification and ex vivo expansion of postnatal human marrow mesodermal progenitor cells. Blood 98: 2615–25

Reyes M, Verfaillie CM (2001) Characterization of multipotent adult progenitor cells, a subpopulation of mesenchymal stem cells. Ann N Y Acad Sci 938: 231–3; discussion 233–5

Reynolds BA, Tetzlaff W, Weiss S (1992) A multipotent EGF-responsive striatal embryonic progenitor cell produces neurons and astrocytes. J Neurosci 12: 4565–74

Reynolds BA, Weiss S (1992) Generation of neurons and astrocytes from isolated cells of the adult mammalian central nervous system. Science 255: 1707–10

Richards M, Fong CY, Chan WK, Wong PC, Bongso A (2002) Human feeders support prolonged undifferentiated growth of human inner cell masses and embryonic stem cells. Nat Biotechnol 20: 933–6

Rideout IW, Eggan K, Jaenisch R (2001) Nuclear cloning and epigenetic reprogramming of the genome. Science 293: 1093–8

Rideout WM, Hochedlinger K, Kyba M, Daley GQ, Jaenisch R (2002) Correction of a genetic defect by nuclear transplantation and combined cell and gene therapy. Cell 109: 17–27

Rietze RL, Valcanis H, Brooker GF, Thomas T, Voss AK, Bartlett PF (2001) Purification of a pluripotent neural stem cell from the adult mouse brain. Nature 412: 736–9

Risau W, Sariola H, Zerwes HG, Sasse J, Ekblom P, Kemler R, Doetschman T (1988) Vasculogenesis and angiogenesis in embryonic-stem-cell-derived embryoid bodies. Development 102: 471–8

Rochat A, Kobayashi K, Barrandon Y (1994) Location of stem cells of human hair follicles by clonal analysis. Cell 76: 1063–73

Roell W, Fan Y, Xia Y, Stoecker E, Sasse P, Kolossov E, Bloch W, Metzner H, Schmitz C, Addicks K, Hescheler J, Welz A, Fleischmann BK (2002) Cellular cardiomyoplasty in a transgenic mouse model. Transplantation 73: 462–5

Rohwedel J, Kleppisch T, Pich U, Guan K, Jin S, Zuschratter W, Hopf C, Hoch W, Hescheler J, Witzemann V, Wobus AM (1998) Formation of postsynaptic-like membranes during differentiation of embryonic stem cells in vitro. Exp Cell Res 239: 214–25

Rohwedel J, Sehlmeyer U, Shan J, Meister A, Wobus AM (1996) Primordial germ cell-derived mouse embryonic germ (EG) cells in vitro resemble undifferentiated stem cells with respect to differentiation capacity and cell cycle distribution. Cell Biol Int 20: 579–87

Rolletschek A, Wiese C, Czyz J, Navarette-Santos A, Kania G, Blyszczuk P, St-Onge L, Boheler KR, Wobus AM (2003) Nestin-positive progenitor cells cultured from mouse intestinal epithelium are multipotent and differentiate into neural, meso-dermal, hepatic and pancreatic cells in vitro Keystone Symposium: From Stem Cells to Therapy, Steamboat Springs, Colorado

Rudolph KL, Chang S, Millard M, Schreiber-Agus N, DePinho RA (2000) Inhibition of experimental liver cirrhosis in mice by telomerase gene delivery [see comments]. Science 287: 1253–8

Sathananthan H, Pera M, Trounson A (2002) The fine structure of human embryonic stem cells. Reprod Biomed Online 4: 56–61

Sauer H, Rahimi G, Hescheler J, Wartenberg M (1999) Effects of electrical fields on cardiomyocyte differentiation of embryonic stem cells. J Cell Biochem 75: 710–23

Schaller RT, Jr., Stevenson JK (1965) Induction of transplantation tolerance by allogeneic thymus grafts. Surg Forum 16: 213–5

Schmitt RM, Bruyns E, Snodgrass HR (1991) Hematopoietic development of embryonic stem cells in vitro: cytokine and receptor gene expression. Genes Dev 5: 728–40

Schneider AI, Maier-Reif K, Graeve T (1999) Constructing an in vitro cornea from cultures of the three specific corneal cell types. In Vitro Cell Dev Biol Anim 35: 515–26

Schuldiner M, Eiges R, Eden A, Yanuka O, Itskovitz-Eldor J, Goldstein RS, Benvenisty N (2001) Induced neuronal differentiation of human embryonic stem cells. Brain Res 913: 201–5

Schuldiner M, Yanuka O, Itskovitz-Eldor J, Melton DA, Benvenisty N (2000) Effects of eight growth factors on the differentiation of cells derived from human embryonic stem cells. Proc Natl Acad Sci U S A 97: 11307–12

Schwartz RE, Reyes M, Koodie L, Jiang Y, Blackstad M, Lund T, Lenvik T, Johnson S, Hu WS, Verfaillie CM (2002) Multipotent adult progenitor cells from bone marrow differentiate into functional hepatocyte-like cells. J Clin Invest 109: 1291–302

Schwedes U, Kaul S, Wdowinski J, Klempa I, Bastert G, Usadel KH (1983) The use of human fetal pancreas for transplantation: experimental and clinical results. Horm Metab Res Suppl : 87–90

Sedivy JM (1998) Can ends justify the means?: telomeres and the mechanisms of replicative senescence and immortalization in mammalian cells. Proc Natl Acad Sci U S A 95: 9078–81

Seshi B, Kumar S, Sellers D (2000) Human bone marrow stromal cell: coexpression of markers specific for multiple mesenchymal cell lineages. Blood Cells Mol Dis 26: 234–46

Shamblott MJ, Axelman J, Littlefield JW, Blumenthal PD, Huggins GR, Cui Y, Cheng L, Gearhart JD (2001) Human embryonic germ cell derivatives express a broad range of developmentally distinct markers and proliferate extensively in vitro. Proc Natl Acad Sci U S A 98: 113–8

Shamblott MJ, Axelman J, Wang S, Bugg EM, Littlefield JW, Donovan PJ, Blumenthal PD, Huggins GR, Gearhart JD (1998) Derivation of pluripotent stem cells from cultured human primordial germ cells. Proc Natl Acad Sci U S A 95: 13726–31

Sharpe AH, Freeman GJ (2002) The B7-CD28 superfamily. Nature Rev Immunol 2: 116–26

Shi Q, Rafii S, Wu MH, Wijelath ES, Yu C, Ishida A, Fujita Y, Kothari S, Mohle R, Sauvage LR, Moore MA, Storb RF, Hammond WP (1998) Evidence for circulating bone marrow-derived endothelial cells. Blood 92: 362–7

Shin T, Kraemer D, Pryor J, Liu L, Rugila J, Howe L, Buck S, Murphy K, Lyons L, Westhusin M (2002) A cat cloned by nuclear transplantation. Nature 415: 859

Slavin S, Kedar E (1988) Current problems and future goals in clinical bone marrow transplantation. Blood Rev 2: 259–69

Sodian R, Hoerstrup SP, Sperling JS, Martin DP, Daebritz S, Mayer JE, Vacanti JP (2000) Evaluation of biodegradable, three-dimensional matrices for tissue engineering of heart valves. Asaio J 46: 107–10

Solter D, Knowles BB (1975) Immunosurgery of mouse blastocyst. Proc Natl Acad Sci U S A 72: 5099–102

Soria B, Roche E, Berna G, Leon-Quinto T, Reig JA, Martin F (2000) Insulin-secreting cells derived from embryonic stem cells normalize glycemia in streptozotocin-induced diabetic mice. Diabetes 49: 157–62

Spangenberg KM, Farr MM, Roy AK (1998) Tissue Engineering of tracheal epithelium: a model of isolation and culture in pluronic F127NF. Tissue Engineering 4: 476

Spencer DD, Robbins RJ, Naftolin F, Marek KL, Vollmer T, Leranth C, Roth RH, Price LH, Gjedde A, Bunney BS, et al. (1992) Unilateral transplantation of human fetal mesencephalic tissue into the caudate nucleus of patients with Parkinson's disease. N Engl J Med 327: 1541–8

Stamm C, Westphal B, Kleine HD, Petzsch M, Kittner C, Klinge H, Schumichen C, Nienaber CA, Freund M, Steinhoff G (2003) Autologous bone-marrow stem-cell transplantation for myocardial regeneration. Lancet 361: 45–6

Steindler DA, Pincus DW (2002) Stem cells and neuropoiesis in the adult human brain. Lancet 359: 1047–54

Steinhoff G, Stock U, Karim N, Mertsching H, Timke A, Meliss RR, Pethig K, Haverich A, Bader A (2000) Tissue engineering of pulmonary heart valves on allogenic acellular matrix conduits : In vivo restoration of valve tissue. Circulation 102: III50–5

Stemple DL, Anderson DJ (1992) Isolation of a stem cell for neurons and glia from the mammalian neural crest. Cell 71: 973–85

Stock UA, Vacanti JP (2001) Tissue engineering: current state and prospects. Annu Rev Med 52: 443–51

Stoye JP, Coffin JM (1987) The four classes of endogenous murine leukemia virus: structural relationships and potential for recombination. J Virol 61: 2659–69

Stranzinger GF (1996) Embryonic stem-cell-like cell lines of the species rat and Bovinae. Int J Exp Pathol 77: 263–7

Strauer BE, Brehm M, Zeus T, Gattermann N, Hernandez A, Sorg RV, Kogler G, Wernet P (2001) Intracoronary, human autologous stem cell transplantation for myocardial regeneration following myocardial infarction. Dtsch Med Wochenschr 126: 932–8

Suemori H, Tada T, Torii R, Hosoi Y, Kobayashi K, Imahie H, Kondo Y, Iritani A, Nakatsuji N (2001) Establishment of embryonic stem cell lines from cynomolgus monkey blastocysts produced by IVF or ICSI. Dev Dyn 222: 273–9

Summer R, Kotton DN, Sun X, Ma B, Fitzsimmons K, Fine A (2003) SP (Side Population) Cells and Bcrp1 Expression in Lung. Am J Physiol Lung Cell Mol Physiol 7: 7

Taylor G, Lehrer MS, Jensen PJ, Sun TT, Lavker RM (2000) Involvement of follicular stem cells in forming not only the follicle but also the epidermis. Cell 102: 451–61

Temenoff JS, Mikos AG (2000) Review: tissue engineering for regeneration of articular cartilage. Biomaterials 21: 431–40

Temple S (1989) Division and differentiation of isolated CNS blast cells in microculture. Nature 340: 471–3

Terada N, Hamazaki T, Oka M, Hoki M, Mastalerz DM, Nakano Y, Meyer EM, Morel L, Petersen BE, Scott EW (2002) Bone marrow cells adopt the phenotype of other cells by spontaneous cell fusion. Nature 416: 542–5

Theise ND, Nimmakayalu M, Gardner R, Illei PB, Morgan G, Teperman L, Henegariu O, Krause DS (2000) Liver from bone marrow in humans. Hepatology 32: 11–6

Thivolet J, Faure M, Demidem A, Mauduit G (1986) Long-term survival and immunological tolerance of human epidermal allografts produced in culture. Transplantation 42: 274–80

Thomas M, Yang L, Hornsby PJ (2000) Formation of functional tissue from transplanted adrenocortical cells expressing telomerase reverse transcriptase [see comments]. Nat Biotechnol 18: 39–42

Thomson JA, Itskovitz-Eldor J, Shapiro SS, Waknitz MA, Swiergiel JJ, Marshall VS, Jones JM (1998a) Embryonic stem cell lines derived from human blastocysts. Science 282: 1145–7

Thomson JA, Kalishman J, Golos TG, Durning M, Harris CP, Becker RA, Hearn JP (1995) Isolation of a primate embryonic stem cell line. Proc Natl Acad Sci U S A 92: 7844–8

Thomson JA, Kalishman J, Golos TG, Durning M, Harris CP, Hearn JP (1996) Pluripotent cell lines derived from common marmoset (Callithrix jacchus) blastocysts. Biol Reprod 55: 254–9

Thomson JA, Marshall VS, Trojanowski JQ (1998b) Neural differentiation of rhesus embryonic stem cells. Apmis 106: 149-56; discussion 156–7

Toma C, Pittenger MF, Cahill KS, Byrne BJ, Kessler PD (2002) Human mesenchymal stem cells differentiate to a cardiomyocyte phenotype in the adult murine heart. Circulation 105: 93–8

Toma JG, Akhavan M, Fernandes KJ, Barnabe-Heider F, Sadikot A, Kaplan DR, Miller FD (2001) Isolation of multipotent adult stem cells from the dermis of mammalian skin. Nat Cell Biol 3: 778–84

Tomita S, Li RK, Weisel RD, Mickle DA, Kim EJ, Sakai T, Jia ZQ (1999) Autologous transplantation of bone marrow cells improves damaged heart function. Circulation 100: II247–56

Tomita S, Mickle DA, Weisel RD, Jia ZQ, Tumiati LC, Allidina Y, Liu P, Li RK (2002) Improved heart function with myogenesis and angiogenesis after autologous porcine bone marrow stromal cell transplantation. J Thorac Cardiovasc Surg 123: 1132–40

Torok E, Pollok JM, Ma PX, Vogel C, Dandri M, Petersen J, Burda MR, Kaufmann PM, Kluth D, Rogiers X (2001) Hepatic tissue engineering on 3-dimensional biodegradable polymers within a pulsatile flow bioreactor. Dig Surg 18: 196–203

Tosh D, Slack JM (2002) How cells change their phenotype. Nat Rev Mol Cell Biol 3: 187–94

Touraine JL (1992) In-utero transplantation of fetal liver stem cells into human fetuses. Hum Reprod 7: 44–8

Travis J (1993) The search for liver stem cells picks up. Science 259: 1829

Tropepe V, Coles BL, Chiasson BJ, Horsford DJ, Elia AJ, McInnes RR, van der Kooy D (2000) Retinal stem cells in the adult mammalian eye. Science 287: 2032–6

Tzanakakis ES, Hess DJ, Sielaff TD, Hu WS (2000) Extracorporeal tissue engineered liver-assist devices. Annu Rev Biomed Eng 2: 607–32

Uchida N, Buck DW, He D, Reitsma MJ, Masek M, Phan TV, Tsukamoto AS, Gage FH, Weissman IL (2000) Direct isolation of human central nervous system stem cells. Proc Natl Acad Sci U S A 97: 14720–5

Uchida N, Fujisaki T, Eaves AC, Eaves CJ (2001) Transplantable hematopoietic stem cells in human fetal liver have a CD34(+) side population (SP)phenotype. J Clin Invest 108: 1071–7

Vassilopoulos G, Wang PR, Russell DW (2003) Transplanted bone marrow regenerates liver by cell fusion. Nature 30: 30

Wagers AJ, Sherwood RI, Christensen JL, Weissman IL (2002) Little evidence for developmental plasticity of adult hematopoietic stem cells. Science 297: 2256–9

Wakayama T, Perry AC, Zuccotti M, Johnson KR, Yanagimachi R (1998) Full-term development of mice from enucleated oocytes injected with cumulus cell nuclei. Nature 394: 369–74

Wakitani S, Goto T, Pineda SJ, Young RG, Mansour JM, Caplan AI, Goldberg VM (1994) Mesenchymal cell-based repair of large, full-thickness defects of articular cartilage. J Bone Joint Surg Am 76: 579–92

Wakitani S, Saito T, Caplan AI (1995) Myogenic cells derived from rat bone marrow mesenchymal stem cells exposed to 5-azacytidine. Muscle Nerve 18: 1417–26

Waldmann H (1999) Transplantation tolerance-where do we stand? Nat Med 5: 1245–8

Walsh CJ, Goodman D, Caplan AI, Goldberg VM (1999) Meniscus regeneration in a rabbit partial meniscectomy model. Tissue Eng 5: 327–37

Wang JS, Shum-Tim D, Galipeau J, Chedrawy E, Eliopoulos N, Chiu RC (2000) Marrow stromal cells for cellular cardiomyoplasty: feasibility and potential clinical advantages. J Thorac Cardiovasc Surg 120: 999–1005

Wang X, Willenbring H, Akkari Y, Torimaru Y, Foster M, Al-Dhalimy M, Lagasse E, Finegold M, Olson S, Grompe M (2003) Cell fusion is the principal source of bone-marrow-derived hepatocytes. Nature 30: 30

Watt FM (1998) Epidermal stem cells: markers, patterning and the control of stem cell fate. Philos Trans R Soc Lond B Biol Sci 353: 831–7

Watt FM (2000) Epidermal stem cells as targets for gene transfer. Hum Gene Ther 11: 2261–6

Watt FM (2001) Stem cell fate and patterning in mammalian epidermis. Curr Opin Genet Dev 11: 410–7

Watt FM (2002) The stem cell compartment in human interfollicular epidermis. J Dermatol Sci 28: 173–80

Weiss RA (1998) Transgenic pigs and virus adaptation. Nature 391: 327–8

Wernet P, Koegler G, Fischer J, Mueller HW, Rosenbaum C, Degistirici O, Knipper A, Beerheide W, Udolf G, Hengstler J Pluripotent stem cells indentified in human cord blood Keystone Symposium, Stem cells: Origins, fates and functions, Keystone Resort 2002, pp 445

Wiles MV, Keller G (1991) Multiple hematopoietic lineages develop from embryonic stem (ES) cells in culture. Development 111: 259–67

Wobus AM, Rohwedel J, Maltsev V, Hescheler J (1995) Development of cardiomyocytes expressing cardiac-specific genes, action potentials, and ionic channels during embryonic stem cell-derived cardiogenesis. Ann N Y Acad Sci 752: 460–9

Woo SL, Hildebrand K, Watanabe N, Fenwick JA, Papageorgiou CD, Wang JH (1999) Tissue engineering of ligament and tendon healing. Clin Orthop 19: S312–23

Wright NA (1997) Stem cell repertoire in the intestine. In: Potten CS (ed) Stem Cells. Academic Press, London, pp 315–30

Xu C, Inokuma MS, Denham J, Golds K, Kundu P, Gold JD, Carpenter MK (2001) Feeder-free growth of undifferentiated human embryonic stem cells. Nat Biotechnol 19: 971–4

Xu RH, Chen X, Li DS, Li R, Addicks GC, Glennon C, Zwaka TP, Thomson JA (2002) BMP4 initiates human embryonic stem cell differentiation to trophoblast. Nat Biotechnol 20: 1261–4

Yamada T, Yoshikawa M, Kanda S, Kato Y, Nakajima Y, Ishizaka S, Tsunoda Y (2002) In vitro differentiation of embryonic stem cells into hepatocyte-like cells identified by cellular uptake of indocyanine green. Stem Cells 20: 146–54

Yang J, Chang E, Cherry AM, Bangs CD, Oei Y, Bodnar A, Bronstein A, Chiu CP, Herron GS (1999) Human endothelial cell life extension by telomerase expression. J Biol Chem 274: 26141–8

Yang L, Li S, Hatch H, Ahrens K, Cornelius JG, Petersen BE, Peck AB (2002) In vitro trans-differentiation of adult hepatic stem cells into pancreatic endocrine hormone-producing cells. Proc Natl Acad Sci U S A 99: 8078–83

Ying QL, Nichols J, Evans EP, Smith AG (2002) Changing potency by spontaneous fusion. Nature 416: 545–8

Yoo J, Ashkar S, Atala A (1998) Excretion of urine-like fluid in vivo from engineered functional kidney structures. Tissue Eng. 4

Yoo JJ, Park HJ, Atala A (2000) Tissue-engineering applications for phallic reconstruction. World J Urol 18: 62–6

Zappone MV, Galli R, Catena R, Meani N, De Biasi S, Mattei E, Tiveron C, Vescovi AL, Lovell-Badge R, Ottolenghi S, Nicolis SK (2000) Sox2 regulatory sequences direct expression of a (beta)-geo transgene to telencephalic neural stem cells and precursors of the mouse embryo, revealing regionalization of gene expression in CNS stem cells. Development 127: 2367–82

Zhang SC, Wernig M, Duncan ID, Brustle O, Thomson JA (2001) In vitro differentiation of transplantable neural precursors from human embryonic stem cells. Nat Biotechnol 19: 1129–33

Zhao LR, Duan WM, Reyes M, Keene CD, Verfaillie CM, Low WC (2002) Human bone marrow stem cells exhibit neural phenotypes and ameliorate neurological deficits after grafting into the ischemic brain of rats. Exp Neurol 174: 11–20

Zhou S, Schuetz JD, Bunting KD, Colapietro AM, Sampath J, Morris JJ, Lagutina I, Grosveld GC, Osawa M, Nakauchi H, Sorrentino BP (2001) The ABC transporter Bcrp1/ABCG2 is expressed in a wide variety of stem cells and is a molecular determinant of the side-population phenotype. Nat Med 7: 1028–34

Zuk PA, Zhu M, Mizuno H, Huang J, Futrell JW, Katz AJ, Benhaim P, Lorenz HP, Hedrick MH (2001) Multilineage cells from human adipose tissue: implications for cell-based therapies. Tissue Eng 7: 211–28

Zulewski H, Abraham EJ, Gerlach MJ, Daniel PB, Moritz W, Muller B, Vallejo M, Thomas MK, Habener JF (2001) Multipotential nestin-positive stem cells isolated from adult pancreatic islets differentiate ex vivo into pancreatic endocrine, exocrine, and hepatic phenotypes. Diabetes 50: 521–33

Zwaka TP, Thomson JA (2003) Homologous recombination in human embryonic stem cells. Nat Biotechnol 21: 319–21

3   The Regulation of Embryo Research in Europe: Situation and Prospects

Bartram C R et. al. (2000) Humangenetische Diagnostik, Wissenschaftsethik und Technikfolgen-beurteilung, vol. 7, Springer, Berlin, Heidelberg, New York

Beckmann JP (2003) Ethik nach Vorgaben des Gesetzes? – Überlegungen zur Aufgabe der Ethik gem. §§ 5 und 6 Stammzellgesetz (StZG). In: Amelung/Beulke/Lilie/Rosenau/Rüping/Wolfs-last (eds.), Strafrecht – Biorecht – Rechtsphilosophie, Festschrift für Hans-Ludwig Schreiber, C. F. Müller, Heidelberg, pp. 593–602

Benda E (2001) Verständigungsversuche über die Würde des Menschen, NJW (Neue Juristische Wochenschrift) C. H. Beck, Munich & Frankfurt, pp. 2147–2148

Beyleveld B, Pattinson S (2001) Embryo Research in the UK: Is Harmonisation in the EU Needed or Possible? In: Friele MB (ed.) Embryo Experimentation in Europe. Europaeische Akademie, Bad Neuenahr-Ahrweiler, Grey Series Nr. 24. pp. 58–74

von Bülow D (2001) Embryonenschutzgesetz. In: Winter/Fenger/Schreiber, Genmedizin und Recht, C. H. Beck, Munich. Margin. 300–380

von Bülow D (1997) Dolly und das Embryonenschutzgesetz. In: DtÄBl. (Deutsches Ärzteblatt) Deutscher Ärzteverlag, Köln, A-718–A-725

Brazier M (1999) Regulating the Reproduction Business? In: Medical Law Review 7. pp. 166–193

Charlesworth M (1990) Community Control of IVF and Embryo Experimentation. In: Singer P, Khuse H, Buckle S, Dawson K & Kasimba P (eds.) Embryo Experimentation. University of Cambridge, Cambridge. pp.147–152

Clothier C (1986) Introduction: Research on Early Human Embryos. In: Bock G & O'Connor M (eds.) Human Embryo Research – Yes or No? Tavistock Publications, London. pp.1–4

Dederer H-G (2002) Menschenwürde des Embryos in vitro? In: AöR (Archiv des öffentlichen Rechts), Mohr Siebeck, Tübingen, pp. 1–26

Deutsch E (1991) Embryonenschutz in Deutschland. In: NJW (Neue Juristische Wochenschrift) C.H. Beck, Munich & Frankfurt, pp. 721–725

Department of Health (1997) Advisory Group on the Ethics of Xenotransplantation, U.K. Animal Tissue into Humans. The Stationery Office, London

Faßbender K (2001) Präimplantationsdiagnostik und Grundgesetz. In: NJW (Neue Juristische Wochenschrift) C. H. Beck, Munich & Frankfurt, pp. 2745–2753

Fikentscher W (1976) Methoden des Rechts, vol. 3, J. C. B. Mohr (Paul Siebeck) Tübingen

Gauthier D (1986) Morals By Agreement. Clarendon Press, Oxford

Gethmann F, Thiele F (2001) Moral arguments against the cloning of humans, Poisis & Praxis 1, pp. 35–45

Graumann S (2001) Zehn Thesen zur Präimplantationsdiagnostik in: Chancen und Grenzen der Biomedizin. Symposium des Diakonischen Werkes der Evangelischen Kirche in Deutschland e. V., Stuttgart, Oktober 2001, published in: Informationen und Materialien aus dem Diakonis-chen Werk der EKD 05/02, pp. 39–50

Gutmann T (2001) Strafbarkeit des Klonens von Menschen. In: Roxin C, Schroth U, Medizin-strafrecht, 2nd edition, Boorberg, Stuttgart

Herdegen M (2001) Menschenwürde im Fluß des bioethischen Diskurses. In: JZ (Juristenzeitung) Mohr Siebeck, Tübingen, pp. 773–779

Herzog F (2001) Präimplantationsdiagnostik – Im Zweifel für ein Verbot? In: ZRP (Zeitschrift für Rechtspolitik) C. H. Beck, Munich, pp. 339–347

Heun W (2002) Embryonenforschung und Verfassung – Lebensrecht und Menschenwürde des Embryos. In: JZ (Juristenzeitung) Mohr Siebeck, Tübingen, pp. 517–524

HFEA (1994) Ovarian tissue in embryo research & assisted conception: public consultation docu-ment. Human Fertilisation & Embryology Authority, London

Hilgendorf E (2001) Klonverbot und Menschenwürde – Vom Homo sapiens zum Homo xerox? Überlegungen zu § 6 Embryonenschutzgesetz. In: Arndt/Geis/Lorenz (Hrsg.) Staat-Kirche-Verwaltung, Festschrift für Hartmut Maurer zum 70. Geburtstag, C. H. Beck, München. pp. 1147–1164

Höfling W (2001) Verfassungsrechtliche Aspekte der Verfügung über menschliche Embryonen und „humanbiologisches Material" Gutachten für die Enquete-Kommission des Deutschen Bun-destages, Recht und Ethik in der modernen Medizin, http://www.bundestag.de/gremien/medi/medi_gut_hoe.pdf

Ipsen J (2001) Der verfassungsrechtliche Status des Embryos in vitro. In: JZ (Juristenzeitung) Mohr Siebeck, Tübingen, pp. 989–996

Jähnke B (1989) In: Leipziger Kommentar zum Strafgesetzbuch (Commentary on the Criminal Code), 10th edition, Walter de Gruyter, Berlin & New York, section 219

Keller R, Günther H-L, Kaiser P (1992) Embryonenschutzgesetz, Kommentar, W. Kohlhammer, Stuttgart, Berlin, Köln

Kirejczyk M (1999) Parliamentary Cultures and Human Embryos: The Dutch and British Debates Compared. In: Social Studies of Science 29/6 (December 1999) pp. 889–912

Kloepfer M (2002) Humangentechnik als Verfassungsfrage. In: JZ (Juristenzeitung), Mohr Siebeck, Tübingen, pp. 417–428

Köcher R (2001) Menschenwürde – Leitbild der modernen Gesellschaft? In: Katholische Akademie (Catholic Academy) Freiburg (editor), Leben als Gottes Bild – Die Bedeutung ethischer Ressourcen, pp. 51–63

Köcher R (1996) Emotionen – ein Standortfaktor. Eine Dokumentation des Beitrags in der Frankfurter Allgemeinen Zeitung Nr. 164 vom 17. Juli 1996, Institut für Demoskopie Allensbach

Larenz, K (1995) Methodenlehre, 3rd edition, Springer, Berlin

Laufs A (1998) Arzt und Recht – Fortschritte und Aufgaben. In: NJW (Neue Juristische Wochenschrift), C. H. Beck, Munich & Frankfurt, pp. 1750–1761

Laufs A (1999) Arzt, Patient und Recht am Ende des Jahrhunderts, NJW (Neue Juristische Wochenschrift) C. H. Beck, Munich & Frankfurt, pp. 1758–1769

Laufs A (2000) Nicht nur der Arzt allein muß bereit sein, das Notwendige zu tun. In: NJW (Neue Juristische Wochenschrift). C. H. Beck, Munich & Frankfurt, pp. 1757–1769

Lee RG, Morgan D (2001) Human Fertilisation and Embryology: Regulating the Reproductive Revolution, Blackstone Press Limited

Lema Añón C (2000a) Los problemas pendientes de la regulación jurídica española sobre reproducción assistida: la sentencia de Tribunal Constitucional y el I. informe de la Comisión Nacional de Reproducción Humana Asistida (Parte I). In: Revista de Derecho y Genoma Humano (Law and the Human Genome Review), No. 12, Bilbao, p. 47–66

Lema Añón C (2000b) Los problemas pendientes de la regulación jurídica española sobre reproducción assistida: la sentencia de Tribunal Constitucional y el I. informe de la Comisión Nacional de Reproducción Humana Asistida (Parte II). In: Revista de Derecho y Genoma Humano (Law and the Human Genome Review), No. 13, Bilbao, p. 103–118

Lilie H (2003) Biorecht und Politik am Beispiel der internationalen Stammzelldiskussion. In: Amelung/Beulke/Lilie/Rosenau/Rüping/Wolfslast (eds), Strafrecht-Biorecht-Rechtsphilosophie, Festschrift für Hans-Ludwig Schreiber, C. F. Müller, Heidelberg, p. 729–740

Lilie H, Albrecht D (2001) Strafbarkeit im Umgang mit Stammzellinien aus Embryonen und damit im Zusammenhang stehender Tätigkeiten nach deutschem Recht. In: NJW (Neue Juristische Wochenschrift) C. H. Beck, Munich & Frankfurt, pp. 2774–2776

Merkel R (2002) Forschungsobjekt Embryo – verfassungsrechtliche und ethische Grundlagen der Forschung an menschlichen embryonalen Stammzellen, Deutscher Taschenbuchverlag, München

Mildenberger EH (2002) Der Streit um die Embryonen: Warum ungewollte Schwangerschaften, Embryoselektion und Embryonenforschung grundsätzlich unterschiedlich behandelt werden müssen. In: MedR (Medizinrecht) Springer, Berlin, pp. 293–300

Neidert R (1998) Brauchen wir ein Fortpflanzungsmedizingesetz? In: MedR (Medizinrecht) Springer, Berlin, pp. 347–353

Neidert R (2000) Präimplantationsdiagnostik: Zunehmendes Lebensrecht In: DÄBl. (Deutsches Ärzteblatt) Deutscher Ärzteverlag, Köln, 2000, pp. A3483–A3486

Neidert R (2002) Das überschätzte Embryonenschutzgesetz – was es verbietet und nicht verbietet. In: ZRP (Zeitschrift für Rechtspolitik), C. H. Beck, Munich, pp. 467–471

Pattinson SD (2002) Influencing Traits Before Birth. Ashgate, Aldershot

Renzikowski J (2001) Die strafrechtliche Beurteilung der Präimplantationsdiagnostik, NJW (Neue Juristische Wochenschrift). C. H. Beck, Munich & Frankfurt, pp. 2753–2758

Romeo Casabona CM (1996) Del Gen al Derecho, Bogota

Romeo Casabona CM (2001) Embryonic Stem Cell Research and Therapy at European Level: Is a common legal framework needed? Revista de Derecho y Genoma Humano (Law and the Human Genome Review), No. 15, Bilbao, p. 121–138

Romeo Casabona CM (2002) Los llamados delitos relativos a la manipulación genetica, Escuela Judicial – Consejo General del Poder Judicial/Ilustre Colegio Oficial de Médicos de Madrid, Genética y derecho, Separata, 36, Madrid, p. 331–400

Rosenau H (2002) Das Verbot des Klonens menschlichen Lebens: Reproduktives und therapeutisches Klonen, Vortrag (unpublished paper), Kyoto 14. September 2002

Sandor J (1998) Reproductive Rights in Hungary. Central European University, Brussels

Schick PJ (1991) Der Entwurf eines Fortpflanzungshilfegesetzes (FHG) – eine kritische Wertungsanalyse, In: Bernat (ed.) Fortpflanzungsmedizin, Wertung und Gesetzgebung, Verlag-Österreichische Staatsdruckerei, Wien. pp. 3–43

Schlink B (2002) Aktuelle Fragen des pränatalen Lebensschutzes, series of articles of the Juristischen Gesellschaft zu Berlin (Jurists Association of Berlin), De Gruyter Recht, Berlin, Heft 172

Schneider S (2000) Auf dem Weg zur gezielten Selektion – Strafrechtliche Aspekte der Präimplantationsdiagnostik. In: MedR (Medizinrecht) Springer, Berlin, pp. 360–364

Schneider S (2002) Präimplantations- und Pränataldiagnostik (unpublished paper), Kyoto 14. September 2002

Schreiber H-L (2000) Von den richtigen rechtlichen Voraussetzungen ausgehen. In: DÄBl. (Deutsches Ärzteblatt). A1135–A1136

Schroth U (2002) Forschung mit embryonalen Stammzellen und Präimplantationsdiagnostik im Lichte des Rechts. In: JZ (Juristenzeitung) Mohr Siebeck, Tübingen, pp. 170–179

Sendler H (2001) Menschenwürde, PID und Schwangerschaftsabbruch. In: NJW (Neue Juristische Wochenschrift) C. H. Beck, Munich & Frankfurt, pp. 2148–2149

Starck C (2002) Verfassungsrechtliche Grenzen der Biowissenschaft und Fortpflanzungsmedizin. In: JZ (Juristenzeitung). Mohr Siebeck, Tübingen, pp. 1065–1072

Taupitz J (2001) Der rechtliche Rahmen des Klonens zu therapeutischen Zwecken. In: NJW (Neue Juristische Wochenschrift) C. H. Beck, Munich & Frankfurt, pp: 3433–3440

Taupitz J (2002) Der Status des Embryos, insbesondere die Produktion und Verwendung von Embryonen zur Forschung (unpublished paper) Kyoto, 14. September 2002

Taupitz J (2003) Rechtliche Regelung der Embryonenforschung im internationalen Vergleich. Springer, Berlin, Heidelberg, New York

Tröndle H (1999) In: Tröndle/Fischer, Kommentar zum Strafgesetzbuch, 49. edition, C. H. Beck, Munich

Warnock M (1984) Report of the Committee of Inquiry into Human Fertilisation and Embryology. HMSO London, Cmnd 9314. As republished in: Warnock M (1985) A Question of Life. The Warnock Report on Human Fertilization and Embryology. Basil Blackwell, Oxford

Zuck R (2002) Wie führt man eine Debatte? Die Embryonennutzung und die Würde des Menschen (Art. 1 Abs. 1 GG). In: NJW (Neue Juristische Wochenschrift) C. H. Beck, Munich & Frankfurt, p. 869

4   Attitudes to Embryo Research in Europe

Beck U (1992) The Risk Society: Toward a new Modernity. London: Sage

Beck U (1999) World Risk Society. Oxford: Blackwell Publishers

Bonfadelli H, Dahinden U, Leonarz M (2002) Biotechnology in Switzerland: high on the public agenda, but only moderate support. In: Public Understanding of Science 11 (2002), 113–130

Campbell A, Converse PE, Miller WE, Stokes DE (1976) The American Voter. Chicago and London: The University Chicago Press (original edition 1960)

Converse P (1964) The Nature of Belief Systems in Mass Publics. In: David E. Apter (ed.) Ideology and Discontent. New York: Free Press. Pp. 206–61

Converse P (1970) Attitudes and Non-Attitudes: Continuation of a Dialogue. In: Tufte ER (ed.) The Qualitative Analysis of Social Problems. Reading, MA: Addison-Wesley

Daamen, Dancker DL, Van der Lans IA, Midden CJH (1990) Cognitive Structures in the Perception of Modern Technologies. In: Science, Technology, & Human Values, Vol. 15 No. 2, Spring 1990, 202–225

Einsiedel EF, Jelsøe E, Breck T (2001) Publics at the technology table: the consensus conferences in Denmark, Canada, and Australia. In: Public Understanding of Science 10, 83–98

Evans G, Durant J (1995) The relationship between knowledge and attitudes in the public understanding of science in Britain. In: Public Understanding of Science 4, 57–74

Gaskell G et al. (1997) Europe ambivalent on biotechnology". In: Nature, vol. 387, 845–847

Gaskell G, Bauer M, Durant J, Allum NC (1999) Worlds Apart? The Reception of Genetically Modified Foods in Europe and the U.S.. In: Science, Vol. 285, 16 July 1999, 384–387

Gaskell G et al. (2002) Europeans and Biotechnology in 2002. Eurobarometer 58.0. A report of the EC Directorate General for Research from the Project 'Life Sciences in European Society' GLG7-CT-1999-00286.
http://europa.eu.int/comm/public_opinion/archives/eb/ebs_177_en.pdf

Guston DH, Keniston K (1994) Introduction: The Social Contract for Science. In: Guston DH, Keniston K (eds.) The Fragile Contract. University Science and the Federal Government. Cambridge, MA: The MIT Press

Heise UK (forthcoming) Science, Technology and Postmodernism. The Cambridge Companion to Postmodernism. Ed. Steven Connor. Cambridge: Cambridge University Press

Hornig Priest S (2001) A Grain of Truth. The Media, the Public, and Biotechnology. Lanham-Boulder-New York-Oxford: Rowman & Littlefield Publishers, Inc.

House of Lords (Select Committee on Science and Technology) (2000), Science and Society. London: The Stationery Office

INRA (Europe) and Report International (1993) Europeans, Science and Technology. Public Understanding and Attitudes. Brussels: European Commission – Directorate General XII

Joss S, Durant J (1995) Public participation in science. The role of consensus conferences in Europe. London: Science Museum

Levitt Norman (1999) Science and the Contradictions of Contemporary Culture. New Brunswick: Rutgers University Press

Lévy-Leblond J-M (1992) About Misunderstandings about Misunderstandings. In: Public Understanding of Science 1, 17–21

Lyotard J-F (1984) The Postmodern Condition: A Report on Knowledge. Minneapolis: University of Minnesota Press

Marx L (1998) The Domination of Nature and the Redefinition of Progress. In: Leo Marx y Bruce Mazlish (eds.) Progress. Fact or Illusion? Ann Arbor: The University of Michigan Press

Mertig AG, Dunlap RE (1995) Public Approval of Environmental Protection and Other New Social Movement Goals in Western Europe and the United States. In: International Journal of Public Opinion Research Vol. 7, No. 2., pp. 145–156

Miller JD (1983) The American People and Science Policy. New York: Pergamon Press

Miller JD, Pardo R (2000) Civic Scientific Literacy and Attitude to Science and Technology: A Comparative Analysis of the European Union, the United States, Japan, and Canada. In: Dierkes M, von Grote C (eds.) Between Understanding and Trust. The Public, Science, and Technology. Australia-Canada-France: Harwood Academic Publishers, 81–130

National Science Board (2002) Science and Engineering Indicators-2002. Arlington, VA: National Science Foundation

Nelkin D (1992) Controversy: politics of technical decisions (3rd edition). Newbury Park, CA : Sage

Nelkin D, Lindee MS (1998) Cloning in the Popular Imagination. In: Cambridge Quarterly of Healthcare Ethics, vol. 7 no. 2, 145–149

Pardo R (2001) La cultura cientifico-tecnologica de las sociedades de modernidad tardia. In: Treballs de la Societat Catalana de Biologia, Volum 51, 64–86

Pardo R, Midden C, Miller JD (2002) Attitudes toward biotechnology in the European Union, Journal of Biotechnology 98, 9–24

Pardo R, Calvo F (2002) Attitudes toward science among the European public: a methodological analysis, Public Understanding of Science 11, 155–195

Patzig G (2001) Moral Problems of Preimplantation Diagnostics. In: Friele MB (ed.) Embryo Experimentation in Europe. Bio-medical, Legal, and Philosophical Aspects. Bad-Neuenahr-Ahrweiler: Europäische Akademie, Graue Reihe vol. 24, 105–115

Perrow C (1999) Normal Accidents. Princeton: Princeton University Press

Prewitt K (1983) Scientific Illiteracy and Democratic Theory. In: Daedalus, Spring, 49–64

Shamos MH (1995) The Myth of Scientific Literacy. New Brunswick, NJ: Rutgers University Press

Slovic P (1987) Perception of Risk. In: Science, vol. 236, 280–285

Turney J (1998) Frankenstein's Footsteps. Science, Genetic, and Popular Culture. New Haven and London: Yale University Press

Spencer RW (1989) Nuclear Fear: A History of Images. Cambridge, MA: Harvard University Press

Weimann G (1994) The Influentials. People Who Influence People. Albany, NY: State University of New York Press

Wellcome Trust (1998) Public Perspectives on Human Cloning. A Social Research Study. London: The Wellcome Trust

Worcester RM (1993) Public and Élite Attitudes to Environmental Issues. In: International Journal of Public Opinion Research, Vol. 5, No. 4, 315–334

5   Toward a Rational Debate on Embryo Research

Arnst C, Kerry C (2001) "A Ban Won't Make Cloning Go Away", Business Week, 10 December

Beasley S (1999) Contraception and the Moral Status of the Early Human Embryo. In: Evans D (ed.) (1996) Conceiving the Embryo: Ethics, Law and Practice in Human Embryology; The Hague, etc.: Martinus Nijhoff Publishers

Bettelheim A (2001) House Bill to Ban Cloning. CQ Weekly 28 July, vol 57, #30: 1857

Beyleveld D, Brownsword R (2001) Human Dignity in Bioethics and Biolaw, Oxford: Oxford University Press

Boukhari S (1999) Religion, Genetics and the Embryo, UNESCO Courier, Sept. 99: 24

Breyer SG (2000) Genetic Advances and Legal Institution. In: Journal of Law, Medicine & Ethics, Winter Supplement, Vol. 28, Issue 4

Brody AB (1998) Ethics of Biomedical Research. An International Perspective, Oxford: University Press

Campbell PW (1999) Fight over Stem-Cell Research Is Unlikely to Tie up Spending Bill. Chronicle of Higher Education, 15 October, vol: 46, #8

Carney TP (2001) Bioethics Guru Unclear of When to Protect Life, Human Events 10 September, vol 57, #33: 4

Carey J, Licking E (2002) "The Stem-Cell Debate Just Got Thornier", Business Week, 11 February

Coulter A (2001) Embryos in the Rain Forest. Human Events, 30 July, vol 57, #28: 1

Daar JF (2001) "Frozen Embryo Disputes Revisited", Journal of Law, Medicine & Ethics, vol 29, #3: 197

Douglas M (1966) Purity and Danger, Hammondsworth: Penguin Books

Dworkin R (1978) Taking Rights Seriously. Revised edition, London: Duckworth

Dyson A, Harris J (1994) Ethics and Biotechnology, London and New York: Routledge

Economist, "The Facts of Life", 8/17/96 vol 340 #7979: 47

Economist, "The Politics of Cloning", 12/1/2001, vol 361, #8250: 34

Economist, "The Cutting Blob of Ethical Politics," 7/7/2001, vol 360, #8229: 31

Eder K (1996) The Social Construction of Natrue; London, etc: SAGE Publications

Evans D (ed.) 1996: Conceiving the Embryo: Ethics, Law and Practice in Human Embryology; The Hague, etc.: Martinus Nijhoff Publishers

Evans D (1996a) Conceiving the Embryo. In: Evans D (ed.) (1996) Conceiving the Embryo: Ethics, Law and Practice in Human Embryology

Evans D (1996b) Pro-attitudes and Pre-embryos. In: Evans D (ed.) (1996) Conceiving the Embryo: Ethics, Law and Practice in Human Embryology

Evans D (1996c) Procuring Gametes for Research and Therapy. In: Evans D (ed.) (1996) Conceiving the Embryo: Ethics, Law and Practice in Human Embryology

Finnis J (1980) Natural Law and Natural Rights; London: Clarendon Press

Fischer JS (2000) Copies upon Copies, US News and World Report, 02/07/2000, vol: 128, #5: 52

Ford Norman: When Did I Begin? Ref?

Freud S (1938) Totem and Taboo, London: Pelican Books

Gethmann CF, Thiele F Moral Arguments against Cloning of Humans. In: Poiesis & Praxis. 2001 #1: 35–46

Gerrand N (1993) Creating Embryos for Research. In: Journal of Applied Philosophy, vol 10 #2

Giles TS, Ferranti J (1995) Test-Tube Wars. In: Christianity Today, 1/9/95, vol 39, # 1: 38

Habermas J (1970) Toward a Theory of Communicative Competence. In:H. P. Dreitzel. (ed.) Recent Sociology No. 2, 114–148: New York: Macmillan

Hare RM (1999) Abortion and the Golden Rule. In: Kuhse and Singer: Bioethics

Harris J (1992) Wonderwoman and Superman: The Ethics of Human Psychology, Oxford: Oxford University Press

Harris J (1994) Biotechnology, Friend or Foe? Ethics and Control, in Dyson and Harris: Ethics and Biotechnology

Häyry M (1994) Categorical Objections to Genetic Engineering – A Critique. In: Dyson and Harris: Ethics and Biotechnology

Henderson CW (2000) UK Doctors Defend Embryo Research- In: Blood Weekly: 11/23/2000-11/30/2000: 11

Holm S (1999) The Moral Status of the Pre-Personal Human Being. In: Evans D (ed.) (1996) Conceiving the Embryo: Ethics, Law and Practice in Human Embryology; The Hague, etc.: Martinus Nijhoff Publishers

Hursthouse R (1987) Beginning Lives; Oxford: Blackwell & the Open University

Jefferey TP (2001) Think Again, Mr. President. In: Human Events, 8/20/2001, vol. 57, 31

John Paul II (1996) Encykliki Ojca ?wi?tego Jana Paw?a II [Encyclicals of His Holiness John Paul II], Kraków: Wydawnictwo Znak (in Polish)

Kelsen H (1967) Pure Theory of Law. Translated by Max Knight form the 2nd revised and enlarged German edition. Berkeley: California University Press

Kischer CW (2001) Why Hatch is Wrong on Human Life. In: Human Events, 7/16/2001, vol 57, #26: 1

Kolakowski L (2000) Kultura i fetysze [in Polish], Warszawa: PWN

Kuhse H, Singer P (1999) Bioethics; An Anthology, Malden Mass: Blackwell Publishers

Lamb D (1998) Down the Slippery Slope: Arguing in Applied Ethics; London: Croom Helm

Lancet (1999) First Principles in Cloning, 01/09/99, vol 353 #9147: 81

Lauritzen P (2001) Neither Person nor Property , America 03/26/2001, vol: 184, #10: 20

Mason JK, McCall Smith RA, Laurie GT (1999) Law and Medical Ethics, London, Edinburgh, Dublin: Butterworths

Mayor S (2001) Common Votes for Human Embryo Stem Cell Research", British Medical Journal, 01/06/2001, vol 322 #7277: 7

McCarthy TA (1973) A Theory of Communicative Competence. In: Philosophy of the Social Sciences, vol. 3 135–156

McLaren A, Marinho da Silva P, Schroten E (2000) Ethical Aspects of Research Involving the Use of Human Embryo in the Context of the 5th Framework Programme. Opinion of the European Group on Ethics in Science and New Technologies

McNeill PM (1993) The Ethics and Politics of Human Experimentation, Cambridge, University Press

Norman F (1992) When Did I Begin? Conception of the Human Individual in History, Philosophy and Science, Cambridge University Press

Mead M (1970) Culture and Commitment; A Study of the Generation Gap; Garden City, NY: Natural History Press

Meyers MJ, Nelson LJ (2001) Respecting What We Destroy. Hastings Center Report, Jan/Feb 2001, vol 31 #1: 16

New Perspectives Quarterly, 2000 Special Issue, vol 17 #5: 58

van Niekirk A, van Zyl L (1996) Embryo Experimentation, Personhood and Human Rights. In: South African Journal of Philosophy, Nov. 1996, vol 15, #4: 139

Novak M (2001) "The Stem-Cell Slide", National Review, 9/3/2001, vol 53, #17

O'Leary D (1999) Embryo Research Contested. Christianity Today 05/24/1999, vol. 43, #6: 26

Orellana C (2002) Germany's Parliament Approves Imports of Embryonic Stem Cells. Lancet, 2/9/2002, vol. 359 #9305: 506

Peters T (1994) Designer Children: The Market World of Reproductive Choice. Christian Century, 12/14/1994: vol. 111, #36: 1193

Pulford C (2001) Human Embryo Cloning Legalized. Christianity Today, 03/05/2001, vol 45, #4: 32

Resnik DB (1998) The Commodification of Human Reproductive Materials. Journal of Medical Ethics, Dec. 1998, vol: 24, #6: 388

Robertson J (1994) Children of Choice: Freedom and the New Reproductive Technologies, Princeton University Press

Rogers A (1998) European Union Approves Gene-Patent Directive, Lancet, 05/16/98, Vol. 351, Issue 9114

Shannon TA (1998) Remaking Ourselves? Commonweal, 12/04/98, vol: 125, #21: 9

Steinbock B (1994) The Moral Status of Extracorporeal Embryos, in Dyson and Harris: Ethics and Biotechnology

Tesler P (2001) Cell Mates. Current Science, 09/14/2001, vol 87, #2: 8

Thompson JJ (1999) A Defense of Abortion. In: Kuhse and Singer: Bioethics

Tooley M (1999) Abortion and Infanticide, in Kuhse and Singer: Bioethics

Vorraut J-M (1989) Le possible et l'interdit. Les devoirs du droit, Paris 1989: Edition de la Table Ronde

Wessels U (1994) Genetic Engineering and Ethics in Germany. In: Dyson and Harris: Ethics and Biotechnology

Wood-Harper J (1994) Manipulation of the Germ-Line. In: Dyson, Harris: Ethics and Biotechnology

Zuckerman MB (2001) "A Rare Gift of Life", US News and World Report 7/23/2001, vol: 131 #3: 116

List of Authors

Beyleveld, Deryck, Professor Dr., Professor of Jurisprudence at the University of Sheffield and Founding Director of the Sheffield Institute of Biotechnological Law and Ethics (SIBLE). He is currently Vice-Chair of Trent Multi-Centre Research Ethics Committee. He studied Biochemistry at the University of the Witwatersrand, Philosophy and Social and Political Science at the University of Cambridge, and obtained his PhD in Philosophy and Social Theory from the University of East Anglia. His research and publications span criminology, legal and moral philosophy, contract law, product liability law, international human rights law, patents in biotechnology, and many topics in bioethics and biolaw (in both medical and non-medical fields), as well as public perception of biotechnology. He is currently co-ordinator of an EC concerted action on implementation of the Data Protection Directive (95/46/EC) in relation to medical research and the role of ethics committees (PRIVIREAL).

Friele, Minou Bernadette, cand. Phil., Study of Philosophy at the Heinrich-Heine-University Düsseldorf, where she also was a research asstistant at the Philosophical Department from 1998-1999, in autum 2000 she was visiting research fellow at the Centre for Human Bioethics at Monash University, Melbourne. Since September 1999 she has been member of the scientific staff of the Europäische Akademie. She writes a thesis on "The Legitimacy of Supranational Law in Europe" supervised by Professor Dr. Dieter Birnbacher. She acted as the co-ordinator of the project.

Holówka, Jacek, Professor Dr. habil., Professor of Analytic Philosophy at Warsaw University. Lecturer at Postgraduate School of Medical Law and Bioethics at Warsaw University. Editor of Przegląd Filozoficzny. He taught courses at Indiana University (Bloomington), Notre Dame, the Viadrina (Frankfurt/Oder and gave guest lectures at the universities of Oxford, London, Copenhagen, Moscow and Tübingen. His many publications are mostly related to questions of moral philosophy. He is member of the Philosophical Committee of the Polish Academy of Sciences, the Europäische Akademie, and of the Managing Board of the Zdrowie Foundation – an organization that promotes new methods of health care management in Poland.

Lilie, Hans, Professor Dr. jur., Professor of Criminal Law, Criminal Procedure and Medical Law at the Martin-Luther-Universität Halle-Wittenberg and Director of the Interdisciplinary Center Medicine – Ethics – Law at the same University. He studied law at the University of Göttingen and UCLA. His doctor theses deal with the patients rights to access to medical records comparing the German and American legal situation. His research and publications are related to questions of criminal law and criminal procedure. His main research subjects are to be found in questions of bioethics and biolaw as well as classical medical law fields like informed consent, transplantation law, euthanasia. He also works in the fields of embryo research and patients rights. He is currently member of the German delegation at the United Nations Ad Hoc Committee on a ban against reproductive cloning.

Lovell-Badge, Robin, PhD, FRS, Head of the Division of Developmental Genetics at the MRC National Institute for Medical Research, Mill Hill in London, UK. He is also a Visiting Professor in the Biochemistry Department of the University of Hong Kong and an Honorary Professor in the Department of Anatomy and Developmental Biology at University College, London. He obtained his PhD in embryology at UCL in 1978, before carrying out postdoctoral research in the Department of Genetics, University of Cambridge and at the Institute Jacques Monod in Paris. He was then appointed Staff Scientist at the MRC Mammalian Development Unit in London, Directed by Anne McLaren, before moving to the NIMR in 1988. He is a member of EMBO and was elected a Fellow of the Academy of Medical Sciences in 1999 and a Fellow of the Royal Society in 2001. He received the 1995 Louis Jeantet Prize for Medicine and the Amory Prize for 1996 (Awarded by the American Academy of Arts and Sciences), for his work on sex determination. He has had long-standing interests in the biology and uses of embryonic stem cells, in how genes work in the context of development and how decisions of cell fate are reached during embryogenesis. Major themes of his current work include sex determination, development of the nervous system and stem cells in the embryo. He has also been actively involved in measures to increase public appreciation and understanding of science and participated in the UK parliamentary debates on stem cell research and therapeutic cloning that led to changes in the HFEA Act.

Mandla, Christoph, Ass. iur., Study of law at the Martin-Luther-University of Halle-Wittenberg and at the Faculté d'Administration et Echanges (A.E.S.) de l'Université Paris XII (Val de Marne), research assistant to Professor Dr. Hans Lilie. He currently writes a thesis on principles of criminal procedure especially main hearing, its interruption and finding the truth.

Martin, Ulrich, Priv. Doz. Dr. rer. nat., Dipl.-Biol., Study of Microbiology, Genetics, Biochemistry and Immunology from 1988-1994 at the Technical University of Hannover, Ph.D. 1997 at the Hannover Medical School at the Institute of Medical Microbiology on a thesis entitled *"The human C3a anaphylatoxin receptor: Methods for Cloning and Investigation of Expression and Function"*. Since 1997, principal investigator at the Leibniz Research Laboratories for Biotechnology and Artificial Organs (LEBAO), Hannover Medical School. Since June 2001 head of the section "Molecular Biotechnology and Stem Cell Research" and laboratory's scientific director of the LEBAO. Venia legendi for molecular biology in November 2001. Main research areas: Xenotransplantation, tolerance induction, viral gene transfer, stem cell differentiation. Address: Leibniz Research Laboratories for Biotechnology and Artificial Organs, Department of Thoracic and Cardiovascular Surgery, Hannover Medical School, Podbielskistr. 380, 30659 Hannover, Germany.

Pardo, Rafael, Professor Dr., General Director of the BBVA Foundation since 2000. From 1996-2000. He was Professor of Research at the National Council for Scientific Research (CSIC), and Professor of Sociology at the Universidad Pública de Navarra from 1993-96. He has held appointments at Stanford University as Visiting Professor (1998) and Visiting Scholar (1990-91, 1996), and as Visiting Scholar at the Massachusetts Institute of Technology (1987, 1988). He has been an advisor to General Directorate XII (Science & Technology) of the European Commission and to Intel Corporation in San José, USA. From 1994-96, he chaired the National Evaluation Commission for Social Science Research Projects, part of the Spanish National Agency for Evaluation and Science Policy. His major research and publications deal with organization studies, innovation, scientific and technological culture, social dimensions of Artificial Intelligence, cloning, and environmental culture and values.

Solter, Davor, Professor Dr., Director of the Max-Planck Institute of Immunobiology in Freiburg. M.D., Ph.D., both from the University of Zagreb, Croatia, Assistant and associate Professor in the Departments of Anatomy and Biology, University Zagreb Medical School 1966-1973. In 1973 moved to the Wistar Institute, Philadelphia and became Member and Professor in 1981 as well as Wistar Professor at the University of Pennsylvania. In 1991 he was appointed Member of the Max-Planck Society and Director of the Max-Planck Institute of Immunobiology in Freiburg. He is also Adjunct Senior Staff Scientist at the Jackson Laboratory, Bar Harbor. He has been a member of numerous editorial and advisory boards and is currently European Editor of *Genes & Development.* He is a member of American Academy of Arts and Sciences, EMBO and Academia Europea. In 1998 he received the March of Dimes Prize in Developmental Biology for pioneering the concept of imprinting. He contributed significantly to many areas of mammalian developmental biology, namely: differentiation of germ layers; role of cell surface molecules in regulating early development; biology and genetics of teratocarcinoma; biology of embryonic stem cells; imprinting and cloning. His current research interest focuses on genetic and molecular control of genome reprogramming and of activation of the embryonic genome. He was Chair-man of this project.